Renewable Resources for Functional Polymers and Biomaterials

RSC Polymer Chemistry Series

Titles in the Series:
1: Renewable Resources for Functional Polymers and Biomaterials

How to obtain future titles on publication:
A standing order plan is available for this series. A standing order will bring delivery of each new volume immediately on publication.

For further information please contact:
Book Sales Department, Royal Society of Chemistry, Thomas Graham House, Science Park, Milton Road, Cambridge, CB4 0WF, UK
Telephone: +44 (0)1223 420066, Fax: +44 (0)1223 420247, Email: books@rsc.org
Visit our website at http://www.rsc.org/Shop/Books/

Renewable Resources for Functional Polymers and Biomaterials

Edited by

Peter A. Williams
Centre for Water Soluble Polymers, Glyndwr University, Wrexham, UK

RSC Publishing

RSC Polymer Chemistry Series No. 1

ISBN: 978-1-84973-245-1
ISSN: 2044-0790

A catalogue record for this book is available from the British Library

Published by The Royal Society of Chemistry,
Thomas Graham House, Science Park, Milton Road,
Cambridge CB4 0WF, UK

Registered Charity Number 207890

For further information see our web site at www.rsc.org

Printed and bound in the United States of America

Preface

Natural polymers, polysaccharides, nucleic acids and proteins, in addition to being the fundamental basis of life, have a wealth of functional properties that can be readily exploited in a broad range of industrial applications. This volume covers some of the most important biomacromolecules, including polysaccharides, nucleic acids, proteins and microbial polyesters, covering their resources, production, structures, properties, and current and potential application in the fields of biotechnology and medicine.

Following this preface, the main chapters begin with a systematic discussion on the general strategies of isolation, separation and characterization of polysaccharides and proteins. Then subsequent chapters are devoted to polysaccharides obtained from various sources, including botanical, algal, animal and microbial. Of the botanical polysaccharides, separate chapters are devoted to the sources, structure, properties and medical applications of cellulose and its derivatives, starch and its derivatives, pectins, and exudate gums, notably gum arabic. Another chapter discusses the potential of hemicelluloses (xylans and xylan derivatives) as a new source of functional biopolymers for biomedical and industrial applications. The algal polysaccharide, alginate, has significant applications in food, pharmaceuticals and the medical field, all of which are reviewed in a separate chapter. With regards to polysaccharides of animal origin, there are separate chapters on the sources, production, biocompatibility, biodegradability and biomedical applications of chitin (chitosan) and hyaluronan.

With the increasing knowledge and applications of genetic engineering, we have also included in the book an introduction to nucleic acid polymers, genome research and genetic engineering.

On proteins and protein conjugates, one chapter provides a general review of structural glycoproteins, fibronectin and laminin, together with their role in the promotion of cell adhesion in vascular grafts, implants and tissue engineering.

RSC Polymer Chemistry Series No. 1
Renewable Resources for Functional Polymers and Biomaterials
Edited by Peter A. Williams

Published by the Royal Society of Chemistry, www.rsc.org

Another chapter discusses general aspects of a number of industrial proteins, including casein, caseinates, whey protein, gluten and soy proteins, with emphasis on their medical applications, and with reference to the potential of bacterial proteins.

Another natural polymer resource, namely microbial polyesters, although small compared with polysaccharides and proteins, is also gaining increasing interest in biomedical technology and other industrial sectors. Thus one chapter is also devoted to microbial polyesters, with comprehensive coverage of their biosynthesis, properties, enzymic degradation and applications.

Contents

RSC Polymer Chemistry Series No. 1
Renewable Resources for Functional Polymers and Biomaterials
Edited by Peter A. Williams

Published by the Royal Society of Chemistry, www.rsc.org

CHAPTER 1

Natural Polymers: Introduction and Overview

PETER A. WILLIAMS

Glyndwr University, Plas Coch, Mold Road, Wrexham LL11 2AW, UK

1.1 Introduction to Biopolymers

Proteins and polysaccharides are found abundantly in nature and are major constituents of plants, animals and micro-organisms serving a number of important functions.[1,2] Proteins are composed of amino acids and although there are hundreds of different proteins they all consist of linear chains of the same twenty L-α-amino acids which are linked through peptide bonds formed by a condensation reaction. Each amino acid contains an amino group and carboxylic acid group with the general formula:

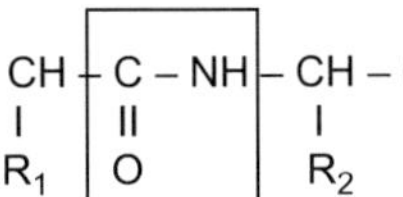

The R group differs for the various amino acids and can impart polar, non-polar, anionic or cationic characteristics. The amino acid units are linked together through a peptide bond to form a polypeptide chain as illustrated below:

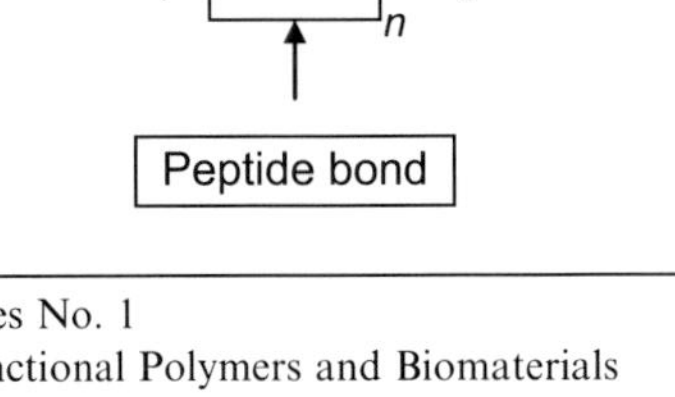

RSC Polymer Chemistry Series No. 1
Renewable Resources for Functional Polymers and Biomaterials
Edited by Peter A. Williams

Published by the Royal Society of Chemistry, www.rsc.org

Proteins consisting of between 15 and 10 000 amino acids are known. Since proteins contain both cationic and anionic charges due to the presence of ionisable groups, notably amine and carboxyl, they have a characteristic isoelectric point which corresponds to the pH at which the molecules have a net zero charge.

There are various levels of protein structure. The protein primary structure is defined by the characteristic sequence of amino acids of the polypeptide chain. Certain amino acids within the chain give rise to local secondary structures such as the α helix and the pleated sheet. There is a tendency for the more hydrophobic amino acids present to reside within the core of the molecule so that they are less exposed to the aqueous environment, and the overall shape of the protein that is formed (the tertiary structure) is stabilised by a range of interactions including hydrogen bonds, disulfide bonds and salt bridges. The protein molecules may self-associate to create a larger assembly referred to as quaternary structures.

Proteins tend to have either a linear or globular conformation and have a unique characteristic molecular mass. Linear proteins function as structural elements in the connective tissue of animals. The polypeptide chains are arranged in parallel forming long fibres. Fibrous proteins are insoluble in aqueous environments and are mechanically strong. Examples include collagen, found in tendons, cartilage and bone, and keratin, found in hair, skin and nails. Other proteins tend to fold into compact spherical or globular conformations. Globular proteins tend to be soluble in aqueous environments and are involved in transport processes or in dynamic functions in the cell. There are also special classes of proteins such as enzymes which are able to selectively cleave covalent bonds in biomacromolecules and antibodies which can attach to specific binding sites. Table 1.1 provides a list of common proteins together with their biological functions.

Table 1.1 List of common proteins and their biological function.

Protein	*Biological function*
Collagen	Fibrous connective tissue (tendons, cartilage, bone)
Elastin	Elastic connective tissue (ligaments)
Keratin	Hair, skin, nails
Sclerotin	Exoskeletons of insects
Fibroin	Spiders web
Myosin	Thick filaments in microfibril
Actin	Thin fibrils in microfibril
Haemoglobin	Transports oxygen in blood
Myoglobin	Transports oxygen in muscle cells
Serum albumin	Transports fatty acids in blood
Ovalbumin	Egg-white protein
Casein	Milk protein
Ferritin	Stores iron in the spleen
Gliadin	Seed protein of wheat
Soy	Seed protein of soya bean
DNA polymerase (enzyme)	Replicates and repairs DNA
Galactosidase (enzyme)	Cleave galactose glycosidic bonds

Proteins are sometimes referred to as 'simple' and 'conjugated'. Simple proteins consist purely of linear chains of amino acids while conjugated proteins have organic or inorganic components linked to the amino acid chain. Examples of the latter include lipoproteins, nucleoproteins and glycoproteins which have lipids, nucleic acids and carbohydrate groups attached respectively. The chapters by Holt and Pham and Mann and Turner review nucleic acid polymers and glycoproteins respectively.

Polysaccharides are composed of sugar units which are connected through so-called glycosidic bonds. The bond occurs by a condensation reaction between the anomeric carbon (C1) of one sugar residue and an oxygen atom from a hydroxyl group on another sugar residue. Since there are hydroxyls present on the C2, C3, C4 and C6 carbons the bonds formed can be $1\rightarrow2$, $1\rightarrow3$, $1\rightarrow4$, or $1\rightarrow6$ linked. Figure 1.1 shows disaccharide residues which are linked $1\rightarrow4$ and $1\rightarrow6$.

The rotation around the glycosidic bond and can be defined by the torsion angles ϕ, ψ, and ω. Since the hydroxyl groups on C1 can be above or below the plane of the ring the linkages are referred to as α or β. The nature of the glycosidic bond leads to a variety of secondary structures. For example, β-($1\rightarrow4$) linkages give rise to extended ribbon-like polysaccharide chains which are relatively stiff and such molecules commonly have a structural role in nature. A typical example is cellulose which consists of linear chains of β-($1\rightarrow4$) linked glucose residues. Amylose molecules on the other hand consist of α-($1\rightarrow4$) linked glucose residues and adopt a helical conformation. A more detailed

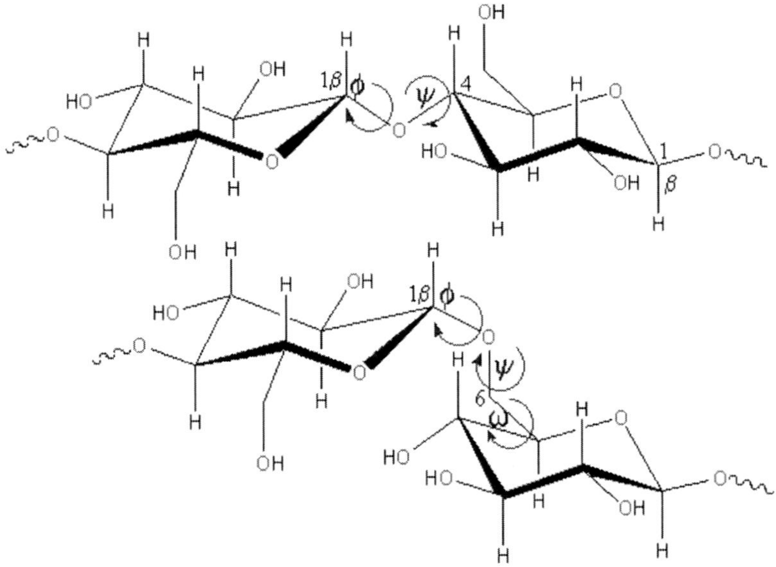

Figure 1.1 Disaccharide repeat units for glucopyranose units linked β-($1\rightarrow4$) (top) and β-($1\rightarrow6$) (bottom). ϕ, ψ and ω define the angles of rotation.

discussion of the structure and properties of cellulose and amylose are given in the chapters by Shibata and by Kennedy *et al.* respectively. Dextran consists of chains of β-(1→6) linked glucose residues and since there are three bonds of rotation between the sugar units the polysaccharide chain is very flexible and the molecules adopt a random coil conformation. It is also possible to have links through more than one hydroxyl group within a single sugar residue resulting in the formation of branched polymers. Gum arabic and amylopectin are classic examples of highly branched polymers and are considered in the chapters by Phillips and Phillips and by Kennedy *et al.* In addition, since there are many different sugar residues that are available as the primary building block, the number of possible polysaccharide structures is infinite. Polysaccharides that are composed of one type of sugar residue are referred to as homopolysaccharides while those composed of two or more types are referred to as heteropolysaccharides. Unlike proteins, polysaccharides do not have a unique molecular mass but tend to have a distribution of molecular mass species present. Polysaccharides perform a range of functions in animals and plants. Linear polysaccharides often have a structural role. Examples include cellulose and xylans (see the chapter by Heinze and Daus), which occur in trees and plants, carrageenans and alginates (see the chapter by Draget and Skjak-Braek) found in seaweeds and chitin (see the chapter by Kean and Thanou) found in the shells of crustacea. Glycogen and amylopectin are the main storage polysaccharides of animals and plants respectively and have highly branched structures containing α-(1→4) linked glucose residues with branches linked α-(1→6). Pectin, which consists of linear chains with branches located periodically along the chain, is found together with cellulose and hemicelluloses such as xyloglucans in the primary cell wall of plants and has a structural role. It is, however, a polyelectrolyte showing a strong affinity for calcium ions and is also involved in transport processes. Details of its production, properties and applications are presented in the chapter by Endress. A list of the common polysaccharides is given in Table 1.2.

1.2 Commercial Applications of Biopolymers

1.2.1 Market Size

Polysaccharides and proteins (often referred to commercially as hydrocolloids) are widely used in a broad range of industrial sectors including food, medical, pharmaceutical, cosmetics, packaging, surface coatings, adhesives, dyes and pigments, paper making, construction, industrial cleaning products, water treatment *etc.*[3–7] They are added to perform a number of functions including thickening and gelling aqueous solutions, acting as emulsifiers and foaming agents, inhibiting the settling and creaming of dispersions and emulsions, preventing ice crystal growth, adhesion and the formation of films and coatings.

Table 1.2 List of common polysaccharides.

Polysaccharide	*Main sources*	*Structure*
Botanical		
Cellulose	Trees, cotton	Linear chains of β-(1→4) linked glucose residues
Starch (composed of amylose and amylopectin)	Cereals, rice, sago	Amylose consists of linear chains of α-(1→4) linked glucose residues Amylopectin consists of chains of α-(1→4) linked glucose residues with extensive branching at C6.
Galactomannans (guar gum, locust bean gum, tara gum)	Seeds of guar, locust bean and tara	Linear chains of β-(1→4) linked mannose residues with galactose residues variably linked at C6 (on average every 2, 3 or 4 mannose units for guar, tara and locust bean gum respectively)
Glucomannan (konjac)	Tuber of *Amorphophallus konjac*	Linear chains of β-(1→4), linked glucose and mannose with some small branches on about every tenth unit in the backbone.
Pectin	Citrus, apple, sugar beet	Linear chains of α-(1→4) linked galacturonic acid (which may be esterified) residues interrupted by a small number of rhamnose residues which are the loci for branches consisting of neutral sugars.
Inulin	Chicory, Jerusalem artichoke	Chains of β-(1→2) linked fructose units terminated by a glucose residue.
Glucan	Oats and barley flour	Linear chains of mixed sequences of β-(1→3), and β-(1→4) glucose residues
Tree gum exudates		
Gum arabic	*Acacia senegal* and *Acacia seyal*	Highly branched with a core of β-(1→3) galactose residues with extensive branching at the C3 and C6 positions. Branches contain arabinose, galactose, rhamnose and glucuronic acid
Gum tragacanth	*Astragalus* species *(microcephalus, gummifer, kurdicus)*	Tragacanthin is a highly branched water soluble component and consists of a core of α-(1→4) galactose residues which ramified branches consisting of arabinose residues Tragacanthic acid is a highly branched water swellable component consisting of linear chains of α-(1→4) galacturonic acid with branches containing, xylose, fucose and galactose
Gum karaya	*Sterculia* species *(urens, villosa, setigera)*	Highly branched with a main chain of rhamnose and galacturonic acid with branches containing galactose and glucuronic acid
Seaweed polysaccharides		
Alginate	Brown seaweeds (*Macrocystis, Ascophyllum, Laminaria* and *Ecklonia* species)	Linear chains consisting of β-(1→4) linked mannuronic and guluronic acid residues which may occur in blocks or randomly.

Table 1.2 (*Continued*)

Polysaccharide	*Main sources*	*Structure*
Carrageenans	Red seaweeds (*Gracilaria*, *Gigartina* and *Euchema* species)	Linear chains of β-(1→3) linked galactose and (1→4) linked 3,6,anhydro αD-galactose repeat unit with varying degrees of sulfation. Forms a double helix.
Agar	Red seaweeds (*Gelidium* species)	Linear chains of β-(1→3) linked galactose and (1→4) linked 3,6,anhydro αL-galactose repeat unit. Forms a double helix.
Animal polysaccharides		
Hyaluronan (also produced by bacterial fermentation)	Rooster comb, bovine vitreous humour	Linear chains of a disaccharide repeat unit consisting of β-(1→4) glucuronic acid and β-(1→3) linked *N*-acetylglucosamine
Heparin	Porcine and bovine mucosal tissues	Linear chains with a disaccharide repeat unit of 2-O-sulfated iduronic acid and N-sulfated glucosamine.
Chitin	Exoskeletons of crustaceans such as crab, lobster and shrimp	Linear chains of β-(1→4) linked *N*-acetylglucosamine residues
Glycogen	Animal cells	Chains of α-(1→4) linked glucose units with extensive branching at C6.
Bacterial polysaccharides		
Xanthan gum	*Xanthamonas campestris* bacterium	Linear chains of β-(1→4) glucose residues with a trisaccharide side chain on every other glucose at C3 consisting of two mannose units with, glucuronic acid inbetween. Forms a double helix.
Gellan gum	*Sphingomonas elodea* bacterium	Linear chains with a tetrasaccharide repeat unit of β-(1→3) glucose, β-(1→4) linked glucuronic acid, β-(1→4) linked glucose, α-(1→4) linked rhamnose. Forms a double helix.
Dextran	*Leuconostoc mesenteroides* bacterium	Linear chains of α-(1→6) linked glucose with 5-35% branches linked α-(1→3) and α-(1→4) depending on the strain.

A major area of application is the food industry where they are employed to control the texture and organoleptic properties of food products. An international numbering system (INS) has been developed by the Codex Committee on Food Additives in order to be able to identify food additives in ingredient lists as an alternative to the declaration of the specific name. The INS is intended as an identification system for food additives approved for use in one or more member countries. It does not imply toxicological approval by Codex. The INS numbers for starch derivatives and other polysaccharide food additives together with their key functions are given in Tables 1.3 and 1.4.

Table 1.3 INS numbers for modified starches.

Modified starch	*INS number*
Dextrin (roasted starch)	1400
Acid treated starch	1401
Alkali treated starch	1402
Bleached starch	1403
Oxidised starch	1404
Monostarch phosphate	1410
Distarch phosphate	1412
Phosphated distarch	1413
Acetylated starch	1414
Starch acetate	1420
Acetylated distarch adipate	1422
Hydroxypropyl starch	1440
Hydroxypropyl distarch phosphate	1442
Starch sodium octenyl succinate	1450
Starch, enzyme treated	1405

The world market for hydrocolloids used in the food industry is of the order of US$ 4 billion *per annum*. Starch and starch derivatives are by far the most abundantly used accounting for some 70% of the total market value followed by gelatine (12%), pectin (5%), carrageenan (5%) and xanthan gum (4%). Locust bean gum, guar gum, alginates, and cellulosics account for most of the remainder. The usage levels are dictated by the cost and ease of supply. Starch is available from a variety of sources, notably cereals, rice and sago, and typically costs less than US$ 1 per kg, while gelatine, which is a form of denatured collagen, is derived from porcine and bovine skins and bones and costs ~US$ 4 per kg. Although xanthan gum is more expensive than starch and other food hydrocolloids (~US$ 10 per kg) it is becoming the thickener of choice in many applications because of its unique rheological properties. It has the ability to suspend particles at rest because of its high viscosity but is highly shear thinning and hence allows dispersions to flow on the application of shear. It is now used as a thickener in many industrial products. It is interesting to compare the cost of these food additives with hyaluronan. This polysaccharide forms highly viscoelastic solutions and, because it is biocompatible, finds application in ophthalmic surgery and in the treatment of arthritic joints (see the chapter by Vercruysse). It costs in the order of US$ 100 000 per kg.

1.2.2 Functional Properties

Biopolymers are finding increased application in a variety of products and processes where they provide a number of key functions, notably their ability to increase viscosity, form gels, stabilise emulsions and dispersions and form films. These are reviewed below.

Table 1.4 INS numbers for hydrocolloids.

Polysaccharide	*INS number*	*Function*
Alginic acid	400	Thickening agent, stabiliser
Sodium alginate	401	Thickening agent, stabiliser, gelling agent
Potassium alginate	402	Thickening agent, stabiliser
Ammonium alginate	403	Thickening agent, stabiliser
Calcium alginate	404	Thickening agent, stabiliser, gelling agent, antifoaming agent
Propylene glycol alginate (propane-1,2-diol alginate)	405	Thickener, emulsifier, stabiliser
Agar	406	Thickener, stabiliser, gelling agent
Carrageenan (including furcelleran)	407	Thickener, gelling agent, stabiliser, emulsifier
Processed Euchema Seaweed	407a	Thickener, stabiliser
Bakers yeast glycan	408	Thickener, gelling agent, stabiliser
Arabinogalactan	409	Thickener, gelling agent, stabiliser
Locust bean gum	410	Thickener, gelling agent
Oat gum	411	Thickener, stabiliser
Guar gum	412	Thickener, stabiliser and emulsifier
Tragacanth gum	413	Emulsifier, stabiliser, thickening agent
Gum arabic (Acacia gum)	414	Emulsifier, stabiliser, thickener
Xanthan gum	415	Thickener, stabiliser, emulsifier, foaming agent
Karaya gum	416	Emulsifier, stabiliser and thickening agent
Tara gum	417	Thickener, stabiliser
Gellan gum	418	Thickener, gelling agent and stabiliser
Gum ghatti	419	Thickener, stabiliser, emulsifier
Curdlan gum	424	Thickener, stabiliser
Konjac flour	425	Thickener
Soybean hemicellulose	426	Emulsifier, thickener, stabiliser, anticaking agent
Pectins	440	Thickener, stabiliser, gelling agent, emulsifier
Cellulose	460	Emulsifier, anticaking agent, texturiser, dispersing agent
Microcrystalline cellulose	460(i)	Emulsifier, anticaking agent, texturiser, dispersing agent
Powdered cellulose	460(ii)	Anticake, emulsifier, stabiliser and dispersing agent
Methyl cellulose	461	Thickener, emulsifier, stabiliser
Ethyl cellulose	462	Binder, filler
Hydroxypropyl cellulose	463	Thickener, emulsifier stabiliser
Hydroxyprpoyl methyl cellulose	464	Thickener, emulsifier stabiliser
Methyl ethyl cellulose	465	Thickener, emulsifier, stabiliser, foaming agent.
Sodium carboxymethyl cellulose	466	Thickener, stabiliser, emulsifier
Ethyl hydroxyethyl cellulose	467	Thickener, stabiliser, emulsifier
Cross-linked sodium carboxymethyl cellulose	468	Stabiliser, binder
Sodium carboxymethyl cellulose, enzymatically hydrolysed	469	Thickener, stabiliser

1.2.2.1 Viscosity Enhancement

Certain biopolymers are able to form viscous solutions at concentrations of 1% or less and are widely used as thickeners in a broad range of products ranging from soups through to paints. These include xanthan gum, carboxymethyl cellulose, non-ionic cellulosics (hydroxyethyl-, methyl, hydroxypropylmethyl-cellulose) and the galactomannans, notably guar gum and locust bean gum. The viscosity of polymer solutions varies as a function of the hydrodynamic volume of the molecules and hence is influenced by the polymer shape and molecular mass. Furthermore, charged polymers tend to expand due to intra-molecular repulsions between the charged groups located along the polymer chain and hence form more viscous solutions than their non-ionic counterparts. At polymer concentrations above the critical overlap concentration, where the molecules are able to interpenetrate, the viscosity is non-Newtonian and solutions display a shear rate dependence. Figure 1.2 shows the viscosity of a number of biopolymer solutions and it is noted that the viscosity–shear rate profiles vary considerably.

Xanthan gum is a very high molecular mass polymer ($>3\times10^6$) consisting of linear β-(1→4) linked glucose chains (cellulose-like) with a trisaccharide unit on every other glucose residue. Under ambient conditions in water it exists in the form of a double helix and hence the molecules are very stiff. As noted above, xanthan gum has a rather unique rheological profile, with a high low-shear viscosity and pronounced shear thinning characteristics, which makes it the thickener of choice in many applications. Carboxymethyl cellulose (CMC) has typical molecular masses of $\sim 3\times10^5$ and the molecules are not as stiff as xanthan. They form solutions which have a lower viscosity than xanthan at low

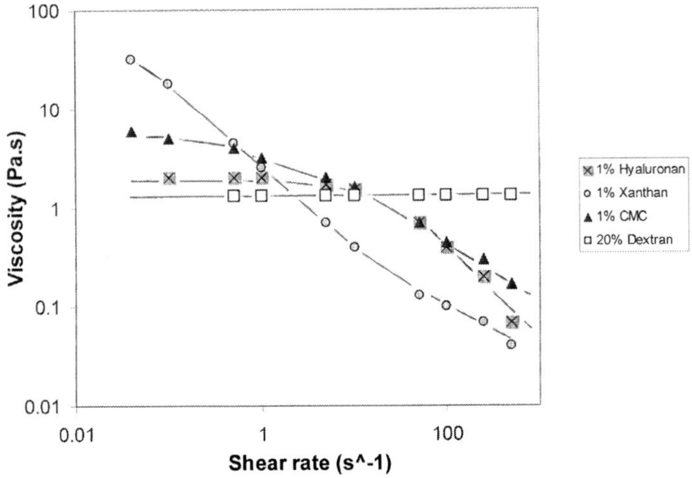

Figure 1.2 Viscosity of polysaccharide solutions as a function of shear rate.

shear rates but are less shear thinning. Dextran consists of mainly α-(1→6) linked glucose chains (with some branching at C3) yielding highly flexible molecules. Dextran does not form viscous solutions even at concentrations of 20%. Hyaluronan consists of linear chains of β-(1→4) and β-(1→3) sugar units with a molecular mass of $\sim 1 \times 10^6$ and produces solutions with lower viscosity than CMC.

1.2.2.2 Gelation

A number of biopolymers are able to form three-dimensional gel structures, at very low concentrations, by physical association of their polymer chains. This results in the creation of stable junction zones which may be formed through various interactions including hydrogen bonding, hydrophobic association and cation-mediated crosslinking. For example, it is argued that amylose molecules tend to self-associate in solution (a process known as retrogradation) through hydrogen bonding, while high-methoxyl pectin self-associates in the presence of high concentrations of sugar probably through both hydrogen and hydrophobic bonding. Low-methoxyl pectin and alginate form gels in the presence of divalent ions, typically calcium, which interact with the carboxylate groups promoting crosslinking of the polysaccharide chains. The nature of the junction zone formed on association of the chains is referred to as an 'egg box' and is discussed later in the chapter by Draget and Skjåk-Braek. For some biopolymers, the gelation process may be triggered by heating the polymer in solution. Typical examples include methylcellulose and hydroxypropylmethyl cellulose. On heating, the polymer molecules reach their lower critical solution temperature and self-associate as a consequence of phase separation thus forming a gel. The process is reversible and so on cooling the gel melts. In contrast a number of biopolymers including agarose, carrageenan, gellan gum and gelatine form gels on cooling. The mechanism of gelation is common for each. At high temperatures they exist in disordered conformations and on cooling undergo a coil–helix transition. The helices then associate to form a three-dimensional gel network (Figure 1.3). Again the process is reversible and so on heating the gel will melt but often the melting temperature is significantly higher than the gelation temperature due to extensive aggregation of the helices.

In contrast to gelatine, which is a linear structural protein, globular proteins will unfold on heating and at extremes of solution pH and ionic strength. The denatured chains will aggregate to form thermally irreversible gels. The properties of the gels will depend on a number of factors including the degree of unfolding of the protein chains and the extent and kinetics of chain aggregation. Association is likely to be driven by interaction between hydrophobic domains along the protein chains which become exposed when the molecules unfold.

Biopolymer gel formation only occurs above the critical minimum concentration required to provide connectivity and is specific for each polymer.

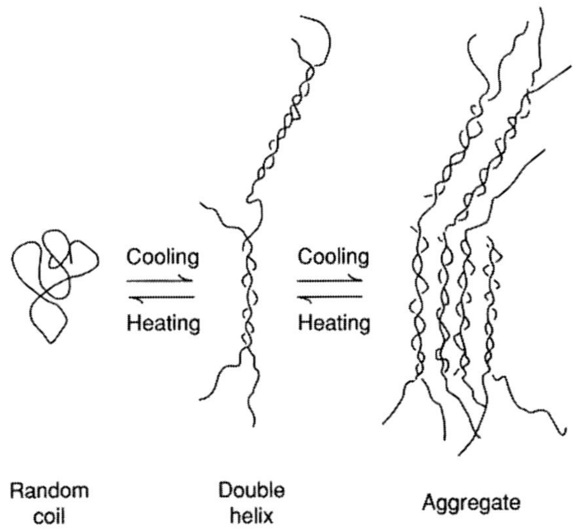

Figure 1.3 Schematic showing the thermally reversible conformational transition of certain polysaccharides which leads to gelation on cooling.

Below this concentration precipitation may result. The properties of biopolymer gels vary considerably in clarity, strength and elasticity due to differences in the degree of chain aggregation, the flexibility of the polymer chains, and the number and nature of the junction zones.

Polysaccharide gels are widely used in the immobilisation of cells. In the case of polysaccharides that form gels on cooling, notably agarose and carrageenan, a typical procedure is to add a solution of the polysaccharide solution containing the cells dropwise into a cold solution resulting in the formation of beads. Alternatively, the polysaccharide solution can be added to a vegetable oil at a temperature above the gelation temperature and sheared to form a water-in-oil emulsion. On cooling to below the gel temperature polysaccharide beads are formed. In the case of cell immobilisation using alginate, the alginate–cell solution is dropped into a solution of calcium chloride which leads to the formation of gel beads. Alternatively an alginate–cell solution containing a sparingly soluble calcium salt such as calcium carbonate can be used to form a water-in-oil emulsion. The calcium carbonate slowly dissolves leading to the formation of alginate gel beads. Polysaccharide gels are also used in tissue engineering to assist in the transport of cells and bioactive molecules to the desired sites within the body of patients who have suffered organ or tissue loss or malfunction.

1.2.2.3 Emulsification

Proteins are commonly used as emulsifiers for oil-in-water emulsions for application in foods and other industries and for encapsulation technologies.

For the latter the active compound is dissolved in the oil phase prior to forming the emulsion. Protein molecules are able to adsorb onto oil droplets as a consequence of their amphiphilic characteristics. On adsorption they often unfold such that the hydrophobic amino acids within their structure reside at the interface while the hydrophilic ones preferentially reside in the aqueous phase. The unfolding process exposes amino acids previously located in the hydrophobic core of the protein molecule which can lead to association with adjacent molecules through hydrophobic bonding and disulfide bond formation. Proteins, therefore, often form thin membranes at the surface of the oil droplets which are highly elastic. Since the thickness of the adsorbed protein layer is very small, typically just a few nanometres, proteins are generally not able to provide a steric barrier for stabilisation. Droplet aggregation can occur at pH values close to the protein isoelectric point and in the presence of electrolyte when electrostatic repulsive forces become inoperative. As a means of overcoming this problem, an anionic polysaccharide can be added to the protein stabilised emulsion forming a secondary layer at the interface through electrostatic interaction with the adsorbed protein. Such systems can then be spray dried to encapsulate the oil with an active compound dissolved within it.

In the food industry, gum arabic is the preferred emulsifier for flavour oils which are used in beverages and details of its functional behaviour are discussed in the chapter by Phillips and Phillips. The gum contains a small proportion of protein (2–3%) but this is present within certain fractions of the gum, notably the high molecular mass arabinogalactan protein fraction (AGP), rather than being uniformly distributed. The AGP fraction has in fact been found to contain ~10% protein. It has been argued that the polypeptide, which has amphiphilic characteristics, anchors the molecules to the surface of the oil droplets while the carbohydrate moieties protrude out into solution and confer stability through both electrostatic and steric mechanisms. Gum arabic stabilised emulsions are stable over a broad range of pH values and in the presence of electrolytes. Further details on the properties and function of gum arabic are given in the chapter by Phillips and Phillips. There is considerable interest nowadays in forming polysaccharide–protein complexes, either through electrostatic interaction or as Maillard conjugates, to match the emulsification performance of gum arabic.

1.2.2.4 Film Formation

Cellulose has been used for a long time for the production of films and membranes and is discussed in detail in the chapter by Shibata. Cellophane was invented almost a hundred years ago and is commercially produced by the extrusion of viscose (cellulose xanthate dissolved in alkali) into an acid bath yielding a transparent film. There has been considerable interest in recent years in the application of other polysaccharides such as starch, carboxymethyl cellulose, carrageenan, chitin and pectin and also proteins, such as whey, soy and wheat gluten for use in biodegradable packaging and as edible films to enhance the shelf life of fruit and vegetables. Edible coatings consisting of a polysaccharide, often with fatty acid esters incorporated, when applied to fruit and vegetables can provide a barrier to moisture and carbon dioxide/oxygen exchange and also improve mechanical handling properties. Some polysaccharides, for example pullulan, are now also being used to form films for administering drugs. The active compound is encapsulated within the film which, when placed on the tongue, dissolves instantaneously.

1.3 Scope of this Book

As noted above, polysaccharides and proteins have a wealth of functional properties that can be readily exploited in a broad range of industrial applications. This book considers some of the most important biomacromolecules that have application in the field of biotechnology and medicine. Consideration is given to polysaccharides obtained from various sources including botanical, algal, animal and microbial. Of the botanical polysaccharides, chapters are included on the medical applications of cellulose and its derivatives, the biotechnology and biomedical applications of starch and its derivatives and the production, properties and applications of pectins and exudate gums, notably gum arabic. A chapter is also included which considers the potential of xylans and xylan derivatives in biomedical and industrial applications and there is also a chapter on the general strategies for the isolation, separation and characterisation of polysaccharides. The algal polysaccharide alginate has significant application in food, pharmaceuticals and the medical field and a review is presented embracing both medical and biotechnological applications. With regards to polysaccharides of animal origin, there are chapters on the biocompatibility, biodegradability and biomedical applications of chitin and hyaluronan. Bacterial sources of the latter are also considered. The book also includes a chapter dedicated to microbial polyesters, with consideration given to their biosynthesis, properties and enzymic degradation.

A general review of the structural glycoproteins, fibronectin and laminin is presented together with their role in the promotion of cell adhesion in vascular grafts, implants and tissue engineering. Included also, is an introduction to nucleic acid polymers and genetic engineering.

References

1. G. E. Schulz and R. H. Schirmer (Eds), *Principles of protein structure*, 4th edn, Springer, 1996.
2. S. Dumitriu (Ed.), *Polysaccharides: Structural diversity and functional versatility*, 2nd edn, Marcel Dekker, New York 2005.

3. A. M. Stephen, G. O. Phillips and P. A. Williams (Eds), *Food Polysaccharides and their applications*, 2nd edn, CRC Taylor and Francis, Boca Raton Florida 2006.
4. G. O. Phillips and P. A. Williams (Eds), *Handbook of hydrocolloids*, CRC Woodhead Publishing Ltd, Boca Raton Florida, 2000.
5. A. Nussinovitch (Ed.), *Hydrocolloid applications*, Blackie Academic and Professional, London UK, 1997.
6. S. E. Hill, D. A. Ledward and J. R. Mitchell (Eds), *Functional properties of food macromolecules*, 2nd edn, Aspen Publishers Inc. Gaithersburg, Maryland, 1998.
7. G. Doxastakis and V. Kiosseoglou (Eds), *Novel macromolecules in food systems*, Elsevier Science B.V. The Netherlands, 2000.

CHAPTER 2

Natural Polymer Resources: Isolation, Separation and Characterization

WERNER PRAZNIK,[a] RENATE LÖPPERT[a] AND ANTON HUBER[b]

[a] Department of Chemistry, University of Natural Resources & Life Science, Vienna, Austria; [b] Institute of Chemistry (IFC), Karl-Franzens University of Graz (KFUG), Graz, Austria

2.1 Introduction

Different sources of natural polymers from plants, algae or animals, as well as increasingly produced by single cell biomass (SCB) are indicated in Figure 2.1. All of these polymers are synthesized by enzymatic processes under differing concentrations and localization of enzymes in different cell compartments *e.g.* cytoplasm, organelles, cell wall or even intracellular.

Likewise, biosynthesis of a given biopolymer is not limited to only one location – an initially formed biopolymer in one location may be transported into a different compartment or organelle of the cell where synthesis may be completed. Biopolymers may be synthesized in different ways: whereas nucleic acids and enzymatically active proteins are formed on the basis of genetic information and are thus strictly fixed in their molecular composition, structural proteins, polysaccharides, polyoxoesters and polyisoprenes, on the other hand, are formed without such genetic template information; their formation is controlled by the mass law, and therefore results in polydisperse materials. This polydispersity or distribution of their molar

RSC Polymer Chemistry Series No. 1
Renewable Resources for Functional Polymers and Biomaterials
Edited by Peter A. Williams

Published by the Royal Society of Chemistry, www.rsc.org

Table 2.1 Molar mass of common natural polymers, as determined by different methods.

Example	*Molar mass*	*Method*	*Ref.*
Homopolysaccharides			
Glucans:			
cellulose (cotton)	10^5–10^7 [a]	LS	2
cellulose (wood, unmodified)	10^5–10^6 [a]	LS	2
amylose (potato)	10^5–10^6 [a]	SEC–LS	3
amylopectin (potato)	1.7×10^8 [a] (M_w)	SEC–LS	3
amylopectin (waxy maize)	8.3×10^8 [a] (M_w)	SEC–LS	3
Fructans:	10^3–10^4	SEC	4
inulin (chicory)	10^3–3×10^4	SEC	4
inulin (globe artichoke)	1.6×10^7–	viscosity	5
levan (bacteria)	2.4×10^7 [a]		
Heteropolysaccharides			
pectin	10^4–2×10^5 [a]	viscosity, LS	6
arabinogalactan (larch)	5×10^4–10^5 [a]	SEC	7
arabinoxylan (rye)	10^5–10^6 [a]	SEC	8
Proteins			
albumin (human serum)	6.85×10^5		9
catalase (horse liver)	2.47×10^5	SDS–PAGE	9
gluten (wheat)	3×10^4–1.2×10^5	UZ	10
gliadin fraction (wheat)	3×10^4		10
Polynucleotides			
chromosomal DNA	10^7–10^8		9
Polyphenols			
lignin (wood)	10^5–10^8 [a]	SEC–LS	11
Polyisoprenoids			
natural rubber (*Hevea brasiliensis* L)	$>10^6$		12

[a]Apparent molecular weight.

masses (degree of polymerization) primarily depends on the polymerization conditions and reaction rates. For practical reasons molar mass distribution (molecular weight distribution) is generally expressed in terms of a mean value, *i.e.* weight average and number average molecular weights (M_w and M_n).[1] (Table 2.1)

Technical utilization of biopolymers from natural sources begins with their isolation or recovery, as illustrated in Figure 2.1, depending on whether deriving from either biomass material such as plants, algae or animals, or, increasingly, from biotechnological intra- and extracellular processing. The fermentative production of biopolymers by means of micro-organisms is limited by cytoplasmic volume, cell density and extraction procedures of the biopolymers produced. Alternative strategies for production of biopolymers are cell-free systems employing isolated enzymes (*e.g.* the application of heat stable DNA polymerases in the polymerase chain reaction, PCR) on the one hand, or fermented production of monomer constituents followed by subsequent polymerization (*e.g.* lactic acid produced by fermentation, followed by chemical polymerization to polylactic acid) on the other. Finally, transgenic

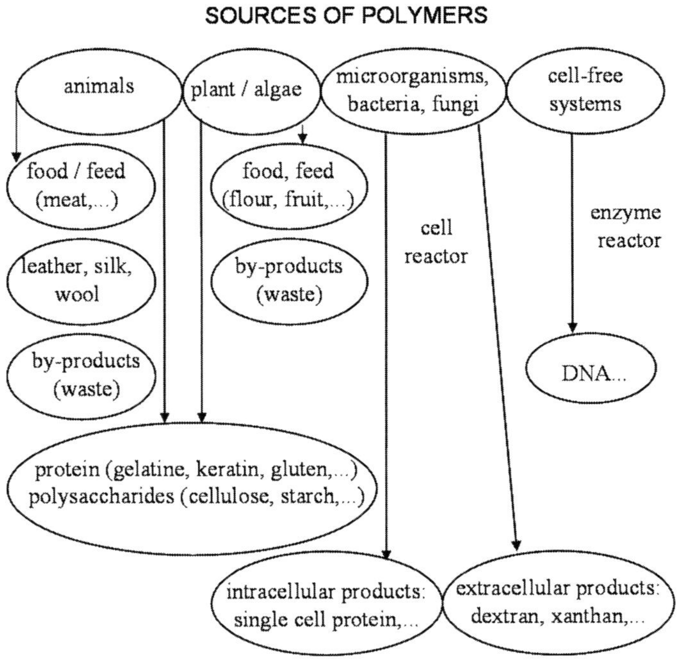

Figure 2.1 Different sources of natural polymers, their inter-relationships, and their main uses.

plants producing specifically required substances might also be an option for tailor-made products in the future.

This chapter focuses on polysaccharides and proteins sourced from native and genetically unmodified plants, algae, animals and selected microorganisms. Due to the wide range of their functionalities, utilization of materials from natural sources is rather tricky as many superimposed heterogeneities often block even the simplest processing steps. The only way to overcome these problems is the comprehensive structural analysis, identification and quantification at every stage of their processing.

A systematic description of various conventional and modern instrumental and enzyme specific methods available for isolation, quantification and characterization of all types of natural polymers, including polysaccharides, proteins, polyesters and polyisoprenes is provided in this chapter. Particular attention is given to plant polysaccharides for providing conclusive information about their molecular background and macroscopic material qualities to provide the full profile of the native raw materials from crops to facilitate their most efficient utilization.

2.2 Established Analytical Techniques in Characterization of Natural Polymers

A number of analytical approaches for identification and quantification of natural polymers have been published within the last decades. Standardized

Table 2.2 Analytical methods used for isolation and characterization of polysaccharides and proteins.

Analytical process	*Method used*	
	Polysaccharides	*Proteins*
Separation and fractionation	Solubility in polar, nonpolar solvents; preparative SEC, HPLC, IEC, HPAEC-PAD	Salt precipitation (hydrophobicity differences); preparative techniques (IEC, HIC, SEC); electrophoresis; preparative UC
Identification and structural analysis	sugar moieties (TLC, HPLC, CE, HPAEC-PAD); methylation analysis (glc/MS/FID); enzymic analysis + LC	amino acid analyzer, HPLC with fluorescence detection; SDS-PAGE, CE
Molecular dimension	LS, viscosity, osmotic pressure; SEC, universal calibration; SEC, multiple detection, molar and mass (apparent and de facto)	2D electrophoresis–MS; UC, LS; SEC
Quantification	photometric techniques, enzymic analysis	photometric techniques, microanalysis, Kjeldahl

and evaluated techniques have been developed and issued by governmental and commercial organizations.[13–16] A general overview about characterization techniques for polysaccharides and proteins is given in Table 2.2. Most of these analytical techniques are rather complex with respect to applied instrumental equipment and time, a fact which in particular causes problems if adjustments/modifications of applied techniques, depending on actually screened natural materials, are necessary.

Elementary analysis of plant materials following their isolation, purification, homogenization and drying provides information about the contents of their nitrogen (N), carbon (C) and sulfur (S). According to standard methods, the content of the mineral elements is obtained by means of atomic absorption spectroscopy (AAS) of ash, giving information about concentrations of sodium (Na), potassium (K), calcium (Ca), magnesium (Mg), iron (Fe) and aluminium (Al) in the investigated samples. After hydrochloric acid extraction of ash, the content of phosphorus (P) is determined in the soluble portion, whereas silica (Si) is determined in the insoluble residue. Finally, the contents of trace elements may be determined according to established extraction procedures combined with AAS or inductively by coupled plasma–mass spectrometry (ICP–MS).[13,16]

The protein content of plant material can be determined by several standard methods that are generally based on quantification of nitrogen (N) as a protein equivalent. The determination of total nitrogen according to Kjeldahl's method[13] as well as nitrogen-specific elementary analyses may be the most appropriate methods. However, these techniques do not allow discrimination of low molecular weight N-compounds such as nitrate, nitrite, amino acids and peptides from proteins. For the classification of nitrogen according to specific classes of compounds, additional preparation steps, sample specific pretreatment and particular analytical techniques must be applied. Soluble proteins, for instance, may be extracted by means of particular buffer systems and quantified by biochemical assays or by photometric methods. Electrophoresis and chromatography, combined with specific staining and labeling, provide detailed information about protein composition and sample specific enzyme patterns.[17–19]

2.3 Characterization of Natural Polymers in Fiber Crops

Fiber plants constitute a wide range of plant species belonging to different taxonomic classes well adapted to the prevalent climatic conditions. In humid and moderate regions such as Central and Southern Europe flax is the main cultivar; cultivation of nettle fibers (great stinging nettle) is possible in these regions as well. In moderate and warmer regions, such as southern areas of Asia like Indonesia, hemp is the predominant fiber plant. A large variety of fiber plants such as cotton, jute, palm tree and agave species (sisal) are also cultivated in tropical and subtropical areas. All these plants contain cellulose, hemicellulose and lignin in different proportions and qualities. Cellulose fiber in plants is present either in the form of seed hairs or is produced by the shoot axis, leaves or the pericarp.

Some applications of plant fibers depend on morphological and technological properties, *e.g.* length and radius of the individual fibers or fiber bundles, tensile strength, breaking elongation, heat resistance, color and density. Plant fibers are rather low in density (1.2–1.4 g cm^{-3}, compared to 2.6 g cm^{-3} for glass fibers, for instance), and thus, are interesting materials for molded parts in the automotive industry.

Analysis of fiber crops (hemp, flax, millet, grasses, *etc.*) first requires representative sampling from selected components (Figure 2.2). For the purpose of reproducibility, materials of interest must be cleaned, pre-selected, weighed and stabilized at the location immediately after harvesting. For subsequent analysis, sample components must be homogenized and dried.

In order to eliminate low molecular weight phenolic compounds and chlorophyll from the fiber material, the initial step in analysis is a non-polar extraction. Subsequent polar extraction of carbohydrates by means of varying ratios of alcohol–water mixtures provides a step by step process to dissolve mono-, di-, oligo- and polysaccharides. The materials remaining after these steps are hemicellulose, cellulose, lignin and other non-aqueous soluble cell wall components. Identification and quantification of these materials, depending on crop variety and available sample quantity, can be achieved by TAPPI methods.[16]

Hemicelluloses predominantly consist of pentoses, *i.e.* xylose and arabinose, and are extracted by alkaline treatment, and identified and quantified by chromatographic methods subsequent to total (acidic) hydrolysis. Fast identification of monomers may be achieved by thin layer chromatography (TLC), quantification by calibrated reverse phase HPLC, anion exchange chromatography combined with pulsed amperometric detection (HPAEC-PAD, Dionex),

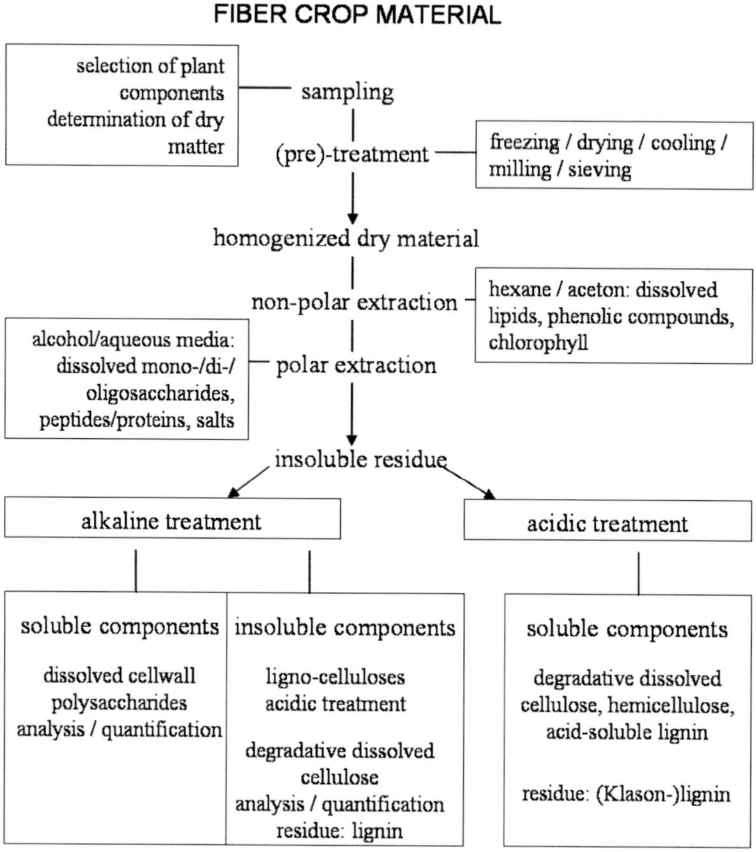

Figure 2.2 Analytical strategy for the separation of cell wall polysaccharides of fiber crops.

or gas–liquid chromatography (GLC) after derivatization.[19–21] The characterization of lignin requires the total hydrolysis of cellulose in lignocellulose (holocellulose) which is achieved by using 72% w/w sulfuric acid (H_2SO_4). This process degrades cellulose to glucose and forms sulfate derivatives of lignin (Klason lignin) which are determined gravimetrically after hot water washing and drying. An optional approach to characterize the cellulose matrix is the oxidative elimination of phenolic lignin (treatment with sodium hyperchlorite or peracetic acid) and gravimetric quantification of the residue (cellulose).[22]

2.4 Characterization of Plant Cell Wall Polysaccharides

There is no general method available for the extraction of cell wall polysaccharides from plants. Polysaccharides, particularly arabinoxylans, arabinogalactans, mixed linked β-glucans and pectins, including associated neutral glycans, are composed of soluble (or extractable) and insoluble constituents. The ratio of soluble to insoluble fractions depends on conditions and treatments, in particular on applied enzymes, temperature and period of solubilization.

Subsequent to the elimination of low molecular weight carbohydrates (by ethanol–water mixtures), of proteins and storage carbohydrates such as starch or inulin (by enzymic hydrolysis), and of pectins (by aqueous chelating agents), the noncellulosic polysaccharides are preferentially extracted with aqueous alkali containing sodium borohydride. Borohydride converts the reducing terminal hemiacetal of polysaccharides into hydroxy groups, and thus decreases alkaline degradation. To allow the required structural analysis of the contained polysaccharides, the alkaline polysaccharide extracts often require a further purification step which can be achieved by, for instance, preparative liquid chromatography.[23]

The primary structure of plant cell wall polysaccharides is determined by an analytical strategy similar to that for fiber crops. Methylation analysis combined with gas liquid chromatography–mass spectrometry (GC-MS), or gas liquid chromatography–flame ionization detector (GS-FID), is applied to identify the kind and position of glycosidic linkages and mean branching characteristics. This approach provides quantitative information about the relative amounts of terminal, intra-molecular linked and branched sugar units, and enables the computation of the mean degree of polymerization (DP) and reconstruction of the mean structural features of the test polymer.[24] More specific methodologies required for cereal, fruits and vegetables are discussed below.

2.5 Characterization of Structural Cereal Polysaccharides

Cell walls of cereal tissues are multicomponent matrices of cross linked polymers with embedded networks of cellulose microfibrils. The matrix is composed of mixed linked β-glucans, arabinoxylans, acidic xylans and cellulose. An additional network formed by structural proteins (*e.g.* glycine-rich), and by threonine- and hydroxyproline-rich glycoproteins, also penetrates this matrix. Minor amounts of lignin and phenolic acids may also be covalently linked to the polysaccharide matrix.[25]

Cellulose is a β-(1→4) linked homopolysaccharide formed by D-glucopyranosyl residues. The symmetry of this molecule is a β-sheet stabilized by intra- and intermolecular hydrogen bonds. Due to the high symmetry and intermolecular stabilization, native cellulose is highly crystalline and more or less insoluble. The more or less pronounced elasticity of cellulosic materials is largely attributed to the alternating regions of crystalline and less ordered amorphous chains.

β-Glucans, occurring in barley and oat grains, are mainly composed of a polymeric backbone consisting of C3-linked cellotriosyl and cellotetrasyl units and minor branches of protein, arabinose and xylose residues. Molecular weights between 20 000 and 40 000 000 are reported for these materials, mainly depending on the applied separation procedure and the preferred analytical technique.[26]

Another major group of cereal polysaccharides is arabinoxylan which consists of a main chain of β-(1→4) linked D-xylopyranosyl residues, substituted at the C2 or C3 position by arabino-furanosyl residues. Arabinoxylans may be substituted by phenolic acids (*e.g.* ferulic acid). Distribution of arabinosyl substituents happens blockwise, rather than homogeneously, *i.e.* some segments carry arabinosyl-substituted xylose clusters, and other segments contain relatively few arabinosyl residues, thus easily being attacked by xylanases. For instance, several classes of rye arabinoxylans are distinguished with respect to their extractability and structure.[27,28]

2.6 Characterization of Pectic Polysaccharides in Fruits and Vegetables

The heterogeneous polysaccharide pectin occurs in the cell walls of plants, and contributes to many physiological aspects related to growth, cell size and shape, integrity and rigidity of tissues, ion transport, water binding capacity, and defense mechanisms against infections and associated damage. The amount and nature of pectin strongly influence the texture of fruits and vegetables during growth, ripening and storage.[29]

Pectins consist of galacturonic acid units with varying amounts of branches consisting of neutral sugars. The building units are composed of "smooth" homogalacturonic segments and rhamnose dominated "hairy" segments.[30] Galacturonans predominantly consist of α-(1→4) linked galacturonic acid residues with minor amounts of symmetry-breaking rhamnose in the homogalacturonan sequence. Citrus and apple pectins typically contain at least 70–100 galacturonic acid residues, 60–90% of which occur as methyl esters.

Depending on the plant species the hairy segments of pectins may consist of the following three different subunits.

1. Rhamnogalacturonan consisting of a backbone of alternating (1→4) linked galacturonic acid and (1→2) linked rhamnose residues partly substituted with galactose residues (1→4) linked to the rhamnose units.
2. Arabinogalacturonan substituted with relatively long chain arabinan branches and predominantly terminal arabinose residues, (1→3) and (1→3,5) linked.
3. Xylogalacturonan subunit (confirmed in apple) consisting of a galacturonan backbone with xylose residues on C3.

Homogalacturonan-rich pectins as obtained by acid extraction of sugar beet pulp are used as a gelling agent following oxidative cross linking. Moreover, if partially deacetylated and demethylated, these pectins even form gels in the presence of multivalent cations, preferably with calcium ions.

The gel formation capacity of pectins primarily depends on their methoxy content, pH, concentration of sugar and Ca^{2+} ions. Pectins are well suited to the production of films which maintain their properties at constant (environmental) conditions but, however, are sensitive to any changes resulting in swelling phenomena, a fact which makes them highly interesting for encapsulation and controlled release formulations.[6]

2.7 Characterization of Chitin and Chitosan

Chitin is a water insoluble polymer with β-(1→4)-linked glucopyranosyl units (GlcNAc) carrying C2-acetamido substitution. It is abundant in the exoskeleton of insects and crustacea (*e.g.* shells of shrimps, lobsters, prawns and crabs) and in the cell walls of many fungi and algae. Chitosan is a copolymer of β-(1→4)-linked D-glucopyranose with varying ratios of random C2-acetamido (NAc) and C2-amino (NH_2) substitution. In nature, it is found only in the cell walls of certain fungi, but is produced by partial alkaline hydrolysis (deacetylation) of chitin for use in various applications.

Chitin and chitosan may be distinguished from each other by their solubility in dilute acidic solutions (formic, acetic, citric acid, *etc.*), wherein chitosan is soluble and chitin is not.[31] The functionality of chitosan primarily depends on its degree of polymerization (DP) and on the proportion and heterogeneity of its NAc groups. Analysis of these qualities may be achieved by several techniques such as IR and UV spectroscopy, size exclusion chromatography, colloid titration, elementary analysis, dye absorption and acid–base titration.[32] An enzymatic method based on total hydrolysis, prior to colorimetric or HPLC determination of GlcNAc and GlcN may also be used.[32] An alternative to total enzymatic hydrolysis is acidic hydrolysis of soluble and insoluble fractions combined with HPLC analysis of N-acetyl groups.[33] Additionally, proton NMR spectroscopy, in particular for the quantification of NAc substituents, is a convenient technique for the analysis of chemical composition of chitosan.[34]

2.8 Characterization of Mucilage and Gums from Plants and Algae

Gums are cell wall components or reserve polysaccharides of plants and algae. They are generally considered as indigestible for humans, and thus useful as dietary fiber. Gums, however, are degraded and metabolized by intestinal microbes, and hence provide added energy value in human nutrition. Primary industrial applications of plant and algal gums in food production are hydrocolloid texture modifiers and viscosity enhancers.[35] Mucilages are sticky polysaccharide films secreted by plants and algae with strong water binding capacity.[7,36]

The most commonly used gum type reserve polysaccharides of plants are galactomannans of legume seeds (guar gum, locust bean gum, tara gum and cassia gum) and glucomannans of *Amorphophallus konjac* tubers. Mixed moieties of xylose, arabinose, rhamnose, mannose, and glucuronic and galacturonic acid are contained in galactans of okra, psyllium, quince seed gum (mucilages), and in plant exudates of gum arabic, karaya gum, gum tragacanth and gum ghatti.

Cell wall mucilages of marine algae and seaweed polysaccharides are excellent sources for hydrocolloids with a high water binding capacity and excellent gel formation capacity.

2.9 Characterization for General Identification and Typing

For the structure analysis of gums and mucilages, the techniques are similar to those applied for cell wall polysaccharides, but analysis of blends of gums (*e.g.* carrageenan–locust bean gum) is a challenge in itself. For instance, glucuronic and mannuronic acids are predominant in the polysaccharide structure of alginates, whereas the main chain in agar, carrageenan and furcelluran is galactan with D-galactose and 3,6-anhydro-D-galactose with different linkages and degrees of sulfate-substitution.

Identification and determination of different carbohydrate ratios are the initial steps. Total hydrolysis and monosaccharide analysis by chromatographic methods (TLC, HPLC or HPAEC-PAD) are additional useful tools.[37]

2.10 Isolation and Identification of Sugar Residues in *Ocimum basilicum* L

The alkali soluble polysaccharides of mucilage from *Ocimum basilicum* L become increasingly important for pharmaceutical applications, in particular as slimming products and laxatives. Their functionality depends on monosaccharide composition, organization of sugar moieties, kind of linkages and molecular weight distribution.

The characterization of these polysaccharides requires a multistep approach – extraction of mucilage by aqueous swelling of *O. basilicum* L seeds, followed by separation of the seeds, drying in hot air and finally grinding to powder. The isolation of the polysaccharide is achieved by treatment with 4 M NaOH at 60 °C overnight, followed by neutralization and precipitation in methanol. Subsequent preparative and calibrated analytical size exclusion chromatography (SEC) provides molecular weight characterization.

Sugar moieties in the freeze-dried polysaccharide fractions are identified by TLC after total hydrolysis with 2 M TFA (1 mg per 100 μL, 80 °C for 6 h), then evaporation under N_2 stream with methanol several times to remove the volatile TFA completely. TLC analysis is typically as follows:

Aluminium sheets with silica gel 60, sample and reference performance with Linomat (CAMAG), eluent $CH_3CN : H_2O$ ratio 17 : 3, developed twice; stained with 0.2% solution of thymol in methanol with 5% H_2SO_4. The subsequent application of several software packages, including scanning, digitization by means of Un-Scan-It program (USI), computation, and visualization (MSExcel and CPCwin32), provides data for the quantification of the carbohydrates present in the mixture.

The SEC profile of alkali soluble mucilage from *O. basilicum* L contains two populations with a bimodal distribution of polysaccharides (PS1 and PS2) in approximately equal amounts, but differing in molecular weight range (Figure 2.3). The applied preparative SEC (Figure 2.4) separates both polysaccharides, but not quantitatively as evidenced by subsequent analytical SEC, with pool 1

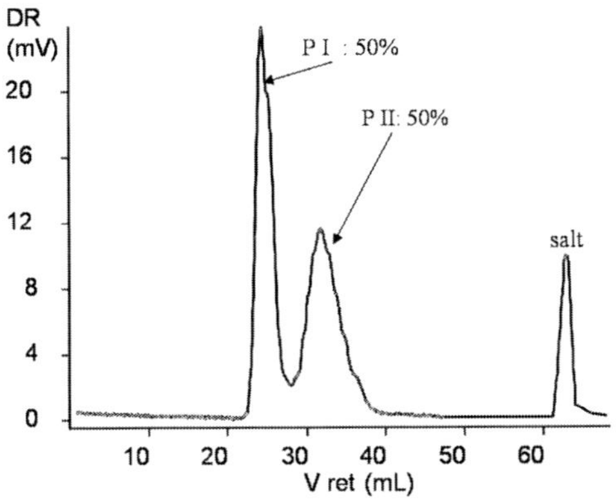

Figure 2.3 Analytical size exclusion chromatography (SEC) of alkali soluble mucilage from *O. basilicum* L. P I – polysaccharide 1, P II – polysaccharide 2. Column: Superose 6 + Superose 12 (high performance); eluent: 0.05 M NaCl; flow rate: 0.6 mL min^{-1}; injected volume of 0.1% carbohydrate: 0.3 mL; DRI detection (mV); calibration with dextran oligosaccharides; polysaccharide PS I: M_w 70 920, M_n 65 910; polysaccharide PS2: M_w 35 670, M_n 18 950.

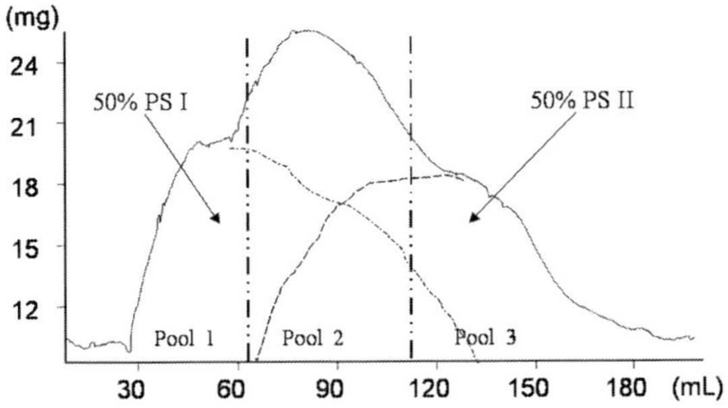

Figure 2.4 Preparative size exclusion chromatography (SEC) of alkali soluble mucilage from *O. basilicum* L. PS I – polysaccharide 1, PS II – polysaccharide 2. Column: Superose 6 prep (column size 15 × 500 mm); eluent: 0.05 M NaCl; flow rate: 0.6 mL min^{-1}; injected volume of 1.5% carbohydrate: 2.2 mL; DRI-detection (calculated in mg); pool 1 containing pure PS1, pool 2 containing 30% PS1 and 70% PS2, pool 3 containing 15% PS1 and 85% PS2.

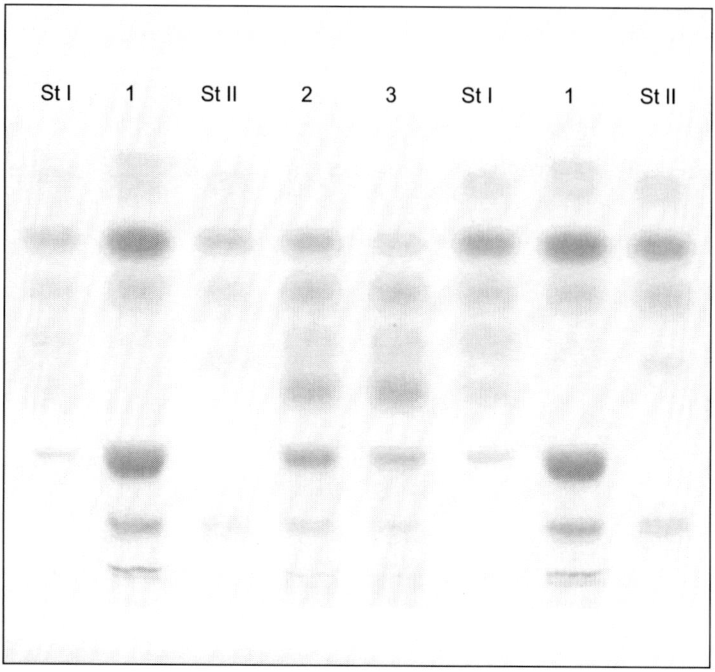

Figure 2.5 TLC analysis of pool 1 (1), pool 2 (2) and pool 3 (3) of prep SEC (see Figure 2.4). Developing system: CH_3CN/H_2O, 17 : 3; visualization with Thymol in CH_3OH with 5% H_2SO_4, standard 1 (St 1 from top to bottom): Rha, Xyl, Ara, Man, Gal, GluUA (each 0.25 mg mL^{-1}); standard 2 (St 2 from top to bottom): Rha, Xyl, Ara, Glu, GalUA (each 0.25 mg mL^{-1}).

containing pure PS1, pool 2 containing 30% of PS1 and 70% of PS2, and pool 3 containing 15% of PS1 and 85% of PS2.

Quantitative TLC analysis subsequent to total hydrolysis of preparative SEC pools enables detailed analysis of monomer composition in the different pools, as follows (Figure 2.5).

Pool 1 (pure PS1) consists of rhamnose (10%), xylose (27%), arabinose (13%), glucuronic acid (36%) and galacturonic acid (14%); pool 3 (85% PS2 and about 15% PS1) consists of xylose (5%), arabinose (18%), mannose (23%), galactose (44%), glucuronic and galacturonic acids (10%) and glucose (trace).

The results suggest that PS1 is an acidic polysaccharide composed of glucuronic and galacturonic acids containing xylose, arabinose and rhamnose, and PS2 is a neutral polysaccharide consisting mainly of galactose and mannose in a ratio of approx. 2 : 1 with additional minor amounts of arabinose units.

2.11 Characterization of Plant Reserve Polysaccharides

For the isolation of starch and fructan, both major reserve polysaccharides in a variety of quite different crops, a number of approaches can be used, all of which include aqueous processing of seeds, stems, tubers or roots. However, details of established processes employed for industrial handling of different crops (potato, maize, wheat, chicory, agave and others) differ from each other to ensure the highest yield and purity of the polysaccharide of interest. There are also worldwide efforts, including the tailoring of crops by genetic modification, to improve the quality of starch- and fructan-containing crops for both food and non-food applications. In both cases, each modification must be identified and evaluated by continuous screening or analysis of individual samples. Identification and evaluation may be applied at two ends – on samples that have been taken directly from the crop source at the location of breeding or harvesting and on samples that have already undergone industrial processing. Irrespective of the source of the sample, careful isolation and preparation is the first crucial step in any laboratory procedure for reliable characterization.

2.12 Characterization of Fructans

Fructans are oligomers and polymers consisting of predominantly β-D-fructofuranosyl residues (Figure 2.6). Naturally occurring fructans contain β2,1) and β2,6) linkages; C3, C4 and C5 are never involved in their glycosidic linkages. In most cases fructans contain a terminal α-D-glycopyranosyl residue from the basic metabolite sucrose which blocks the reducing terminal hemiketal group. β-D-Fructofuranosyl residues linked to C6-position of the α-D-glucopyranoside produce non-terminal glucopyranosyl residues, constituting neo-kestose-type fructans. Contrary to most glucans which contain a reducing terminal hemiacetal, fructans do not have a reducing terminal hemiketal, but a full ketal with a glucopyranosyl residue.[4,38]

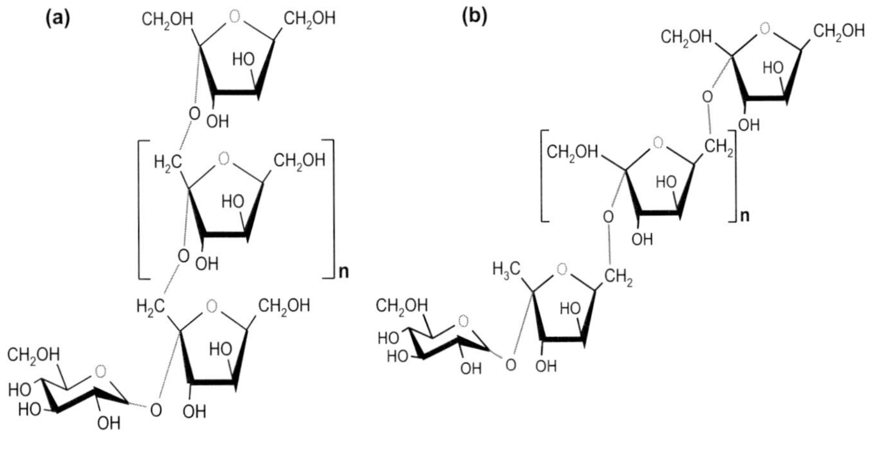

Figure 2.6 Chemical structures of different fructan types: (a) DP13-Inulin sequence (1-kestose-β2,1-Fruf_{10}); (b) DP13-Levan sequence (6-kestose-β2,6-Fruf_{10}).

The most important and commercially exploited fructan crop varieties are dicotyledons, in particular *Asteraceae* (chicory, Jerusalem artichoke, globe artichoke). This plant family contains predominantly an inulin-type fructan (β2,1 linked) with varying degrees of polymerization, genetically imprinted by the diversity of plant species. Levan-type fructan with β2,6 linkages have variable branching characteristics as in *Poaceae* (*i.e.* cereals including wheat, rye, barley, oat and grasses), and complex structured fructans with β2,1 and β2,6 linkages in the same molecule found predominantly in *Liliaceae* (garlic, onion, leek and red squill), *Agavaceae* (blue agave), *Iridaceae* (iris), and *Asparagaceae* (asparagus). The mixed structure fructans with high branching provide high solubility in aqueous media, and do not precipitate under chilly conditions in water, in contrast with inulin solutions which are unstable and precipitate.[4]

2.12.1 Isolation and Polymer Characterization

The efficient use of fructan-containing crops in food, feed and non-food products, as well as in pharmaceuticals, requires the full knowledge of the composition, structural details, degree of polymerization, polydispersity, and the amount of fructan in storage organelles of the plant. To obtain reliable and reproducible information about selected organs of fructan containing crops, representative homogenized samples are first prepared by freeze-drying. Apolar components such as polyphenolic compounds, chlorophyll and others must be eliminated by apolar extraction with hexane–acetone, in particular for samples from the leaves and stem. Then water soluble components such as carbohydrates may be extracted by hot water extraction at 80 °C, leaving hemicellulose, cellulose, lignin and water insoluble cell wall components as residues. Enzymatic approaches, chromatographic techniques, controlled fragmentation combined with methylation, and reductive cleavage and fragment analysis (GC-MS and FID) are useful tools for determination of total fructan content, degree of polymerization distribution and structural features.[4]

Enzymic assays and various modes of liquid chromatography, *e.g.* TLC or anion exchange chromatography combined with pulsed amperometric detection (HPAEC-PAD, Dionex), reversed phase or aminated high performance matrices (rpHPLC, NH2HPLC) with an evaporating light scattering detector (ELSD), are also used for qualitative and quantitative analysis of the composition of mono- and disaccharides and low molecular weight fructans in samples.[39–43] In particular, for mid-range and high DP fructans, calibrated size exclusion chromatography (SEC) with mass detection from refractive index provides quantitative information about molar mass and degree of polymerization distribution. From this distribution, mean molar mass values, number average molar mass M_n (DP_n), weight average molar mass M_w (DP_w) and polydispersity as the ratio of M_w/M_n (DP_w/DP_n) can be calculated relatively easily, providing essential characteristic data for the fructan sample.[44,45]

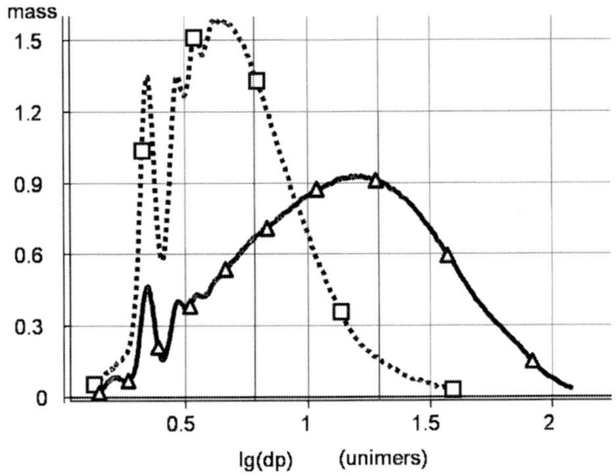

Figure 2.7 Degree of polymerization (DP) distribution of inulin from tubers of Jerusalem artichoke (analysis by calibrated size exclusion chromatography). Triangles: autumn harvest: $DP_w = 19$, $DP_n = 9$ (calculation including mono- and disaccharides). Squares: spring harvest: $DP_w = 6$, $DP_n = 4$ (calculation including mono- and disaccharides).

As an example, DPs of inulin from Jerusalem artichoke by SEC analysis are presented in Figure 2.7. The results show rapid changes in inulin polymer distribution at different harvest terms, indicating that the carbohydrate constitution and polymer distribution of inulin in the tubers are highly dependent on soil profile, climate and weather condition (temperature and rainfall) during the vegetation period. Knowledge of these properties for different plant species allows the specific application of different crops for food and non-food use.

2.12.2 Structural Characterization by Methylation, Acetylation, Controlled Fragmentation and Chromatographic Fragment Analysis

The combination of methylation, reductive hydrolysis (cleavage) and acetylation, followed by identification of obtained fragments by means of gas–liquid chromatography (GLC) with flame ionization detection (FID) or mass spectroscopy (MS) provides a powerful tool for structural analysis of fructans.[46–48]

For permethylation, the well dried sample is dissolved in dimethylsulfoxide [$(CH_3)_2SO$, DMSO], alkalized with thoroughly dried powdered sodium hydroxide (NaOH), stirred briefly, and then methylated with iodomethane (CH_3I) (only absolutely anhydrous conditions enable successful analysis). After the reaction water is added to the mixture for quenching, the resulting derivatives are extracted into dichloromethane (CH_2Cl_2) or chloroform (CH_3Cl); residual water traces are removed with dry sodium sulfate (Na_2SO_4), the organic phase is filtered and evaporated under nitrogen.

For reductive cleavage the dry permethylated polysaccharide is dissolved in a small quantity of dichloromethane (CH_2Cl_2) dried over a molecular sieve; trimethylsilane [$(C_2H_5)_3SiH$] is added as a reducing agent and trimethylsilyl-trifluormethansulfonate [$(CH_3)_3SiOSO_2CF_3$] as a catalyst and the solution stirred at room temperature over night

For acetylation of the hydroxy groups liberated at the cleavage stage, acetic anhydride [$(CH_3CO)_2O$] is added and the solution stirred for 2 h. After adding more CH_2Cl_2 and neutralizing with saturated solution of sodium bicarbonate ($NaHCO_3$), the organic phase containing the methylated derivatives is quenched again with water, dried (removing H_2O) over dry Na_2SO_4, filtered, and the organic phase evaporated under nitrogen.

For GC-FID analysis, the methylated anhydro alditols are redissolved in CH_2Cl_2, and analyzed in a suitable column (*e.g.* Durabond DB-1701), as follows:

Column: L 30 m, ID 0.25 mm, coat 0.25 μm; injected volume 2 μL, split 1 : 5; temperature program 100–250 °C, 4 °C min^{-1}; carrier gas, helium (pressure 1–1.2 bar); temperature: injection 230 °C, detection 300 °C.

The resulting methylated and acetylated 2,5-D-anhydroglucitols and 2,5-D-anhydromannitols provide quantitative information about relative amounts of different kinds of glycosidic linkages and mean DP values which enable the structural modeling of the polysaccharide.

Following methylation and reductive cleavage, aldoses (*e.g.* glucose) involved in glycosidic linkage of fructan show only one kind of alditol, because its glycosidic carbon does not have chirality. However, the glycosidic carbon of ketoses (*e.g.* fructose) involved in such linkages has a chiral carbon atom (anomeric C), and thus shows two anomeric forms of alditol.

Figure 2.8 shows GLC separation of alditol derivatives of a fructan isolated from wild type garlic (grown in Northern Spain), with peak identification as follows.

- Peak 1: 1,5-anhydro-2,3,4,6-tetra-O-methyl-D-glucitol, *i.e.* terminal glucose.
- Peaks 2 and 3: 2,5-anhydro-1,3,4,6-tetra-O-methyl-D-mannitol and 2,5-anhydro-1,3,4,6-tetra-O-methyl-D-glucitol, *i.e.* terminal fructose.

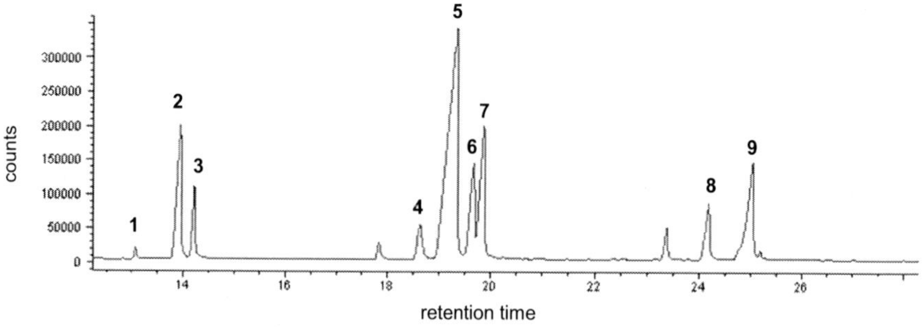

Figure 2.8 GLC separation of alditol derivatives of fructan from wild-type garlic (neo kestose series, peaks 1–9, see text for details). Column HP 5890 Series II Plus (for details see text, retention time in min).

- Peak 4: 6-O-acetyl-1,5-anhydro-2,3,4-tri-methyl-D-glucitol, *i.e.* internal glucose.
- Peak 5 and 6: 1-O-acetyl-2,5-anhydro-3,4,6-tri-O-methyl-D-mannitol and 1-O-acetyl-2,5-anhydro-3,4,6-tri-O-methyl-D-glucitol, *i.e.* internal (2,1) β-D-fructofuranosyl residue.
- Peak 5 and 7: 6-O-acetyl-2,5-anhydro-1,3,4-tri-O-methyl-D-mannitol, and 6-O-acetyl-2,5-anhydro-1,3,4-tri-O-methyl-D-glucitol, *i.e.* internal (2,6) β-D-fructofuranosyl residue.
- Peaks 8 and 9: 1,6-di-O-acetyl-2,5-anhydro-3,4-di-O-methyl-D-mannitol, and 1,6-di-O-acetyl-2,5-anhydro-3,4-di-O-methyl-D-glucitol, *i.e.* branched β-D-fructofuranosyl residue.

Based on these results, a structural model of garlic fructan (garlic fructan as neo kestose series) with a mean DP of 10 may be worked out as shown in Figure 2.9.

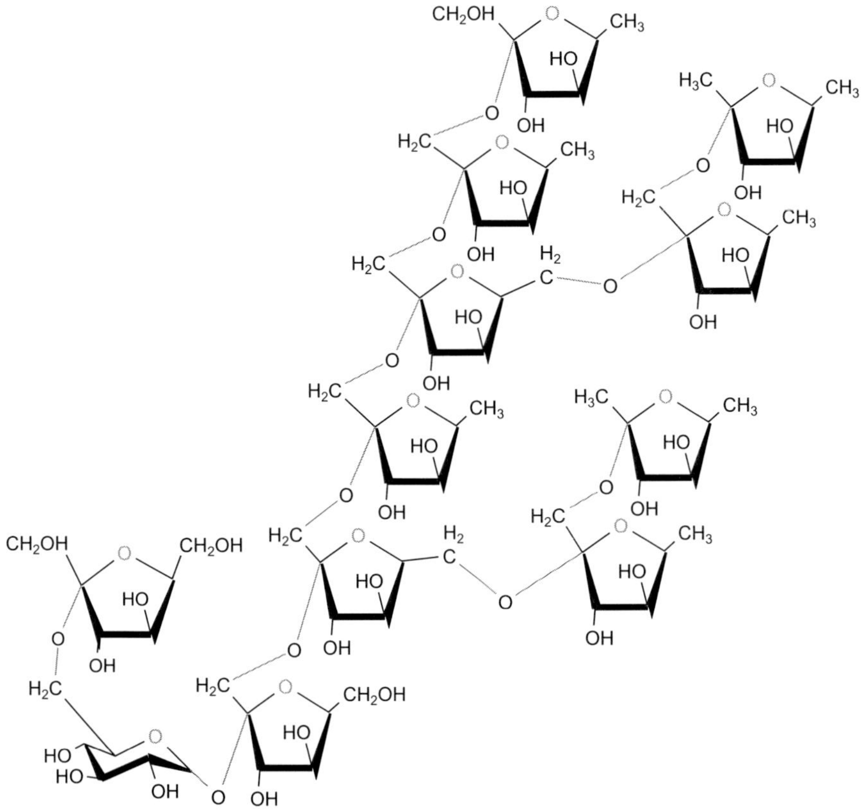

Figure 2.9 Most probable model of fructan from wild-type garlic based on the results of methylation analysis (Figure 2.8).

2.13 Characterization of Starches

Starch is abundantly produced as an energy reservoir in many diverse crops of higher plants. It is a complex homopolymer of glycosyl residues, with a number of chain segments or sub-structures. Isolation of starches from starch-containing crops such as tubers, roots or seeds is achieved by smashing and dispersion in water, and separation from fiber by sedimentation of the granules.

The material called "amylose" was discovered in 1940 by Meyer and his coworkers who found that properties of a material leached from maize starch with hot water at 70–80 °C was different from those of native maize starch.[49] Detailed investigation (methylation, end group analysis, degradation with β-amylase, and osmotic pressure measurement) revealed a non-branched molecule with about 300 glucosyl residues with α-(1→4) links. The rest of the polymer with additional α-(1→6) linked branches is termed "amylopectin". Shortly after, Cori and Cori synthesized the nonbranched polysaccharide amylose *in vitro* using glucose-1-phosphate and malto oligomers as starting materials in presence of muscle phosphorylase, thus describing the first successful *in vitro* synthesis of a biopolymer.[50]

The first quantitative separation of starch polymers was achieved by precipitation of amylose as an n-butanol inclusion complex, and recovering amylopectin from the supernatant.[51] At the same time waxy maize starch consisting of nearly 100% amylopectin, and being devoid of amylose, was commercialized. In the 1950s, commercial production of high amylose starch was started with specially adapted hybrids. These high amylose starches were of interest due to their ability to form gels and films. Today the quality of starch in different crops can be modified by breeding, as well as increasingly by genetic engineering. A wide variety of crop mutants are already known containing non-branched (nb), long chain branched (lcb) glucans (amylose), or short chain branched (scb) glucans (amylopectin), with a variety of different qualities, and thus can be regarded as tailor-made natural compounds.[52,53]

2.13.1 Characterization of Starch Granules

Plant physiology and environmental conditions influence the biosynthesis pathway of starch, resulting in various kinds of starch glucans with different DPs and branching characteristics. For storage purposes the crystalline insoluble form of starch granules is preferred, but the formation and growth of starch granules is a complex process that ends up with each granule as an individual object. The major component of all types of starch granules is glucan with different percentages of protein, lipid, water and charged species (mainly phosphates). A different percentage of crystallinity provides more or less ordered structures in a more or less predominant amorphous background.

The model of ordered double helix glucans in amorphous single chain regions is best supported by ^{13}C cross polarization magic angle spinning NMR (CP-MAS NMR).[54,55] Scb- and lcb-glucans are more or less incompatible, *i.e.* scb-glucans form compact layers of high order, whereas lcb-glucans form amorphous precipitates in less dense domains. The tendency for phase separation in the granule increases with increasing percentage of lcb-glucans; examples of homogeneous scb–lcb-glucan blends inside a starch granule are not known.[56,57] Thus, starch granules are composed of a crystalline scb-glucan (amylopectin) framework in an amorphous lcb-glucan (amylose) background.

Starch granules are characterized by X-ray diffraction and classified as outlined below for typical examples (approximate data).

- A-type spectrum – starch from seeds (*e.g.* from maize, wheat, rye, rice): Reflexes at scattering angle $2\theta \sim 15$–$16°$; $2\theta \sim 18$–$19°$; $2\theta \sim 23$–$24°$; left-handed parallel strand double helix, crystallized in monoclinic space group *B*2; compact packing of glucan chains with low water contents of 9–12%; 12 anhydroglucose units (AGU) with 12 H2O molecules; 6 AGU per helix turn, and 1.04 nm height of each turn.
- B-type spectrum – starch from tubers, *e.g.* from potato, (Figure 2.10) and amylose-rich starch (*e.g.* from amylomaize and wrinkled pea): reflexes at scattering angle $2\theta \sim 18$–$19°$; double helix crystallized in hexagonal space group *P*6; loose packing with high water contents of 12–15%; 12 AGU with 36 H_2O molecules; 6 AGU per helix turn, 1.04 nm height of each turn.
- C-type spectrum – starch (*e.g.* from several rhizomes and beans) can be seen as a mixture of A- and B-type starches or as a different structure of its own.
- V-type spectrum – gelatinized starches giving a diffuse halo spectrum which may indicate an amylose–lipid complex or an amorphous state.[3,58]

Thermal stress on B-type starch results in the loss of water and transformation into A-type, whereas swelling of A-type in aqueous media leads to loss of crystalline structure and yields B-type on recrystallization.[57,58]

Air-dried starch granules swell rather fast in water with 30–40% increased diameter, *i.e.* more than double their initial volume. Increasing the temperature of such starch suspensions may cause irreversible modification of the granules and re-organization of their supramolecular structure and gelatinization pattern.

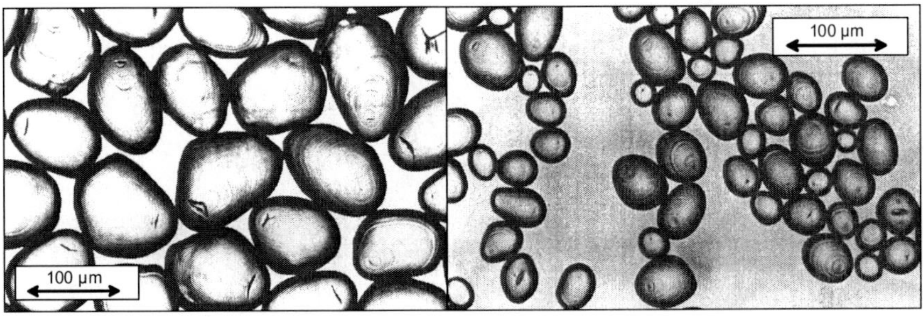

Figure 2.10 Large and small starch granules fractionated from potato tubers by sedimentation.

Continued thermal stress eventually leads to more and more homogeneous suspensions, which can be monitored by their decreasing viscosity. The gelatinization characteristics of starch suspensions are typically monitored by standardized temperature–time programs with a Brabender Viscoamylograph.[59]

Starch granules are also analyzed by means of differential scanning calorimetry (DSC) for their phase transitions, and by conformation analysis. Molecular interactions during the endothermic process show a positive correlation between the amylopectin chain length and transition enthalpy, and a negative correlation between transition temperatures and amylose content. DSC is also an essential tool for understanding starch gelatinization process. For instance, the melting of crystallites is preceded by a glass transition, depending on the water content and temperature profile of the sample.[3,56,60]

2.13.2 Molecular Characterization of Starch Glucans

Key molecular characteristics such as molecular weight distribution molecular weight averages, supermolecular and coherent segment dimensions (*i.e.* dimensions of coherently acting segments, faking bigger segment dimensions than actually present), as well as branching features of starch are essential for the understanding of their bulk properties. A number of techniques including physicochemical, chemical and enzymatic methods can be employed to obtain valuable information about the molecular characteristics of starch polysaccharides.[3,58,59]

After isolation (see above) and defatting of the granules by apolar extraction (*e.g.* with about 75% propanol), they are dissolved in a polar medium such as water, buffer, alkaline (*e.g.* NaOH) or dimethylsulfoxide (DMSO) for non-destructive molecular analysis. Aqueous starch solution can also be fractionated into nb- and lcb-glucans (amylose) and scb-glucans (amylopectin) by precipitation of amylose inclusion complex with higher alcohols, particularly n-butanol, or by preparative size exclusion chromatography (SEC).

Non-destructive molecular analysis provides information about the supramolecular and molecular dimensions and the interactive behavior of starch polysaccharides in aqueous media. Generally the molecular weight distribution of starch glucans is obtained with calibrated SEC with the use of reference glucans (*e.g.* dextrans), which may include peak position calibration or broad standard calibration with a defined material.

Analytical SEC combined with absolute techniques such as light scattering (LS) combined with universal mass detection (SEC–mass–LS), or viscosity combined with universal mass detection (SEC–mass–viscosity) or both combined is a highly sensitive method for high molecular components, in particular for glucan aggregates. Thus, these techniques are primarily used for determination of molecular weight of supra structures, rather than that of the constituting molecules. The study of starch polysaccharides by photon correlation spectroscopy (PCS) also provides information about material diffusion and mobility at molecular and supramolecular levels.

Combining a chemical/analytical technique by means of quantitative derivatization of each glucan molecule with a unique chromophore (labelling on the reducing terminal hemiacetal group) and following analytical SEC combined with specific molar detection (e.g. UV/VIS or fluorescence) and universal mass detection (SEC-mass/molar) represents an additional approach to information regarding the molecular weight of constituting glucans.[61]

For branching analysis, debranching enzymes such as pullulanase and isoamylases for hydrolysis of α-(1$\rightarrow$6) linkages are applied followed by chromatographic techniques such as SEC or HPAEC–PAD. Capillary electrophoresis (CE) can also be applied after labeling the reducing terminal hemiacetal residues of debranched chains, particularly of scb-glucan (amylopectin) with fluorophores such as 8-amino-1,3,6-pyrenetrisulfonic acid (APTS).

Classification of starch polysaccharides into nb-, lcb- and scb-glucans, the basic branching structure of β-limit dextrins (LD), their molar weight averages, and outer and inner chain lengths can be achieved by enzymatic degradation with β-amylases and sometimes by debranching enzymes combined with liquid chromatography.[58,59]

Destructive chemical methods like methylation combined with acid hydrolysis (or reductive cleavage) and acetylation, as well as periodate oxidation followed by reduction of the oxidized polysaccharide, and acid hydrolysis (Smith degradation), are also used for determination of branching points, non-reducing terminal residues, as well as for classification of starch polymers into nb-, lcb- and scb-glucans. For screening of starch molecular weight, reducing end group determination (number average molecular weight) by a Somogyi–Nelson colorimetric procedure, and enzymatic analysis of glucitol after reduction with sodium borohydride and acid hydrolysis, are both useful.[58]

2.13.3 Size Exclusion Chromatography of Starch Glucans

SEC analysis provides information on the distribution of excluded (hydrodynamic or occupying) volume (V_e) of the polymer molecule in an aqueous gel matrix. The excluded volume in aqueous systems is governed primarily by eqn (1):

$$V_e = ip \times md^{mc} \tag{1}$$

where V_e is the excluded volume; *ip* the interactive potential; *md* the molecular dimension; and *mc* the molecular conformation.

The *ip* describes the interaction of the polymer with the aqueous gel matrix mainly *via* H-bonding; *md* is dominated by molecular size, transition states between molecular and coherent supramolecular structures; and *mc* represents molecular symmetries in terms of helixes, branching pattern, crosslinks, oxidation status, structural compatibility and packing density.

The nature of solvent and polymer dissolution significantly influences the ratio of supramolecular/molecular structures, and therefore results in different elution profiles in SEC separations. The term hydrodynamic volume, as often used in literature, is the product of the limiting viscosity and molecular weight, $[\eta]M_w$, expressing approximately the same value as the term excluded volume.

In case of universal calibration, the elution volume in SEC is inversely proportional to log $[\eta]M_w$, and depends on the branching characteristic of the polymer mixture as well as on their monomer composition. Scb-glucans (amylopectin) are more compact, and have smaller $[\eta]$, than nb- and lcb-glucans (amylose) of the same molecular weight, and thus elute later. Molecules with fewer branches elute earlier.[58,59]

Thus, SEC analysis with universal calibration (only mass detection), and with absolute calibration (mass detection with light scattering or viscosity measurement), provides detailed information on molecular weight distribution, branching characteristics, supramolecular dimensions, and coherent segment dimensions.

For native wheat starch (dissolved in DMSO), for example, absolute calibration with light scattering, and with universal calibration with dextrans (with viscosity contribution), provide M_w and apparent M_w (NB these are different from absolute or calibrated weight average molecular weight M_w: apparent values are 'seem-to-be' values, correct from applied experimental equipment and applied equations, but wrong in terms of actual material qualities) of approx. 17×10^6 and 5.5×10^6, respectively. Interestingly, light scattering indicates a rather monodisperse sample, whereas universal calibration shows a polydispersity index of approx. 7. With native potato starch (dissolved in DMSO), LS shows an apparent M_w of approx. 20×10^6.

Thus, more information is required to verify the argument of apparent and real M_w, at least to prove a plausible true molecular weight, as exemplified by the following approach.

Quantitative derivatization with 2-aminopyridine on the free terminal hemiacetal of native glucans (*e.g.* potato starch) produces one fluorophore label per molecule, as measured by a fluorescence detector (FD). The glucan derivative thus obtained can then be fractionated into scb-, nb- and lcb- glucans and analyzed by combined SEC with mass, LS, viscosimetry and FD. Molecular weight analysis by LS for the fractionated scb-glucan shows an apparent M_w of approx. 20×10^6 with high conformity over the whole separation system with a viscosity $[\eta]$ of 80–100 mL g^{-1}, with a *de facto* molar mass (molecular weight; *de facto* molar masses are actual molar masses of investigated materials when all misleading contributions from experimental design and computational approaches are eliminated or avoided) in the range of 20 000–300 000. For lcb-glucans (native amylose) with a high tendency to form supramolecular structures, the apparent M_w by LS is approx. 3×10^6, and *de facto* mean molar masses are M_w 160 000, and $M_n = 89\ 000$, with a M_w/M_n (polydispersity) value of 1.8 (Figure 2.11).

These results indicate that starch glucan molecules interact strongly with each other, hence they have a high potential to form supramolecular structures in various aqueous media. Nb- and lcb-glucans are poorly soluble in aqueous media, and have strong tendency to retrograde, easily gelatinize and form gels and films. Scb-glucans are apparently much better soluble in aqueous media,

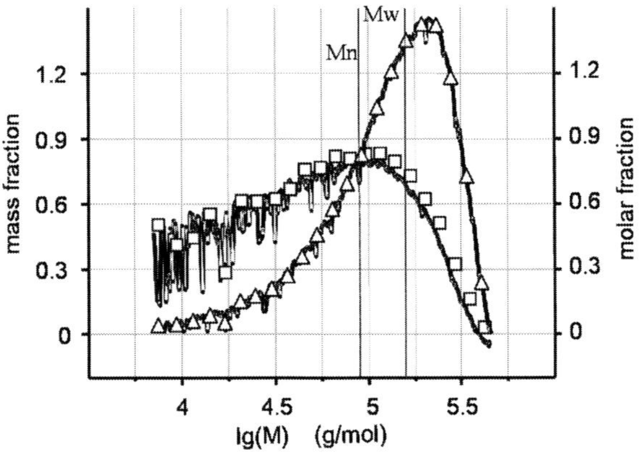

Figure 2.11 Mass (triangle) and molar (square) distribution of 2-aminopyridyl labeled potato lcb-glucans (native amylose). M_w 160 000, M_n 89 000, M_w/M_n 1.8.

associate with relatively large amounts of water, and are less sensitive towards varying environmental conditions.

The *de facto* molecular weights of scb- and lcb-glucans are surprisingly similar despite their derivation from differently structured starch polysaccharides. However, it appears that different starch polysaccharides have similar molecular weights, despite diverse supermolecular features in their granular forms due to different granule growth conditions.

Thus, variation in branching pattern is most probably the key factor in understanding the native structural features of various starch glucans. This means that the percentage and position of branching, and the ratio of lcb/scb-glucans, are highly responsible for the macroscopic quality of starch.

2.14 Characterization of Proteins

From a biochemical point of view, proteins are required for regulation and catalysis, energy modulation and structural functions. Thus it is this background and the potential of proteins for medical applications that have led to the new field of proteomics in recent years. The term proteomics embraces the characterization of proteins,[62,63] as well as the systematic study of proteins in the cells, tissues and organisms in general. Advanced analytical techniques, such as two-dimensional electrophoresis (separation of fluorescence labeled proteins with two-dimensional-sodiumdodecyl sulfate-polyacryl gel electrophoresis, 2D-SDS-PAGE), nano-liquid chromatography and sequence fragment analysis with ion trap mass spectrometer (tandem MS) and data bank identification with validation, provide detailed information about the primary, secondary and tertiary protein structure and their related functions.[64]

Utilization of proteins for food, feed and nonfood applications is wide ranging and dates back to ancient stories. For practical reasons, different proteins are distinguished as animal proteins, plant proteins, and single cell proteins (SCP, *i.e.* of micro-organisms), discussed below. In most plants the protein content with respect to dry mass ranges only between 5 and 10%, although some plants like lucerne and soy bean (legume species) contain up to 40% proteins. Single cell protein (SCP) is produced mainly by yeasts with rather high percentages (30–65% of dry mass); its nutritional value is comparable to soybean and fish.[65] Meat as the primary product of slaughtered animals (including fish) has high nutritional value for the daily diet; depending on the species, only 35–55% of the live animal weight is protein; the rest of the animal body is the meat by-products including bones, hide, skin, fat, hair, blood, horns, hoofs, and so on, each of which has different degrees of utility for food, feed and non-food applications. However, some proteins such as animal hair (wool, especially of sheep or camel), and cocoon (silkworm) fibers are used as raw materials for textile products.[66]

2.14.1 Characterization of Plant Proteins

There is an increasing interest in proteins from plants, particularly proteins isolated from by-products of starch processing (*e.g.* potato, wheat, corn), and from oil-producing seeds (soybean, cotton seed, *etc.*). More recently, new concepts have also been developed for production of biomaterials from grassland (*e.g.* lucerne), the so-called "biorefineries"; one of its main fractions is protein with a very high nutritional value. In this respect the characterization and evaluation of protein composition, in particular the content of essential amino acids is important. Amino acid analysis is carried out by amino acid analyzers and HPLC with precolumn derivatization. The analytical profile of protein quality is developed on the basis of chemical and biological methods for amino acid composition, protein–energy relationship, and immune reactions, as well as sensory analysis.[67]

As proteins are sensitive to water and elevated temperatures, very often there is a need for their stabilization by chemical modification, including crosslinking and incorporation of reactive polar groups causing new hydrophobic–hydrophilic properties. Modification by oxidation (for making strong films without losing elasticity), acetylation (for use as cobinder in paper coating), and deamination (improving foam-forming ability), are also methods employed for the development of protein-based products.

In all of these, detailed information about protein composition and molecular structure, before and after modification, is necessary to match quality norms and achieve effective application.[68]

2.14.2 Characterization of Animal Proteins

The main animal protein is obviously meat used as a major food; proteins from animal waste are less valuable for food use. For preventing the risks of transmissible spongiform encephalopathy (TSE), *e.g.* scarpie and bifidum spongiform encephalopathy (BSE), European legislation strictly forbids the use of specified animal by-products for any food and feed production. This has, for example, resulted in a surplus of animal bone meal. Such materials could be useful for new applications, particularly in polymer production, and could be integrated in future research programs in European countries.

Collagen (gelatin) and keratin are among the main animal proteins. About 30% of animal protein is made up of collagen. This large extracellular protein is composed of three polypeptide chains of about 1000 amino acid units with a repeating sequence of glycine–proline–Y (Y being mostly hydroxyproline). Collagen chains form a triple helix (about 300 nm long), with shorter globular domains (telopeptides) at the ends of the molecule, called tropocollagen. To date 19 different types of collagen have been identified showing different molecular masses and amino acid sequences possessing a high tendency towards self-assembly to fibers with high mechanical strength. These triple helices are resistant to most proteases (trypsin, pepsin, papain), but not to collagenases.[69]

Keratin, the polypeptide of hair, nails and hoofs is rich in cystine (25%), proline (10%), arginine (10%) and glycine. Two main types of insoluble, elastic and fibrous keratin are known.

- Hard keratin with very densely packed filaments and high sulfur content (nails, hoofs, cortex of hair). Hard keratin is resistant to most proteolytic enzymes including pepsin and trypsin, but can be thermally hydrolyzed with water under pressure.
- Soft keratin (horny matrix) which fills the cells of the inner root sheath of hair follicles.

The industrially important product of collagen is gelatin, which is obtained by partial hydrolysis of collagen of hides, mainly from animal skin and bones.[70] The composition of gelatin and its molecular weight – in the range of 10–250 kDa – depend on the source of collagen and the method of hydrolysis. Unlike native collagens, gelatin is easily digested by most common proteases.

The most important characteristic of gelatin is its reversible gel formation in water, hence its name "gelatin". The gel is formed by cooling the gelatin solution prepared in warm (or hot) water. The 3D gel network is formed by ionic interactions between amino and carboxylic groups, hydrogen bonds, as well as hydrophobic interactions. Technical grade gelatin is applied as glue and as carrier for silver halide emulsions (in the photography), and a wide range of other chemical applications including microencapsulation of dyes in the manufacture of carbonless copy paper.[71,72] Highly purified gelatin is used in the pharmaceutical industry, *e.g.* for the production of conventional capsules,[70] microcapsules[71,72] and radiolabeled aggregates for nuclear medicine imaging.[73]

2.14.3 Characterization of Single Cell Proteins

Different kinds of microorganisms are used for the conversion of waste components and by-products into high value products, including protein with high amounts of essential amino acids of potential interest for food and feed applications.

Yeast is an excellent microorganism for the production of single cell biomass (SCB) with high nutritional value with adequate amino acid composition, lipid, vitamins, minerals, and nucleic acid. Due to their fast turnover (1–5 h) yeasts can utilize waste raw materials available in large quantities such as lignocelluloses and starchy materials, whey, molasses, sulfite waste liquor, animal wastes, forest wastes and lipid by-products. SCB can be produced in continuous culture with limited land area and water requirement, and is independent of climate changes; the protein content of yeasts is in the range of 30–65% of dry weight, which is relatively high compared to other natural sources (see above). Generally speaking SCB can replace soy protein cake and fish meal in feed (with addition of 0.1–0.2% methionine). However, the use of yeast in food is possible only if the RNA content is less than 2%.[65]

Recombinant DNA technology has led to the possibility of constructing new strains of baker's yeasts, fermentation yeasts for production of ethanol, and fodder yeasts. It is possible that yeasts can be grown on different renewable materials with high biomass production or with high resistance to ethanol concentration. Thus, yeast biotechnology could offer a profitable use of waste biomass resources such as starch and cellulose wastes and whey from cheese production and at the same time reduce their potential as pollutants. However, consumer resistance to genetically modified organisms, especially for food use, is the main barrier to their development.[74]

2.15 Concluding Remarks

There is no doubt that the efficient utilization of natural polymers requires information from adequate analysis and characterization and this means data from the full range of established polymer-analytical techniques such as elementary analysis, microscopy, staining, spectroscopy, rheology and viscosimetry, calorimetry, controlled fragmentation and fragment analysis, preparative and analytical fractionation combined with off-line and in-line fraction analyses. This chapter outlines the principal strategies for isolation, identification and characterization of fiber-crop polymers with a major focus on fractionation by means of liquid chromatography, as well from a preparative as from an analytical point of view.

Liquid chromatography combined with controlled fragmentation and fragment analysis after reductive methylation and acetylation on the one hand, and multiple detection of size-exclusion-separated sample fractions on the other, provide information about molecular level structural compound details and so resemble a basic background for any subsequent structure–functionality correlation.

A major challenge of natural polymers is their multiple and superimposed heterogeneity: any quality comes as a distribution rather than as a simple figure. Hence, analysis must include fractionation approaches covering the full variety of enthalpy (interaction)-controlled and entropy (geometry/dimension)-controlled separation techniques to obtain the required data. Molecular level information regarding how and when certain structures and functionalities are established, kept and/or disintegrated in obtained fractions may then be correlated with macroscopic material qualities.

Another challenge is the fact of perpetual changes in crop components' compositions. With respect to this fact, information about general activity potentials of investigated crops in their environment is needed, and additionally, information about specific control and response mechanisms upon applied stress in particular are required.

Although the analysis of molecular level composition and structural details of crop materials is complex enough, these data are only pieces in the puzzle of comprehensive understanding of crop-performance. Such biological systems generate conditions for many potential qualities and, depending on demands, actually establish material qualities quasi on-demand. Key processes for switching from potential to actually established qualities are minor conformational modification of oligomeric and polymeric materials in aqueous environment which result in significant changes of interactive properties and, finally, more or less pronounced modified surface/interface- and bodywork-properties of material-clusters. A sensitive, diffuse and permanent transition between architecture/geometry (static properties) and transient coherent supermolecular activities (dynamic properties) advocates and develops actual macroscopic material functionalities.

However, there are limits to the correlation of crop performance and molecular level crop-compound characteristics. As any structure may contribute to many, even rather different, activities, and, on the other hand, any activity may be achieved by different combinations of component structures, structural analysis is an important, however insufficient way of obtaining unambiguous correlations between molecular level structure and macroscopic material functionality. Consequently, reproducible crop performance is an indicator but no guarantee of achieving the desired qualities for crop-produced narrowly-specified compounds.

To summarize, for comprehensive understanding and control of crop-produced material-qualities we need reliable structural information and also information about the dynamics of structural changes upon applied stress to the crops, classification of structural influences on functionalities at different levels, and information about sensitivity and modes of amplification and suppression of control mechanisms. Besides the technical task to gather such multi-dimensional information sets, the major future challenge most probably will be 'data mining': to detect triggers, controlling contributions, threshold levels, critical states, limiting structures and similar features which together establish specific crop-performances.

2.16 List of Abbreviations

Xyl	xylose
UV/VIS	ultra violet/visible
USI	Un-Scan-It program
UC	ultracentrifuge

TAPPI	Technical Association of the Pulp and Paper Industry
TSE	transmissible spongiform encephalopathy
TLC	thin layer chromatography
TFA	trifluoracetic acid
SEC	size exclusion chromatography
SDS–PAGE	sodium dodecylsulfate–polyacrylamide gel electrophoresis
scb	short chain branched
SCB	single cell biomass
Rha	rhamnose
PS	polysaccharide
PCS	photon correlation spectroscopy
PCR	polymerase chain reaction
nb	nonbranched
M_w	weight average molecular mass
MS	mass spectrometer
M_n	number average molecular mass
Man	mannose
lsb	long chain branched
LS	light scattering
LC	liquid chromatography
ID	inner diameter
ICP-MS	inductively coupled plasma–mass spectrometry
IEC	ion exchange chromatography
HIC	hydrophobic interaction chromatography
HPLC	high performance liquid chromatography
HPAEC-PAD	high performance anion exchange chromatography–pulsed amperometric detection
GluUA	glucuronic acid
Glu*p*	glucosylpyranose
Glu	glucose
GlcNAc	2-desoxy-2-acetylamidoglucose
GlcN	2-desoxy-2-aminoglucose
GLC	gas liquid chromatography
GC-MS	gas liquid chromatography–mass spectrometry
GC-FID	gas liquid chromatography–flame ionization detection
GalUA	galacturonic acid
Gal	galactose
Fru*f*	fructosylfuranose
FID	flame ionization detector
FD	fluorescence detector
EPS	exopolysaccharide
ELSD	evaporating light scattering detector
DMSO	dimethylsulfoxide
DP	degree of polymerization
DPD	degree of polymerization distribution
CP-MAS	cross polarization–magic angle spinning

CE	capillary electrophoresis
BSE	bovine spongiform encephalopathy
Ara	arabinose
APTS	8-amino-1,3,6-pyrenetrisulfonic acid
AAS	atomic absorption spectroscopy

References

1. A. Huber and W. Praznik, Identification and quantification of renewable crop materials. In *Renewable Bioresources*, (Eds) C. V. Stevens and R. G. Verhé, Wiley & Sons, Chichester, 2004, pp. 138–159.
2. D. A. I. Goring and T. E. Timell, Molecular weight of native celluloses, *Tappi*, 1962, **45**, 454–460.
3. J. Jane, Starch: structure and properties. In *Chemical and Functional Properties of Food Saccharides*, (Ed.) P. Tomasik, CRC press, Boca Raton, 2004, pp. 81–101.
4. W. Praznik, R. Löppert and A. Huber, Analysis and molecular composition of fructans from different plant sources. In *Recent Advances in Fructooligosaccharides Research*, (Eds) N. Shiomi, N. Benkeblia and S. Onodera, Research Signpost, Kerala (India), 2007, pp. 93–117.
5. S. A. Arvidson, B. T. Rinehart and F. Gadala-Maria, Concentration regimes of solutions of levan polysaccharides from *Bacillus* sp., *Carbohydr. Polym.*, 2006, **65**, 144–149.
6. A. G. J. Voragen, W. Pilnik, J. F. Thibault, M. A. V. Axelos and C. M. G. C. Renard, Pectins. In *Food Polysaccharides and their Applications*, (Ed.) A. M. Stephen, Marcel Dekker, New York, 1995, pp. 237–340.
7. L. Ramsden, Plant and algal gums and mucilages. In *Chemical and Functional Properties of Food Saccharides*, (Ed.) P. Tomasik, CRC press, Boca Raton, 2004, pp. 231–254.
8. M. A. Izydorczyk and C. G. Biliaderis, Cereal arabinoxylans: advances in structure and physicochemical properties, *Carbohydr. Polym.*, 1995, **28**, 33–48.
9. G. Zubay (Ed.), *Biochemistry*, Addison-Wesley Publ. Com., London 1983.
10. A. Seifert, L. Heinevetter, H. Cölfen and S. E. Harding, Characterization of gliadin-galactomannan mixtures by analytical ultracentrifugation – Part I. Sedimentation velocity, *Carbohydr. Polym.*, 1995, **28**, 325–332.
11. B. Cathala, B. Saake, O. Faix and B. Monties, Association behavior of lignins and lignin model compounds by multidetector size exclusion chromatography, *J. Chromatogr. A*, 2003, **1020**, 229–239.
12. K. Cornish, J. Castillon and D. J. Scott, Rubber molecular weight regulation, in vitro, in plant species that produce high and low molecular weights in vivo, *Biomacromolecules*, 2000, **1**, 632–641.
13. AOAC International, Association of Official Analytical Chemists, Official Methods of Analysis, Gaithersburg, Maryland 20877, USA, www.aoac.org
14. ASTM International, American Society for Testing Materials, Standard methods, Headquarters, PA 19428-2959, USA, www.astm.org

15. DIN Methods, Deutsches Institut für Normung, www.din.de
16. TAPPI, Technical Association of the Pulp and Paper Industry, Test Methods & Technical Information Papers, www.tappi.org
17. Y. P. Kalra (Ed.), *Handbook of Reference Methods for Plant Analysis*, Saint Lucie Press, 1997.
18. R. K. Owusu-Apenten (Ed.), *Food Protein Analysis*, Marcel Dekker, New York, 2002.
19. R. E. Wrolstad, T. E. Acree, E. A. Decker, M. H. Penner, D. S. Reid, S. J. Schwartz, C. F. Shoemaker, D. Smith and P. Sporns (Eds), *Handbook of Food Analytical Chemistry*, Wiley & Sons, Hoboken, 2005.
20. J. N. Be Miller, D. J. Manners and R. J. Sturgeon (Eds), Methods in Carbohydrate Chemistry. In *Enzymic Methods*, Vol. 10, John Wiley & Sons, Chichester, 1994.
21. M. F. Chaplin and J. F. Kennedy (Eds), *Carbohydrate Analysis*, Oxford University press, second edition, New York, 1994.
22. H. Haas, W. Schoch and U. Störle, Herstellung von Skelettsubstanzen mit Peressigsäure, *Das Papier*, 1955, **9**, 469–475.
23. R. Andersson, E. Westerlund and P. Åman. Cell wall Polysaccharides: Structural, Chemical, and Analytical Aspects. In *Carbohydrates in Food* (Ed.) A. C. Eliasson, 2nd edn, CRC press, Boca Raton, 2006, pp. 129–166.
24. A. S. Perlin and B. Casu, Spectroscopic methods. In *The Polysaccharides*, (Ed.) G. O. Aspinall, Vol. 1, Academic Press, New York, 1982, pp. 133–185.
25. A. W. Mc Gregor and G. B. Fincher, Carbohydrates of barley grain. In *Barley: Chemistry and Technology, Amer. Assoc. of Cereal Chemists*, (Eds) A. W. Mc Gregor and R. S. Bhatty, St Paul, 1993, pp. 73–130.
26. G. B. Fincher and B. A. Stone, Cell walls and their components in cereal grain technology, *Adv. Cereal Sci. Technol.*, 1986, **8**, 207–295.
27. P. Åman and S. Bengtsson, Perjodate oxidation and degradation studies on the major water soluble arabinoxylan in rye grain, *Carbohydr. Polym.*, 1991, **15**, 405–414.
28. S. Bengtsson, P. Åman and R. E. Andersson, Structural studies on watersoluble arabinoxylans in rye grains using enzymic hydrolysis, *Carbohydr. Polym.*, 1992, **17**, 277–284.
29. B. L. Ridley, M. A. O'Neill and D. Mohnen, Review. Pectins: structure, biosynthesis, and oligogalacturonide related signalling, *Phytochemistry*, 2001, **57**, 929–967.
30. F. Voragen, G. Beldman and H. Schols, Chemistry and enzymology of pectins. In *Advanced Dietary Fiber Technology*, (Eds) B. V. Mc Cleary and L. Prosky, Blackwell Science, Oxford, 2001, pp. 379–398.
31. G. A. F. Roberts (Ed.), *Chitin Chemistry*, Macmillian, Houndmills, 1992.
32. F. Nanjo, R. Katsumi and K. Sakai, Enzymic method for determination of the degree of deacetylation of chitosans, *Anal. Biochem.*, 1991, **193**, 164–167.
33. F. Niola, *et al.*, A rapid method for the determination of the degree of N-acetylation of chitin–chitosan samples by acid hydrolysis and HPLC, *Carbohydr. Res.*, 1993, **238**, 1–9.

34. K. M. Vårum, *et al.*, Determination of the degree of acetylation and the distribution of acetyl groups in partially N-deacetylated chitins (chitinosan) by high field NMR spectrodcopy. Part I: High field NMR spectroscopy of partially N-deacetylated chitin (chitosans), *Carbohydr. Res.*, 1991, **211**, 17–23.
35. J. L. Doublier and G. Cuvelier, Gums and hydrocolloids: Functional aspects. In *Carbohydrates in Food* (Ed.) A. C. Eliasson, 2nd edn, CRC press, Boca Raton, 2006, pp. 233–272.
36. B. Quemener, C. Marot, L. Mouillet, V. Da Riz and J. Diris, Quantitative analysis of hydrocolloids in food systems by methanolysis coupled to reserve HPLC. Part 1. Gelling carrageenans, *Food Hydrocoll.*, 2000, **14**, 9–17.
37. A. Versari, S. Biesenbruch, D. Barbanti, P. J. Farnell and S. Galassi, HPAEC-PAD analysis of oligogalacturonic acids in strawberry juice, *Food Chem.*, 1999, **66**, 257–261.
38. M. Suzuki and N. J. Chatterton (Eds), *Science and Technology of Fructan*, CRC Press, Boca Raton, 1993.
39. L. Prosky and H. Hoebregs, Methods to determine food inulin and oligofructose, *J. Nutr.*, 1999, **129**(Suppl. S), 1418–1423.
40. M. K. Ernst, N. J. Chatterton, P. A. Harrison and G. Matitschka, Characterization of fructan oligomers from species of the genus *Allium* L, *J. Plant Physiol.*, 1998, **153**, 53–60.
41. N. Shiomi, S. Onodera, N. J. Chatterton and P. A. Harrison, Separation of fructooligosaccharide isomers by anion exchange chromatography, *Agric. Biol. Chem.*, 1991, **55**, 1427–1428.
42. J. W. Timmermans, M. B. Van Leeuwen, H. Tournois, D. De Wit and J. F. G. Vliegenhardt, Quantitative analysis of the molecular weight distribution of inulin by means of anion exchange HPLC with pulsed amperometric detection, *J. Carbohyd. Chem.*, 1994, **13**, 881–888.
43. B. Heinze and W. Praznik, Seperation and purification of inulin oligomers and polymers by reversed phase high performance liquid chromatography, *J. Appl. Pol. Sci.: Appl. Pol. Symp.*, 1991, **48**, 207–225.
44. R. H. F. Beck and W. Praznik, Molecular characterization of fructans by high–performance gel chromatography, *J. Chromatogr.*, 1986, **369**, 208–212.
45. A. Huber, Polysaccharides. In *Encyclopedia of Chemical Processing (ECHP)*, (Ed.) K. B. L. Sunggyu, Dekker, London 2005, pp. 2349–2367.
46. N. C. Carpita, T. L. Housley and J. E. Hendrix, New features of plant fructan structure revealed by methylation analysis and ^{13}C NMR spectroscopy, *Carbohydr. Res.*, 1991, **217**, 127–136.
47. T. Spies, W. Praznik, A. Hofinger, F. Altmann, E. Nitsch and R. Wutka, The structure of the fructan sinstrin from *Urginea maritime*, *Carbohydr. Res.*, 1992, **235**, 221–230.
48. S. Baumgartner, T. G. Dax, W. Praznik and H. Falk, Characterization of the high molecular weight fructan isolated from garlic (*Allium sativum* L.), *Carbohydr. Res.*, 2000, **328**, 177–183.

49. K. H. Meyer, W. Bretano and P. Bernfeld, Starch. II. Nonhomogeneity of starch, *Helv. Chim. Acta*, 1940, **23**, 845–853.
50. G. T. Cori and C. F. Cori, Crystalline muscle phosphorylase. IV. Formation of glycogen, *J. Biol. Chem.*, 1943, **151**, 57–63.
51. T. J. Schoch, The fractionation of starch, *Adv. Carbohydr. Chem.*, 1945, **1**, 247–268.
52. C. D. Boyer and L. C. Hannah, Kernel mutants of corn. In *Speciality Corns*, (Ed.) A. Hallauer, CRC Press, Boca Raton, 1994, pp. 2–14.
53. M. Obanni and J. N. Be Miller, Identification of starch from various maize endosperm mutants via ghost structure, *Cereal Chem.*, 1995, **72**, 436–444.
54. D. French, Organisation of starch granules. In *Starch Chemistry and Technology*, (Eds) R. L. Whistler and J. N. BeMiller, E. F. Paschal, Academic press, London 1984, pp. 183–221.
55. M. J. Gidley and S. M. Bociek, Molecular organization in starches: ^{13}C CP/MAS NMR study, *J. Am. Chem. Soc.*, 1985, **107**, 7040–7044.
56. A. C. Eliasson and M. Gudmundson, Starch: physicochemical and functional aspects. In *Carbohydrates in Food*, (Ed.) A. C. Eliasson, 2nd edn, CRC press, Boca Raton, 2006, pp. 391–469.
57. H. F. Zobel, Molecules to granules: a comprehensive review, *Starch/Stärke*, 1988, **40**, 44–50.
58. S. Hizukuri, J. Abe and I. Hanashiro, Starch: analytical aspects. In *Carbohydrates in Food*, (Ed.) A. C. Eliasson, 2nd edn, CRC press, Boca Raton, 2006, pp. 305–390.
59. A. Huber and W. Praznik, Analysis of molecular characteristics of starch polysaccharides. In *Chemical and Functional Properties of Food Saccharides*, (Ed.) P. Tomasik, CRC press, Boca Raton, 2004, pp. 349–369.
60. K. J. Zeleznak and R. C. Hoseney, The glass transition in starch, *Cereal Chem.*, 1987, **64**, 121–124.
61. W. Praznik and A. Huber, De facto molecular weight distributions of glucans by size exclusion chromatography combined with mass/molar detection of fluorescence labeled terminal hemiacetals., *J. Chromatogr.*, 2005, B, **824**, 295–307.
62. C. Delahunty and J. R. Yates III, Proteomics: a shotgun approach without two-dimensional gels. In *Encyclopedia of Life Science*, John Wiley & Sons, 2006, www.els.net
63. W. Blackstock and M. Mann (Eds), *Trends in Proteomics*, Elsevier, London, 2000.
64. P. Roepstorff, Mass spectrometry instrumentation in proteomics. In *Encyclopedia of Life Science*, John Wiley & Sons, 2006, www.els.net
65. W. Praznik and C. V. Stevens, Integral valorization. In *Renewable Bioresources*, (Eds) C. V. Stevens and R. G. Verhé, Wiley & Sons, Chichester, 2004, pp. 46–71.
66. W. Rymowicz, Primary production of raw materials. In *Renewable Bioresources*, (Eds) C. V. Stevens and R. G. Verhé, Wiley & Sons, Chichester, 2004, pp. 72–104.

67. M. Friedman, Nutritional value of proteins from different food sources. *A review, J. Agric. Food Chem.*, 1996, **44**, 6–29.
68. R. G. Verhé, Industrial products from lipids and protein. In *Renewable Bioresources*, (Eds) C. V. Stevens and R. G. Verhé, Wiley & Sons, Chichester, 2004, pp. 208–250.
69. A. J. Bailey and N. D. Light, *Connective tissue in meat and meat products*, Elsevier Applied Science, London, 1989.
70. R. C. Clarke and A. Courts, The chemical reactivity of gelatin. In *The Science and Technology of Gelatin*, (Eds) A. G. Ward and A. Courts, Academic Press, London, 1977, pp. 209–247.
71. R. Arshady (Ed.) Microspheres, Microcapsules & Liposomes, The MML Series. Vol. 1, *Preparation and Chemical Applications*, Citus (now Kentus) Books, London, 1999.
72. R. Arshady and B. Boh (Ed.) Microcapsule Patents and Products, The MML Series, Vol. 6, Citus (Kentus) Books, London, 2003.
73. R. Arshady (Ed.) Microspheres, Microcapsules & Liposomes, The MML Series. Vol. 3, *Radiolabeled and Magnetic Particulates in Medicine and Biology*, Citus (Kentus) Books, London, 2001.
74. A. N. Glazer and H. Nikaido (Eds), *Microbial Biotechnology*, Freeman & Co, New York, 1995.

CHAPTER 3

Cellulose and Its Derivatives in Medical Use

TOHRU SHIBATA

Daicel Chemical Industries, Research Center, 1239 Shinzaike, Aboshi-ku, Himeji, 671-1283, Japan

3.1 Introduction

Cellulosic materials have a long history of utilization by mankind because of their abundance and excellent physical properties. In the medical and pharmaceutical fields, they have also contributed to human welfare in the form of medical devices, pharmaceutical auxiliaries and in related biomedical materials. Cellulose has a number of important merits as a biomaterial resource for medical applications. First, it is a renewable resource, with a natural production of about 10^{11} tons per year. Second, it is a highly pure polymer of glucose, and hence shows good environmental and biological safety. Third, it has an excellent physical strength that is warranted by its role in plants. Fourth, it is biodegradable; while biodegradability may not necessarily apply in case of all cellulose derivatives, cellulose acetate undergoes biodegradation. Furthermore, the carbon source of cellulose is atmospheric carbon dioxide, hence the consumption of cellulose back to carbon dioxide is environmentally carbon neutral. Thus cellulose has an outstanding position among various bioresource materials.

Cellulose is produced mostly in high plants by photosynthesis, and only negligible amounts by microorganisms and animals. The main sources and applications of cellulosic materials are summarized in Table 3.1.[1] The main raw materials used for high performance cellulosics discussed in this chapter are

RSC Polymer Chemistry Series No. 1
Renewable Resources for Functional Polymers and Biomaterials
Edited by Peter A. Williams

Published by the Royal Society of Chemistry, www.rsc.org

Table 3.1 Major cellulose sources and their worldwide annual production and use in 2004.[1]

Source	*Production (million tons)*	*Main uses*
Wood pulps	170.0	Paper, casing, sanitary products
Dissolving wood pulps	2.95	Rayon, cellophane, derivatives
Cotton	26.2	Textile fibers, filler
Cotton linter	About 10% of cotton	Fillers, medical fabrics, sanitary products, membranes, derivatives
Other fiber pulps	19.0	Paper

mostly the so-called "dissolving wood pulps" and cotton and cotton linters. Dissolving wood pulps are highly purified wood pulps, and are more accessible and cheaper, although are less pure and have lower molecular weights than cotton and cotton linter.[2]

Cotton linters are the shorter fibers cropped from cotton seeds after cutting the long fiber (cotton lints). Among these two, the first cut material is used mainly as medical and hygienic supplies, *e.g.* blood absorbing pads and similar products; the second cut is used as the starting material for regenerated cellulose and cellulose derivatives.

There are many other differences between these two raw materials, which can affect the quality of their final products. Thus, one should be especially careful when the product is intended as a biomaterial and for medical use. Processing conditions should also be suitably adjusted depending on the raw material and the intended application.

Cellulose produced by micro-organisms (bacterial cellulose) attracts interest for its high molecular weight, high purity, and microfibrillar nature; it is not commercially supplied (a sample may be available from Ajinomoto Corp).

Cellulose synthesis has also been attained both chemically and enzymically, but neither of these appears industrially feasible so far, though synthetic methodologies may be useful in the future as a pathway to certain variants of natural cellulose.

This chapter provides an introduction to general cellulose technology and the manufacture of cellulose derivatives, followed by a comprehensive review of cellulosic membranes and sorbents in blood purification and a wide range of other separations. The use of cellulosics in pharmaceuticals and related applications is also highlighted.

3.2 Chemistry of Cellulose and Its Derivatives[3]

3.2.1 Chemical Structure of Cellulose

Cellulose consists of anhydroglucose units bound by β-(1→4) linkages (Figure 3.1). Both bonds carrying the neighboring glucose residues are in the equatorial conformation; this helps cellulose to take a well extended linear conformation.

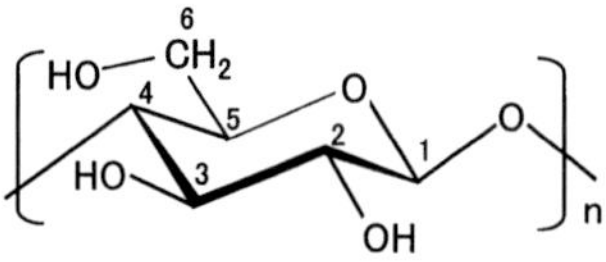

Figure 3.1 Chemical structure of the anhydro glucose unit in cellulose, indicating the equatorial conformation of the β-1,4 linkages, and the numbering for the carbons.

The two hydroxy groups at C2 and C3, and the hydroxymethyl group at C5 also adopt the equatorial conformation, and this gives cellulose an equatorial hydrophilic zone and an axial lipophilic zone. All three hydroxy groups contribute to the physical strength of cellulose through intra- and intermolecular hydrogen bond networks as well as the intermolecular hydrophobic interaction. The hydroxyl groups are also important as convenient anchors for chemical derivatization, as briefly described in 3.2.3 below.

3.2.2 Microcrystalline and Regenerated Celluloses

The fibrous raw cellulose can be transformed into powdery forms by mechanical cutting or milling, and by heterogeneous acid hydrolysis. The latter chemical process produces so-called microcrystalline cellulose (MCC), in which the uneven dissolution is accompanied by molecular weight reduction to a degree of polymerization (DP) of 100–300.[4] On the other hand, mechanical impact in water results in an axial split of the fibers to microfibrils of submicron thickness, *i.e.* microfibrillar cellulose (Celish, Daicel Chemical) which has a gel-like property and occludes a quantity of water.[5]

Cellulose can also be dissolved and regenerated into various forms, *e.g.* fiber, film or porous material.[6] While many dissolution processes have been developed, and some have been commercially applied, the cuprammonium process (also called the Bemberg process) is used for preparation of medical cellulose products. The solvent in this process is an aqueous solution of tetraammonio copper(II) dihydroxide. Another common dissolution process is the viscose process. Here treatment of cellulose with aqueous solution of sodium hydroxide and then with carbon disulfide results in the formation of sodium cellulose xanthate which dissolves in water, and subsequent acid or thermal treatment leads to elimination of carbon disulfide and regeneration of cellulose. Hydrolysis of cellulose derivatives is also another route to regenerated cellulose; for instance, the first hollow fiber membrane was prepared from cellulose acetate. Trimethylsilyl ether of cellulose is also a potentially interesting intermediate.

3.2.3 Cellulose Ethers and Esters

Chemical derivatization endows cellulose with new properties suitable for a variety of applications. Conventional and novel methodologies of cellulose

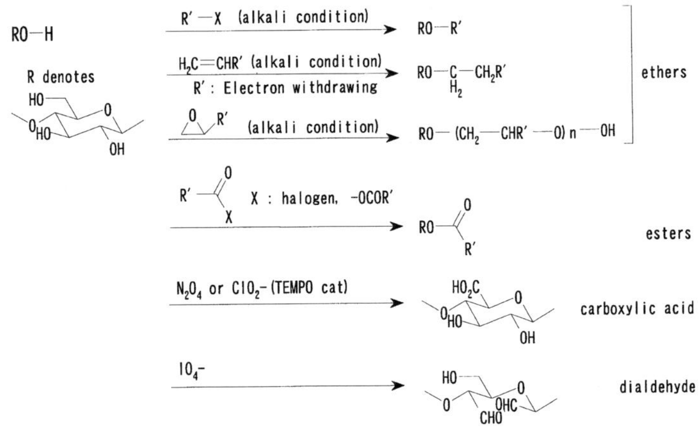

Figure 3.2 Some typical cellulose derivatives in which R denotes a glucopyranose ring and *n* depends on the reaction conditions.

derivatization are reviewed by Klemm *et al.*,[7] and some of the derivatives discussed in this chapter are outlined in Figure 3.2.

It should be emphasized in connection with the derivatization of cellulose that products with apparently the same chemical composition and molecular weight may perform differently, depending on the method of derivatization or reaction conditions. This may be due to different distribution of the substituents on the three chemically different glucosidic positions, in different segments of the polymer chain, and among different molecules.

The most important reactions are etherification, esterification and oxidation. Etherification is industrially applied to produce a variety of water soluble cellulosics. It usually includes cellulose treatment with concentrated aqueous sodium hydroxide, leading to swelling and the formation of cellulose alkoxides, which then reacts with an alkylating agent such as alkyl halide, epoxide or Michael acceptor. Thus, methylcellulose (MC) and carboxymethyl cellulose (CMC, also called carmelose) are derived from methyl chloride and sodium chloroacetate respectively.

Reaction with epoxides results in the formation of a hydroxyalkyl ether, *e.g.* hydroxyethylcellulose (HEC) or hydroxypropylcellulose (HPC). As the hydroxy group derived from the epoxide can also be etherified, the number of bound epoxides per glucose (molar substitution, MS) is usually larger than the number of the substituted hydroxy groups of anhydro glucose (degree of substitution, DS). Reaction with epoxides can also be quenched with a different alkylating agent to give a mixed ether such as hydroxyethylmethylcellulose (HEMC) and hydroxypropylmethylcellulose (HPMC). As HPC of a high MS and HPMC are moderately lipophilic, they are soluble in organic solvents such as alcohols. Cellulose ethers heavily substituted with hydrophobic groups, *e.g.* ethylcellulose of DS 2.5 and benzylcellulose, are no longer soluble in water, but soluble in organic solvent.

Several kinds of cellulose ester, namely nitrate, acetate, acetate propionate, and acetate butyrate, are commercially supplied, among which cellulose acetate is particularly important for medical applications. Cellulose acetate is prepared from raw cellulose by reaction with acetic anhydride in acetic acid as solvent in presence of sulfuric acid as catalyst. After complete dissolution of the raw material (*i.e.* acetylation), the product is partially hydrolyzed by adding water to produce a DS about 2.8–2.9, described as cellulose triacetate, or a DS around 2.4 as cellulose diacetate (CDA). The diacetate can be dissolved in a variety of common solvents and processed, *e.g.* dissolution in acetone for cigarette filter spinning. The triacetate can be dissolved in a rather limited number of solvents, *e.g.* in dichloromethane for the manufacture of photo and optical films. General information on cellulose acetate is available in ref. 7–12.

3.2.4 Other Cellulose Derivatives

Cellulose urethanes, also called cellulose carbamates, can be easily obtained by treatment of cellulose with an isocyanate in pyridine. While this reaction is often adopted in laboratory work, cellulose carbamate itself is not commercially supplied.

Oxidation of the $-CH_2OH$ to $-CO_2H$ is industrially performed with N_2O_4. The product, which is called cellouronic acid, is soluble in aqueous alkaline media.

3.3 Cellulosic Membranes

Membrane separation technology is rapidly growing as a low energy separation and purification process. Cellulosics have been the core material throughout the development of membrane separation technology and still have an important position in every category of membrane separation. Table 3.2 provides a

Table 3.2 Overview of membrane separation technology with indicative membrane pore sizes and respective applications.

Category[a]	*Rough pore size (A)*	*Application example*
Gas permeation	10^0	Separation of H_2 from fuel gas, O_2 enrichment,
Pervaporation	10^0 to 10^1	Alcohol–water separation
Reverse osmosis	10^1 to 10^2	Desalination, Concentration of fruit juice, Production of ultrapure water
Nanofiltration	10^1 to 10^2	Protein purification, Recovery of proteins from cheese whey
Dialysis	10^1 to 10^2	Hemodialysis Sausage production
Ultrafiltration	10^2 to 10^3	Drinking water purification Removal of microorganism from fermentation products and pharmaceuticals
Microfiltration	10^3 to 10^5	Removal of microorganism Waste water treatment

[a]The categories are not strictly defined. Separation takes place not only through the pores, but also by affinity between membrane material and the particle (or molecules). Driving force of transportation, *e.g.* diffusion, pressure, and evaporation, is also considered in the categorization.

summary of the major categories of membrane separation technology for cellulose characterized mainly by the membrane pore size (sieving size).

3.3.1 Outline of Membrane Separation

Presswood has reviewed the history of membrane separation technology.[13] Membrane research first started with natural materials such as animal organs. Fick made a membrane from cellulose nitrate which had been invented by Schönbein in 1845, and used it in his study on dialysis in 1855. Then membrane formation from collodion (cellulose nitrate solution in diethyl ether–ethanol) was thoroughly studied by researchers, and the basic technique of phase separation was established.

Phase separation (or gelation) of a solution by contact with a non-solvent before complete solvent evaporation was found to be the key process in producing membranes with larger pore sizes and good permeability. The addition of a small amount of low volatile liquid to the original solution also increased permeability. Bechhold prepared a membrane with pore diameters ranging from 1 to 5 μm, used it for filtration by pressure, and termed the filtration process "ultrafiltration" (UF). Then cellulose acetate membrane was first prepared by Brown in the 1910s. In Germany during the Second World War, membrane technology was studied to inspect drinking water by collecting and cultivating bacteria on the membrane. This idea was then brought to the US by the US Army and was developed as a filtration method. Thus was developed the membrane technology for the removal of micro-organisms, which is nowadays adapted to drinking water supplies.

Hemodialysis, which requires smaller pore sizes, was also tried with a collodion membrane first by Abel in 1912 on a rabbit. As mentioned later, a remarkable clinical success was also attained with cellophane tube around 1945.

Another important field of membrane separation is desalination by reverse osmosis. This technology was developed under the promotion of the US government (Saline Water Act 1952) to secure pure water from sea water or saline. Such a membrane that can afford pure water of reasonable purity at a reasonable rate was attained by Loeb and Sourirajan in 1960 with cellulose acetate.[14] The key to the success was the very thin separation active layer on a porous support layer (asymmetric membrane) to attain a higher permeation rates.

Cellulose acetate is also used for gas permeation, in which the thinness of the separation layer of the asymmetric acetate membrane attains a large flux.

As mentioned above, cellulosic materials, especially cellulose acetate, which was called "a separating medium without equal",[15] can form membranes with a wide range of pore sizes. Thus, cellulose acetate has been the key material for membrane technology, and is utilized for a variety of purposes, including recent developments in hemodialysis, blood product preparation, and water purification, as discussed in the rest of this section.

The physicochemical processes involved in the manufacture of semipermeable membranes has been extensively studied, and a great deal of literature is available on the subject.[16–18]

3.3.2 Cellulosic Membranes in Hemodialysis and Related Technologies

Treatment of an ailment by removing its detrimental cause(s) from the blood is generally called blood purification, and includes a variety of different methods as follows:

- Hemodialysis is the removal of low molecular weight waste matters from blood plasma by a membrane.
- Plasmapheresis is used for separation of plasma from the blood. The separated plasma is substituted with donated plasma (plasma exchange, plasmafusion), or subjected to further treatment to selectively remove a target material(s) by a secondary membrane filtration (double filtration), filtration by cooling (cryofiltration), selective adsorption (plasma adsorption), or salting out, and then returned to the blood. It is indicated for the treatment of autoimmune diseases, poisoning, endotoxemia and others.
- Separation of blood plasma from a healthy donor is also called plasmapheresis.
- Hemoadsorption and plasmadsorption are used for direct removal of target material(s), respectively, from the patient's blood or plasma by adsorption, as discussed later.
- Hemofiltration is the partial removal and replenishment of blood plasma across the membrane.
- Hemodiafiltration is a combination of hemodialysis and hemofiltration.

Cellulosic membranes have an important position in hemodialysis. They are also used in the second step of double filtration, and also for the fractionation of donated blood plasma. While centrifugation is the most commonly applied technique for plasmapheresis, membrane separation is also used.

3.3.3 History of Hemodialysis

Hemodialysis[19] is indicated mainly for kidney insufficiency, and is used for the treatment of about 600 000 patients worldwide. This is apparently one of the most important fields where membrane technology makes a remarkable contribution to human welfare.

The first success of a hemodialysis remedy was attained for a 67 year old lady suffering from acute renal failure in 1945. The dialyzer, Kolff's rotary drum artificial kidney, consists of a 20 m cellophane tube of sausage casing rolled spirally on a rotary drum. In the Second World War and the Korean war, the

Table 3.3 Materials and major suppliers of dialysis membranes (adapted from ref. 20).

Material	*Spinning*	*Manufacturer*[a]
Cellulose (cuprammonium)	Wet Dry–wet	Asahi Medical (Jpn), Membrana (Ger)
Cellulose (saponified cellulose acetate)	Melt	Cordis Dow, Althin, Teijin (Jpn.)
Modified cellulose (see text)	Wet Dry–wet	Asahi Medical, Membrana
Cellulose acetates	Dry–wet Melt	Toyobo (Jpn.), Cordis Dow Gambro (Swed), Althin
PAN (acrylonitrile based copolymers)	Wet	Hospal (Fr), Asahi Medical
PMMA (complex of isotactic- and syndiotactic polymers)	Dry–wet	Toray (Jpn)
EVA (saponified ethylene vinyl acetate copolymer)	Wet	Kuraray (Jpn)
Polysulfone modified with polyvinylpyrrolidone	Dry–wet	Fresenius Medical Care (Ger) Asahi Medical, Toray
Polyethersulfone modified with polyvinylpyrrolidone	Dry–wet	Gambro, Membrana
PEPA (blend of polyacrylate and polyethersufone)	Dry–wet	Nikkiso (Jpn)
Aromatic polyamide	Wet	Gambro

[a]The proprietorship of the products is changing considerably. On the ANZDATA website (www.anzdata.org.au/forms/2009SurveyDialyser/Codes.pdf), the following brands (not manufacturer) and their materials are cited: Baxter (CDA, CTA, Polysynthane, PES), Fresenius (PS, PS-Helixone, SAVH/CUVH), Gmbro (Haemophan, Polyamix, AN/Na methallyl sulfonate copolymer), Terumo (Cuprophane, Exebrane), Nipro (CTA, PES).

efficacy of hemodialysis for acute renal failure from crash syndrome was established. Since then cuprammonium cellulose has become the standard material for the manufacture of dialysis membranes.

The structure of the dialyzer depends on the form of the membrane. The coil type dialyzer contains a tubular membrane; the parallel flow type (or the Kiil type) contains laminated flat membranes. A hollow fiber dialyzer was first developed by Cordis Dow around 1968, which has become dominant nowadays because of its compactness and minimal blood residence.

The strong point about cellulose as a hollow fiber material, in addition to its permeating and physiological inertness, is said to be its wet strength and lower swelling due to its partial crystallinity. However, a variety of synthetic dialysis membranes have recently been developed (Table 3.3). According to the survey by The Japanese Society for Dialysis Therapy, cellulose, cellulose acetate, and synthetics had even shares in 2000 but polysulfone 50%, CTA 20%, polyethersulfone 11% modified cellulose (cellulosynthetics as mentioned later) 8% in 2008.[20] Thus the share of synthetics is getting larger in the West. As the exact comparison of the medical materials under the same condition is very difficult in a clinical field and the sales also depend on the strategy of the maker, the author cannot find the conclusive reason for this change of the share.

The materials, production, and characteristics of cellulosic dialysis membranes will be outlined in the following part.

3.3.4 Cellulosic Hollow Fibers

The methods of spinning dialyzing hollow fibers are broadly classified into wet spinning (including dry–wet spinning) and melt spinning which is often applied to synthetic polymers. In either case, the hollow structure is formed with a double tube spinneret (Figure 3.3). The bore-forming fluid is chosen from a coagulating liquid to form the inside skin layer, or inert liquid or gas to avoid the formation of an inside skin layer.

Spinning conditions for different cellulosic hollow fibers are shown in Table 3.4. Cuprammonium fibers are spun by the wet or dry–wet method. The polymer solution is extruded from the spinneret and treated with a coagulant such as concentrated aqueous sodium hydroxide without stretch.[21] The caustic coagulant causes the gelation of polymer solution to result in a gel with the molar composition Glu : Cu : Na of 2 : 1 : 2. The coagulation condition is in contrast to that for textile filaments, where coagulation is performed with water under strong stretching (several hundred times) to attain rapid coagulation with high orientation. Then the hollow fiber is washed with dilute sulfuric acid to remove the copper ions. Finally a plasticizer, typically glycerol, is applied to avoid brittleness and the collapse of pores during drying. Sketches of one common cuprammonium spinning machine, and one designed with an inert contacting bath to prevent skin formation, are given in Figure 3.4.[22]

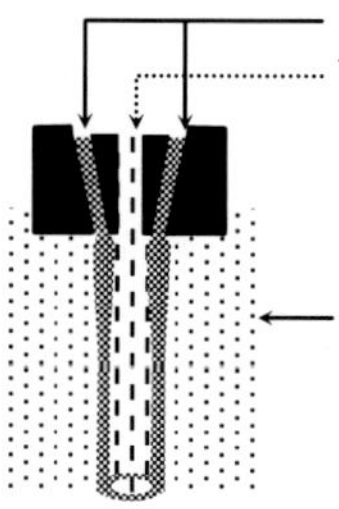

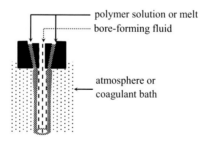

Figure 3.3 Conceptual sketch of a concentric tubular spinneret.

Table 3.4 Examples of spinning conditions for cellulosic hollow fibers.

Material	*Solution (or melt) and porogen*	*Core fluid*	*Coagulant*
Cellulose	Aq. $Cu(NH_3)_4(OH)_2$	2-Propyl myristate	Aq. NaOH
Diacetate[a]	DMF–PEG	Liquid paraffin	Water
Diacetate	DMF	?	Water
Diacetate	NMP	Glycerol	Water?
Triacetate	NMP–PEG	Water–PEG–NMP	Water–NMP
Diacetate	Melt (240–280) polyalcohol	Inert gas?	Water?

[a]Saponified after spinning (saponified cellulose acetate, SCE).

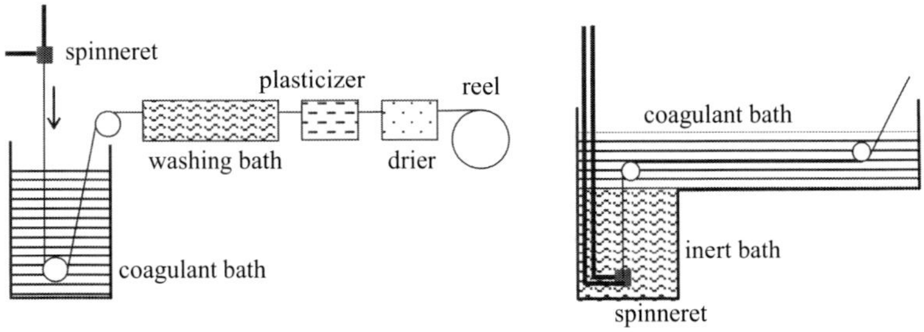

Figure 3.4 A typical spinning machine (left, dry–wet method[23]), and the coagulant bath of a modified system (right) for cuprammonium spinning.[22]

Cellulose di- and triacetates are dry–wet spun.[23a] PEG is sometimes added to the starting solution to control pore formation. Cellulose diacetate is spun also by melt spinning. A polyhydric alcohol like glycerol is also added to plasticize cellulose acetate and then undergo phase separation by cooling to form pores.[23b] The first cellulose hollow fiber module was produced by Cordis-Dow by saponification of preformed diacetate hollow fibers (saponified cellulose ester, SCE).

3.3.5 Recent Developments in Hemodialysis Membranes

Although the technology of hemodialysis is well established, what should actually be removed from the patient's blood by hemodialysis is still a controversial issue. Initially low M_w compounds such as creatinine, urea, and ureic acid were targeted, and this achieved a remarkable increase in survival rate. However, the success of long term hemodialysis then elicited so called hemodialysis-associated complications, and how to avoid them and improve the quality of life (QOL) of the patients have become the focus of recent R&D.

3.3.5.1 Fractionation Molecular Weight

Some of these complications, a class of diseases called hemodialysis-associated amyloidosis, the typical one of which is carpal tunnel syndrome, was supposed to be caused by the deposition of the protein called β-amyloid on tissues. The protein called β_2-microglobulin (β_2-MG), which is believed to have a role in immune response and metabolized in a kidney, was suspected as the precursor for β-amyloid[24] and the removal of it from blood was proposed. The molecular weight of β_2-microglobulin is 11 700, and that of serum albumin which should be retained in the blood is 66 000. Therefore the development of ultrafiltration (UF) membranes with a fractionating size between these two proteins was designed and the dialyser equipped with such a membrane was called "the high-performance membranes". As the transport of such a high M_W material by pure diffusion is very slow, some dialyzers are so designed that hemodialysis and hemofiltration are combined to achieve what is called hemodiafiltration.

While high-performance dialysers were successful against amyloidosis, they sometimes appear to be less effective against some other complications such as anemia and psychosomatization, which were assumed to be caused by the pathogenic factors with molecular weights near that of serum albumin. Thus, efforts to further increase the fractionating point was made. On the other hand, some people thought that only the sieving mechanism was not satisfactory and its combination with other mechanisms such as selective adsorption, *i.e.* extracorporeal therapy[25] was of particular interest (see later).

3.3.5.2 Biocompatibility Issues

Biocompatibility poses one of the major challenges in biomedical technology in general,[26a] particularly in hemodialysis.[25,26] Some of the main biocompatibility issues are as follows:

- Blood coagulation. Blood coagulates by contact with a foreign material. While blood coagulation can be suppressed with heparin or other anti-coagulants, it is preferable to reduce the anticoagulant dose from the viewpoints of both safety and cost.
- Complement activation. Complement activation covers a part of the immune system, involving several factors constituting two cascades. It was suggested that repeated complement activation induced by the contact with membrane is a possible cause of some hemodialysis-associated diseases and also induce the deposition of β-amyloid. Cellulose is assumed to activate the complement system by binding of one of the factor in the cascade on its OH groups.[25–27] Thus the avoidance of complement activation has been one of the most important working hypotheses in developing new membranes.
- Cytokine induction. The compliment activation may also induce interleukins and tumor necrosis factors which may lead to muscle atrophy, arteriosclerosis and related problems, and the possible mechanism is called the cytokine hypothesis.
- Endotoxin reverse diffusion. Lipopolysaccharides (LPSs), also called endotoxins or pyrogens (in a narrow sense, but inaccurately), are a constituent of the cell wall of Gram-negative bacteria, and are very commonly present in our living environment. Once penetrating a mammal's blood, LPSs show a strong pyrogenic (fevering) and immunoadjuvant activity even at quantities below μg. LPSs have an amphiphilic structure composed of a polysaccharide and a lipid A components with α-(1$\rightarrow$6) chitobiose skeleton carrying ester-bonded phosphoric acids and long chain fatty acids. While it is usually present in highly aggregated micellar forms, it is pointed out that the reverse diffusion of oligomeric LPs into the blood through a high-performance membrane can occur, causing complications as the M_w of monomeric LPSs is about 4000. To reduce such potential risks, the dialyzate must be kept free from LPS (also see later for adsorptive removal of LPS).

3.3.6 Recently Developed Cellulosic Hemodialysis Membranes

To overcome or reduce these and other possible risks, the development of a new membrane, often called the New High-Performance Membrane is extensively continued. It includes some based on cellulose acetate, improved regenerated cellulose, chemically modified cellulose (cellulosynthetic), and synthetics. Some accomplishments on cellulosic membranes will be introduced below.

3.3.6.1 Cellulose Acetate Membrane

A membrane with large pore size and wet strength made of cellulose triacetate (CTA) was introduced by Toyobo; that was the thinnest hemodialysis membrane (wet wall thickness below 15 μm) of the day, thus contributed to the compactness of the dialyzer. The Toyobo FB-F series, which is highly permeable by β_2-MG, is reported to mitigate joint pain and numbness; reduction of complement activation is also reported. According to Ohno, the permeaselectivity between β_2-MG and serum albumin is enhanced by the reversible adsorption of albumin onto the CTA surface.[28] The CTA membrane has furthermore improved to remove α_1-microglobulin (M_w 33 000). As sieving out the protein from HSA (M_w 66 000) with a single sieving layer (also called screen filtration) is impossible, it is designed to be a depth filter having the efficiency of multistep filtration. The structure was attained by initial gelation not by contact with coagulant but by thermally induced phase separation.[28b]

3.3.6.2 Vitamin E-Coated Cellulose Membrane

The blood of hemodialyzed patients is said to be possibly oxidative because of the release of active oxygen by neutrocytes, an immune reaction. Thus the idea of combining the membrane with vitamin E came from the expectation of mitigating the oxidative state. Such a vitamin E-modified membrane is made as follows.

An acrylic polymer carrying fluorocarbon and oleyl alcohol moieties capped with epoxide groups is added to the bore-forming fluid (2-propyl myristate) used for spinning the cellulose hollow fiber. The hydrophobic moiety of the modifying agent localizes in the boundary of the bore-forming fluid and the cuprammonium solution, hence the chemical modification of the inner surface of the hollow fiber at a high efficiency (Figure 3.5). Then vitamin E is adsorbed on the modified surface by hydrophobic interaction. Such a regenerated cellulose based membrane, Excebrane (Terumo, later transferred to Asahi Medical) is reported to alleviate leukopenia (a transient reduction of leukocytes from the blood under hemodialysis), and induce less IL-6 and less active oxygen *in vitro* than plain regenerated cellulose, with clinical results indicating its improved biocompatibility.[29]

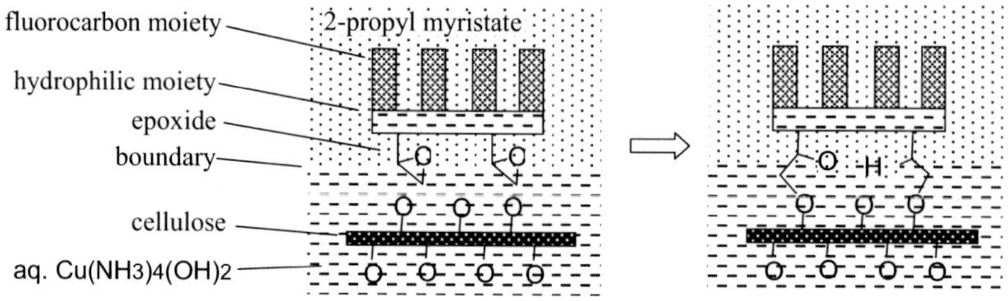

Figure 3.5 Schematic illustration of the hydrophobic modification of cellulose surface. Oleyl alcohol and vitamin E are physically carried between the fluorocarbon moieties by hydrophobic interaction.[22]

3.3.6.3 Poly(ethylene glycol) Grafted Cellulose Membrane

PEG grafted membranes, AM-PC and AM-BC (also Asahi Medical), were developed on the assumption of Ikada that a surface with a high molecular mobility (diffusive layer) like that of living tissue should have a higher biocompatibility, and thus lower protein adsorption.[30] Here a post spinning modification is applied, *i.e.* regenerated cellulose hollow fibers are treated with carboxylic anhydride derivative of PEG.[31] This membrane is reported to reduce leukopenia, platelet adhesion, and other problems in *in vitro* and clinical *in vivo* reactions, and enable a reduction of the heparin dose in some cases.

3.3.6.4 DEAE-Modified Regenerated Cellulose Membrane

A membrane made by the cuprammonium process from cellulose slightly etherified with diethylaminoethyl (DEAE) group is named Hemophane (Akzo Nobel, transferred to Membrana). It is reported that modification by only 1 wt% DEAE attained a considerable reduction of complement activation.[32] While the mechanical strength and permeability of urea and creatinine remained essentially the same as that of unmodified cuprammonium cellulose, the permeation of phosphates increased by about 10%. Many indicators of biocompatibility except TNF, *e.g.* leukopenia, granulocyte elastase, C3a (one of complements), and P-selectin, were also improved.[33] P-Selectin is a protein involved in blood coagulation, and this membrane is reported as the strongest anticoagulant. It is suggested that adsorption of anionic heparin on the cationic cellulose may contribute to this property. The adsorption of acidic albumin is also suggested as another mechanism of improving biocompatibility.

3.3.6.5 Symmetric Gradient Cellulose Membrane

A high-performance membrane based on plain regenerated cellulose, but characterized by a smaller pore size distribution on both surfaces, is referred to as a symmetric gradient membrane. It is expected that if filtration is performed under absolute equilibrium, the permeation rate of particles should be the same in both directions, but in practice absolute equilibrium is impossible. If the pore size distribution across the membrane is like a funnel (*i.e.* dead end filtration), it may happen that the particle to be rejected cannot diffuse backward, and so stays around the funnel's bottom to give a higher local concentration and eventually a higher possibility of permeation. The BCX membrane (Asahi Medical) was so designed that its pore size is minimal in both surfaces to make reverse diffusion efficient and it is reported to reduce reverse diffusion of LPS by half.[34]

3.3.6.6 Various Chemically Modified Cellulosic Membranes

The chemical modification of cellulose has been the subject of continuous research in membrane development and a variety of chemically modified membranes such as cellulose esters,[32,35] cellulose ethers,[36–38] or those carrying ion exchange groups,[39–41] or carbamates and acetate carbamates,[42] have been reported. According to Diamantoglou *et al.*,[42] *in vitro* evaluation revealed that there is an optimal balance of hydrophilic and lipophilic moieties. Thus, the generation of C5a, a factor in the complement cascade, could be completely suppressed by the introduction of 0.1 mole/Glu butyl- or methylbenzyl carbamate onto cellulose diacetate (DS 2.5) without much effect on platelet adhesion. The authors suggested that the interaction of the membrane with certain unidentified protein(s) in a physiological cascade is optimized at an optimal hydrophilic–lipophilic balance.[42]

Thus a variety of chemically different membranes have been developed, some based on cellulose and some on synthetic polymers, as summarized in Table 3.3. However, because of the complexity of biological systems, and the difficulty of clinical evaluation, it will take some time to fully assess the physiological and clinical characteristics of these membranes.

3.3.7 Removal of Pathogens with Cellulosic Membranes

Cellulosic UF and microfiltration membranes have been commercially available since Sartorius started production of an acetate membrane in 1927. Even today membranes made of regenerated cellulose, cellulose acetate, cellulose nitrate, and cellulose acetate propionate are the most popular laboratory filters for collecting or removing microorganisms and cells. However, as far as the author is aware, synthetic, rather than cellulosic, membranes tend to be employed for industrial processes presumably because of their higher durability against thermal sterilization. In this section, the membranes used industrially to remove HIV from blood products and pathogenic microorganisms from drinking water are discussed.

3.3.7.1 Cellulose Membrane Filters for Virus Removal

A cuprammonium membrane was developed which targeted the removal of HIV from blood products; its importance is now growing as one of several successful medical applications of cellulosic membrane. The particle size of many of the essential blood proteins is about 10 nm, while those of viruses are 20 to 200 nm. Therefore, membranes used to remove viruses from blood require a very high accuracy of sieving. According to Manabe *et al.*,[43–45] such a separation is impossible by screen filtration and homogeneous membranes in which the pore size is depth-independent (*i.e.* depth filters), like laminated sieves, are rather ideal for the purpose. Such a membrane, named the Bemberg Microporous Membrane (BMM) is attained by a modified cuprammonium process,[43] and is supplied as Planova filters (Asahi Medical). According to the patent, a water–acetone–ammonia mixture is employed as the coagulant in the cuprammonium process.[44] Acetone is said to modulate the surface property of cellulose. It is also stated that any stress to the fiber in the spinning process should be strictly avoided. For this reason, a long and linear coagulation bath was designed to avoid the bending of the fiber line. This filter seems to be widely used in the preparation of blood products and its possible application for the removal of the Creutzfeldt–Jakob disease pathogen (abnormal prion) is mentioned.[45]

3.3.7.2 Acetate Membrane for Drinking Water Purification

The market for reverse osmosis desalination is shared between cellulose acetate and polyamide membranes. However, more recently ultrafiltration technology is being employed to remove turbidity and pathogenic micro-organisms from surface water.

The conventional method of water purification comprises the sequence of flocculation, precipitation and sand filtration processes, and thus requires a large floor space. The introduction of membrane filtration into water purification reduces the size of filtration facilities, simplifies the operation, and increases the accuracy of purification. The outbreak of massive infections by cryptosporidium, a commonly distributed pathogen, which is resistant to chlorine sterilization and can cause diarrhea, has boosted the development of membrane technology for drinking water purification. The rejection efficiency of 5 μm particles (*e.g.* oocysts of cryptosporidium parvum) by UF membranes is estimated at about 100 times higher than that of conventional systems. Thus membrane filtration is recognized as one of the promising technologies of water purification.

However, it is ultrafiltration, rather than reverse osmosis, that dominates efforts to minimize the cost, energy and facilities at the desired purification level. Nakatsuka *et al.* have compared UF membranes based on cellulose acetate, polyacrylonitrile and polyethersulfone with similar sieving sizes and fluxes (flow rate per unit area and pressure drop); they conclude that cellulose acetate membranes retain the highest level of flux, and resist fouling (loss of permeability by deposited material) (Figure 3.6).[46,47]

Presently up to 3 million cubic meters of surface water are purified daily worldwide by membrane filtration, and this volume is rapidly growing.

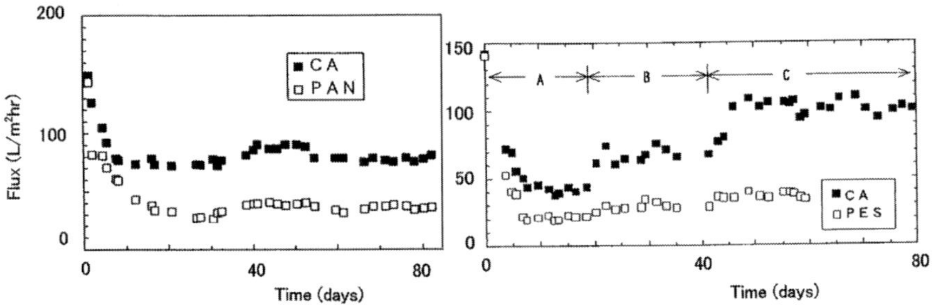

Figure 3.6 Flux of cellulose triacetate (CA), polyacrylonitrile (PAN, left), and polyethersulfone (PES, right) membranes used in water purification. Water of Ibo River in Hyogo, Japan filtered at 50 kPa. A, B, and C indicate different operating conditions.[47]

Among a variety of membrane materials, *e.g.* polypropylene, polyacrylonitrile, polysulfone, polytetrafluoroethylene and others, cellulose acetate is estimated to occupy about one quarter of the total market.

3.4 Cellulosics in Chromatography and Related Technologies

Chromatography is a technology for separating materials on the basis of differences in their thermodynamic partition coefficients in solid–liquid or liquid–liquid systems, usually in a multistep separation system. The separation mechanism is clearly different from the kinetic mechanism of membrane filtration. Since the discovery of paper chromatography, cellulosics have been used as the chromatographic stationary phase in increasingly more refined forms and diversified settings. The major applications of chromatographic cellulosics outlined below include the isolation of bioproducts. Nowadays, the development of so-called biopharmaceuticals including antibodies, hormones, vaccines is a definite trend in the field of pharmaceuticals and that exhibits the potential applications of cellulosics as well as other materials. Other topics are the resolution of optical isomers (chiral separation), and the removal of a specific type of leukocytes as a medical treatment. First a general description of cellulosic stationary phases (gels) is made.

3.4.1 Cellulose-Based Chromatography Gels to Separate Biomaterial

3.4.1.1 Matrix and Ligands

Most of the production cost of biopharmaceuticals is said to be of purification. Thus the development of suitable separating media is crucially important. The purification of such biomaterials should carefully avoid denaturation or irreversible adsorption, and this is one of the reasons why polysaccharide matrices are often employed in bioseparation. While Sephadex and Sepharose, developed in Pharmacia and aquired by GE Healthcare are the established brands of such media, cellulosic materials have also been utilized for long and seem to have a good potential. The advantage of cellulosic beads are their mechanical strength, which enables a higher flow rate,[54] and the minimized non-specific interactions achievable by reductive removal of their acidic groups (carboxylate and sulfate).

Powdered cellulose, *e.g.* cut fibers and microcrystalline cellulose, have been traditionally used as packing materials for column chromatography; the particles may be plain cellulose, or carrying ionic groups (ion exchange ligands), and various ion exchange cellulosics are supplied from many suppliers.[48a] Although they are relatively inexpensive, these cellulose powders are not optimized as to their particle shape, size and porosity. So suspension cross-linking technology has been applied for producing a variety of beaded cellulose particles with controlled pore size and functionality to improve separation efficiency and capacity,[48b,c.] some of which are commercially applied (see below). A sponge-like monolithic support made from regenerated cellulose has also proposed to resolve the problems facing granular particles.[49] Examples of various commercially available cellulosic adsorbents are summarized in Table 3.5.

Iontosorb supplies beaded plain cellulose named Perloza[50] and that carrying a ligand named Iontosorb. The key step in preparation of these materials is the thermal gelation of a water-in-oil (w/o) suspension of a viscose cellulose solution in a water-immiscible liquid like chlorobenzene.[48b,c,51]

Chisso Corporation also supplies another brand, Cellulofine or Cellufine. It adopts two processes of preparation. One has its base in Hirayama *et al.*,[52] solvent evaporation from the o/w suspension of a dichloromethane solution of cellulose triacetate containing a porogen (an additive to make a porous structure by phase separation) such as 1-octanol, a suspension stabilizer (*e.g.* starch derivative) in water is stirred near the boiling point of dichloromethane. This produces a suspension of polymer solution droplets in water; drawing a draft in the vessel results in slow removal of the solvent from the polymer droplets, and the formation of cellulose triacetate beads in water, followed by extraction of the porogen, hydrolysis of cellulose triacetate to afford the corresponding cellulose beads. Another process is based on the report by Kuga,[53] including the gelation of cellulose solution in conc. aq. calcium thiocyanate and commented that it gave a higher mechanical strength for their pore size.

Pall Corporation also supplies a series of cross-linked cellulose beads, HyperCel, but the author has no information on the production process.

Cellulose beads were also obtained *via* trimethylsilylcellulose and the properties have been compared with other supports.[54] The molecular weight range covered by total Perloza grades seems to be higher than that by Cellulofine. In general crosslinking treatment can improve mechanical strength.

A variety of ligands are bound to the plain cellulosic particles, including ion exchange groups, hydrophobic groups, protein binding groups with a specific affinity, chelating groups to trap heavy metal ions, dyes for group specific

Table 3.5 Representative examples of commercially available cellulosic adsorbents.

Powdered natural cellulose	*Regenerated cellulose beads*	*Functional groups*
CC, CF (Whatman[a]) Microcrystalline cellulose (E Merck)	Perloza MT (Iontosorb) Cellulofine GH (Chisso)	None (plain cellulose)
DE (Whatman) Express-Ion D (Whatman)	Iontosorb DEAE (Iontosorb) DEAE Cellulofine (Chisso)	$-CH_2-CH_2-N(C_2H_5)_2$
QA (Whatman) Express-Ion Q (Whatman)	Iontosorb TMAHP QA Cellulofine	Quaternary amine
CM (Whatman) Express-Ion C (Whatman)	Iontosorb CM CM Cellulofine (Chisso)	$-CH_2-CO_2^-$
P (Whatman)	Iontosorb P	$-PO(OH)_2$
SE (Whatman) Express-Ion S (Whatman)	Sulfate-Cellulofine (Chisso)	$-CH_2CH_2SO_3^-$
	Iontosorb BUTYL, Octyl, Phenyl Phenyl-Cellulofine, Butyl-Cellulofine (Chisso)	Hydrophobic groups (alkyl, aryl)
	Iontosorb AV, CNC, TS, A-Urea Formyl-Cellulofine (Chisso)	Protein binding groups (affinity ligands)
	Iontosorb OXIN, SALICYL, DETA, DTTA, DITHIZON	Chelating groups
	Iontosorb Blue 2, Red, Blue 4, Green 5, Green 19, Green 19a, Yellow, Brown	Dye (group selective affinity ligands)
	Anhydrothrombin-Cellulofine (Chisso)	Protein
	ET Clean-PL (Chisso)	ε-Polylysine
	MEP HyperCel (Pall) HEA	4-mercaptoethylpyridine
	HyperCel (Pall) PPA	hexylamine
	HyperCel (Pall) IMAC	phenylpropylamine
	HyperCel (Pall)	iminodiacetic acid

[a]Whatman is a part of GE Healthcare.

chromatography, and so on. For instance, the polylysine ligand is designed for selective removal of LPS (see 3.3.5.2). Observations in our lab show that cellulose beads carrying primary amino groups at a high density are effective in the removal of LPS from protein solution.[55] A cellulosic depth filter, Zeta Plus, is supplied by Cuno for the same purpose, but its technical details are not clear.[56]

3.4.1.2 *Application of Cellulose-Based Packing Materials*

Many reports on research applications of stationary phases are given on suppliers' websites. Some work aiming at industrial production of biomaterials is reviewed

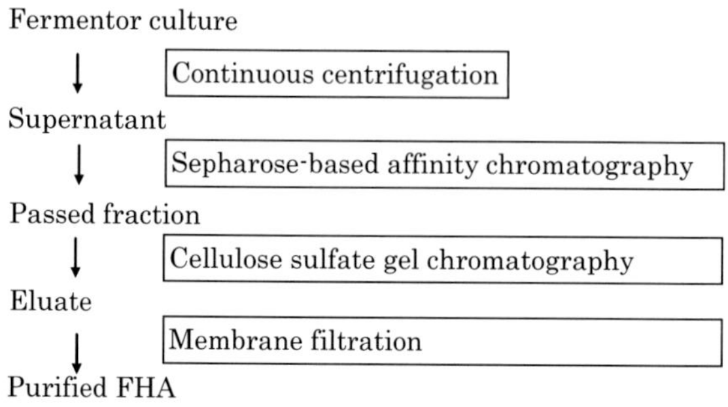

Figure 3.7 General chart for the purification of a component of Pertussis vaccine. Chromatography on sulfated cellulose is reported to give highly pure endotoxin-free material by a simple process with a high yield.[58]

in this section. For example, recombinant human serum albumin can be purified by a process including chromatography on Phenyl-Cellulofine,[57] and Sulfate-Cellulofine has been used to purify the Pertussis component vaccine,[58] Japanese encephalitis vaccine on the same support,[59] recombinant adeno-associated virus vectors,[60,61] and the rabies virus.[62] For industrial applications, loading capacity and durability are crucially important.[63] Figure 3.7 shows a chart of purification process for a protein component of Pertussis (whooping cough) vaccine.

3.4.2 Adsorbents for Hemoadsorption

The term hemoadsorption means the removal of a harmful substance from blood by adsorption on particulate or fibrous materials, including cellulosics, as exemplified below.

3.4.2.1 Granulocytapheresis with Cellulose Acetate Beads

Leukocytes perform an important role in immune response. However, their unusual activation is linked to autoimmune diseases, and thus they are sometimes partially removed from the blood of such patients by adsorption. When a particular class of leukocytes is targeted, the process is called leukocytapheresis.

Cellulose acetate selectively was examined for adsorption of granulocytes in the early 1990s, and a column packed with cellulose diacetate beads (2 mm, Adacolumn) has been developed for this application (Japan Immunoresearch Laboratory Co Ltd). The process, especially important for steroid resisting patients, has been officially approved as a treatment for ulcerative colitis,[64,65] and is also reported to mitigate rheumatoid arthritis and cachexia in advanced cancer patients. The remedial function of CDA is rationalized by the adsorption of C3b (a factor in the complement cascade) and IgG for which granulocytes have a receptor site(s).

Another adsorbent (Cellsorba, Asahi Medical) commercialized for similar treatments in Japan is a non-woven filter made of polyester fiber carrying polyethyleneimine as its active ligand.

A related blood treatment is the partial removal of leukocytes from blood used for transfusion. The method is claimed to reduce transfusion-associated side effects, but it has pros and cons. Some filter devices are commercially supplied, one of which (named Immuguard) contains cotton as the adsorbent.[30]

3.4.2.2 Other Hemoadsorbents

Several selective sorbents for hemoadsorption or plasmadsorption are commercially available, several of which are based on cellulose, presumably like those mentioned in 3.4.1 (Table 3.6).[66] Both Selesorb and Liposorber are cellulose beads with dextran sulfate groups, but designed for different targets. The selectivity of these matrixes is determined by their pore sizes (larger in Liposorber).[67] For instance, Lixelle (hexadecylated microporous cellulose beads) is designed particularly to adsorb β2-microglobulin from blood, and is used by connecting in series with an artificial kidney. The smaller pore size of these beads allows access to its ligands only by low M_w proteins such as the targeted β2-microglobulin. The preparation of uniform beads by a vibrating nozzle has also been described for this sorbent.[68] Other sorbents have been designed with ligands for antigen–antibody interactions.[69]

Table 3.6 Commercially available selective hemoadsorbents (adapted from ref. 66).

Ligand/support	*Indication/target*	*Trade name/Supplier*
Dextran sulfate/cellulose	Systemic lupus erythematosus/Anticardiolipin, Anti-DNA antibodies	Selesorb/Kaneka
Dextran sulfate/cellulose	Hyperlipidemia/LDL	Liposorber/Kaneka
Acetylcholine receptor peptide/(synthetic) cellulose	Myasthenia gravis/acetylcholine receptor blocking antibody	Medisorba/Kuraray
Sheep polyclonal antibody/cellulose	Many autoimmune diseases/antigens	Ig-Therasorb/Therasorb Med Systems
Hexadecyl groups/cellulose	Hemodialysis- associated amyloidosis/β2-microglobulin	Lixelle/Kaneka
Protein A/agarose	Myasthenia gravis and other autoimmune diseases/	Immunosorba/Excorim
Protein A/silica	Rheumatoid arthritis/IgG	Prosorba/Cypress Bioscience
Phenylalanine/PVA	Autoimmune diseases/	Immusorba PH/Asahi Medical
Tryptophan/PVA	Autoimmune diseases/	Immusorba TR/Asahi Medical
Polymixin B/polystyrene	Septicemia/Lipopolysaccharide	Toraymixin/Toray

Hemoadsorption is also used for poisoning treatment. Active charcoal is the most commonly used adsorbent,[25] which may be coated with cellulose, poly(2-hydroxyethyl methacrylate) or other polymers to improve biocompatibility and suppress the adsorption of essential blood proteins. A variety of other beaded adsorbents have also been described.[70]

3.4.3 Chromatographic Chiral Separation

Although a pair of enantiomers are identical in their chemical composition and the make up of their bonding, their biological activities are sometimes dramatically different from each other. One may reasonably speculate that life, which is dominated by the L-form of amino acids and the D-form of glucose, must discriminate between enantiomeric molecules. However, the importance of chirality in life science, which has been conceptually known, is seriously recognized with increasing concrete knowledge especially in pharmacology.[71] As a result the development of racemic drugs, which was once very common, is nowadays avoided as much as possible. Thus, the methodology of enantiomer analysis and preparative separation has become essential. Among them, liquid chromatography has substituted for the previous methods of analysis and is developing its application to preparation of a pure enantiomer. Nowadays a fair number of adsorbents for enantiomer separation have become available but the cellulose and amylose-based ones are far more often used than others.

3.4.3.1 Chromatographic Sorbents for Chiral Separation

Chiral separation requires a homochiral stationary phase, a packing material of virtually pure enantiomer. In addition various other operating parameters must be adjusted to enable an effective interaction between the stationary phase and the analyte enantiomer to be separated, thus making the development of a chiral stationary phase (CSP) difficult. For instance, sugars including cellulose have been regarded as a potential CSP for a long time, and exploratory chiral separation on lactose, cellulose (paper and powder), starch, and cellulose 2.5-acetate was reported up to the 1960s,[72a] but the separations were not effective. The first complete separation of enantiomers by liquid chromatography was reported on microcrystalline cellulose triacetate (MCT) by Hesse and Hagel in 1973.[72b] Although MCT, the product of heterogeneous acetylation of microcrystalline cellulose, performed many good chiral separations, it required long separation times and a lot of solvent.

In about 1980, several novel CSPs were reported, including crosslinked polyacryl- and polymethacrylamides carrying a homochiral amine residue, immobilized homochiral amides carrying π-acidic benzoyl groups, immobilized amino acid, immobilized bovine serum albumin, and homochiral isotactic poly(triarylmethyl methacrylate).[71] While many of these CSPs were commercialized, many racemates still remained inseparable.

The CSPs mentioned above were designed as high performance stationary phases in which the chiral ligands were attached chemically or physically to silica gel beads. It was considered impossible to make CTA into a high performance phase because MCT substantially loses its chiral recognition after dissolution in solvents,[72] generally assumed to be due to the loss of its crystalline structure. Furthermore, once the crystal lattice of MCT belonging to CTA I is dissolved, it can never be reconstructed because of the irreversible transformation into the more stable CTA II lattice. Therefore we sought a methodology to re-establish the potential chiral recognition ability of CTA II, as outlined below.[73,74]

1. CTA I and CTA II have completely different retention properties, and they often have the opposite enantioselectivity (*i.e.* which enantiomer they retain more strongly).
2. Increasing crystallinity as measured by X-ray diffraction reduces adsorption and eventually chiral discrimination.
3. Coating CTA on porous supports is a convenient method for obtaining durable high performance stationary phases which are easy to handle.
4. Chiral discrimination ability sometimes show a strong hysteretic effect of the coating solvent related to the hysteretic conformational regularity of the resulting CTA detected by solid state NMR,[74] and also shown in Figure 3.8.

Based on these observations, a wide range of cellulose and other polysaccharide derivatives were evaluated by us and by Okamoto and his colleagues,[75–78] and the knowledge gained by both groups may be summarized as follows:

1. Generally speaking, cellulose ethers showed only poor chiral discrimination, and the tribenzoate, tricinnamate, and triphenylcarbamate were better than triacetate.
2. The secondary substituent(s) on the aromatic rings of cellulose tribenzoate and triphenylcarbamate had a profound effect on their selectivity. Strongly polar secondary substituents had a negative effect, presumably because of their non-chiral adsorption.
3. Most polysaccharide derivatives have more or less chiral discrimination, each with its own range of resolvable enantiomeric pairs (referred to here as analyte selectivity); cellulose derivatives cover the widest range, followed by amylose derivatives.

Following this work, the cellulose and amylose phases (or the respective packed columns) indicated in Figure 3.9 and Table 3.7 have been commercialized.

The abovementioned adsorbents comprise polysaccharide derivatives coated on a silica gel support which can be damaged by applying solvents which dissolve or strongly swell the polymer. The applicable solvents were limited to an alkane–alcohol mixture and aqueous acetonitrile or alcohol. To overcome this shortcoming, immobilized adsorbents were developed. Generally to say, methods are available to afford the polymers with solvent resistance,

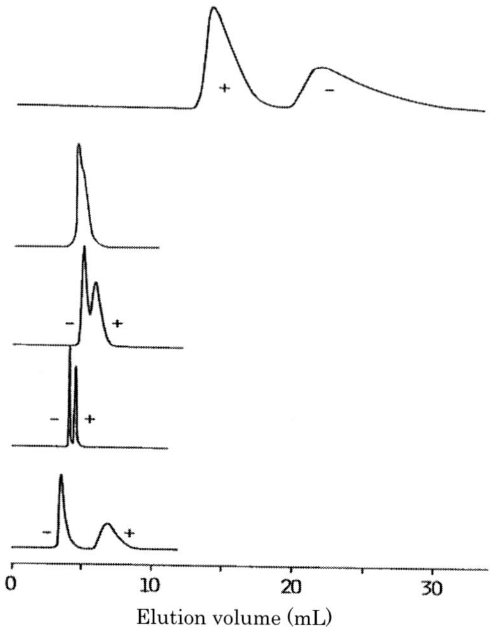

Figure 3.8 Dependence of chromatogram of *trans*-1,2-diphenyloxirane on the form of CTA stationary phase. +/– denotes the polarimetric sign. Chromatograms from top to bottom are numbered 1–5. 1, MCT (mobile phase, ethanol); 2, CTA II high crystallinity (mobile phase, ethanol); 3, CTA II low crystallinity (mobile phase, ethanol); 4, CTA II coated on porous silica beads with dichloromethane (mobile phase, hexane–2-propanol 9 : 1 v/v); 5, CTA II coated on porous silica beads with phenol–dichloromethane (8 : 1 v/v) (mobile phase, hexane–2-propanol 9 : 1 v/v).

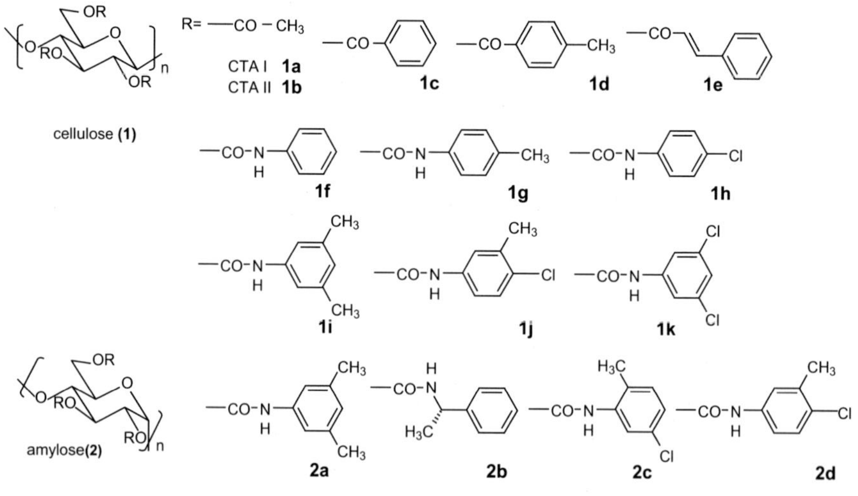

Figure 3.9 Commercially available cellulose- and amylose-based chiral selectors.

Table 3.7 Commercially available cellulose and amylose phases.

Chiral selector	*Commercial name*	*Supplier*
MCT 1a	Chiralcel CA-1	Daicel
	Chiral Triacel	Macherey-Nagel
Coated CTA 1b	Chiralcel OA	Daicel
Plain 1c	Chiral Tribencel	Macherey-Nagel
Coated 1c	Chiralcel OB	Daicel
Coated 1d	Chiralcel OJ	Daicel
Coated 1e	Chiralcel OK	Daicel
Coated 1f	Chiralcel OC	Daicel
Coated 1g	Chiralcel OG	Daicel
Coated 1h	Chiralcel OF	Daicel
Coated 1i	Chiralcel OD	Daicel
Coated 2a	Chiralpak AD	Daicel
Coated 2b	Chiralpak AS	Daicel
Coated 2c	Chiralpak AY	Daicel
Coated 2d	Chiralpak AZ	Daicel
Immobilized 2a	Chiralpak IA	Daicel
Immobilized 1i	Chiralpak IB	Daicel
Immobilized ij	Chiralpak IC	Daicel

intermolecular cross-linking, chemical bonding on the support, and so on. While immobilization slightly changes the selectivity from that of the corresponding coating one, a wide variety of mobile phases can be tried without anxiety. It is also a strong merit that a reaction mixture in a reaction solvent can be subjected to analysis without purification.

However, a much wider variety of derivatives is also potentially available. For example, a family of cycloalkyl carbamates is of interest for thin layer chromatography with UV light detection.[78] Cyclodextrins chemically bonded to silica gel and then derivatized (Cyclobond, ASTEC) are also commercially supplied.[79]

3.4.3.2 *Practice and Impact of Chiral Separation*

According to Okamoto and Yashima, chromatography has become the major method of chiral analysis and the polysaccharide CSPs are the most used.[78a] Francotte suggested that about 70% of the CSPs used for preparative-scale chromatography are derived from polysaccharides.[78b]

The choice of an effective phase for a racemate is empirical, and data tables are published to help this choice.[80,81] Group selectivity is often found, some of which are seen in Table 3.8, and two chromatogram examples are shown in Figure 3.10.

The chemical stability of polysaccharide CSPs allows a good variety of chromatographic conditions. For the separation of analytes carrying carboxyl or amino group(s) under normal phase conditions, a strong carboxylic acid (*e.g.* formic or trifluoroacetic acid), or strong amine (*e.g.* diethylamine), respectively, can be added to improve the separation by suppressing non-specific adsorption. A reversed phase condition is also applicable with suitable choice of the mobile phase.[82] Thus, the appearance of polysaccharide phases dramatically extended the application range of chiral liquid chromatography.[71] Chiral analysis has become possible with much smaller amount of samples at

Table 3.8 Example of group selectivity of polysaccharide chiral selectors.

Chiral selector	*Group of compounds*
1c	Sulfoxides, esters of aliphatic alcohols, α-Aryloxypropionic derivatives
1d	α-Arylpropionic derivatives (anti-inflammatory)
	Barbiturates and cyclic amides and imides carrying aromatic group (central nervous depressant)
1e	Similar to above
1i	3-Aryloxy-2-hydroxypropylamine (β2-blockers)
	Applicable to many compounds
2a	Applicable to many compounds
2b	β-Lactams (penem and carbapenem antibiotics)

much improved accuracy and convenience. This contributes to improvement in reliability of production control and R&D. General aspects of chiral liquid chromatography are compared with polarimetry in Table 3.9.

A pharmacodynamic study of chiral drugs, which should be conducted with microgram amounts of the sample recovered from the blood or urine of humans or animals, would be almost impossible without this kind of technology. Many reports of such studies appear in the literature.[83]

The preparation of homochiral materials has also become easier by chromatographic methods. The advantages of cellulosic phases over preparative chromatography, *i.e.* their performance (chiral discrimination, durability, and loading capacity) and availability, also inspired the engineering of an automated single column process, and then a simulated moving bed process (SMB). Thus a variety of preparation system became available depending on the amount of material needed.

Thus, semi-preparative columns (1–2 cm diameter), preparative columns (5–10 cm ϕ), and the SMB system (4–12 columns of 2–5 cm ϕ connected with 3-way valves) can be applied to obtain a target in milligrams per day to tons per year. Thus it has become much easier to acquire a preclinical test sample, and this has shortened the time of development. The production of a homochiral pharmaceutical or intermediate by chromatography has also been realized. The application of the SMB process to the production of a hypolipidemic is shown in Figure 3.11.[84] The yields for the desired (3R,5S)-form (the first elute) on a single column (20 cm ϕ × 50 cm) and on an SMB with 12 columns (3 cm ϕ × 15 cm) show that the SMB process enabled about twice the productivity, and reduction of mobile phase between 66% to 75%. The alternative route without chiral chromatography to the homochiral NK-104 Ca consists of hydrolysis and resolution with homochiral 1-phenylethylamine, which suffers from low recovery.

While chiral discrimination is the distinctive function of the cellulosic phase, it is often useful in separating similar (but not enantiomeric) compounds which could hardly be separated on plain silica gel or octadecylsilylated silica gel (ODS). Figure 3.12 shows the separation of three stereoisomers of 1,3-pentanediyl diacetate on Chiralcel OB-H (cellulose tribenzoate), where the meso form is separated from each chiral form much better than on silica gel.

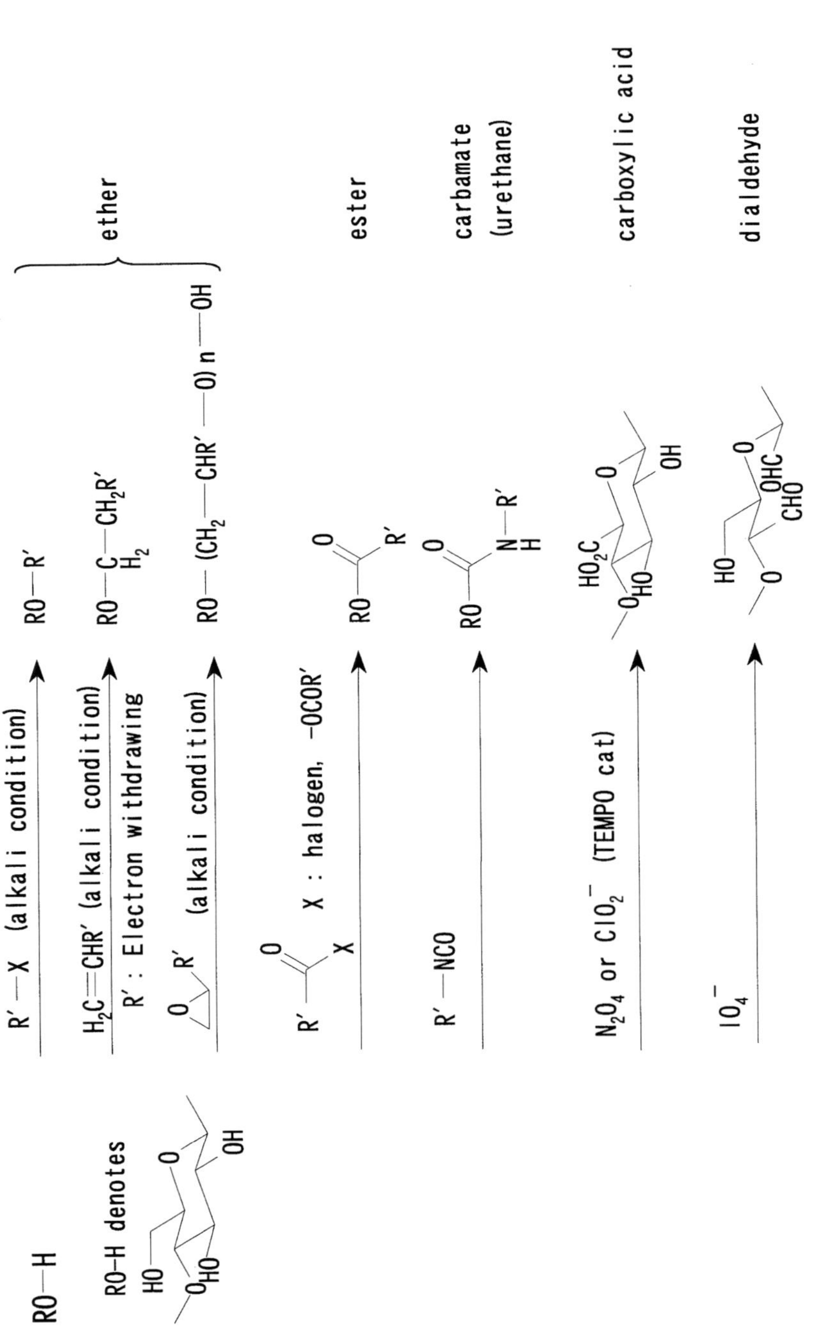

Figure 3.10 Two examples of chiral separation on cellulosic phases listed in Table 3.7. Top: methaqualone (a central nervous depressant) on **1d**. Bottom: pindolol (a β2-blocker) on **1i** (Chiralcel OD). The extreme separations are rather unconvenient for analytical purposes but can be made moderate by control of chromatographic condition.

Table 3.9 Comparison of chiral liquid chromatography with polarimetry.

Analytical feature	*Chiral HPLC*	*Polarimetry*
Sample amount	>1 μg	>1 mg
Accuracy	0.1%	1%
Contamination	Generally no effect	Causes error
Automation	Easy	Not easy

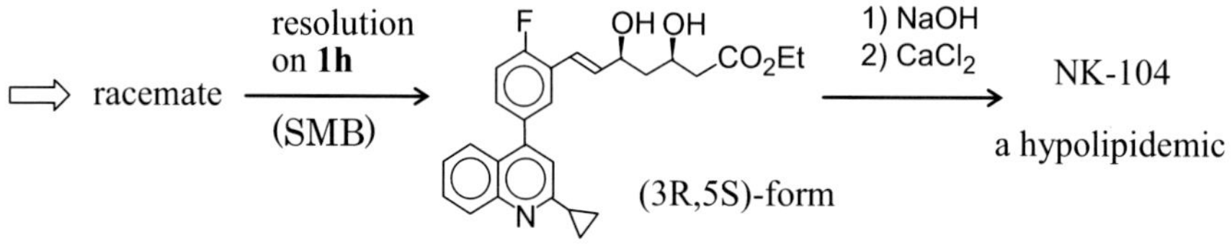

Figure 3.11 Preparative chiral separation of racemic intermediate of NK-104 Ca (a hypolipidemic agent). SMB: simulated moving bed.

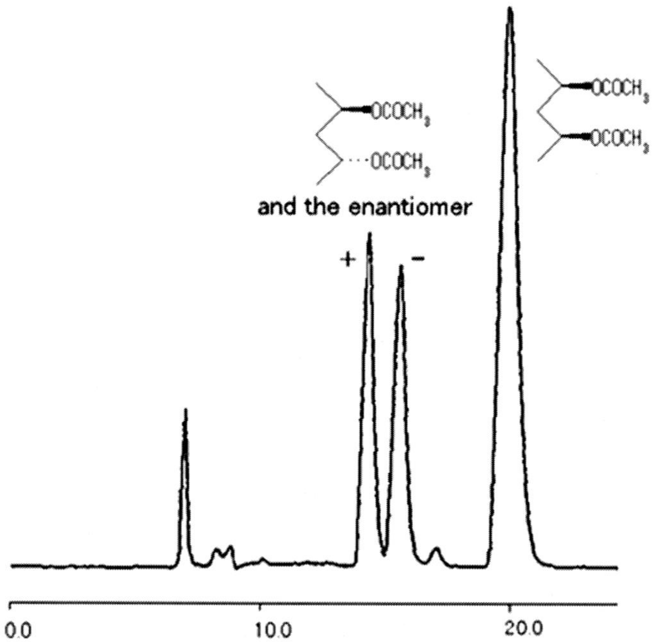

Figure 3.12 Separation of the three stereoisomers of 2,4-diacetoxypentane on "CHIRALCEL OB-H". Mobile phase: hexane–2-propanol (9 : 1 v/v); flow rate: 0.5 ml min^{-1}; temperature: 25°C. The earlier eluting two peaks of the same strength are assigned to each enantiomer of the racemic *trans*-forms (+/– denotes the polarimetric sign) and the later eluting larger peak, to the achiral cis-form.

The retention ratio of triphenylene and o-diphenylbenzene is sometimes discussed as an indicator of planarity preference of stationary phases, which is about 10 for cellulosic phases and about 2 for ODS. This higher planarity preference is apparently one of the selection rules.

3.4.3.3 Mechanism of Chiral Separation

The mechanism of chiral separation has attracted much interest as it would provide a guide to improving polysaccharide CSP and to designing a novel type of CSPs. Hesse and Hagel hypothesized the inclusion of a benzene ring in a pocket between two triacetylglucose residues of MCT.[72b] While this mechanism does not seem to be based on a close examination, the authors well reflected the hydrophobic nature of the adsorption on MCT.

Okamoto and his colleagues paid attention to hydrogen bonding between cellulose phenyl carbamate and the analyte. They rationalized that the secondary substituent(s) on the aromatic ring of cellulose phenyl carbamates controls their retention property by affecting proton donor–acceptor strength of the urethane.[85] Then they proposed a specific interaction mechanism based on computer modeling and a two-dimensional NMR technique.[77] The study was based on the interaction between cellulose tris(2-methyl-5-fluorophenylcarbamate) and 1,1′-binaphthol, as sketched by the partial molecular structure in Figure 3.13, where only two adjacent glucose residues and two phenylcarbamoyl groups at C3′ and at C2 are drawn in a 3/2 helix. Note that the two carbonyl groups arrowed are nearly parallel in a twisted 2/1 helix. This twisted configuration allows a better synchronous hydrogen bonding with the twisted two hydroxy groups of (S)-binaphthol.

While this is the only close study describing the mechanism of a cellulosic phase, it does not reveal all the possible interactions. For example, it is possible

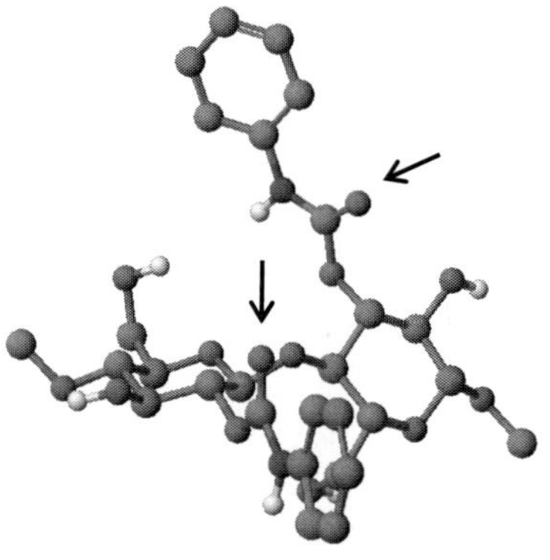

Figure 3.13 The partial model of the left handed 3-folded cellulose phenyl carbamate. The twisted alignment of the two carbonyl groups (arrows) allows the better simultaneous hydrogen bonding interaction with the phenolic protons of (S)-binaphthol.[77] Figure remade by T. Shibata.

that dipole–dipole interaction may be involved in some cases, or a stacking interaction between the aromatic moieties of the cellulosic support and the analyte in others. Interaction between the α-face of the glucose ring and the aromatic ring is also suggested by an NMR study.[86] Various other types of molecular interactions may also be involved which remain to be discovered.

Consideration of the possible function of the remaining hydroxy groups of partially substituted cellulose benzoate indicates an enhanced retention, not for a H-bonding molecule, but for a condensed aromatic ring. The voids left by incomplete substitution may offer a receptive site, and may suggest a novel approach to the design of molecular recognition.[87]

3.5 Cellulosics in Pharmaceutical Formulations

Cellulosics are commonly used in pharmaceutical formulations, particularly in tablets. Tablets should have a suitable swallowability, physical and chemical stability, and releasing property in the digestive system. To meet these requirements, the active ingredient is mixed with a main excipient, and a variety of auxiliary materials, variously called expander, stabilizer, binder or disintegrator, according to their functions (Table 3.10), but the terminology may be somewhat arbitrary.

Table 3.10 Cellulosics and other materials used as excipient and auxiliary for pharmaceutical preparations.

Function	*Cellulosic*	*Other materials*
Filler or binder	MCC, MC, HPC, HPMC HEMC, Na-CMC,	Minerals (Ca, Mg salts), starch, HP starch, lactose, sucrose
Disintegrant	Ca-CMC, H-CMC, L-HPC	Na-CM starch, Amberlite
Binder for wet tableting	Na-CMC, MC, HPC (for organic solvent) HPMC	Gum arabic, starch, gelatin, PVA, Na alginate
Binder for sugar coating	Na-CMC, MCC	Gelatin, PVP, Na-CM starch, gum arabic, Na alginate
Film coating	HEC, HPC	
Enteric (film) coating	CMEC, HPC succinate, CA succinate, CA phthalate	Methacrylic acid copolymers
Thickener, dispersion stabilizer, emulsifier for liquid preparations	MC, Na-CMC, MCC, HPC, HEC	Gum arabic, tragacant gum, Na alginate, pectin, gelatin, casein, PVA, bentonite
Poultice plaster	Na CMC	Na polyacrylate
Ointment base	HPMC	

[a]Abbreviations: MCC, microcrystalline cellulose; MC, methyl cellulose; HPC, hydroxypropyl cellulose; L-HPC, water insoluble HPC with a low DS; HPMC, hydroxypropylmethylcellulose; HEMC, hydroxyethylmethylcellulose; Na-CMC, Ca-CMC, H-CMC, sodium, calcium and acid form of carboxymethylcellulose; HEC, hydroxyethylcellulose; CMEC, carboxymethylethylcellulose.

One of the merits of MCC as an excipient in comparison to starch and lactose is its larger aspect ratio which increases tablet porosity, hence rapid disintegration in water.[88–90] This property seems more or less common in all cellulosic excipients. Cellulose ethers, *e.g.* MC and HPC, are also important as excipients.[91] L-HPC is designed to swell with water but not dissolve, and to enable overall quicker disintegration.

The surface of the tablet is often film-coated to mask the taste, to make swallowing easier, or to protect the active ingredient from moisture or oxygen. Water- or alcohol-soluble polymers are used for the film coating. Polymers carrying carboxyl groups are used to protect the tablet from acidic gastric juice and disintegrate it in the alkaline condition of the intestine (enteric coating).

Cellulosics are also used in external preparations. For instance, CMC is used in poultice plaster as a gel-forming agent, and HPMC in oral ointments for adhesion on mucous membrane. Liquid preparations also often require water-soluble polymers as dispersion stabilizer or thickener. Steroid eye lotion is one such example.

Some formulations are designed to attain delivery at a suitable time and location. For instance it is common practice to make a hardly soluble ingredient dissolve more rapidly. Simple grinding is the first choice, and even molecular dispersion can be attained by grinding it with an excipient such as MCC or MC.[92] As the excipient molecules hinder the aggregation of the ingredient molecules, a metastable condition such as supersaturation can result.[93]

Dispersion in a solid matrix can also be attained from a melt, solution or a melt–solution mixture. For instance, tacrolimus–HPMC was developed by this method.[94] The hydrophobic powder of hexobarbital can be treated with MC to accelerate dissolution by hydrophilizing the surface.[95] A time-controlled disintegration system was designed by doubly coating the active ingredient particles with an inner L-HPC film and an outer water insoluble polymer film. The expansion of L-HPC by water absorption causes the "explosion" of the outer coat, and the time of "explosion" could be controlled by the thickness of the coating.[96]

The application of cellulose esters to controlled release and other uses is reviewed in detail by Edgar *et al.*[97] Acidic derivatives, CA phthalate and HPMC phthalate are used in enteric coatings with a suitable plasticizer. The rather hydrophobic property and good compatibility with moderately polar organic materials make cellulose acetate and its mixed esters a suitable matrix material for sustained release. A variety of methods to prepare such matrixes are available, including direct compression, co-precipitation with the active, impregnation of preformed ultramicroporous CTA (Proplastic) with liquid actives, spray drying, *etc.* This methodology is applied not only in medical fields, but also in agriculture and other fields. Osmotic pump delivery is an important technology for controlled drug delivery and at high rates, and CDA is commercially used as a semipermeable shell material for this purpose.

While these and related uses of cellulosics in pharmaceutical formulations are commonplace, R&D activity also continues to fine tune some of the existing formulations and create new ones.

3.6 Other Medical Applications of Cellulosics

Cellulosics are used in a wide range of other medically related applications, and some offer similar functions in other fields. For instance, CMC itself is approved as the active ingredient of a laxative. CMC solutions gel by γ-ray irradiation at high concentrations (20–30%), but degrade at lower concentrations.[98] The gel thus formed is utilized as packing for beds for long-term medical care.

While much medical gauze (pack) is made of cotton fibers, more specialized materials are also supplied from different cellulosics. For instance, regenerated Bemberg fibers are made into spinbonded nonwoven Bemliese (Asahi Medical). Here cuprammonium fibers are loaded on a conveyer before complete coagulation to form bonds between the fibers. The sheet thus made is less prone to dust or irritate the diseased part.[99] Regenerated oxidized cellulose fibers are also made into gauze (Surgecell, Johnson & Johnson). It swells up by contact with blood, blocks bleeding, and decomposes in tissue.[100] Cellulose acetate fibers are also claimed to be suitable for wound dressing.[101]

MCC, microfibrillar cellulose, and possibly microfibrous regenerated cellulose, are used as food additives to improve water preservation and elasticity of processed food. They are also known to help bowel movement, increase fecal amount, and reduce the risk of intestinal tumors and hyperlipidemia in animals.[102,103] This kind of effect as a dietary fiber may become more important in future eating habits.

In another application, an HEC derivative is claimed to suppress the activity of tick and *Cryptomeria japonica* (Japanese cedar pollen) allergens, and a spray for clothes polluted with pollen has been formulated (Figure 3.14).[104]

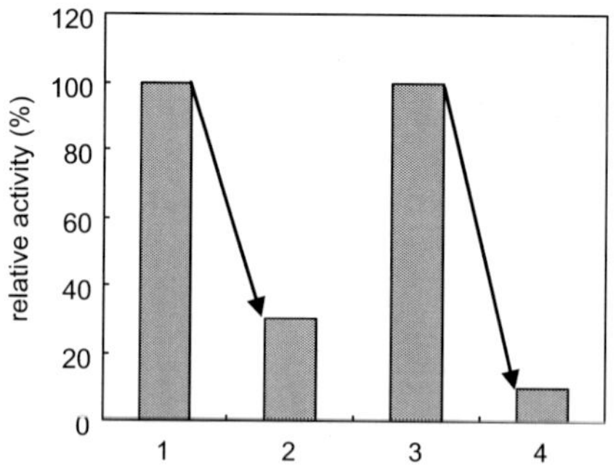

Figure 3.14 Reduction of allergenic activity with a hydroxyethylcellulose (HEC) derivative, as measured by ELISA. 1: tick extract; 2: 1+ HEC derivative; 3: cryptomeria pollen extract; 4: 3+ HEC derivative (adapted from ref. 104).

3.7 Concluding Remarks

The widespread medical uses of cellulosics derive from its ready availability, physical properties and physiological inertness. As the use of cellulosics by humans dates back to ancient history, some may often be regarded as stale. Yet we still understand cellulose only partially. But the needs of our time are changing, and many untapped potential applications of cellulosics may yet serve our future needs. Cellulose is an abundant, renewable and gentle resource, and almost a necessity for the future human living. If cellulose were, like β-(1 → 3)-glucan, a cell component of a micro-organism, it might have an outstanding bioactivity, yet much less widely usable. However, it is conceivable that advanced technologies of chemical modification would make it possible to endow cellulose with a particular bioactivity.[105,106]

3.8 List of Abbreviations

PVA	poly(vinyl alcohol)
HEC	hydroxyethylcellulose
DEAE	N,N-diethylaminoethyl
DP	degree of polymerization
MC	methyl cellulose
CMC	carboxymethyl cellulose
HEC	hydroxyethyl cellulose
HPC	hydroxypropyl cellulose
MS	molar substitution
DS	degree of substitution
HEMC	hydroxyethylmethyl cellulose
CDA	cellulose diacetate
UF	ultrafiltration
SCE	saponified cellulose ester
QOL	quality of life
MW	molecular weight
LPS	lipopolysaccharide
CTA	cellulose triacetate
MG	microglobulin
TNF	tumor necrosis factor
HPLC	high-performance liquid chromatography
CSP	chiral stationary phase
MCT	microcrystalline cellulose triacetate
SMB	simulated moving bed
ODS	octadecylsilylated silica gel
MCC	microcrystalline cellulose
HPMC	hydroxypropylmethyl cellulose

References

1. FAOSTAT Database Results 2004, http://faostat.fao.org/faostat/servlet.
2. S. Saka and H. Matsumura, in *The raw materials of CA*, Reference 12, pp. 7–48.
3. D. Klemm, B. Philipp, T. Heinze, U. Heinze and W. Wagenknecht, Comprehensive Cellulose Chemistry, In *Fundamentals and Analytical Methods*, Vol. 1, Wiley-VCH, Weinheim, Germany, 1998.
4. O. A. Battista, *Microcrystal Polymer Science*, McGraw-Hill Inc, New York, NY, 1975.
5. T. Hatakeyama, H. Hatakeyama, *Cellulosics: Chemical Biochemical and Materials Aspects*, (Eds) J. F. Kennedy, G. O. Phillips and P. A. Williams, Ellis Horwood, Chichester, UK, 1993, p. 225.
6. D. Klemm, B. Philipp, T. Heinze, U. Heinze and W. Wagenknecht, Comprehensive Cellulose Chemistry, In *Fundamentals and Analytical Methods*, Vol. 1, Wiley-VCH, Weinheim, Germany, 1998, pp. 60–82.
7. D. Klemm, B. Philipp, T. Heinze, U. Heinze and W. Wagenknecht, Comprehensive Cellulose Chemistry, In *Functionalization of Cellulose*, Vol. 2, Wiley-VCH, Weinheim, Germany, 1998.
8. E. Ott and H. M. Spurlin, M. W. Grafflin (Ed.) High Polymers In *Cellulose and Cellulose Derivatives, Part I, II, and III*, Vol. V, Interscience Publishers Inc, New York, NY, 1954.
9. Encyclopedia of Polymer Science and Engineering, Vol. 3, John Wiley & Sons, US, 1985.
10. K. Balser and L. Hoppe (Ed.) *Ullman's Encyclopedia of Industrial Chemistry*, Vol. A5, VCH Verlagsgesellschaft mbH, Germany, 1986.
11. *Encyclopedia of Chemical Technology*, 4th edn, Vol. 5, John Wiley & Sons, New York, NY, 1993.
12. P. Rustemeyer (Ed.) *Cellulose Acetates: Properties and Applications*, WILEY-VCH Verlag GmbH & Co. Weinheim, Germany, 2004.
13. W. G. Presswood, The membrane filter: Its history and characteristics. In *Membrane Filtration, Applications, Techniques, and Problems*, (Ed.) B. D. Dutka, Marcel Dekker, New York, 1981, pp. 1–17.
14. S. Sourirajan and T. Matsuura, *Reverse Osmosis/Ultrafiltration, Process and Principles*, National Research Council Canada, Ottawa, Canada, 1985.
15. P. D. Randall, Cellulose acetate – a separating medium without equal. In *Cellulosics: Materials for selective separations and other technologies*, (Eds) J. F. Kennedy, G. O. Phillips and P. A. Williams, Ellis Horwood, Chichester, West sussex, 1993, pp. 127–136.
16. T. Matsuura, S. Sourirajan, Material Science of Reverse-Osmosis-Ultrafiltration Membranes. In *Reverse Osmosis and Ultrafiltration*, (Ed.) M. J. Comstock, (ACS symposium series, ISSN 0097-6156, 281), American Chemical Society, 1985, pp. 1–19.
17. R. E. Kesting, *Synthetic Polymeric Membranes*, John Wiley & Sons Inc, New York, 1985, pp. 237–286.

18. R. W. Baker, Membrane technology, (Ed.) M. Howe-Grant, *Kirk-Othmer Encyclopedia of Chemical Technology*, Vol. 16, 4th edn, 1997, pp. 135–193.
19. M. J. Lysaght, U. Bauermeister, Dyalysis, *Kirk-Othmer Encyclopedia of Chemical Technology*, (Ed.) M. Howe-Grant, Vol. 8, 4th edn, 1997, pp. 58–74.
20. H. Sakurai, Cellulosic membrane processing for homodialysis, *Cellulose Commun.*, 2003, **10**, 8–11.
21. U. Baurmeister, J. Vienken and M. Pelger, Cellulosic membranes: high adaptability to changing demands, *Blood Purif.*, 1986, **4**, 205–206.
22. M. Sasaki, Conventional regenerated cellulose membrane and VE modified cellulose membrane. In *The New High Performance Dialyzer*, (Ed.) S. Takezawa, Tokyo Igakusha, Tokyo, Japan, 1998, pp. 13-27(Title translated by the author).
23. (a) H. Sakurai, M. Ohno, T. Masuda, Cellulose acetate. In *The New High Performance Dialyzer*, (Ed.) S. Takezawa, Tokyo Igakusha, Tokyo, Japan, 1998, pp. 46–53 (Title translated by T Shibata); (b) M. J. Kell and R. D. Mahoney (to Cordis Dow Corporation), Cellulose acetate hollow fiber and method for making same, *US. Patent* 4276173, 1981.
24. F. Gejyo, T. Yamada and S. Odani, *et al.*, A new form of amyloid protein associated with chronic hemodialysis was identified as beta 2-microglobulin, *Biochem. Biophys. Res. Commun.*, 1985, **129**, 701–706.
25. S. V. Mikhalovsky, Microparticles for hemoperfusion and extracorporeal therapy, Microspheres, Microcapsules and Liposomes, The MML Series, Vol. 2. In *Medical and Biotechnology Applications*, (Ed.) R. Arshady, Citus Books, London, 1999, pp. 133–169.
26. (a) R. Arshady, Polymeric biomaterials: Chemistry, concept, criteria. In *Introduction to Polymeric Biomaterials*, (Ed.) R. Arshady, Kentus Books, London, 2003, pp. 11–64; (b) R. M. Hakim, Clinical implication of hemodialysis membrane biocompatibility, *Kidney Int.*, 1994, **44**, 484–494.
27. (a) P. R. Craddock, J. Fehr, A. P. Dalmasso, K. L. Brigham and H. S. Jacob, Hemodialysis leukopenia: Pulmonary vascular leukostasis resulting from complement activation by dialyzer cellophane membranes, *J. Clin. Invest.*, 1977, **59**, 879–888; (b) G. Pertosa, E. A. Tarantino, L. Gesualdo, V. Montinaro and F. P. Schena, C5b-9 generation and cytokine production in hemodialyzed patients, *Kidney Int.*, 1993, (Suppl. 41), S221–S225.
28. (a) M. Ohno, M. Suzuki, M. Miyagi, T. Yagi, H. Sakurai and T. Ukai, CTA hemodialysis membrane design for β2-microglobulin removal. In *Cellulosics: Chemical, Biochemical and Material Aspects*, (Eds) J. F. Kennedy, G. O. Phillips and P. A. Williams, Ellis Horwood Limited., Chichester, West Sussex, U.K., 1993, pp. 415–420; (b) T. Sagara, K. Tomita and K. Mabuchi, Newly developed cellulose triacetate hemodialysis membrane, designed for enhanced separation of low molecular weight proteins, *Cellulose Commun.*, 2008, **16**, 93.
29. M. Sasaki, Vitamin-E modified cellulose membrane (Excebrane^{TM}), *Membrane*, 1995, **20**, 239–241.

30. Y. Ikada, Biochemical Applications of cellulose membranes. In *Cellulose: Structural and Functional Aspects*, (Eds) J. F. Kennedy, G. O. Phillips and P. A. Williams, Ellis Horwood Limited., Chichester, West Sussex, U.K., 1989, pp. 447–455.
31. Y. Ikada and K. Imamura (to Asahi Chemical Ind.), Improved regenerated cellulose film and its manufacture, *Jpn. Patent* 297103, 1989.
32. M. Diamantoglou, H. D. Lemke and J. Vienken, Cellulose-ester as membrane materials for hemodialysis, *Int. J. Artif. Organs*, 1994, **17**, 585–591.
33. W. P. Oosterhuis, M. de Metz, A. Wadham, M. R. Daha and R. H. Go, *In vivo* evaluation of four hemodialysis membranes: Biocompatibility and clearances, *Dial. Transplant.*, 1995, **24**, 450–458.
34. M. Fukuda, T. Hiyoshi, K. Sakai and K. Kokubo, Anisotropic Differences in Solute Transfer Rate Through Asymmetric Membrane (BIOREX® AM-BC-X), *Jpn. J. Artif. Organs*, 2000, **29**, 411–418.
35. O. P. Ivanovich, D. Chenoweth and R. Schmidt, Symptoms and activation of granulocytes and complement activation with two dialysis membranes, *Kidney Int.*, 1983, **24**, 758–763.
36. U. Baurmeister, W. Brodowski, M. Diamantohlou, G. Dünweg, W. Henne, M. Pelger and H. Schulze, Dialysis membranes of modified cellulose with improved biocompatibility, *US Patent* 4668396, 1987.
37. D. Falkenhagen, T. Bosch, G. Brown, B. Schmidt, M. Holtz, U. Baurmeister, H. Gurland, H. Klinkmann and A. clinical study on different cellulosic dialysis membranes, *Nephrol. Dial. Transplant*, 1987, **2**, 537–545.
38. J. Toufik, M. Carreno, M. Josefowicz and D. Labarre, Activation of the complement system by polysaccharide surfaces bearing carboxymethyl, carboxymethyl benzylamide and carboxymethylbenzylamide sulphonate groups, *Biomaterials*, 1995, **16**, 993–1002.
39. M. Carreno, D. Labarre, F. Maillet, M. Josefowicz and M. Kazatchkine, Regulation of the human alternative complement pathway: Formation of a ternary complex between factor H, surface bound C3b and chemical groups on nonactivating surfaces, *Eur. J. Immunol.*, 1989, **19**, 2145–2159.
40. E. Rauterberg, H. Schulze and E. Ritz, Limited derivatisation of Cuprophan increases facter H binding and diminishes complement activation, *23rd Congress European Renal Association*, Budapest, 1986, a 146.
41. R. Johnson, M. Lelah, T. Sutliff and D. Boggs, A modification of cellulose that facilitstes the control of complement activation, *Blood Purif.*, 1990, **8**, 318–328.
42. M. Diamantoglou, J. Platz and J. Vienken, Cellulose carbamates and derivatives as hemocompatible membrane materials for hemodialysis, *Artif. Organs*, 1999, **23**, 15–22.
43. T. Tsurumi, T. Sato, N. Osawa, H. Hitaka, T. Hirasaki, K. Yamaguchi, Y. Hamamoto, S. Manabe, T. Yamashiki and N. Yamamoto, Structure and filtration performances of improved cuprammoniun regenerated cellulose hollow fiber (Improved BMM hollow fiber) for virus removal, *Polymer J.*, 1990, 1085–1100.

44. S. Manabe and M. Satani (to Asahi Kasei Kogyo Kabushiki Kaisha), Porous hollow fiber membrane and a method for the removal of a virus by using the same, *US Patent* 4808315, 1989.
45. J. Tateishi, T. Kitamoto, G. Ishikawa and S. Manabe, Removal of causative agent of Creutzfeldt-Jakob disease (CJD) through membrane filtration method, *Membrane*, 1993, **18**, 357–362.
46. I. Nakate and S. Nakatsuka (to Daicel Chemical Ind. Ltd.), Cellulose acetate hollow-fiber separation membrane and its production, *Jpn. Patent* POH08-108053, 1996.
47. S. Nakatsuka, Development of a high flux hollow fiber membrane, New Membrane Technology Symposium '98, March 1998, (The Membrane Society of Japan), part 1-3-1.
48. (a) P. R. Levison, Cellulosics as ion-exchange materials. In *Cellulosics: Materials for selective separations and other technologies*, (Eds) J. F. Kennedy, G. O. Phillips and P. A. Williams, Ellis Horwood, Chichester, West sussex, UK, 1993, pp. 25–36; (b) R. Arshady, Beaded polymer supports and gels: 1. Manufacturing techniques, *J. Chromatogr.*, 1991, **586**, 181–197; (c) R. Arshady, Beaded polymer supports and gels: 2. Physicochemical criteria and functionalization, *J. Chromatogr.*, 1991, **586**, 199–219.
49. R. Noel, A. Sanderson and L. Spark, A monolithic ion-exchange material suitable for downstream processing of bioproducts, in *Cellulosics: Materials for selective separations and other technologies*, (Eds) J. F Kennedy, G. O. Phillips and P. A. Williams, Ellis Horwood, Chichester, West Sussex, UK, 1993, pp. 17–24.
50. I. Køepelka, Perloza-products made of regenerated beaded cellulose. In *Cellulosics: Materials for selective separations and other technologies*, (Eds) J. F. Kennedy, G. O. Phillips and P. A. Williams, Ellis Horwood, Chichester, Wrest sussex, UK, 1993, pp. 9–16.
51. J. Peška, J. Štamberg and Z. Pelzbauer, Regenerated cellulose in the bead form. Aftertreatments and their effects on the porous structure of cellulose, *Cellul. Chem. Technol.*, 1978, **21**, 419–428.
52. C. Hirayama, T. Sonoi, K. Matsumoto and Y. Motozato, *Kobunshi Ronbunshu*, 1982, **39**, 597–604.
53. S. Kuga, New cellulose gel for chromatography, *J. Chromatogr.*, 1980, **195**, 221–230.
54. W. H. Velander, J. A. Kaster, G. Kumar, K. Van Cott, W. de Oliviera and W. G. Glasser, Structural and Chromatographic characteristics of beaded cellulose supports. In *Cellulosics: Materials for selective separations and other technologies*, (Eds) J. F. Kennedy, G. O. Phillips and P. A. Williams, Ellis Horwood, Chichester, Wrest sussex, UK, 1993, pp. 1–8.
55. S. Nagamatsu, Y. Tanaka, T. Shibata (to Daicel Chemical Ind. Ltd., Tanabe Seiyaku Co. Ltd.), Separation method and separation agent, EP P:901233163.
56. Y. Planques, N. Bendris, Lipids removal from protein solutions by a selective adsorption with a depth filter. In *Cellulosics: Materials for*

selective separations and other technologies, (Eds) J. F. Kennedy, G. O. Phillips and P. A. Williams, Ellis Horwood, Chichester, Wrest sussex, UK, 1993, pp. 63–69.
57. A. Sumi, K. Okuyama, K. Kobayashi, W. Ohtani and T. Ohmura, Purification of recombinant human serum albumin, *PDA J. GMP and validation in Japan*, 2000, **2**, 28–33.
58. A. Ginnaga, K. Morokuma, K. Aihara, M. Sakou, A. Imaizumi, Y. Suzuki, H. Sato, Y. Sato, K. Ueda, H. Kuno-Sakai and M. Kimura, Characterization and clinical study on the acellular pertussis vaccine produced by a combination of column purified pertussis toxin and filamentous hemagglutinin, *J. Exp. Clin. Med.*, 1988, **13**(Suppl.), 59–69.
59. K. Sugawara, K. Nishiyama, Y. Ishikawa, M. Abe, K. Sonoda, K. Komatsu, Y. Horikawa, K. Takeda, T. Honda, S. Kuzuhara, Y. Kino, H. Mizokami, K. Mizuno, T. Oka and K. Honda, Development of vero cell-derived inactivated Japanese encephalitis vaccine, *Biologicals*, 2002, **30**, 303–314.
60. C. R. O'Riordan, A. L. Lachapelle, K. A. Vincent and S. C. Wadsworth, Scaleable chromatographic purification process for recombinant adeno-associated virus (rAAV), *J. Gene Med.*, 2000, **2**, 444–454.
61. K. Tamayose, Y. Hirai and T. Shimada, A new strategy for large-scale preparation of high-titer recombinant adeno-associated virus vectors by using packing cell lines, sulfonated cellulose column chromatography, *Human Gene Therapy*, 1996, **7**, 507–513.
62. K. Sakamoto, T. Kawahara, K. Okuma, M. Sakawa and A. Yamada, *The Clinical Report*, 1990, **24**, 3155–3159.
63. J. Moscariello, G. Purdom, T. W. Root and E. N. Lightfoot, Characterizing the performance of industrial-scale columns, *J. Chromatogr. A*, 2001, **908**, 131–141.
64. S. Asakura and M. Adachi (to Japan Immuno Res. Lab., Sekisui Chemical Co. Ltd.), Method of treating inflammatory diseases, *US Patent* 5567443A, 1996.
65. N. Kashiwagi, K. Sugimura, II. Koiwai, II. Yamamoto, T. Yoshikawa, A. R. Saniabadi, M. Adachi and T. Shimoyama, Immunomodulatory Effects of granulocyte and monocyte adsorption apheresis as a treatment for patients with ulcerative colitis, *Digestive Diseases and Sciences*, 2002, **47**, 1334–1341.
66. S. Nakaji, Current topics on immunoadsorption therapy, *Therapeutic Apheresis*, 2001, **5**, 301–305.
67. S. Daimon, T. Saga, M. Nakayama, Y. Nomura, H. Chikaki, K. Dan and I. Koni, Dextran sulphate cellulose columns for the treatment of nephritic syndrome due to inactive lupus nephritis, *Nephrol. Dial. Transplant*, 2000, **15**, 235–238.
68. T. Eguchi and M. Tsunomori (to Kanegafuchi Chemical Ind.), Method of making uniform polymer particles, *US Patent* 5015423 A, 1991.
69. S. Nakaji, K. Oka, M. Tanihara, K. Takakura and M. Takamori, Development of a specific immunoadsorbent containing immobilized

synthetic peptide of acetylcholine receptor for treatment of myasthenia gravis, *Therapeutic Plasmapheresis, VSP*, 1993, **XII**, 573–576.
70. S. Margel, S. Sturchak, E. Ben-Bassat, *et al.*, Functional microspheres for biomedical applications. In *Microspheres, Microcapsules and Liposomes, The MML Series*, Vol. 2, (Ed.) R. Arshady, Medical and Biotechnology Applications, Citus (now Kentus) Books, London, 1999, pp. 1–42.
71. (a) E. J. Ariens, Racemates–An impediment in the use of drugs and agrochemicals. In *Chiral Separations by HPLC. Applications to pharmaceutical compounds*, (Ed.) A. M. Krstulovic, Ellis Horwood Limited, Chichester, UK, 1989, pp. 31–68; (b) Y. MachidaH. Nishi, Chiral purity in drug analysis. In *Encyclopedia of Analytical Chemistry*, (Ed.) R. A. Meyers, John Wiley & Sons Ltd, Chichester, UK, 2000, pp. 7076–7100.
72. (a) A. Lüttringhaus, U. Hess and H. J. Rosenbaum, Conformational enantiomerism. I. Optically active 4,5,6,7-dibenzo-1,2-dithiacyclooctadiene, *Z. Naturforschung*, 1967, **B22**, 1296–1300; (b) G. Hesse and R. Hagel, Complete separation of a racemic mixture by elution chromatography on cellulose triacetate, *Chromatographia*, 1973, 277–280.
73. T. Shibata, I. Okamoto and K. Ishii, Chromatographic optical resolution on polysaccharides and their derivatives, *J. Liq. Chromatogr.*, 1986, **9**, 313–340.
74. T. Shibata, T. Sei, H. Nishimura and K. Deguchi, Hysteretic effect of the coating solvent on chiral recognition by cellulose derivatives, *Chromatographia*, 1987, **24**, 552–554.
75. H. Namikoshi, T. Shibata, H. Nakamura, I. Okamoto, K. Shimizu and Y. Toga, Chromatographic optical resolution on cellulose and other polysaccharide derivatives. In *Wood and Cellulosics*, (Eds) J. F. Kennedy, G. O. Phillips and P. A. Williams, Ellis Horwood, Chichester, UK, 1987, pp. 611–617.
76. Y. Okamoto, M. Kawashima and K. Hatada, Useful chiral packing materials for high-performance liquid chromatographic resolution of enantiomers: phenylcarbamates of polysaccharides coated on silica gel, *J. Amer. Chem. Soc.*, 1984, **106**, 5357–59.
77. E. Yashima, C. Yamamoto and Y. Okamoto, NMR studies of chiral discrimination relevant to the liquid chromatographic enantioseparation by a cellulose phenylcarbamate derivative, *J. Amer. Chem. Soc.*, 1996, **118**, 4036–4048.
78. (a) Y. Okamoto and E. Yashima, Polysaccharide derivatives for Chromatographic Separation of Enantiomers, *Angew. Chem. Int. Ed.*, 1998, **37**, 1020–1043; (b) E. Francotte, Preparation of drug enantiomers by chromatographic resolution on chiral stationary phases. In *The Impact of Stereochemistry on Drug Development and Use*, (Eds) H. Y. Aboul-Enain and I. W. Wainer, Wiley, New York, NY, USA, 1997, pp. 633-683.
79. D. W. Armstrong, A. M. Stalcup, M. L. Hilton, J. D. Duncan, J. R. Faulkner and S. C. Chang, Derivatized cyclodextrins for normal-phase liquid chromatographic separation of enantiomers, *Anal. Chem.*, 1990, **62**, 1610–1615.

80. B. Koppenhoefer, R. Graf, H. Holtzschuh, A. Nothdurft, U. Trettin, P. Piras and C. Roussel, Chirbase, a molecular database for the separation of enantimers by chromatography, *J.Chromatogr. A.*, 1994, **666**, 557–563.
81. Application Guid for Chiral Column Selection, Crownpak, Chiralcel, Chiralpak Chiral HPLC Columns for Optical Resolution, 3rd edn, Daicel Chemical Industries Ltd, Tokyo, Jpn.
82. A. Ishikawa and T. Shibata, Cellulosic chiral stationary phase under reversed-phase condition, *J. Liq. Chromatogr.*, 1993, **16**, 859–878.
83. I. W. Wainer, The use of HPLC chiral stationary phases in pharmacokinetic and pharmacodynamic studies: putting a new technology to work. In *Chiral Separation by HPLC: Application to Pharmaceutical Compounds*, (Ed.) A. M. Krsturovic, Ellis Horwood, Chichester UK, 1989, pp. 194–207.
84. S. Nagamatsu, K. Murazumi and S. Makino, Chiral separation of a pharmaceutical intermediate by a simulated moving bed process, *J. Chromatogr. A*, 1999, **832**, 55–65.
85. Y. Okamoto, M. Kawashima and K. Hatada, Chromatographic resolution. XI. Controlled chiral recognition of cellulose triphenylcarbamate derivatives supported on silica gel, *J. Chromatogr.*, 1986, **363**, 173–186.
86. T. Shibata, Chiral recognition by cellulose derivatives. In *Wood and Cellulosics*, (Eds) J. F. Kennedy, G. O. Phillips and P. A. Williams, Ellis Horwood, Chichester, UK, 1990, pp. 291–298.
87. Y. Toga, K. Hioki, H. Namikoshi and T. Shibata, Dependance of chiral recognition on the degree of substitution of cellulose benzoate, *Cellulose*, 2004, **11**, 65–71.
88. K. Obae, H. Iijima and K. Imada, Morphological effect of microcrystalline cellulose particles on tablet tensile strength, *Int. J. Pharm.*, 1999, **182**, 155–164.
89. S. Westermark, A. M. Juppo, L. Kervinen and J. Yliruusi, Microcrystalline cellulose and its microstructure in pharmaceutical processing, *Eur. J. Pharm. Biopharm.*, 1999, **48**, 199–206.
90. E. Hirosawa, K. Danjo and H. Sunada, Influence of granulating method on physical and mechanical properties, compression behavior, and compactibility og lactose and microcrystalline cellulose granules, *Drug Dev. Ind. Pharm.*, 2000, **26**, 583–593.
91. R. C. Rowe, The molecular weight of methyl cellulose used in pharmaceutical formulation, *J. Pharm.*, 1982, **11**, 175–179.
92. Y. Nakai, Mechanochemical effect of mixing and grinding to organic crystals, *Funtai Kogaku Kaishi (J. Soc. Powder Technol. Jpn.)*, 1979, **16**, 473–481.
93. K. Yamamoto, M. Nakano, T. Arita and Y. Nakai, Dissolution behavior and bioavailability of phenitoin from ground mixture with microcrystalline cellulose, *J. Pharm. Sci.*, 1976, **65**, 1484–1488.
94. T. Honbo, M. Kobayashi, K. Hane, T. Hata and Y. Ueda, The oral dosage form of FK-506, *Transplant. Proc.*, 1987, **XIX**, 17–22.

95. C. F. Lerk, M. Lagas, J. T. Fell and P. Nauta, Effect of hydrophilization of hydrophobic drugs on release rate from capsules, *J. Pharm. Sci.*, 1976, **65**, 1702–1704.
96. T. Hata and S. Ueda, Rationality of oral rate-controlled drug delivery system – Introduction of concept of time-controlled explosion system (TES) as a novel oral rate controlled drug delivery system, *Pharm. Tech. Japan*, 1988, **4**, 1415–1422.
97. K. J. Edgar, C. M. Buchanan, J. S. Debenham, P. A. Rundquist, B. D. Seiler, M. C. Shelton and D. Tindall, Advances in cellulose ester performance and application, *Prog. Polym. Sci.*, 2001, **26**, 1605–1088.
98. B. Fei, R. A. Wach, H. Mitomo, F. Yoshii and T. Kume, Hydrogel of biodegradable cellulose derivatives. I. Radiation-induced crosslinking of CMC, *J. Appl. Polymer Sci.*, 2000, **78**, 278–283.
99. K. Nishiyama, Status of nonwoven wipers and market development of "Bemliese", *Nonwovens Review*, 1998, **9**, 1–6.
100. P. M. Watt, Harvey W. E. Lorimer, D. Wiseman (to Johnson & Johnson Medicl Inc), Use of oxidized cellulose and complexes thereof for chrinic wound healing, PCT Int Appl. 1998, WO 9800180 A1.
101. J. C. Chen, L. J. Deutsch and P. M. Garrett (to Hoechst Celanese Corporation) Cellulose ester wound dressing, *US Patent* 5685832, 1997.
102. T. Oku, F. Konishi and N. Hosoya, Effect of various unavailable carbohydrates and administrating periods on several physiological functions of rats, *Eiyou to Shokuryou (Nutrition and Food)*, 1981, **34**, 437–443.
103. H. J. Freeman, Effect of differing purified cellulose, pectin, and hemicellulose fiber diets on fecal enzymes in 1,2-dimethylhydrazine-induced rat colon carcinogenesis, *Cancer Res.*, 1986, **46**, 5529–5532.
104. K. Hori, H. Nojiri, M. Nonomura, M. Nonomura, F. Okuda, H. Yanagida (to K Hori, Kao Corp, H Nojiri, M Nonomura), Allergen inactivator, *US Patent* 197319 AA, 2005 .
105. K. Kamide, K. Okajima, T. Matsui, M. Ohnishi and H. Kobayashi, Roles of molecular characteristics in blood anticoagulant activity and acute toxicity of sodium cellulose sulfate, *Polym. J.*, 1983, **15**, 309–321.
106. T. Matsuzaki, I. Yamamoto (to Daicel Chemical Ind. Ltd.), Antitumor agent, *Jpn. Patent* POS61-15836, 1986.

CHAPTER 4

Xylan and Xylan Derivatives – Basis of Functional Polymers for the Future

THOMAS HEINZE* AND STEPHAN DAUS

Centre of Excellence for Polysaccharide Research, Friedrich Schiller University of Jena, Humboldtstraße 10, D-07743 Jena, Germany

4.1 Introduction

Hemicelluloses, accounting on average for up to 50% of the biomass of annual and perennial plants, have emerged as an immense renewable resource of biopolymers. Their application potential, emphasized many times by leading polysaccharide scientists, has not yet been exploited on an industrial scale. During the last decade, increasing interest has been noticeable in research into biopolymers from renewable resources among scientists from academia and industry. A future shortage of the required natural resources, the replacement of petroleum-based products that is connected with the solution of the worldwide environmental problems, the design of novel products and materials based on structure "synthesized by nature", and demands for healthy food and alternative medicines are the main driving forces of the immense activity in research into polysaccharides including hemicelluloses.

The term hemicellulose resulted from the assumption that these polysaccharides are precursors of cellulose and was originally proposed by Schulze.[1] In contrast to cellulose, hemicelluloses can be extracted from higher plants by aqueous alkaline solutions. Today, the term hemicellulose is still commonly

RSC Polymer Chemistry Series No. 1
Renewable Resources for Functional Polymers and Biomaterials
Edited by Peter A. Williams

Published by the Royal Society of Chemistry, www.rsc.org

used to refer to non-starch polysaccharides found in association with cellulose in the cell walls of higher plants. The term does not include pectic polysaccharides, as these are extractable by hot water, weak acids, or chelating agents.

As recently reviewed,[2] hemicelluloses can be divided into four general classes of structurally different cell wall polysaccharides, *i.e.*,

(i) xylans (xyloglycans),
(ii) mannans (mannoglycans),
(iii) β-glucans with mixed linkages, and
(iv) xyloglucans.

They occur in structural variations which differ in side chains, their distribution and localization as well as the types of glycoside linkages in the macromolecular backbone. Xylans are the main hemicellulose components of secondary cell walls constituting about 20–30% of the biomass of dicotyl plants (hardwoods and herbaceous plants). In some tissues of monocotyl plants (grasses and cereals) xylans even comprise up to 50%.[3] Xylans are thus available in huge amounts as by-products from forestry, the agriculture, wood, and pulp and paper industries. Nowadays, xylans of some seaweed represent a novel biopolymer resource. The diversity and complexity of xylans suggest that great potential for novel products exists and that these polysaccharides are considered as possible raw materials for various exploitations. As a renewable resource, xylans are being rediscovered not only in wood but also in all kinds of biomass.

There are many excellent reviews published in the field of xylan research. The aim of this review is to highlight typical aspects concerning the occurrence, isolation and characteristics of xylan. Special attention is given to the topic of xylan dervatives accompanied by selected results of our own work in this field.

4.2 Occurrence and Structural Diversity of Xylans

Xylans can be grouped into several structural subclasses, namely,

(i) homoxylans,
(ii) glucuronoxylans,
(iii) (arabino)glucuronoxylans and (glucurono)arabinoxylans,
(iv) arabinoxylans, and
(v) heteroxylans.

Homoxylans are linear polysaccharides composed of D-xylopyranosyl (Xyl*p*) residues linked by β-(1→3) linkages (Figure 4.1a), β-(1→4)-linkages (Figure 4.1b) and/or mixed β-(1→3, 1→4) linkages (Figure 4.1c). They occur in seaweeds of the *Palmariales* and *Nemaliales* consisting of Xyl*p* residues linked by β-(1→3) or mixed β-(1→3, 1→4) glycosidic linkages. They are assumed mainly

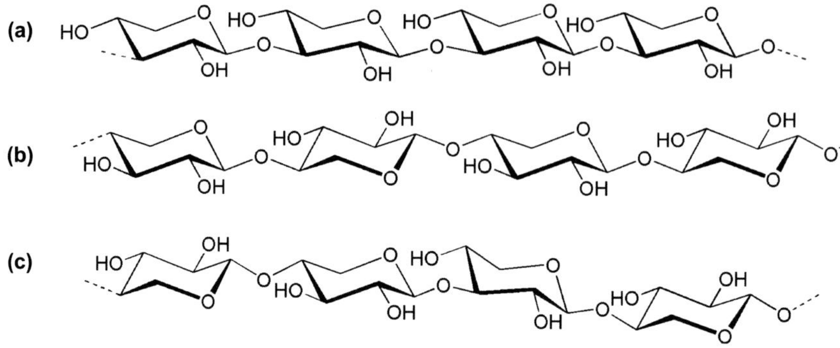

Figure 4.1 Primary structure of (a) β-(1→3)-D-xylan, (b) β-(1→4)-D-xylan and (c) β-(1→3, 1→4)-D-xylan.

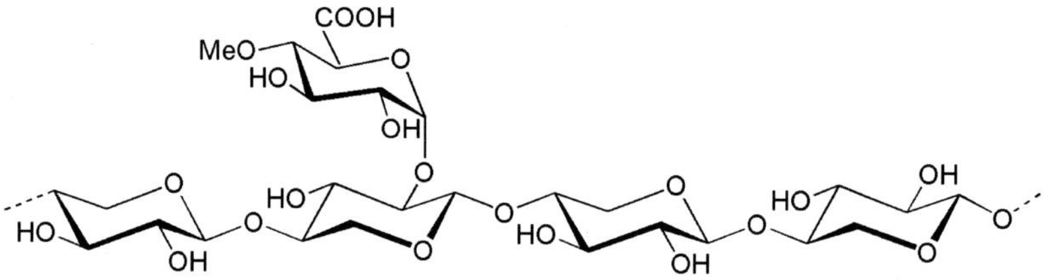

Figure 4.2 Primary structure of 4-*O*-methyl-D-glucurono-D-xylan (MGX).

to have a structural function in the cell wall architecture; however, a reserve function cannot be ruled out.[4] From the microfibrils of green algae (*Siphonales*) such as *Caulerpa* and *Bryopsis* sp., β-(1→3) linked homoxylan was isolated and the structure was confirmed by methylation analysis and ^{13}C-NMR spectroscopy,[5] as well as by mass spectrometry of enzymatic released linear oligosaccharides up to a degree of polymerization (DP) of 25.[6] Recently, β-(1→3, 1→4) linked homoxylan was isolated from edible red seaweed *Palmaria palmata*.[7,8] The regular distribution of (1→3) linkages idealized in a pentameric structure was analysed by HPAEC-PAD, ESI-MS and NMR spectroscopy of oligomeric fragments produced by enzymatic hydrolysis. However, minor amounts of phosphate, sulphate, and galactosylated and xylosylated peptides have been found in some xylan fractions as well.[8,9]

The glucuronoxylans are mostly substituted with a 4-*O*-methyl-α-D-glucopyranosyl uronic acid residue (MeGlcA) attached at position 2 of the main chain (Figure 4.2). This structure type is usually named as 4-*O*-methyl-D-glucurono-D-xylan (MGX). However, the non-methylated glucuronic acid side chain (GlcA) may appear as well. MGX represents the main hemicellulose component of hardwoods showing Xyl : MeGlcA ratios from 4 : 1 to 16 : 1 depending on extraction conditions.[3]

In the native state the xylan is supposed to be *O*-acetylated. The content of acetyl groups of MGX isolated from hardwoods of temperate zones varies between 3–13%.[4] The acetyl groups are split during the alkaline extraction conditions. The acetyl groups may be at least partly preserved under hot water or steaming treatment conditions. Water-soluble acetylated MGXs have been isolated from $NaClO_2$-delignified wood of birch and beech with DMSO.[10] Water-soluble acetylated MGXs of low molecular weight were isolated from aspen wood[11] and from bast fibres of flax[12] by microwave treatment showing a degree of acetylation in the range from 0.3 to 0.6 acetyl groups per repeating unit. MGX was also isolated from fruits and storage tissues such as the pericarp seed of the *Opuntia ficus-indica* pear,[13] luffa (*Luffa cylindrica*) fruit fibres,[14] date seed fibres (*Phoenix dactylifera*),[15,16] sugar beet pulp,[17] grape skin,[18,19] hulls of Jojoba (*Simmondsia chinensis*),[20] and olive fruits at different ripening stages.[21] In the olive fruit cell walls[22] and leaf cell walls of the tree *Argania spinoza*,[23] the MGXs were isolated in the form of xylan–xyloglucan complexes.

Alkali-soluble hemicelluloses of hardwood dissolving pulps have been investigated.[24] Their composition and molecular properties depended on the pulp origin and steeping conditions. The MGX of the β-fraction from press lye had a low uronic acid content (the ratio of MeGlcA to Xyl is about 1 : 20). The molecular weight of the hemicellulose fractions varied between 5000 and 10 000 g mol^{-1}.

(Arabino)glucurono- and (glucurono)arabinoxylan (AGX and GAX) have single MeGlcA and α-L-Ara*f* residues attached at positions 2 and 3 of the β-(1→4)-D-xylopyranose backbone (Figure 4.3a). AGX occurs in appreciable quantities in coniferous species.[3] The AGX backbone contains 5–6 Xyl residues per uronic acid while the hardwood MGX possesses 10 on average.

AGX are the dominant hemicelluloses in the cell walls of lignified supporting tissues of grasses and cereals. They were isolated from sisal, corn cobs and the straw from various wheat species.[4] Corn cob xylans[25] contain a linear, water-insoluble polymer (wis-AGX, ~95% of the backbone is non-substituted) and a water-soluble xylan (ws-AGX) with more than 15% substitution at the backbone.[25,26] The uronic acid content was lower in the wis-AGX (~4%) than in ws-AGX (about 9%). A small proportion of the Xyl*p* residues of the backbone are di-substituted by α-L-Ara*f* residues. A peculiar structural feature of the ws-AGX is the presence of disaccharide side-chains (Figure 4.3b). This sugar moiety has been found, usually esterified by ferulic acid (FA), at position 5 of the Ara*f* unit and occurs as a widespread component of grass cell walls.[27] The ester-linked FA is lost during the alkaline extraction of AGX.[28]

GAX consists of an arabinoxylan backbone containing an amount of uronic acid about ten times lower than that of α-L-Ara*f*. Some Xyl*p* residues are substituted with both sugars. GAX are located in the non-endospermic tissues of cereal grains (wheat, corn, rice bran). The degree and pattern of substitution of GAX vary depending on the source. These differences are reflected in the ratio of Ara to Xyl, the content of MeGlcA, and the presence of disaccharide[3,4] and dimeric arabinosyl side chains.[27]

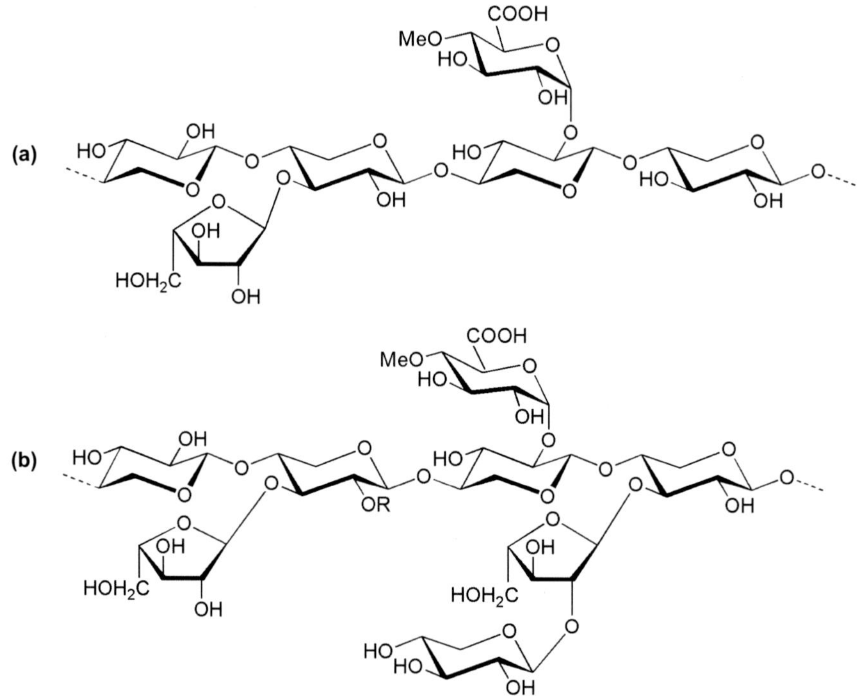

Figure 4.3 Primary structure of (a) (L-arabino)-4-*O*-methyl-D-glucurono-D-xylan (AGX) and (b) water-soluble AGX (ws-AGX).

Arabinoxylan (AX) has been identified in various cereals of commerce (wheat, rye, barley, oat, rice, corn, sorghum) and in other plants such as rye grass, bamboo shoots, and pangola grass. They represent the major hemicellulose component of cell walls of the starchy endosperm (flour) and outer layers (bran) of the cereal grain. AX contents vary from 0.15% in rice endosperm to ~13% in whole grain flour from barley and rye, and up to 30% in wheat bran.[29] AXs occur as neutral polymers as well as slightly acidic ones, the last being usually included into the GAX group. AX has a linear backbone in part substituted by α-L-Ara*f* residues positioned either on *O*-3 or *O*-2 (mono-substitution) or on both *O*-2 and *O*-3 (di-substitution) of the Xyl*p* units (Figure 4.4). In addition, phenolic acids like ferulic and coumaric acid have been found to be esterified to *O*-5 of some Ara*f* residues of AX.[4,30–40]

Heteroxylans (HX) are structurally complex.[29] They have a β-(1→4)-D-xylopyranose backbone decorated with various mono- and oligoglycosyl side-chains except the single uronic acid and arabinosyl residues (Figure 4.1b). Reinvestigations of the HX isolated from corn bran[41] have confirmed that the xylan backbone is heavily substituted (at both positions 2 and 3) with β-D-Xyl*p*, β-L-Ara*f*, α-D-Glc*p*A residues and oligosaccharide side chains (Figure 4.5a–c).

Several HX were isolated from the leaves and barks of tropical dicots such as *Litsea* species.[29] The mucilage-forming seeds of *Plantago* sp. contain very

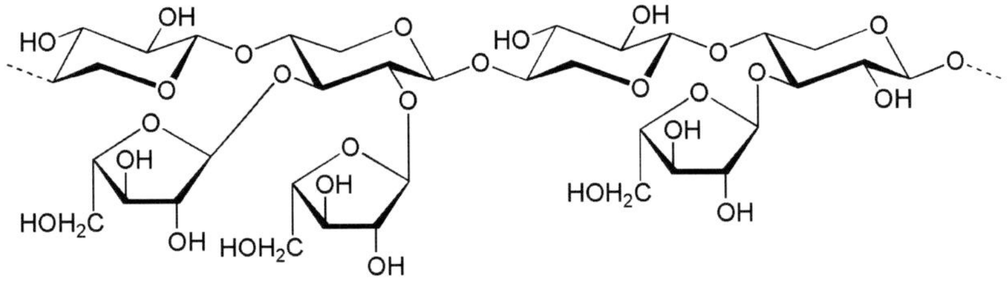

Figure 4.4 Primary structure of water-soluble L-arabino-D-xylan (AX).

β–D-Xyl*p*-(1→2)-α-l-Ara*f*-(1→ (a)

β-D-Gal*p*-(1→5)-α-L-Ara*f*-(1→ (b)

L-Gal*p*-(1→4)-β-D-Xyl*p*-(1→2)-α-L-Ara*f*-(1→ (c)

α-L-Ara*f*-(1→3)-β-D-Xyl*p*-(1→ (d)

α-D-Glc*p*A-(1→3)-α-L-Ara*f*-(1→ (e)

α-L-Ara*f*-(1→3)-β-D-Xyl*p*-(1→)-α-L-Ara*f*-(1→ (f)

Figure 4.5 Structures of side chains of xylans.

complex heteroxylans.[42,43] For the HX from *Plantago major* seeds,[42] a (1→3, 1→4) mixed-linkage xylan backbone has been suggested, possessing short side chains attached to position 2 or 3 of some (1→4) linked D-Xyl*p* residues. The side chains consist of β-D-Xyl*p* and α-L-Ara*f* residues, and disaccharide moieties (Figure 4.5d and e).

Further structural analysis of oligosaccharides generated by partial acid hydrolysis of the polysaccharide[44] revealed the presence of (1→4) linked xylotrisaccharide and (1→3) linked xylooligosaccharides with DP 6–11. These oligosaccharides were suggested to be building blocks in the backbone of HX. In addition, the presence of single β-D-Xyl*p* 2-linked to the backbone, as well as of the acidic disaccharide (Figure 4.5e), was confirmed.

In contrast to the former *Plantago* complex heteroxylan (CHX),[44] the gel-forming HX from psyllium (*Plantago ovata*) husks[44] was found to be a neutral, highly branched arabinoxylan with the (1→4)-β-D-xylopyranose backbone substituted at position 2 with single Xyl*p* units and at position 3 with the trisaccharide moiety (Figure 4.5f).

4.3 Resources and Isolation of Xylans

Potential resources of xylans are by-products of the pulp and paper industry (forest chips, wood meal and shavings), where GX and AGX comprise 25–35%

of the biomass and annual crops (straw, stalks, husk, hulls, bran) containing 25–50% of AX, AGX, GAX, and CHX.[4] Moreover, xylans were isolated from flax fiber,[12,45] abaca fiber,[46] wheat straw,[47,48] sugar beet pulp,[17,49] sugarcane bagasse,[50] rice straw,[51] wheat bran,[52,53] jute bast fibre,[14] and from vetiver grasses.[54]

The extractability of xylan is restricted due to its strong interactions with other cell wall constituents. Different covalent linkages appear including ester bonds formed by uronic acid moieties and hydroxyl groups of lignin.[29,55–57] Moreover, phenolic moieties provide some potential for covalent interactions of the xylan with other cell wall polymers such as feruloylated, arabinose-containing pectic polysaccharides[27] or lignin.[58–60] The existence of covalent linkages between AX and protein has been reported.[59]

A variety of extraction methods and conditions were elaborated to isolate the polymeric xylan from plant cell walls. The effects of conditions and alkali type on the yield and composition of xylan-rich hemicelluloses from wheat straw,[47] sugar beet pulp,[49] and barley straw[60] have been extensively investigated. The isolation procedure and material properties of the MGX from aspen wood were described.[61] For the isolation of xylans from hardwood, in particular, two-step procedures with a $NaOH/H_2O_2$ delignification step were shown to be more acceptable in practice than the hazardous delignification with sodium chlorite. The application of ultrasound was shown to be very effective.[28,50,62–65] Due to the sonomechanical effect of ultrasound, the cell walls were disintegrated achieving higher yields of xylans without substantial modification of their structural and molecular properties.

It is known that disintegration of lignified cell walls can be achieved by steam explosion treatments resulting in soluble, partly depolymerized hemicelluloses.[66,67] The application of this method to wheat bran yielded feruloylated GAX with a different ferulic acid content.[68] Partly depolymerized water-soluble, acetylated AGX was obtained from aspen wood by employing a microwave treatment.[69]

Large-scale isolation procedures have been developed for the isolation of xylans from wheat flour,[70] rye whole-meal[71] and de-starched wheat bran.[52] Optimum extraction conditions were elaborated[72] for the isolation of the CHX (known as corn gum) from corn hulls (corn fibre), a by-product from the wet-milling isolation of cornstarch. This xylan represents a further valuable co-product next to fuel alcohol prepared from the corn fibre. Oxygen-aided alkaline extractions of oat spelt, containing from 22 to 39% of xylan, have been studied and the results evaluated with regard to the yield, purity, and molecular weight of the isolated arabinoxylans.[73] Using a twin-screw extruder, xylans were isolated from poplar wood[74] and co-extracted from a mixture of wheat straw and wheat bran.[75] This procedure might be a technical approach to design a high-throughput process.

Xylans were selectively removed from paper-grade pulps by an extraction with tris(2-aminoethyl)amine (Nitren) to produce dissolving pulp (Figure 4.6).[76,77] Nitren strongly interacts with the hydroxyl groups of the xylan forming coordinative bonds (a discussed de-protonating of the polymer is

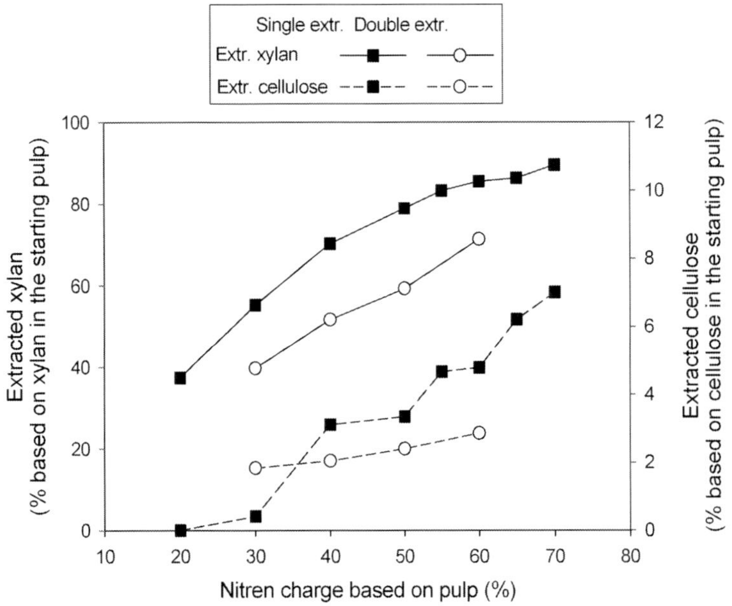

Figure 4.6 Extracted xylan and cellulose from birch kraft pulp after single and double extraction (liquor : pulp ratio = 10 : 1, 1 h, 30 °C).

questionable) at a concentration of 3% Nitren.[78] Higher concentrations of Nitren yield to dissolution of cellulose as well.

4.4 Characteristics

4.4.1 Molecular Mass

An essential problem in determining the molecular mass of polysaccharides is the solubility and the aggregation in solution. It is known that the solubility and aggregation behaviour of dissolved xylan depends on various structural features as described in the previous sections. The type and degree of substitution of xylans by the glycosyl side chains and their distribution pattern, which is responsible for aggregation tendencies, acetyl groups, chemical linkages with lignin, and cross-links through phenolic acids, have to be pointed out in this context. A study on the alkali-extracted, lignin-containing MGX from aspen wood indicates that the lignin component, present in 'bound' and 'unbound' forms, is the main contributor to aggregate formation.[79] In addition, the solubility of xylans is affected by the patterns of intra- and intermolecular hydrogen bonds that may change during the isolation and drying processes of the preparations and by storing.[80] For splitting the hydrogen bonds, aqueous, aprotic, and complexing solvents (such as cuoxam, cadoxen, and FeTNa) have also been applied in combination with ultrasonication and autoclave heating.[81,82]

The values of molecular mass reported show considerable variations and may vary depending on the method applied.[3,4,29] For cereal AX and CHX, the M_w values ranged between 64 000 and 380 000 g mol^{-1}. The M_w values for AGX and MGX ranged from 30 000 to 370 000 g mol^{-1} and 5000 to 130 000 g mol^{-1}, respectively. Comparable M_w values (150 000 and 162 000 g mol^{-1}) were obtained for the water-soluble corn cob AGX by ultracentrifugation and viscometry coupled to static light scattering of fractions from gel chromatography, respectively.[83,84] The molecular mass determination of very low-branched xylans is still an unsolved problem, because they are poorly soluble even in complexing solvents commonly used for the dissolution of cellulose.[80]

Most frequently, size exclusion chromatography (SEC) with dextran-, pullulan-, or polystyrene-calibration standards has been used to characterize the molecular properties of xylans. A sufficient ionic strength of solvents is a prerequisite for valuable SEC measurements of charged polysaccharides, glucuronoxylans included.[85–87] An advantage of the SEC technique is that the presence of protein and phenolic components or oxidative changes can be detected by simultaneous UV-detection.

4.4.2 Interaction of Xylans with other Polysaccharides

The interactions of xylan with starch have been studied because they affect the bread-making quality of flours.[88–90] The addition of ws-AXs shows that no effect on gelation kinetics occurs.[91] However, an enhanced storage modulus G′ appears and starch retrogradation is influenced. The results suggest that AX increased the effective starch concentration in the continuous phase. Well-defined AX samples of low molecular weight (LMW-AX) and of high molecular weight (HMW-AX) have been used as additives.[92] Dough characteristics were affected to a larger extent by HMW-AX than by LMW-AX as assessed by mixography. Small-scale bread-making tests showed that neither the water-holding capacity nor the oxidative gelation or viscosity was the main factor governing the LMW-AX functionality.

The effect of xylan rich ground husks from the seeds of *Plantago ovata* (isabgol) on the bread-making properties of various flours was studied in order to substitute gluten in non-wheat bread.[93] Addition of isabgol alone or in combination with hydroxypropylmethylcellulose at 2% and 1% replacement levels, respectively, significantly enhanced the loaf volumes of bread prepared from rice flour.

Interactions of GAX fractions of wheat bran with xanthan show that for lowly substituted GaMs a synergy in viscosity was observed at low total polymer concentrations yielding a maximum of the relative viscosity at nearly equal proportions of both polysaccharides.[94,95]

The assembly of hemicelluloses including xylans and their interactions with cellulose has been intensively studied. Methods using spectral fitting of CP-MAS ^{13}C-NMR spectra[96] as well as AFM and QCM-D[97] were used to monitor interactions between cellulose I and xylan. Interactions are possible because the linear xylan backbone allows a partial alignment and formation of hydrogen bonds to cellulose microfibrils. In autoclave experiments of cellulose–GX mixtures under controlled conditions, the amount and localization of the globular-shaped xylan assemblies were visualized by immunolabelling and confocal laser microscopy (ref. 98 and references therein). The xylans retained on the cellulose surfaces formed nano- and micro-sized particles. This is of practical importance as xylan-modified lignocellulosic fibres show improved wetting and liquid spreading properties.

4.4.3 Thermal Behaviour

The thermal behaviour has been characterized,[4,96,99,100] indicating that xylans show no distinct thermal transitions. Several MGX fractions obtained by fractional extraction of steam-exploded poplar wood[101] as well as of AGX isolated under various alkaline extraction conditions from wheat straw[99] show a broad onset of thermal degradation near 200 °C. For lignin-rich hemicellulose fraction, the DSC curves showed a prominent effect at 250–540 °C with three maxima around 280, 435, and 500 °C corresponding to about 38, 20, and 13% of the total weight losses. The weight loss was interpreted as being due to decarboxylation in addition to dehydration and oxidation reactions of the carbohydrates and of less condensed structures of the lignin component.[45] In the lignin-poor hemicellulose fraction, the peak maximum at 435 °C was dominating.

A fraction isolated from delignified wheat straw contains about 75% of the hemicellulosic material mainly AGX and a minor amount of residual lignin.[102] The degradation of this fraction started at 220 °C and was completed at around 395 °C. The glass transition temperature T_g of dry xylan was estimated to be in the range of 167–180 °C.[103]

The DTA of a flexible film prepared from birch *O*-acetyl-MGX after conditioning at room humidity displayed endothermic behaviour over a broad temperature range from 60 to 225 °C with a maximum at 150 °C (ref. 104) resulting from water removal of the xylan hydrate non-crystalline state. Thermoplasticity was also displayed by AGX film from pinewood, the latter being far more extensible.

4.5 Application Potential of Xylans

Cross-linked branan ferulate has been marketed as a 'super gel' for wound dressing.[105] Recently, the production of branane ferulate/alginate fibres by the wet spinning technology has been reported.[106] The application of xylans as an additive in bread-making or drug carrier seems to be a possibility as well. Xylans from beech wood, corn cobs and the alkaline steeping liquor of the viscose process have been shown to be applicable as pharmaceutical auxiliary aids.[3] Micro- and nanoparticles were prepared by a coacervation method from xylan isolated from corn cobs, which may be used in drug delivery systems.[107]

The process is based on the neutralization of an alkaline solution in the presence of a surfactant.

Xylo-oligosaccharides are at the centre of interest, as are oligosaccharides in general. There are several reasons for the permanent research going on in this area. One reason is that oligosaccharides have been suggested as new functional food ingredients modifying food flavour and physicochemical characteristics.[108] Moreover, they may possess properties beneficial to health. They show non-carcinogenicity, a low caloric value, and the ability to stimulate the growth of beneficial bacteria in the colon, thus representing potential prebiotics.[109] Acidic xylo-oligosaccharides produced by family 10 and 11 endoxylanases were reported to exhibit antimicrobial activity.[110]

Xylo-oligosaccharides of defined structure are very important substrates that serve as model compounds for the optimization of hydrolytic processes and in enzyme assays. The enormous development in enzyme methodology for chemical modification and structural analysis[111–114] and in physical methods for the isolation, separation, and structural characterization of carbohydrates[115–121] offered the preparation of value-added oligomers of xylans and other hemicelluloses.

Classical acid hydrolysis,[122] autohydrolysis at high temperatures based on acetic acid formed from the released of acetyl groups,[123] endoxylanase treatment alone[110] or in combination with steaming and microwave heating[124] were used to prepare xylo-oligosaccharides from various xylan-rich materials. The hydrothermal treatment is of interest[125,126] because it enables us to fractionate hemicellulose from cellulose and lignin, and separate the formed oligosaccharides as demonstrated on *Eucalyptus* wood and brewery's spent grain.[115,127]

The functionalization of synthetic polymers by carbohydrates has become a current subject of research. It is aimed at preparing new bioactive and biocompatible polymers capable of exerting a temporary therapeutic function. The large variety of methods of anchoring carbohydrates onto polymers as well as the current and potential applications of the functionalized polymers has been recently discussed in a critical review.[128]

4.6 Biological Activity of Xylans and their Derivatives

The various physiological effects of xylan-rich dietary fibres and xylans isolated from various cereal bran and red algae are well known and have been reviewed (see ref. 4, 29, 129 and references therein). Heteroxylans from various plants show a dose-dependent antimutagenic activity tested by the antibleaching effect of these polysaccharides against acridine orange- and ofloxacin-induced mutagenicity in the *Euglena* assay.[130] The feruloylated xylan fragments from corn bran were reported to possess antioxidative activity that might contribute to some physiological functions of cereal dietary fibre.[129]

Medicinal plants are known to be a potential source of biologically active polysaccharides.[4,25] Anticomplementary activities have been reported for the complex heteroxylans of the seeds of *Plantago major*[42] and immunomodulating effects for the MGX from *Rudbeckia fulgida*.[131]

Heteroxylan modified by enzyme treatment from rice bran (commercially known as MGN-3) possesses potent anti-HIV activity without any notable side effects.[132,133] The ws-AGX from corn cobs[134] exhibited significant immunostimulatory effects in the *in vitro* rat thymocyte tests. Its activity was comparable to those of the commercial immunomodulator Zymosan, a fungal β-glucan. The responses in the comitogenic thymocyte test of the cob ws-AGX and MGXs prepared from beech wood and the herbs *Rudbeckia*, *Altheae*, and *Mahonia* were compared in terms of the molecular weight and the content and distribution of MeGlcA substituents[135] (Figure 4.7).

However, no unequivocal relation either to the content of MeGlcA or to its distribution pattern was found for the MGX samples. The immunostimulatory effect of ws-AGX seems to be affected less by the molecular weight than by the presence of the disaccharide side-chain (Figure 4.4a) supposed to be important for the expression of the observed biological response.[26] Antitussive effects (tested on cats) have been reported for MGXs from *Rudbeckia*[136] and mahony.[137]

Xylan sulphates are permanently studied with regard to their biological activities.[29,138–141] Usually, sulphuric acid, sulphur trioxide, or chlorsulphonic acid are employed as sulphating agents alone or in combination with alcohols, amines or chlorinated hydrocarbons as reaction media.[142,143] Recently, xylan sulphate has been prepared from oat spelt xylan.[144] More extensive studies on

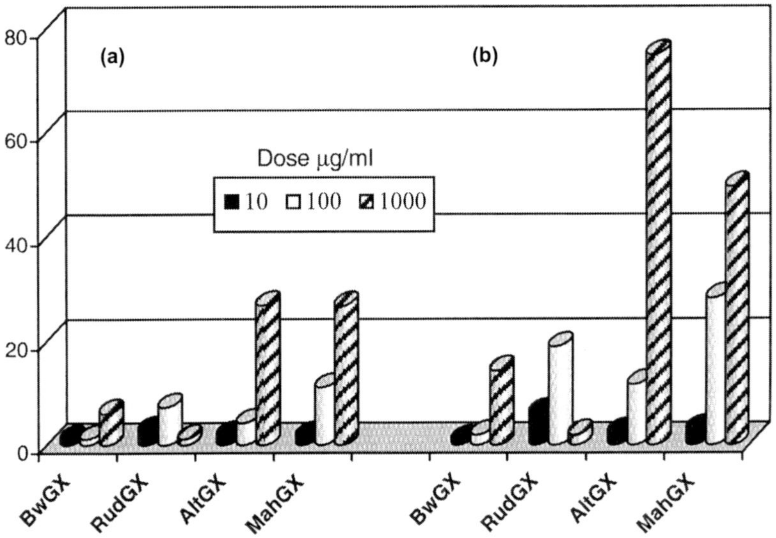

Figure 4.7 Immunostimulatory activity of MGX from beech wood (BwGX), *Rudbeckia fulgida* (RudGX), *Altheae officinalis* (AltGX), and *Mahonia aquifolium* (MahGX);[137] (a) mitogenic activity, SI_{mit} and (b) comitogenic activity, SI_{comit}.

the biological activity of beech wood xylan sulphate, known as pentosan polysulphate (PPS), have been carried out. The highly sulphated xylan exhibits remarkable anticoagulant activity and thus is useful in the prevention of vessel diseases. PPS was recently approved by the US Food and Drug Administration (FDA) for its action as an orally active urological compound.[145]

4.7 Chemical Modification of Xylans

Due to the lack of commercial supply as well as their usually low molecular weight and poor solubility, interest in the chemical modification of xylans has been rather low compared to polysaccharides like cellulose or starch which are commercially available at different qualities. Various chemical modifications have been investigated during the past decade under lab scale conditions as summarized in recent reviews.[29,146] The chemical modification of xylans will be the most important path to novel products with defined properties.

4.7.1 Xylan Ethers

Etherification of the beech wood MGX with *p*-carboxybenzyl bromide in aqueous alkali yielded fully water-soluble xylan ethers with a degree of substitution (DS) of up to 0.25 without significant depolymerization; the M_w determined by sedimentation velocity was 27 000 g mol^{-1}.[147,148] The derivatives exhibited remarkable emulsifying and protein foam-stabilizing activities. Another approach to amphiphilic xylan derivatives is the reaction of the beech wood xylan and its sulphoethyl ether with 1-bromododecane in DMSO at moderate reaction temperatures. The products may be used as biosurfactants.[149]

Polymeric xylan was reacted with propylene oxide (PO) in aqueous alkali homogeneously. The hydroxypropyl xylan obtained could be peracetylated in formamide solution leading to a water-insoluble acetoxypropyl xylan (APX) that is thermoplastic. The glass transition temperature (T_g) of APX varies in relation to the DS of hydroxypropyl groups declining from 160 to 70 °C.[150]

4.7.1.1 Carboxymethylation

The carboxymethylation of polysaccharides is one of the most versatile transformations leading to products that swell or dissolve in water depending on the DS. Moreover, the carboxylic groups may form various interactions with inorganic materials, metal cations, polycations and others leading to supramolecular interactions of practical interest.[151] Xylans from various resources like birch, beech, and eucalyptus wood and from oat husk, rye bran, and corn cob were used to investigate the important method of carboxymethylation in detail.[152]

Different activation procedures were elaborated to synthesize carboxymethyl xylans (Figure 4.8). One-step reactions lead to products with a degree

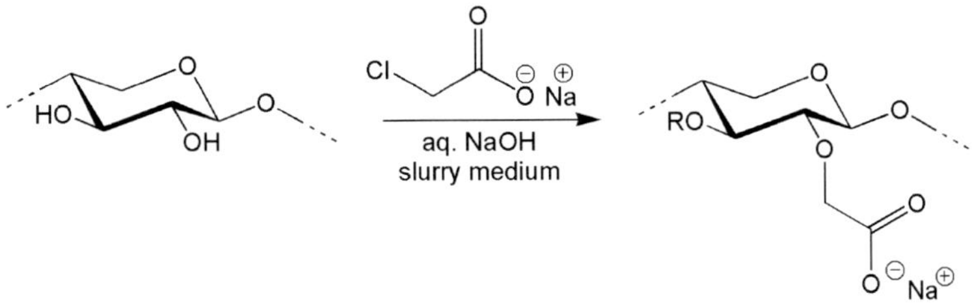

Figure 4.8 Carboxymethylation of an idealized xylan structure.

of substitution (DS) from 0.13 to 1.22 depending on the molar ratio of anhydroxylose unit (AXU) to reagent. Two-step syntheses yielded DS values up to 1.65. Carboxymethyl xylans are water soluble at a DS of 0.3. The solutions have different clearness depending on the provenance. NMR spectroscopy and HPLC were applied to characterize the carboxymethyl xylans in detail.

The DS values obtained under totally heterogeneous conditions depend on the slurry medium. A mixture of ethanol–toluene (1 : 1) or ethanol as slurry media lead to comparable DS value of about 0.6 while 2-propanol yields a significant increase of the DS to 0.91. An alternative synthesis procedure starts with xylan dissolved in 25% aqueous NaOH followed by the addition of the slurry medium. Ethanol–toluene (1 : 1) and ethanol as slurry media, and the reaction without any slurry medium, lead to comparable DS values of 0.6 while a conversion in 2-propanol results an increase of the DS to 0.7 at a molar ratio of 1.0 : 2.0 : 4.1 (AXU : SMCA : NaOH). An increase of the molar ratio leads to an increase of the DS in the slurry medium studied up to 1.3 (reaction time: 70 min at 65 °C). A consecutive reaction, *i.e.*, carboxymethylated xylan was reacted again with SMCA, gave DS values up to 1.66. After 3 reaction steps, nearly a complete functionalized CMX is obtained.

^{1}H NMR spectroscopy of CMX after chain degradation with 25% D_2SO_4 is useful to analyse the substituent distribution. The signals for the protons of the CM substituents in position 2 and 3 can be identified. Additionally, the signals of the protons at C1 (H-1α and H-1β) can be detected depending on the sugar unit which is useful for the DS calculation (Figure 4.9).

After hydrolysis, the sugar units obtained can be separated from the un-, mono-, and di-*O*-substituted units by employing a cation exchange column. The mole fractions fit with the theoretically calculated curves.[153,154] In contrast, the mole fractions of the CMX prepared by a complete heterogeneous procedure do not follow the calculated curves. The polymers contain more unsubstituted and di-*O*-carboxymethylated units and fewer mono-*O*-carboxymethylated units, indicating a non-uniform distribution of the CM functions within the polymer chains.

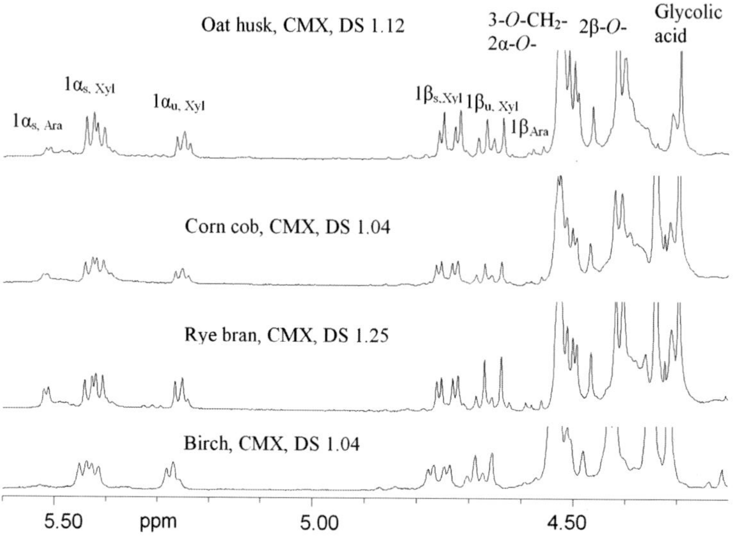

Figure 4.9 ^{1}H NMR spectra of carboxymethyl xylan (CMX) from different types in D_2SO_4 after depolymerization (s means substituted with carboxymethyl groups in position 2, u means unsubstituted, Ara means arabinose, Xyl means xylose).

4.7.1.2 Cationic Xylan Ethers

Investigations were directed towards xylan-rich waste materials such as hardwood sawdust, corn cobs, and sugar cane bagasse by reaction with 3-chloro-2 hydroxypropyltrimethyl-ammonium chloride (CHTMAC).[146] Subsequent extraction steps with water and dilute alkali lead to fractionation into 2-hydroxypropyltrimethylammonium (HPMA)-cellulose, HPMA-hemicellulose, and lignin. The cationization of xylans, isolated from beech wood, corn cobs, rye bran and the viscose spent liquor, was investigated.[155] The results indicated that the DS depends on the molar ratios of CHTMAC : xylan and NaOH : CHMTAC as well as on the xylan type used. The functionalization pattern of the cationized xylans with DS 0.25–0.98 was characterized by ^{13}C-NMR spectroscopy after the hydrolytic chain degradation.[156] Monosubstitution of xylose units was found mainly to occur at lower DS, where position 2 is preferred, whereas at higher DS the regioselectivity is lost.

The HPMA-MGX isolated from cationized aspen sawdust was reported to be applicable as a beater additive; it increased significantly the tear strength of bleached spruce organosolv pulp.[29] The HPMA-derivatives prepared from isolated xylans were shown to improve the papermaking properties and act as flocculants for pulp fibres at very low additions ($\sim$0.25%), very probably due to irreversible adsorption onto cellulose fibres.[157] The derivatives exhibit antimicrobial activity against some Gram-negative and Gram-positive bacteria depending on the DS and xylan type.[158]

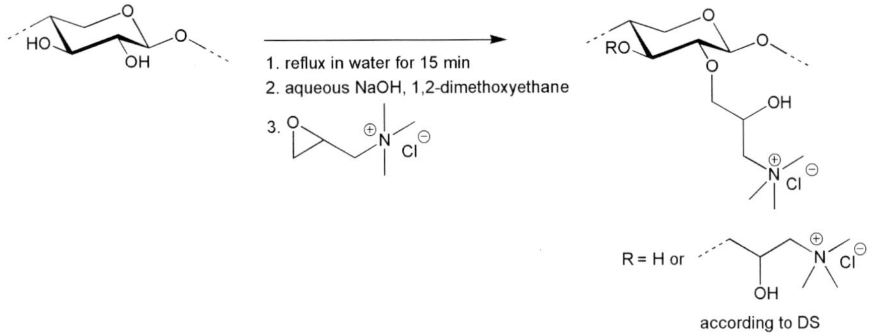

Figure 4.10 Reaction scheme of the cationization of xylan with 2,3-epoxypropyl-trimethyl-ammonium chloride (EPTA).

Cationic 2-hydroxypropyltrimethylammonium (HPMA)-xylan is well studied.[159–161] The synthesis was carried out in an alkaline medium with 3-chloro-2-hydroxypropyltrimethylammonium chloride. Depending on the molar ratio of xylan to reagent and of reagent to NaOH, derivatives with different degrees of substitution (DS) up to 0.98 were obtained.[155] Products with a DS from 0.03 to 0.25 possess microbiological activity.[162]

Recently, the reaction of the birch wood xylan with 2,3-epoxypropyl-trimethylammonium chloride in 1,2-dimethoxyethane as slurry medium has been shown to yield water-soluble, cationic 2-hydroxypropyltrimethyl-ammonium xylan derivatives with high DS values (Figure 4.10). The DS values up to 1.6 can be controlled by adjusting the molar ratio in a one-step synthesis. The structure of the cationic xylan derivatives was confirmed by means of DEPT(135) NMR spectroscopy.[163]

The DEPT(135) spectra of samples with a DS of 0.36 and 1.64 are given in Figure 4.11. Peaks of the CH_2-groups at positions 6, 8 and 5 show negative intensities and can easily be distinguished from the CH moieties. The signals 8 and 7 of the 2-hydroxypropyltrimethylammonium chloride function were determined at δ 68.7 and 65.7 ppm. The two signals (6) observed at δ 74.9 and 71.7 ppm might result from the substitution of the hydroxyl group at position 2 and 3 of the xylan residue. The three methyl groups of the ammonium moieties show an intensive signal at δ 55.0 ppm. The C-1 signal of the 4-linked AXU appeared for an unmodified position 2 and/or 3 at δ 102.2 ppm (1) and for the modified position 2 and/or 3 and at δ 101.7 ppm (1′). The remaining carbons of the AXU can be assigned as follows: the signal for a substituted position 2 shifted to δ 82.1 ppm, for an unsubstituted position 2, the signal can be identified at δ 77.0 ppm. The peak area of 2 s increases with increasing DS, as can be clearly seen in Figure 4.6. The signals of C-4 and C-3 can be detected at δ 76.0 and 73.7 ppm. At δ 64.2 ppm, the negative signal for the CH_2-group corresponds to C-5.

The cationic xylan derivatives form films from aqueous solution which can be stored in ethanol. SEM measurements show a dependence of the morphology

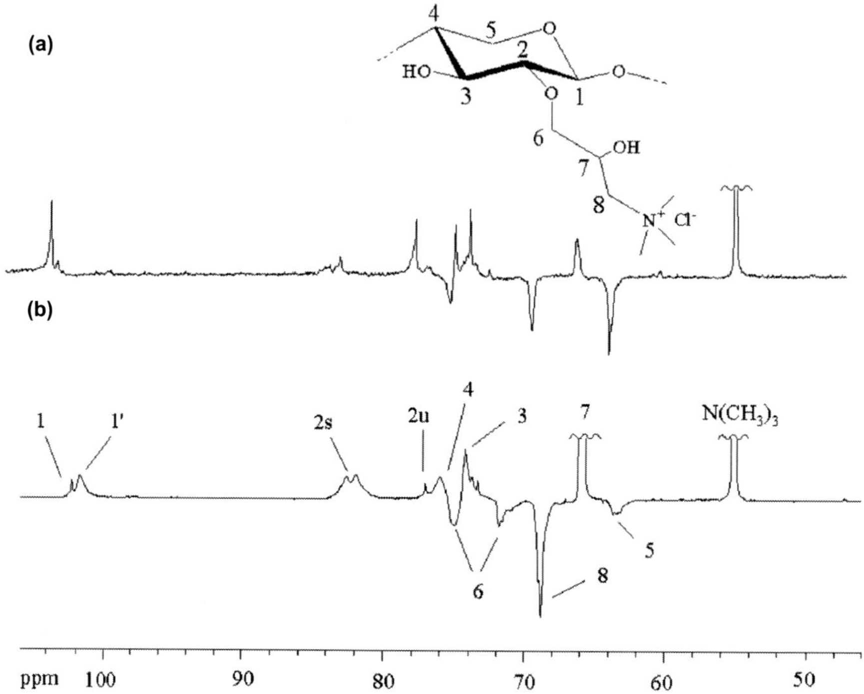

Figure 4.11 DEPT(135) NMR spectra of 2-hydroxypropyltrimethylammonium (HPMA) xylans with different degree of substitution (DS) in D_2O: (a) DS = 0.36 and (b) DS = 1.64, s means substituted, u means unsubstituted in position 2, ′ means influenced by substitution at position 2.

of the films on thickness and on the concentration of the HMPA xylan solution used (Figure 4.12). Remarkably, the films with a thickness of 100 μm possess bigger pores than the films with a thickness of 50 μm independent from the concentration of the original solution. Films formed from 18.5% (w/w) solution possess pores in the range from 250 to 500 nm (50 μm) and at 100 μm the pore size ranges between 500 and 1200 nm. The pores of the films from the 27.3% (w/w) solution exhibit a size of 300–1000 nm (50 μm) and 600–1500 nm (100 μm). No pores can be observed for the film obtained from the 33.3% (w/w) solution with a thickness of 50 μm. The film of 100 μm thickness has a pore size of 100–650 nm. Summarizing, the film morphology can be tailored by the concentration of the solution and the thickness.

Understanding the central role of hemicelluloses at the interface between cellulose-rich and lignin-rich domains may enable the design of better interfaces in wood plastic composites. The self-assembly of hydroxypropyltrimethylammonium-xylan (HPMA) with different degree of substitution onto a model cellulose surface is examined by surface plasmon resonance (SPR) spectroscopy.[164] Results indicate that more HPMA xylan binds onto cellulose than on hydroxyl-terminated alkane thiol self-assembled monolayers (SAM).

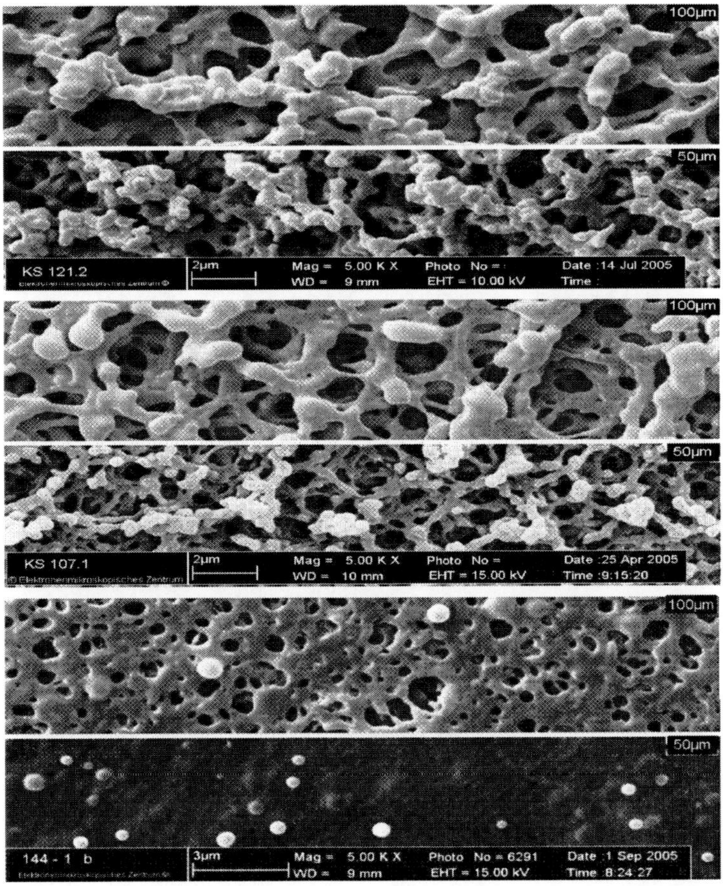

Figure 4.12 SEM studies of 2-hydroxypropyltrimethylammonium (HPMA) xylan films (degree of substitution = 0.36) with different thickness from 50 µm (lower pictures), 100 µm (upper pictures). The films were made from (1) 18.5% (w/w), (2) 27.3% (w/w) and (3) 33.3% (w/w) polymer solution.

The observations yield the surprising conclusion that hydrophobic interactions are at least as important as elecrostatic interactions for driving self-assembly, and that hydrogen bonding may play a minor role.[165]

Alternatively, *N*,*N*-dialkylaminoethyl xylans were prepared by a convenient heterogeneous procedure applying a mixture of aqueous NaOH and 1,2-dimethoxyethane as medium (Figure 4.13).[166] Prior the reaction, the xylan was refluxed in water and dissolved in aqueous NaOH. As reagents, 2-chloro-N,N-dialkylethylammine-hydrochlorides, with methyl, ethyl and diisopropyl as alkyl substituents were used. DS values of the products up to 1.54 could be achieved controlled by adjusting the molar ratio of biopolymer to reagent. The novel xylan ethers are soluble in water depending on the DS, alkyl chain type, and pH value.

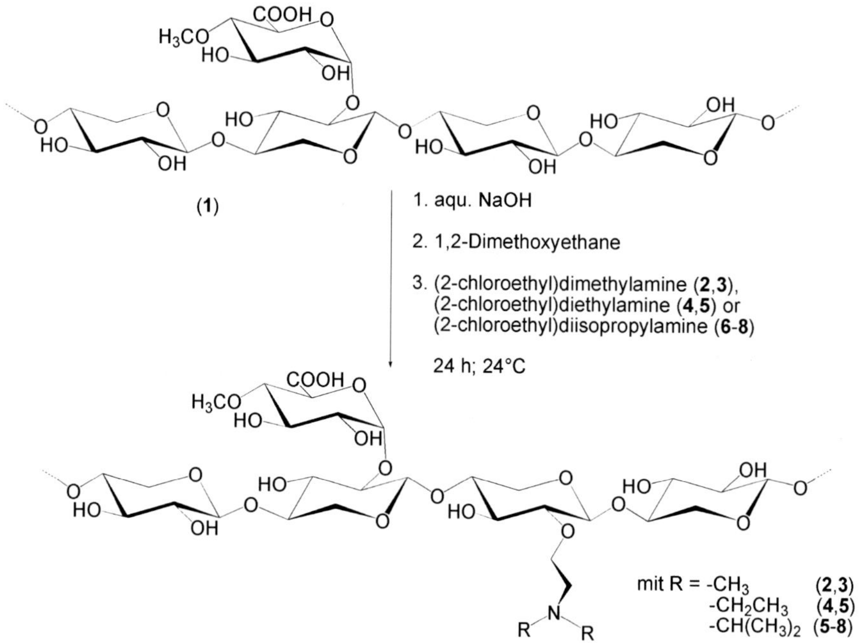

Figure 4.13 Reaction scheme for the preparation of *N,N*-dimethylaminoethyl (DMAE)-, *N,N*-diethylaminoethyl (DEAE)-, and *N,N*-diisopropylaminoethyl (DIAE)-xylan in aqueous NaOH and 1,2-dimethoxyethane.

4.7.2 Xylan Esters

Carboxylic acid esters of xylan are prepared under typical conditions used for polysaccharide esterification, *i.e.*, activated carboxylic acid derivatives are allowed to react with the polymer both heterogeneously and homogeneously (Figure 4.14). Heterogeneous esterification of oak wood sawdust and wheat bran hemicelluloses with excess of octanoyl chloride (without solvent) was described.[167] The separated liquid fraction contained, except for lignin, degraded esterified hemicelluloses.

Homogeneous acylation of xylan-rich hemicelluloses has been mostly carried out in DMF in combination with LiCl.[168] Under moderate reactions conditions, acylated derivatives with DS ranging from 0.18 to 1.71 were prepared from MGX of poplar wood and wheat straw AGX.[169–172] SEC measurements revealed that polymer degradation was low at reaction temperatures below 80 °C. The conversion of wheat straw and bagasse hemicelluloses with succinic anhydride in aqueous alkaline solutions yielded carboxyl groups containing derivatives with $DS < 0.26$.[173] Applications as thickening agents and metal ion binders were proposed for these derivatives. Various catalysts have been used to accelerate acetylation, succinoylation and oleoylation of wheat straw and bagasse hemicelluloses.[174,175]

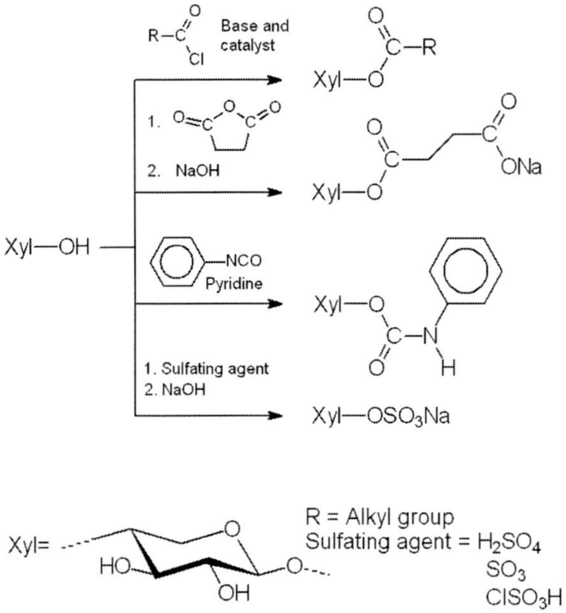

Figure 4.14 Introduction of ester groups into xylan.

By the treatment of oat spelt xylan with phenyl- or tolyl isocyanate in pyridine the fully functionalized corresponding carbamates were prepared.[176] Xylan-3,5-dimethylphenylcarbamte showed higher recognition ability for chiral drugs compared to that of the same cellulose derivative.[177]

The preparation of corn fibre arabinoxylan esters by reaction of the polymer with C_2–C_4 anhydrides using methanesulphonic acid as a catalyst is described.[178] The water-insoluble derivatives of high molecular weight showed glass transition temperatures from 61 to 138 °C depending on the DS and substituent type. The products were thermally stable up to 200 °C.

4.7.3 Thermoplastic and Unconventional Xylan Derivatives

Thermoplastic xylan derivatives have been prepared by "in-line" modification with propylene oxide of the xylan present in the alkaline extract of barley husks.[179,180] Carrying out peracetylation of the hydroxypropylated xylan in formamide solution yielded the water-insoluble acetoxypropyl xylan. The thermal properties of the derivative qualify this material as a potential biodegradable and thermoplastic additive to melt-processed plastics.

AFM and SEM were used to elucidate superstructural features, in particular macro-pore distribution and surface roughness of films of unconventional polysaccharide furan-2-carboxylic acid esters (furoates). The polysaccharides studied are cellulose, curdlan, dextran, starch, and xylan. Furan-2-carboxylic imidazolide in DMA or DMSO was simply added to the polysaccharide solution and the reaction proceeded at 60 °C for 24 h (Figure 4.15). Pure

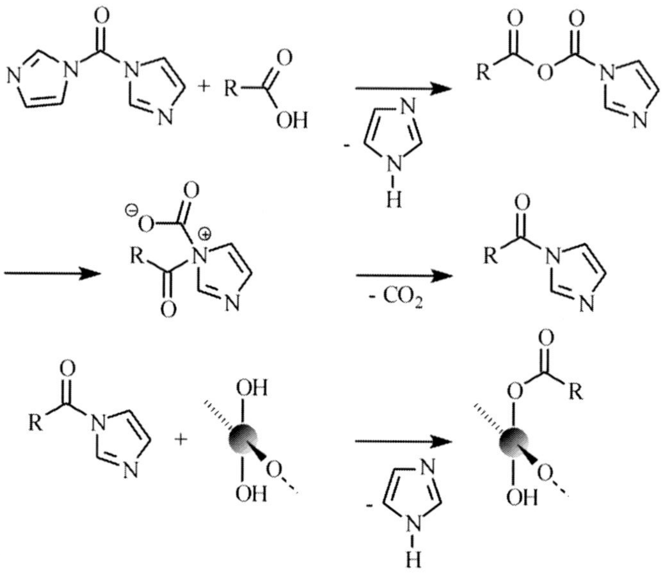

Figure 4.15 General method for esterification of polysaccharides with furan-2-carboxylic acid.[182]

products obtained by a usual work-up procedure were soluble in organic solvents depending on the DS values. In case of xylan furoate it was shown that the DS was controlled by increasing the molar ratio of the furan-2-carboxylic acid to repeating unit. Xylan furoate with a DS of 0.86 was completely soluble in solvents like DMSO, DMA, or DMF, and thus can be processed to films.[181]

AFM on polymer-coated mica surfaces resulted in significantly different qualities of the polysaccharide ester films, *i.e.*, the surface roughness of furan-2-carboxylic acid esters determined by AFM strongly depends on the backbone structure of the polysaccharides. Representative surface characteristics of the furan-2-carboxylic acid esters of cellulose, curdlan, dextran, starch, and xylan on the mica surfaces are shown in Figure 4.16.

For the roughness analysis (Table 4.1), different AFM figures were utilized, which were taken at a Topometrix instrument with a scanner size of 130 to 130 μm and a *z*-range of about 12 μm.

The films of cellulose furoate (DS 1.81) on mica surfaces exhibited the highest roughness values (R_a, approx. 170 nm) and maximum heights (R_{max}) of about 1700 nm, followed by xylan furoate (DS 0.86) with $R_a \approx 150$ nm and $R_{max} \approx$ 1500 nm. Further characterization of the film's surfaces was carried out by means of SEM. For the SEM measurements, the wet polymer films were dried by critical point drying and gold sputtered. The scanning electron micrographs of the xylan furoate (DS 0.86) film's rear side exhibited quite porous structures with a macro-pore size predominantly smaller than 250 nm (Figure 4.17a, upper picture). In contrast, the up side of xylan furoate films looked like a tender meshwork with pore structures of about 400 nm size (Figure 4.17b). In some

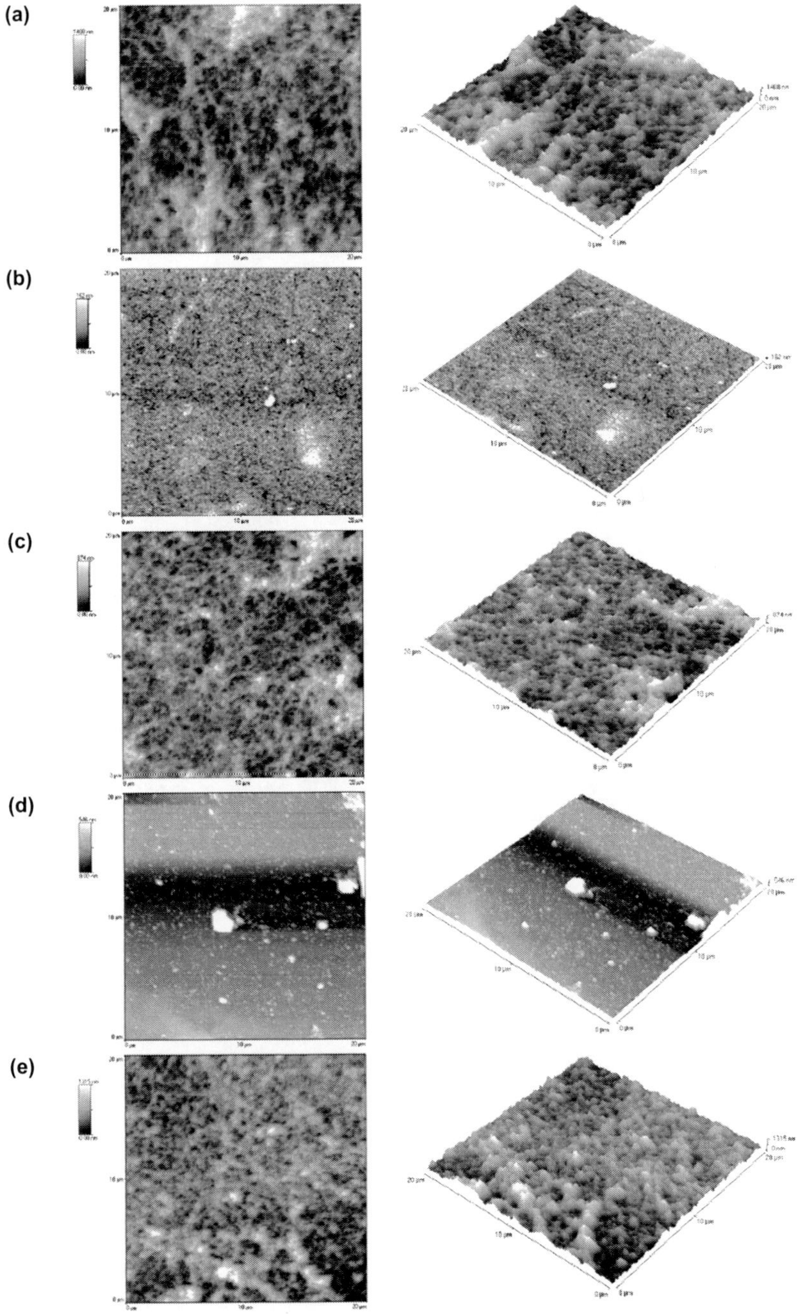

Figure 4.16 AFM images (contact mode; scan range 20×20 μm) of the supermolecular structures of furan-2-carboxylic acid ester films on mica surfaces: (a) cellulose (DS 1.81), (b) curdlan (DS 1.59), (c) dextran (DS 1.93), (d) starch (DS 0.98), and (e) xylan (DS 0.86).

Table 4.1 Roughness analysis of the AFM images and macro-pore sizes of films based on polysaccharide furan-2-carboxylic acid esters.

		Characteristics/nm			
Furan-2-carboxylic acid ester of	*DS*	*Mean roughness* (R_a)	*Root mean square (RMS)*	*Maximum height* (R_{max})	*Pore size*[a]
Cellulose	1.81	174.0 ± 14.2	220.1 ± 17.4	1725 ± 240	300 to 800
Curdlan	1.59	25.2 ± 8.7	33.4 ± 11.6	372 ± 190	200 to 550
Dextran	1.93	102.8 ± 10.4	132.0 ± 14.1	1010 ± 161	-
Starch	0.98	67.3 ± 23.6	86.2 ± 31.7	776 ± 499	-
Xylan	0.86	152.5 ± 20.1	193.7 ± 26.5	1535 ± 424	250 to 750

[a]Macro-pore size of self-supporting films determined by SEM.

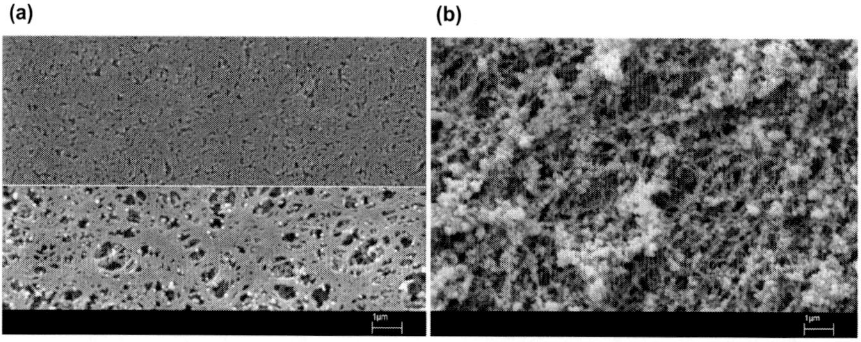

Figure 4.17 Scanning electron micrographs of a self-supporting film of xylan furan-2-carboxylic acid ester (DS 0.86): (a) glass–film contact side, (b) film–solvent contact side.

rear side regions, however, sections of the meshwork were observed through extended macro-pores. In these regions, the pore structures were characterized by sizes up to 800 nm (Figure 4.17a, lower picture). This fact implied that the high-density rear side of a film would not control its separation properties, *e.g.*, in activity as a membrane or filter material, and could be removed quite easily. Its tightness apparently depended on the feasibility of precipitation. As described previously,[182] the polysaccharide furoate can be modified specifically by UV irradiation yielding solvent-resistant, stable films. If samples were treated with light having a wavelength in the range of the absorption maximum of the furan moiety (290 nm), cross-linking succeeded presumably *via* a pericyclic reaction and yielded insoluble polymers. It was confirmed that the irradiation did not lead to a change in the surface roughness or the pore size.[183,184]

In recent years, several studies have been devoted to the formation and characterization of nanoparticles based on polymers. Xylan furan- and pyroglutamic acid esters, efficiently synthesized by conversion of the biopolymer with furan- and pyroglutamic acid and *N,N′*-carbonyldiimdazole as the activating agent form spherical nanoparticles of a size down to 60 nm and a narrow particle size distribution by applying a simple dialysis process.[185]

4.7.4 Oxidation of Xylans

Oxidation is an important tool for the introduction of carbonyl and carboxyl functions into biopolymers. The tendency of the oxidation of polysaccharides depends substantially on the nature of the oxidants and the conditions. Most of the oxidants known from the low-molecular organic chemistry produce both carbonyl and carboxyl functions to a varying extent depending on the experimental conditions.[142] Moreover, even so-called selective oxidation reactions will result in more or less depolymerization of the macromolecules. In principle, xylan as polyhydroxy compounds bearing secondary hydroxyl groups can be oxidized to 2-keto-, 3-keto- and 2,3-diketo xylans neglecting the transformation of the end groups. Moreover, 2,3-dialdehyde xylan may be obtained in the well-known glycol cleavage oxidation of vicinal diol units with sodium periodate, which can be further oxidized to give 2,3-dicarboxyl xylan. Figure 4.18 summarizes the structures of the possible oxidized repeating units.

The biodegradation of various carboxylic group containing polymers was studied by Matsumura *et al.*[186–188] The aim of this work was to develop compounds which are biologically degradable detergents. Thus, xylan from oat spelt (Sigma Chemical) as well as cellulose and starch were oxidized to 2,3-dicarboxylic polysaccharides in a two-step procedure by using HIO_4/$NaClO_2$ as oxidants. Xylan was treated with periodic acid for 6 h at 4 °C to obtain 2,3-dialdehyde xylan with 39 mole% dialdehyde groups. Subsequent treatment with sodium chloride for 24 h at 20 °C and 1 h at 50 °C leads to sodium 2,3-dicarboxylic xylan with 39 mole% dicarboxylate moieties. A set of oxidized samples with a content of dicarboxylate functions in the range from 29 to 53 mole% was synthesized by varying the reaction time.

The biodegradability of these polyanions was compared with artificial poly(carboxylic acids). An important result was the fact that the biodegradability is improved if the polymer chain contains functionalities like hydroxyl, ether, ester or carbonyl groups as well as sugar residues. The biodegradability and detergency is improved by the introduction of carboxylic groups.[186] Poly(sodium acrylate) was used as reference in a five-day biochemical oxygen

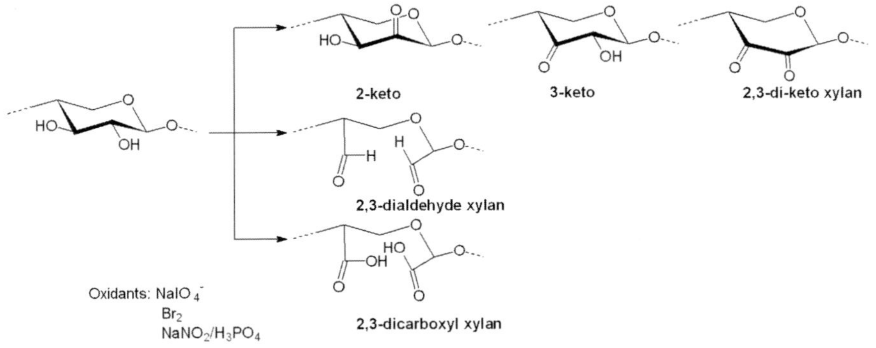

Figure 4.18 Typical repeating units of oxidized xylan.

demand measurement (BOD_5). No oxygen consumption was measured for the reference substance. It was found that the BOD_5 decreases with increasing content of dicarboxylic units in the polymers used.

Andersson *et al.* investigated the oxidation of xylan and cellulose with sodium nitrite in orthophosphoric acid.[189] This oxidant is known to attack primary hydroxyl groups yielding carboxylic acid groups, *i.e.*, cellulose 6-carboxyl cellulose is formed.[190,191] Consequently, an oxidation of xylan should not occur due to the absence of a primary hydroxyl group. NMR measurements of the oxidized xylan indicate that the polymer is preferentially oxidized to 2-ulose residues. Oxidation at C-3 occurs only to a very low extent. The peak assignment could be performed using 2D-COSY experiments. The result is in agreement with the fact that esterification of the *O*-2 is faster compared to *O*-3. In addition, the low oxidation at C-3 was revealed by means of the more sensitive sugar analysis after reduction with sodium borohydride compared to NMR measurements.

The neutral xylan fraction from *Palmaria decipiens* was treated with aqueous bromine solution.[192] This xylan contains (1→4) as well as (1→3) linked sugar units. Carbonyl groups are formed which was revealed by means of FTIR (ν_{CO} at 1741 cm^{-1}) and NMR spectroscopy. The peaks found in the ^{13}C-NMR spectrum were assigned (Table 4.2). The presence of C=O was also revealed chemically by reactions with *p*-chloroaniline or bovine serum albumin followed by reduction to the corresponding amine with sodium cyanoboranate. The oxidation occurs at C-2 only as determined by means of GLC after reductive cleavage, reduction and peracetylation.

The sodium periodate oxidation of xylan isolated from *Palmaria decipiens* yields 2,3-dialdehyde xylan.[193] The dialdehyde moieties can be converted into a Schiff base type compound by the reaction with *p*-chloroaniline (Figure 4.19). Imine structures are known to be ligands for the complexation of metal ions. Thus, the Schiff base type compounds obtained from 2,3-dialdehyde xylan coordinates copper(II) in a complex. The complexes were characterized by means of spectroscopic methods (IR, UV/VIS, cyclic voltammetry).

Table 4.2 Chemical shift assignments for the ^{13}C-NMR spectra of xylan and its bromine-oxidized derivative.[192]

	Chemical shift/ppm						
	C-1		*C-2*		*C-3*	*C-4*	*C-5*
β-D-Xylopyranosyl residue	*1→3*	*1→4*	*1→3*	*1→4*			
Xylan							
3-O-substituted	104.08	102.50	73.39	73.39	84.39	68.61	65.78
4-Ovc-substituted	104.08	102.50	74.02	73.64	74.62	77.29	63.80
Br_2 oxidized derivative							
3-O-substituted	104.05	102.48	73.56	73.56	84.28	68.51	65.75
4-O-substituted	104.05	102.48	74.15	74.15	74.54	77.22	63.80
CO	203.73						

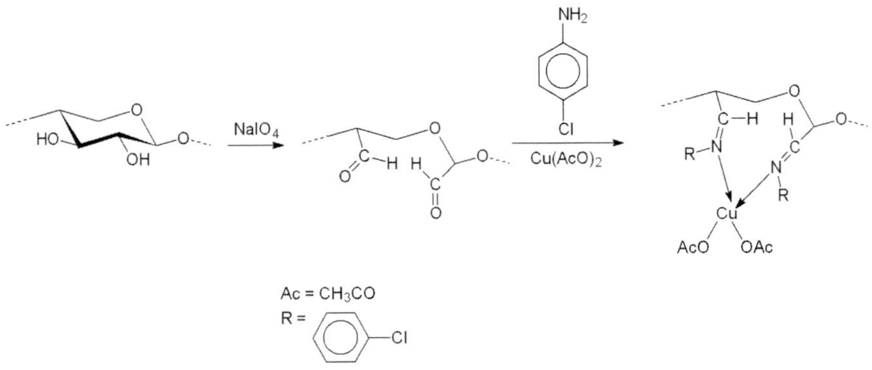

Figure 4.19 Complexation of copper(II) by a ligand synthesized from xylan.[192]

4.8 Concluding Remarks

The studies in the field of hemicelluloses, which have been covered in this review, indicate the significantly increased importance of this valuable class of biopolymers as plant constituents and isolated polymers. It was the aim to describe typical aspects and not to give a complete picture. There are many excellent reviews available that are also cited in this chapter. The main problem using hemicellulose in large quantities and as a chemical feedstock is the fact that still no large-scale commercial availability exists. It may be assumed that hemicelluloses become available in the future as products obtained by alternative pulping of wood, by extraction of hemicellulose from paper pulp, and by direct isolation from various plants due to their outstanding properties including bioactivity in the unmodified form (or just easily modified by the extraction procedure) and as raw material for the chemical conversion. There are different issues regarding chemical modification: on one hand, the polymeric hemicellulose can be transformed to new polymeric materials by reactions at the polymer backbone using typical polysaccharide chemistry as reviewed. This is an obvious trend in polysaccharide research in general. On the other hand, degradation of the hemicelluloses to the corresponding sugars (including oligosaccharides) or to simple organic molecules is the basis of the technical organic chemistry at present.

The usefulness of hemicellulose-based or derived materials in an industrial and biomedical context is now beyond dispute. It will stimulate further research into the development of isolation and purification processes from the natural sources, and the development of analytical tools for structural characterization of the isolated hemicelluloses and their derivatives.

4.9 List of Abbreviations and Symbols

AFM Atomic force microscopy
AGX (L-Arabino)-4-*O*-methyl-D-glucurono-D-xylan
APX acetoxypropyl xylan

AX	L-arabino-D-xylan
AXU	Anhydroxylose unit
BOD5	biochemical oxygen demand measurement (BOD_5)
Cadoxen	Cadminium complex with ethylenediamine
CHX	Complex heteroxylan
CHTMAC	3-chloro-2 hydroxypropyltrimethyl-ammonium chloride
CM	Carboxymethyl
CMX	Carboxylmethyl xylan
COSY	Correlated Spectroscopy
CP-MAS	Cross polarization magic angle spinning ^{13}C-NMR
Cuoxam	Copper complex with ammonia
DEAE	diethylaminoethyl
DIAE	diisopropylaminoethyl
DMA	NN-Dimethylacetamide
DMF	N,N-Dimethylformamide
DMSO	Dimethyl sulfoxide
DP	Degree of polymerization
DS	Degree of substitution
DSC	Differential scanning calorimetry
DTA	Differential thermal analysis
ESI-MS	Electrospray ion trap mass spectrometry
EPTA	2,3-epoxypropyltrimethyl-ammonium chloride
FA	Ferulic acid
FDA	Food and Drug Administration
FeTNa	Ferric tatric acid complex in alkaline aqueous solution
FTIR	fourier transform infra-red
G'	Storage modul
GaM	D-Galacto-D-mannan
GAX	(4-*O*-Methyl-D-glucurono)-L-arabino-D-xylan
GX	Glucuronoxylan
HMW-AX	High molecular weight arabinoxylan
HPAEC	High performance anion exchange chromatography
HPLC	High pressure liquid chromatography
HPMA	2-hydroxypropyltrimethylammonium
HX	Heteroxylans
IR	infra-red
LMW-AX	Low molecular weight arabinoxylan
MGX	4-*O*-Methyl-D-glucurono-D-xylan
M_w	weight average molecular weight
NMR	nuclear magnetic resonance
PAD	Pulsed amperometric detection
ppm	parts per million
PPS	Pentosan Polysulphate
PO	Propylene oxide
QCM-D	Quartz crystal microbalance with dissipation
Ra	Mean roughness

RMS	Root mean square
R_{max}	maximum height
SAM	self-assembled monolayers
SEC	Size exclusion chromatography
SEM	scanning electron microscopy
SMCA	sodium monochloroacetate
SPR	surface plasmon resonance
Tg	Glass transition temperature
US	United States
UV	ultra violet
UV/VIS	ultra violet/visual
wis	Water insoluble
ws	Water soluble

Sugars

Ara*f*	L-Arabinofuranose
GlcA	D-Glucuronic acid
Glc*p*	D-Glucopyranose
MeGlcA	4-*O*-Methyl-D-glucuronic acid
NaOH	sodium hydroxide
Xyl*p*	D-Xylopyranose

Acknowledgements

The authors acknowledge the financial support of the research in the field of hemicellulose by the German BWMA/AiF–DGfH (project No. 13698 BR), by the Fachagentur Nachwachsende Rohstoffe (FNR)/BMVEL (project No. 22011401), by the European Community program COST D29 (project No. D29/0008/03), and COST D28 (project No. D28/006/03), the European Polysaccharide Network of Excellence (EPNOE) and the Fonds der Chemischen Industrie of Germany. The general financial support of Borregaard Chemcell, Hercules, Dow Chemicals, Wolff Cellulosics, ShinEtsu, and Rhodia is gratefully acknowledged. The authors thank Dr Katrin Petzold for proof reading the manuscript.

References

1. E. Schulze, Zur Kentniss der chemischen Zusammensetzung der pflanzlichen Zellmembranen, *Ber. Dtsch. Chem. Ges.*, 1891, **24**, 2277–2287.
2. A. Ebringerova, Structural Diversity and Application Potential of Hemicelluloses, *Macromol. Symp.*, 2006, **232**, 1–12.
3. A. M. Stephen, Other plant polysaccharides. In *Polysaccharides*, (Ed.) G. O. Aspinall, Vol. 2, Academic Press, New York, 1983, pp. 97–193.
4. A. Ebringerová and T. Heinze, Xylan and xylan derivatives – biopolymers with valuable properties: 1. Naturally occurring xylans: structures, isolation, procedure and properties, *Macromol. Rapid Commun.*, 2000, **21**, 542–556.

5. T. Yamagaki, M. Meada, K. Kanazawa, Y. Ishizuka and H. Nakanishi, Structural clarification of *Caulerpa* cell wall β-1,3-xylan by NMR spectroscopy, *Biosci., Biotechnol., Biochem.*, 1997, **61**, 1077–1080.
6. T. Yamagaki, M. Meada, K. Kanazawa, Y. Ishizuka and H. Nakanishi, NMR spectroscopic analysis of sulfated β-1,3-xylan and sulfation stereochemistry, *Biosci., Biotechnol., Biochem.*, 1997, **61**, 1281–1285.
7. E. Deniaud, B. Quemener, J. Fleurence and M. Lahaye, Structural studies of the mix-linked β-(1→3)/β-(1→4)-D-xylans from the cell wall of *Palmaria palmata* (Rhodophyta), *Int. J. Biol. Macromol.*, 2003, **33**, 9–18.
8. E. Deniaud, J. Fleurence and M. Lahaye, Interactions of the mix-linked β-(1→3)/β-(1→4)-D-xylans in the cell walls of *Palmaria palmata* (Rhodophyta), *J. Physiol.*, 2003, **39**, 74–82.
9. E. Deniaud, J. Fleurence and M. Lahaye, Preparation and chemical characterization of cell wall fractions enriched in structural proteins from *Palmaria palmata* (Rhodophyta), *Bot. Mar.*, 2003, **46**, 366–377.
10. A. Teleman, M. Tenkanen, A. Jacobs and O. Dahlman, Characterization of O-acetyl-(4-O-methylglucurono)xylan isolated from birch and beech, *Carbohydr. Res.*, 2002, **337**, 373–377.
11. A. Teleman, J. Lundqvist, F. Tjerneld, H. StÅlbrand and O. Dahlman, Characterization of acetylated 4-O-methylglucuronoxylan isolated from aspen employing ^{1}H and ^{13}C NMR spectroscopy, *Carbohydr. Res.*, 2000, **329**, 807–815.
12. A. Jacobs, M. Palm and G. Zacchi, Dahlman, Isolation and characterization of water-soluble hemicelluloses from flax shive, *Carbohydr. Res.*, 2003, **338**, 1869–1876.
13. Y. Habibi, M. Mahrouz and M. R. Vignon, Isolation and structure of D-xylans from pericarp seeds of *Opuntia ficus-indica* prickly pear fruits, *Carbohydr. Res.*, 2002, **337**, 1593–1598.
14. M. R. Vignon and C. Gey, Isolation, ^{1}H and ^{13}C NMR studies of (4-O-methyl-D-glucurono)-D-xylans from luffa fruit fibers jute bast fibers, and mucilage of quince tree seeds, *Carbohydr. Res.*, 1998, **307**, 107–111.
15. O. N. Haq and J. Gomes, Studies on xylan from date fruits (*Phoenix dactylifera*), *Bangladesh J. Sci. Ind. Res.*, 1977, **12**, 76–80.
16. O. Ishurd, Y. Ali, W. Wei, F. Bashir, A. Ali, A. Ashour and Y. Pan, An alkali-soluble heteroxylan from seeds of *Phoenix dactylifera* L, *Carbohydr. Res.*, 2003, **338**, 1609–1612.
17. E. Dinand and M. Vignon, Isolation and NMR characterisation of a (4-O-methyl-D-glucurono)-D-xylan from sugar beet pulp, *Carbohydr. Res.*, 2001, **330**, 285–288.
18. J. M. Igartuburu, E. Pando, F. Rodiguez-Luis and A. Gil-Serrano, Hemicellulose A Fraction in Dietary Fiber from the Seed of Grape Variety Palomino (*Vitis vinifera* cv. Palomino), *J. Nat. Prod.*, 1998, **61**, 876–880.
19. J. M. Igartuburu, E. Pando, F. Rodiguez-Luis and A. Gil-Serrano, A Hemicellulose B Fraction from Grape Skin (*Vitis vinifera*, Palomino Variety), *J. Nat. Prod.*, 2001, **64**, 1174–1178.

20. T. Watanabe, Y. Mitsuishi and Y. Kato, Isolation and characterization of an acidic xylan from jojoba (*Simmondsia chinensis*) hulls, *J. Appl. Glycosci.*, 1999, **46**, 281–284.
21. E. Vierhuis, H. A. Schols, G. Beldman and A. G. J. Voragen, Isolation and characterisation of cell wall material from olive fruit (*Olea europaea* cv koroneiki) at different ripening stages, *Carbohydr. Polym.*, 2000, **43**, 11–21.
22. E. Vierhuis, H. A. Schols, G. Beldman and A. G. J. Voragen, Structural characterization of xyloglucan and xylans present in olive fruit (*Olea europaea* cv koroneiki), *Carbohydr. Polym.*, 2001, **44**, 51–62.
23. B. Ray, C. Loutelier-Bourhis, C. Lange, E. Condamine, A. Driouich and P. Lerouge, Structural investigation of hemicellulosic polysaccharides from *Argania spinosa*: Characterization of a novel xyloglucan motif, *Carbohydr. Res.*, 2004, **339**, 201–208.
24. U. Mais and H. Sixta, Characterization of alkali-soluble hemicelluloses of hardwood dissolving pulps. In *ACS Symposium Series, Volume 864 (Hemicelluloses), Science and Technology*, (Eds) P. Gatenholm and M. Tenkanen, American Chemical Society, Washington DC, 2004, pp. 94–107.
25. A. Ebringerová, A. Kardošová, Z. Hromádková and V. Hříbalová, Mitogenic and comitogenic activities of polysaccharides from some European herbaceous plants, *Fitoterapia*, 2003, **74**, 52–61.
26. A. Ebringerová, Z. Hromádková, J. Alföldi and V. Hříbalová, The immunologically active xylan from ultrasound-treated corn cobs: extractability, structure and properties, *Carbohydr. Polym.*, 1998, **37**, 231–239.
27. T. Ishii, Structure and functions of feruloylated polysaccharides, *Plant Sci.*, 1997, **127**, 111–127.
28. Z. Hromádková, J. Kovačiková and A. Ebringerová, Study of the classical and ultrasound assisted extraction of the corn cob xylan, *Ind. Crop. Prod.*, 1999, **9**, 101–109.
29. A. Ebringerová and Z. Hromádková, Xylans of industrial and biomedical importance. In *Biotechnology and Genetic Engineering Reviews*, (Ed.) S. E. Harding, Intercept Ltd., Vol. 16, England, 1999, pp. 325–346.
30. J. A. Delcour, H. V. Win and P. J. Grobet, Distribution and structural variation of arabinoxylans in common wheat mill streams, *J. Agric. Food Chem.*, 1999, **47**, 271–275.
31. G. Dervilly, C. Leclercq, D. Zimmermann, C. Roue, J.-F. Thibault and L. Saulnier, Isolation and characterization of high molar mass water-soluble arabinoxylans from barley and barley malt, *Carbohydr. Polym.*, 2002, **47**, 143–149.
32. C. J. A. Vinkx and J. A. Delcour, Rye (*Secale cereale* L.) arabinoxylans: A critical review, *J. Cereal Sci.*, 1996, **24**, 1–14.
33. A. G. J. Voragen, H. Gruppen, M. A. Verbruggen and R. J. Vietor, Characterization of cereal arabinoxylans. In *Volume 7 (Xylans and Xylanases) Progress in Biotechnology*, (Eds) J. Visser, G. Beldmann, A. S. Kusters van

Someren and A. G. J. Voragen, Elsevier Science Publishers, Amsterdam, 1992, pp. 51–67.
34. M. S. Izydorczyk and C. G. Biliaderis, Cereal arabinoxylans: advances in structure and physiochemical properties, *Carbohydr. Polym.*, 1995, **28**, 33–48.
35. M. S. Izydorczyk, L. J. Macri and A. W. MacGregor, Structure and physicochemical properties of barley non-starch polysaccharides. I. Water-extractable β-glucans and arabinoxylans, *Carbohydr. Polym.*, 1998, **35**, 249–258.
36. M. S. Izydorczyk, L. J. Macri and A. W. MacGregor, Structure and physicochemical properties of barley non-starch polysaccharides-II. Alkali-extractable β-glucans and arabinoxylans, *Carbohydr. Polym.*, 1998, **35**, 259–269.
37. J. P. Roubroeks, R. Andersson and P. Åman, Structural features of (1→3), (1→4)-β-D-glucan and arabinoxylan fractions isolated from rye bran, *Carbohydr. Polym.*, 2000, **42**, 3–11.
38. J.-Y. Han, Structural characteristics of arabinoxylan in barley, malt, and beer, *Food Chem.*, 2000, **70**, 131–138.
39. C. D. Nandini and P. V. Salimath, Structural features of arabinoxylans from sorghum having good roti-making quality, *Food. Chem.*, 2001, **74**, 417–422.
40. M. V. S. S. T. S. Rao and G. Muralikrishna, Non-starch polysaccharides and bound phenolic acids from native and malted finger millet (ragi, *Eleusine coracana*, Indaf - 15), *Food Chem.*, 2001, **72**, 187–192.
41. L. Saulnier, C. Marot, E. Chanliaud and J.-F. Thibault, Cell wall polysaccharide interactions in maize bran, *Carbohydr. Polym.*, 1995, **26**, 279–287.
42. A. B. Samuelson, I. Lund, J. M. Djahromi, B. S. Paulsen, J. K. Wold and S. H. Knutsen, Structural features and anti-complementary activity of some heteroxylan polysaccharide fractions from the seeds of *Plantago major* L., *Carbohydr. Polym.*, 1999, **38**, 133–143.
43. M. H. Fischer, N. Yu, G. R. Gray, J. R. Ralph, L. Anderson and J. A. Marlett, The gel-forming polysaccharide of psyllium husk (*Plantago ovata* Forsk), *Carbohydr. Res.*, 2004, **339**, 2009–2017.
44. A. B. Samuelsen, E. H. Cohen, B. S. Paulsen, L. P. Brull and J. E. Thomas-Oates, Structural studies of a heteroxylan from *Plantago major* L. seeds by partial hydrolysis, HPAEC-PAD, methylation and GC-MS, ESMS and ESMS/MS, *Carbohydr. Res.*, 1999, **315**, 312–318.
45. J. M. van Hazendonk, E. J. M. Reinerink, P. de Waard and J. E. G. van Dam, Inter-glycoside acetals. Part 3. Synthesis and structure determination of cyclic monobenzylidene acetals of cyclodextrin derivatives bridging between two continuous D-glucopyranosyl residues, *Carbohydr. Res.*, 1996, **291**, 53–62.
46. R. C. Sun, J. M. Fang, A. Goodwin, J. M. Lawther and A. J. Bolton, Fractionation and characterization of polysaccharides from abaca fiber, *Carbohydr. Polym.*, 1998, **37**, 351–359.

47. J. M. Lawther, R. C. Sun and W. B. Banks, Effects of extraction conditions and alkali type on yield and composition of wheat straw hemicellulose, *J. Appl. Polym. Sci.*, 1996, **60**, 1827–1837.
48. R. Sun, J. M. Lawther and W. B. Banks, Fractional and structural characterization of wheat straw hemicelluloses, *Carbohydr. Polym.*, 1996, **29**, 325–331.
49. R. C. Sun and S. Hughes, Fractional extraction and physicochemical characterization of hemicelluloses and cellulose from sugarbeet pulp, *Carbohydr. Polym.*, 1998, **36**, 293–299.
50. J.-X. Sun, R. C. Sun, X.-F. Sun and Y. Q. Su, Fractional and physicochemical characterization of hemicelluloses from ultrasonic irradiated sugarcane bagasse, *Carbohydr. Res.*, 2004, **339**, 291–300.
51. R. C. Sun, J. Tomkinson, P. L. Ma and S. F. Liang, Comparative study of hemicelluloses from rice straw by alkali and hydrogen peroxide treatments, *Carbohydr. Polym.*, 2000, **42**, 111–122.
52. M. Bataillon, P. Mathaly, A. P. N. Cardinali and F. Duchiron, Extraction and purification of arabinoxylan from destarched wheat bran in a pilot scale, *Ind. Crop. Prod.*, 1998, **8**, 37–43.
53. M. E. F. Schooneveld-Bergmans, A. M. C. P. Hopman, G. Beldman and A. G. J. Voragen, Extraction and partial characterization of feruloylated glucuronoarabinoxylans from wheat bran, *Carbohydr. Polym.*, 1998, **35**, 39–47.
54. P. Methacanon, O. Chaikumpollert, P. Thavorniti and K. Suchiva, Hemicellulosic polymer from Vetiver grass and its physicochemical properties, *Carbohydr. Polym.*, 2003, **54**, 335–342.
55. O. Eriksson and B. O. Lindgren, About the linkage between lignin and hemicelluloses in wood, *Svensk Papperstidn.*, 1997, **80**, 59–63.
56. G. Wallace, W. R. Russel, J. A. Lomax, M. C. Jarvis, C. Lapierre and A. Chesson, Extraction of phenolic-carbohydrate complexes from graminaceous cell walls, *Carbohydr. Res.*, 1995, **272**, 41–53.
57. C. Lapierre, B. Pollet, M.-C. Ralet and L. Saulnier, The phenolic fraction of maize bran: evidence for lignin-heteroxylan association, *Phytochemistry*, 2001, **57**, 765–772.
58. T. B. T. Lam, K. Kadoya and K. Iiyama, Bonding of hydroxycinnamic acids to lignin: ferulic and p-coumaric acids are predominantly linked at the benzyl position of lignin, not the β-position, in grass cell walls, *Phytochemistry*, 2001, **57**, 987–992.
59. L. Saulnier, R. Andersson and P. Åman, A study of the polysaccharide components in gluten, *J. Cereal Sci.*, 1997, **25**, 121–127.
60. R. C. Sun and V. F. Sun, Fractional and structural characterization of hemicelluloses isolated by alkali and alkaline peroxide from barley straw, *Carbohydr. Polym.*, 2002, **49**, 415–423.
61. M. Gustavsson, M. Bengtsson, P. Gatenholm, W. Glasser, A. Teleman and O. Dahlman, Isolation, characterisation and material properties of 4-O-methylglucuronoxylan from aspen. In *Biorelated Polymers: Sustainable Polymer Science and Technology*, (Eds) E. Chiellini, H. Gil, G. Braunegg,

J. Burchert, P. Gatenholm and M. van der Zee, Kluwer Academic, Plenum Publishers, Dordrecht, pp. 41–52.
62. Z. Hromádková and A. Ebringerová, Ultrasonic extraction of plant materials – investigation of hemicellulose release from buckwheat hulls, *Ultrason. Sonochem.*, 2003, **10**, 127–133.
63. R. C. Sun and J. Tomkinson, Characterization of hemicelluloses isolated with tetraacetylethylenediamine activated peroxide from ultrasound irradiated and alkali pre-treated wheat straw, *Eur. Polym. J.*, 2003, **39**, 751–759.
64. R. C. Sun and J. Tomkinson, Characterization of hemicelluloses obtained by classical and ultrasonically assisted extractions from wheat straw, *Carbohydr. Polym.*, 2002, **50**, 263–271.
65. R. C. Sun, X. F. Sun and X. H. Ma, Effect of ultrasound on the structural and physicochemical properties of organosolv soluble hemicelluloses from wheat straw, *Ultrason. Sonochem.*, 2002, **9**, 95–101.
66. W. G. Glasser, W. E. Kaar, R. K. Jain and J. E. Sealey, Isolation options for non-cellulosic heteropolysaccharides (HetPS), *Cellulose*, 2000, **7**, 299–317.
67. K. Shimizu, K. Sudo, H. Ono, M. Ishihara, T. Fujii and S. Hishiyama, Integrated process for total utilization of wood components by steam-explosion pre-treatment, *Biomass Bioenerg.*, 1998, **14**, 195–203.
68. M. E. F. Schooneveld-Bergmans, M. J. W. Dignum, J. H. Grabber, G. Beldman and A. G. J. Voragen, Studies on the oxidative crosslinking of feruloylated arabinoxylans from wheat flour and wheat bran, *Carbohydr. Polym.*, 1999, **38**, 309–317.
69. A. Jacobs, J. Lundqvist, H. StÅlbrand, F. Tjerneld and O. Dahlman, Characterization of water-soluble hemicelluloses from spruce and aspen employing SEC/MALDI mass spectroscopy, *Carbohydr. Res.*, 2002, **337**, 711–717.
70. A.-L. Faurot, L. Saulnier, S. Berot, Y. Popineau, M.-D. Petit, X. Rouau and J.-F. Thibault, Large-scale isolation of water-soluble and water-insoluble pentosans from wheat flour, *Food Sci. Technol.*, 1995, **28**, 436–441.
71. J. A. Delcour, N. Rouseu and I. P. Vanhaesendonck, Pilot-scale isolation of water-extractable arabinoxylans from rye, *Cereal Chem.*, 1999, **76**, 1–2.
72. R. B. Hespell, Extraction and Characterization of Hemicellulose from the Corn Fiber Produced by Corn Wet-Milling Processes, *J. Agr. Food Chem.*, 1998, **46**, 2615–2619.
73. B. Saake, N. Erasmy, Th. Kruse, E. Schmekal and J. Puls, Isolation and characterization of arabinoxylan from oat spelts. In *ACS Symposium Series, Volume 864 (Hemicelluloses), Science and Technology*, (Eds) P. Gatenholm and M. Tenkanen, American Chemical Society, Washington DC, 2004, pp. 52–65.
74. S. N'Diaye and L. Rigal, Factors influencing the alkaline extraction of poplar hemicelluloses in a twin-screw reactor: correlation with specific

mechanical energy and residence time distribution of the liquid phase, *Bioresour. Technol.*, 2000, **75**, 13–18.

75. P. Maréchal, J. Jorda, P.-Y. Pontalier and L. Rigal, Twin screw extrusion and ultrafiltration for xylan production from wheat straw and bran. In *ACS Symposium Series, Volume 864 (Hemicelluloses), Science and Technology*, (Eds) P. Gatenholm and M. Tenkanen, American Chemical Society, Washington DC, 2004, pp. 38–51.
76. R. Janzon, J. Puls and B. Saake, Upgrading of paper-grade pulps to dissolving pulps by nitren extraction: optimisation of extraction parameters and application to different pulps, *Holzforschung*, 2006, **60**, 347–354.
77. J. Puls, N. Schroeder, A. Stein, R. Janzon and B. Saake, Xylans from oat spelts and birch kraft pulp, *Macromol. Symp.*, 2006, **232**, 85–92.
78. G. Kettenbach and A. Stein (to Rhodia Acetow GmbH), Verfahren zum Abtrennen von Hemicellulosen aus hemicellulosehaltiger Biomasse sowie die mit dem Verfahren erhältliche Biomasse und Hemicellulose, *Ger. Patent* 10109502, 2001.
79. J. P. Roubroeks, B. Saake, W. Glasser and P. Gatenholm, Contribution of the molecular architecture of 4-O-methyl glucuronoxylan to its aggregation behaviour in solution, in P. Gatenholm. In *ACS Symposium Series, Volume 864 (Hemicelluloses), Science and Technology*, (Ed.) M. Tenkanen, American Chemical Society, Washington DC, 2004, pp. 167–183.
80. A. Ebringerová, Z. Hromádková, W. Burchard, R. Dolega and W. Vorwerg, Solution properties of water-insoluble rye-bran arabinoxylan, *Carbohydr. Polym.*, 1994, **24**, 161–169.
81. D. R. Picout, S. B. Ross-Murphy, N. Errington and S. E. Harding, Pressure cell assisted solubilization of xyloglucans: tamarind seed polysaccharide and detarium gum, *Biomacromolecules*, 2003, **4**, 799–807.
82. Q. Wang, P. R. Ellis, S. B. Ross-Murphy and W. Burchard, Solution characteristics of the xyloglucan extracted from *Detarium senegalense* Gmelin, *Carbohydr. Polym.*, 1997, **33**, 115–124.
83. R. Dhami, S. E. Harding, N. J. Elizabeth and A. Ebringerová, Hydrodynamic characterization of the molar mass and gross conformation of corncob heteroxylan AGX, *Carbohydr. Polym.*, 1995, **28**, 113–119.
84. A. Ebringerová, Z. Hromádková, J. Alföldi and G. Berth, Structural and solution properties of corn cob heteroxylans, *Carbohydr. Polym.*, 1992, **19**, 99–105.
85. T. E. Eremeeva and O. E. Khinoverova, Exclusion liquid chromatography of 4-O-methylglucuronoxylan in dipolar aprotic solvents, *Cellul. Chem. Technol.*, 1990, **24**, 439–444.
86. T. E. Eremeeva and T. O. Bykova, High-performance size-exclusion chromatography of wood hemicelluloses on a poly(2-hydroxyethyl methacrylate-co-ethylene dimethacrylate) column with sodium hydroxide solution as eluent, *J. Chromatogr.*, 1993, **639**, 159–164.
87. T. Bykova and T. Arnis, Solubility and molecular weight of hemicelluloses from *Alnus incana* and *Alnus glutinosa*. Effect of tree age, *Plant Physiol. Bioch.*, 2002, **40**, 347–353.

88. C. G. Biliaderis, M. S. Izydorczyk and O. Rattan, Effect of arabinoxylans on bread-making quality of wheat flours, *Food. Chem.*, 1995, **53**, 165–171.
89. C. M. Courtin and J. A. Delcour, Arabinoxylans and Endoxylanases in Wheat Flour Bread-making, *J. Cereal Sci.*, 2002, **35**, 225–243.
90. T. Sasaki, T. Yasui and J. Matsuki, Influence of non-starch polysaccharides isolated from wheat flour on the gelatinization and gelation of wheat starches, *Food Hydrocolloids*, 2000, **14**, 295–303.
91. M. Gudmundsson, A.-C. Eliasson, S. Bengtsson and P. Åman, The effects of water-soluble arabinoxylan on gelatinization and retrogradation of starch, *Starch/Staerke*, 1991, **43**, 5–10.
92. C. M. Courtin and J. A. Delcour, Physicochemical and Bread-Making Properties of Low Molecular Weight Wheat-Derived Arabinoxylans, *J. Agr. Food Chem.*, 1998, **46**, 4066–4073.
93. A. Haque, E. R. Morris and R. K. Richardson, Polysaccharide substitutes for gluten in nonwheat bread, *Carbohydr. Polym.*, 1994, **25**, 337–344.
94. M. E. F. Schooneveld-Bergmans, Y. M. Van Dijk, G. Beldman and A. G. J. Voragen, Physicochemical characteristics of wheat bran glucurono-arabinoxylans, *J. Cereal Sci.*, 1999, **29**, 49–61.
95. P. A. Williams and G. O. Phillips, Interactions in Mixed Polysaccharide Systems. In *Food Polysaccharides and their Applications*, (Ed.) A. M. Stephen, Marcel Dekker Inc., New York, 1995, pp. 463–500.
96. P. T. Larrson, Interaction between cellulose I and hemicelluloses studied by spectral fitting of CP/MAS ^{13}C-NMR spectra. In *ACS Symposium Series, Volume 864 (Hemicelluloses), Science and Technology*, (Eds) P. Gatenholm and M. Tenkanen, American Chemical Society, Washington DC, 2004, pp. 254–268.
97. A. Paananen, M. Osterberg, M. Rutland, T. Tammellin, T. Saarinen, K. Tappura and P. Stenius, Interaction between cellulose and xylan: An atomic force microscope and quartz crystal microbalance study. In *ACS Symposium Series, Volume 864 (Hemicelluloses), Science and Technology*, (Eds) P. Gatenholm and M. Tenkanen, American Chemical Society, Washington DC, 2004, pp. 269–290.
98. A. Linder and P. Gatenholm, Effect of cellulose substrate on assembly of xylans. In *ACS Symposium Series, Volume 864 (Hemicelluloses), Science and Technology*, (Eds) P. Gatenholm and M. Tenkanen, American Chemical Society, Washington DC, 2004, pp. 236–253.
99. R. C. Sun, J. M. Fang, P. Rowlands and J. Bolton, Physicochemical and Thermal Characterization of Wheat Straw Hemicelluloses and Cellulose, *J. Agr. Food Chem.*, 1998, **46**, 2804–2809.
100. B. Xiao, X. F. Sun and R. C. Sun, Chemical, structural, and thermal characterizations of alkali-soluble lignins and hemicelluloses, and cellulose from maize stems, rye straw, and rice straw, *Polym. Degrad. Stabil.*, 2001, **74**, 307–319.
101. N. Rauschenberg, K. Dhara, J. Palmer and W. Glasser, Xylan derivatives from steam-exploded lignocellulosic resources – structure and properties, *Polymer Prepr. (Am. Chem. Soc. Div. Polym. Chem.)*, 1990, **31**, 650–652.

102. J. M. Fang, R. C. Sun, J. Tomkinson and P. Fowler, Acetylation of wheat straw hemicellulose B in a new non-aqueous swelling system, *Carbohydr. Polym.*, 2000, **41**, 379–387.
103. G. M. Irvine, The glass transitions of lignin and hemicellulose and their measurement by differential thermal analysis, *Tappi J.*, 1984, **67**, 118–121.
104. R. H. Marchessault, Isolation and properties of xylan: Rediscovery and renewable resource. In *ACS Symposium Series, Volume 864 (Hemicelluloses), Science and Technology*, (Eds) P. Gatenholm and M. Tenkanen, American Chemical Society, Washington DC, 2004, pp. 158–166.
105. L. L. Lloyd, J. F. Kennedy, P. Methacanon, M. Paterson and C. J. Knill, Carbohydrate polymers as wound management aids, *Carbohydr. Polym.*, 1998, **37**, 315–322.
106. M. Miraftab, Q. Qiao, J. F. Kennedy, S. C. Anand and M. R. Groocock, Fibres for wound dressings based on mixed carbohydrate polymer fibres, *Carbohydr. Polym.*, 2003, **53**, 225–231.
107. R. B. Garcia, T. Nagashima, A. K. C. Praxedes, F. N. Raffin, T. F. A. L. Moura and E. S. T. do Egito, Preparation of micro and nanoparticles from corn cobs xylan, *Polym. Bull.*, 2001, **46**, 371–379.
108. R. G. Crittenden and M. J. Playne, Production, properties and applications of food-grade oligosaccharides, *Trends Food Sci. Technol.*, 1996, **7**, 353–361.
109. L. J. Fooks, R. Fuller and G. R. Gibson, Prebiotics, probiotics and human gut microbiology, *Int. Dairy J.*, 1999, **9**, 53–61.
110. P. Chirstakopolulos, P. Katapodis, E. Kalogeris, D. Kekos, B. J. Macris, H. Stamatis and H. Skaltsa, Antimicrobial activity of acidic xylo-oligosaccharides produced by family 10 and 11 endoxylanases, *Int. J. Biol. Macromol.*, 2003, **31**, 171–175.
111. R. P. de Vries and J. Visser, Aspergillus enzymes involved in degradation of plant cell wall polysaccharides, *Microbiol. Mol. Biol. Rev.*, 2001, **65**, 497–522.
112. A. C. E. Gregory, A. P. O'Connell and G. P. Bolwell, *Xylans, Biotechnol. Genet. Eng. Rev.*, 1998, **15**, 439–455.
113. P. Biely, Xylanolytic enzymes. In *Handbook of Food Enzymology*, (Eds) P. Whitaker, A. G. J. Voragen and H. Wong, Marcel Dekker Inc., New York, 2003, pp. 879–915.
114. A. Lappalainen, M. Tenkanen and J. Pere, Specific antibodies for immunochemical detection of wood-derived hemicelluloses. In *ACS Symposium Series, Volume 864 (Hemicelluloses), Science and Technology*, (Eds) P. Gatenholm and M. Tenkanen, American Chemical Society, Washington DC, 2004, pp. 140–156.
115. M. A. Kabel, H. A. Schols and A. G. J. Voragen, Complex xylo-oligosaccharides identified from hydrothermally treated *Eucalyptus* wood and brewery's spent grain, *Carbohydr. Polym.*, 2002, **50**, 191–200.
116. I. Tanczos, C. Schwarzinger, H. Schmidt and J. Balla, THM-GC/MS analysis of model uronic acids of pectin and hemicelluloses, *J. Anal. Appl. Pyrolysis*, 2003, **68**, 151–162.

117. M. Tenkanen and M. Siika-aho, An α-glucuronidase of Schizophyllum commune acting on polymeric xylan, *J. Biotechnol.*, 2000, **178**, 149–161.
118. M. A. Kabel, P. de Waard, H. A. Schols and A. G. J. Voragen, Location of O-acetyl substituents in xylo-oligosaccharides obtained from hydrothermally treated *Eucalyptus* wood, *Carbohydr. Res.*, 2002, **337**, 69–77.
119. M. Kačuráková and R. H. Wilson, Developments in mid-infrared FT-IR spectroscopy of selected carbohydrates, *Carbohydr. Polym.*, 2001, **44**, 291–303.
120. A. Reis, M. A. Coimbra, P. Domingues, A. J. Ferrer-Correia and M. R. M. Domingues, Fragmentation pattern of underivatised xylo-oligosaccharides and their alditol derivatives by electrospray tandem mass spectrometry, *Carbohydr. Polym.*, 2004, **55**, 401–409.
121. A. Rydlund and O. Dahlman, Oligosaccharides obtained by enzymatic hydrolysis of birch kraft pulp xylan: analysis by capillary zone electrophoresis and mass spectrometry, *Carbohydr. Res.*, 1997, **300**, 95–102.
122. H.-J. Sun, S. Yoshida, N.-H. Park and I. Kusakabe, Preparation of (1→4)-β-D-xylooligosaccharides from an acid hydrolysate of cotton-seed xylan: suitability of cotton-seed xylan as a starting material for the preparation of (1→4)-β-D-xylooligosaccharides), *Carbohydr. Res.*, 2002, **337**, 657–661.
123. F. Carvalheiro, M. P. Esteves, J. C. Parajo and H. Pereira, Production of oligosaccharides by autohydrolysis of brewery's spent grain, *Bioresour. Technol.*, 2004, **91**, 93–100.
124. M. Palm and G. Zacchi, Extraction of hemicellulosic oligosaccharides from spruce using microwave oven or steam treatment, *Biomacromolecules*, 2003, **4**, 617–623.
125. G. Garrote, H. Dominguez and J. C. Parajo, Mild autohydrolysis: an environmentally friendly technology for xylooligosaccharide production from wood, *J. Chem. Technol. Biotechnol.*, 1999, **74**, 1101–1109.
126. E. G. Koukios, A. Pastou, D. P. Koullas, V. Sereti and F. Kolosis, New green products from cellulosics. In *Biomass: a growth opportunity in green energy and value-added products*, (Eds) R. P. Overend and E. Chornet, Pergamon Press, Oxford, 1999, pp. 641–647.
127. M. A. Kabel, F. Carvalheiro, G. Garotte, E. Avgerinos, E. Koukios, J. C. Parajo, F. M. Girio, H. A. Schols and A. G. J. Voragen, Hydrothermally treated xylan rich by-products yield different classes of xylo-oligosaccharides, *Carbohydr. Polym.*, 2002, **50**, 47–56.
128. A. J. Varma, J. F. Kennedy and P. Galgali, Synthetic polymers functionalized by carbohydrates: a review, *Carbohydr. Polym.*, 2004, **56**, 429–445.
129. T. Ohta, S. Yamasaki, Y. Egashira and H. Sanada, Antioxidative Activity of Corn Bran Hemicellulose Fragments, *J. Agr. Food Chem.*, 1994, **42**, 653–656.
130. A. Belicová, L. Ebringer, J. Krajčovič, Z. Hromádková and A. Ebringerová, Antimutagenic effect of heteroxylans, arabinogalactans, pectins and mannans in the euglena assay, *World J. Microb. Biot.*, 2001, **17**, 293–299.

131. M. Bukovský, A. Kardošová, H. Koščová and D. Koštálová, Immunomodulating activity of 4-O-methyl-D-glucurono-D-xylan from *Rudbeckia fulgida var. sullivantii*, *Biologia (Bratislava)*, 1998, **53**, 771–775.
132. M. H. Ghoneum, Anti-HIV activity in vitro of MGN-3, an activated arabinoxylane from rice bran, *Biochem. Bioph. Res. Comm.*, 1998, **243**, 25–29.
133. M. Ghoneum and A. Jewett, Production of tumor necrosis factor-α and interferon-γ from human peripheral blood lymphocytes by MGN-3, a modified arabinoxylan from rice bran, and its synergy with interleukin-2 in vitro, *Cancer Detect. Prev.*, 2000, **24**, 314–324.
134. A. Ebringerová, Z. Hromádková and V. Hřibalová, Structure and mitogenic activities of corn cob heteroxylans, *Int J. Biol. Macromol.*, 1995, **17**, 327–331.
135. A. Ebringerová, A. Kardošová, Z. Hromádková, A. Malovíková and V. Hřibalová, Immunomodulatory activity of acidic xylans in relation to their structural and molecular properties, *Int. J. Biol. Macromol.*, 2002, **30**, 1–6.
136. G. Nosál'ová, A. Kardošová and S. Fraňová, Antitussive activity of a glucuronoxylan from *Rudbeckia fulgida* compared to the potency of two polysaccharide complexes from the same herb, *Pharmazie*, 2000, **55**, 65–68.
137. A. Kardošová, A. Malovíková, V. Pätoprstý, G. Nosál'ová and T. Matáková, Structural characterization and antitussive activity of a glucuronoxylan from *Mahonia aquifolium* (Pursh) Nutt., *Carbohydr. Polym.*, 2002, **47**, 27–33.
138. S. J. Elliot, L. J. Striker, W. G. Stetler-Stevenson, T. A. Jacot and G. E. Striker, Pentosan polysulfate decreases proliferation and net extracellular matrix production in mouse mesangial cell, *J. Am. Soc. Nephrol.*, 1999, **10**, 62–68.
139. J. Giedrojc, P. Radziwon, M. Klimiuk, M. Bielawiec, H. K. Breddin and J. Kloczko, Experimental studies on the anticoagulant and antithrombotic effects of sodium and calcium pentosan polysulfate, *J. Physiol. Pharmacol.*, 1999, **50**, 111–119.
140. W. D. Figg, J. M. Pluda and O. Sartor, Pentosan polysulfate: A polysaccharide that inhibits angiogenesis by binding growth factors. In *Antiangiogenic Agents in Cancer Therapy*, (Ed.) B. A. Teicher, Humana Press Inc., Totowa NJ, 1999, pp. 371–383.
141. P. Ghosh, The pathobiology of osteoarthritis and the rationale for the use of pentosan polysulfate for its treatment, *Semin. Arthritis Rheu.*, 1999, **28**, 211–267.
142. D. Klemm, B. Philipp, T. Heinze, U. Heinze and W. Wagenknecht, *Comprehensive Cellulose Chemistry*, Wiley-VCH, Weinheim, 1998.
143. D. Papy-Garcia, V. Barbier-Chassefiere, V. Rouet, M. Kerros, C. Klochendler, M. Tournaire, D. Barritault, J. Caruelle and E. Petit, Nondegradative Sulfation of Polysaccharides. Synthesis and Structure Characterization of Biologically Active Heparan Sulfate Mimetics, *Macromolecules*, 2005, **38**, 4647–4654.

144. K. Hettrich, S. Fischer, N. Schröder, J. Engelhardt, U. Drechsler and F. Loth, Derivatization and Characterization of Xylan from Oat Spelts, *Macromol. Symp.*, 2006, **232**, 37–48.
145. V. R. Anderson and C. M. Perry, Pentosan polysulfate: a review of its use in the relief of bladder pain or discomfort in interstitial cystitis, *Drugs*, 2006, **66**, 821–835.
146. Th. Heinze, A. Koschella and A. Ebringerová, Chemical functionalization of xylan: A short review. In *ACS Symposium Series, Volume 864 (Hemicelluloses), Science and Technology*, (Eds) P. Gatenholm and M. Tenkanen, American Chemical Society, Washington DC, 2004, pp. 312–325.
147. A. Ebringerová, Z. Hromádková, A. Malovíková, V. Sasinková, J. Hirsch and I. Sroková, Structure and properties of water-soluble p-carboxybenzyl polysaccharide derivatives, *J. Appl. Polym. Sci.*, 2000, **78**, 1191–1199.
148. A. Ebringerová, J. Alföldi, Z. Hromádková, G. M. Pavlov and S. E. Harding, Water-soluble p-carboxybenzylated beechwood 4-O-methyl-glucuronoxylan: structural features and properties, *Carbohydr. Polym.*, 2000, **42**, 123–131.
149. A. Ebringerová, I. Sroková, P. Talába, M. Kačuráková and Z. Hromádková, Amphiphilic beechwood glucuronoxylan derivatives, *J. Appl. Polym. Sci.*, 1998, **67**, 1523–1530.
150. K. Rajesh, M. Sjostedt and W. G. Glasser, Thermoplastic xylan derivatives with propylene oxide, *Cellulose*, 2001, **7**, 319–336.
151. Th. Heinze, New ionic polymers by cellulose functionalization, *Macromol. Chem. Phys.*, 1998, **199**, 2341–2364.
152. K. Petzold, K. Schwikal, W. Günther and Th. Heinze, Carboxymethyl xylan - Control of properties by synthesis, *Macromol. Symp.*, 2006, **232**, 27–36.
153. Th. Heinze, U. Erler, I. Nehls and D. Klemm, Determination of the substituent pattern of heterogeneously and homogeneously synthesized carboxymethyl cellulose by using high-performance liquid chromatography, *Angew. Makromol. Chem.*, 1994, **215**, 93–106.
154. U. Heinze, T. Heinze and D. Klemm, Synthesis and structure characterization of 2,3-O-carboxymethyl cellulose, *Macromol. Chem. Phys.*, 1999, **200**, 896–902.
155. A. Ebringerova, Z. Hromadkova, M. Kacurakova and M. Antal, Quaternized xylans: synthesis and structural characterization, *Carbohydr. Polym.*, 1994, **24**, 301–308.
156. A. Ebringerová and Z. Hromádková, Substituent distribution in cationic xylan derivatives, *Angew. Makromol. Chem.*, 1996, **242**, 97–104.
157. M. Antal, A. Ebringerová, Z. Hromádková, I. Pikulík, M. Laleg and M. M. Micko, Structure and papermaking properties of (aminoalkyl)xylans, *Papier*, 1997, **51**, 223–226.
158. A. Ebringerová, A. Belicová and L. Ebringer, Antimicrobial activity of quaternized heteroxylans, *J. Microbiol. Biotechn.*, 1994, **10**, 640–644.
159. M. Antal, A. Ebringerová and I. Simkovic, New aspects in cationization of lignocellulose materials. I. Preparation of lignocellulose materials

containing quaternary ammonium groups, *J. Appl. Polym. Sci.*, 1984, **29**, 637–642.

160. M. Antal, A. Ebringerová and I. Simkovic, New aspects in cationization of lignocellulose materials. II. Distribution of functional groups in lignin, hemicellulose, and cellulose components, *J. Appl. Polym. Sci.*, 1984, **29**, 643–650.
161. A. Ebringerová, M. Antal and I. Simkovic, New aspects in cationization of lignocellulose materials. III. Influence of delignification on reactivity and extractability of TMAHP-hemicelluloses, *J. Appl. Polym. Sci.*, 1986, **31**, 303–308.
162. A. Ebringerová, A. Belicova and L. Ebringer, Antimicrobial activity of quaternized heteroxylans, *World J. Microbiol. Biotechnol.*, 1994, **10**, 640–644.
163. K. Schwikal, Th. Heinze, A. Ebringerova and A. Petzold, Cationic xylan derivatives with high degree of functionalization, *Macromol. Symp.*, 2006, **232**, 49–56.
164. K. Schwikal, Dissertation, FSU Jena, 2007.
165. D. A. Drazenovich, A. Kaya, W. G. Glasser, K. Schwikal, Th. Heinze, A. R. Esker, Hydroxypropyltrimethylammonium-xylan adsorption studies onto model surfaces via surface plasmon resonance spectroscopy, 233rd ACS National Meeting 2007, March 25–29.
166. K. Schwikal, Th. Heinze, Dialkylaminoethyl xylans: polysaccharide ethers with pH-sensitive solubility, *Polym. Bull.*, in press.
167. S. Thiebaud and M. E. Borredon, Analysis of the liquid fraction after esterification of sawdust with octanoyl chloride – Production of esterified hemicelluloses, *Bioresour. Technol.*, 1998, **63**, 139–145.
168. J. M. Fang, R. C. Sun, P. Fowler, J. Tomkinson and C. A. S. Hills, Esterification of wheat straw hemicelluloses in the N,N-dimethylformamide/lithium chloride homogeneous system, *J. Appl. Polym. Sci.*, 1999, **74**, 2301–2311.
169. R. C. Sun, J. M. Fang, J. Tomkinson and C. A. S. Hill, Esterification of hemicelluloses from poplar chips in homogeneous solution of N,N-dimethylformamide/lithium chloride, *J. Wood Chem. Technol.*, 1999, **19**, 287–306.
170. R. C. Sun, J. M. Fang and J. Tomkinson, Stearoylation of hemicelluloses from wheat straw, *Polym. Degrad. Stabil.*, 2000, **67**, 345–353.
171. R. C. Sun, J. M. Fang, J. Tomkinson and G. L. Jones, Acetylation of wheat straw hemicelluloses in N,N-dimethylacetamide/LiCl solvent system, *Ind. Crop. Prod.*, 1999, **10**, 209–218.
172. R. C. Sun, J. M. Fang, J. Tomkinson, Z. C. Geng and J. C. Liu, Fractional isolation, physicochemical characterization and homogeneous esterification of hemicelluloses from fast-growing poplar wood, *Carbohydr. Polym.*, 2001, **44**, 29–39.
173. R. C. Sun, X. F. Sun and X. J. Bing, Succinylation of wheat straw hemicelluloses with a low degree of substitution in aqueous systems, *J. Appl. Polym. Sci.*, 2002, **83**, 757–766.

174. X. F. Sun, R. C. Sun, J. Tomkinson and M. S. Baird, Preparation of sugarcane bagasse hemicellulosic succinates using NBS as a catalyst, *Carbohydr. Polym.*, 2003, **53**, 483–495.
175. X. F. Sun, R. C. Sun and J. X. Sun, Oleoylation of sugarcane bagasse hemicelluloses using N-bromosuccinimide as a catalyst, *J. Sci. Food Agr.*, 2004, **84**, 800–810.
176. M. Vincedon, Xylan derivatives: aromatic carbamates, *Makromol. Chem.*, 1993, **194**, 321–328.
177. Y. Okamoto, J. Noguchi and E. Yashima, Enantioseparation on 3,5-dichloro- and 3,5-dimethylphenylcarbamates of polysaccharides as chiral stationary phases for high-performance liquid chromatography, *React. Funct. Polym.*, 1998, **37**, 183–188.
178. C. M. Buchnanan, N. L. Buchanan, J. S. Debenham, P. Gatenholm, M. Jacobsson, M. C. Shelton, T. L. Watterson and M. D. Wood, Preparation and characterization of arabinoxylan esters and arabinoxylan ester/cellulose ester polymer blends, *Carbohydr. Polym.*, 2003, **52**, 345–357.
179. R. K. Jain, M. A. Sjöstedt and W. G. Glasser, Thermoplastic xylan derivatives with propylene oxide, *Cellulose*, 2001, **7**, 319–336.
180. W. G. Glasser, R. K. Jain and M. A. Sjöstedt, 5430142 Thermoplastic pentosan-rich polysaccharides from biomass, *Biotechnol. Adv.*, 1996, **14**, 605.
181. St. Hesse, T. Liebert and Th. Heinze, Studies on the film formation of polysaccharide based furan-2-carboxylic acid esters, *Macromol. Symp.*, 2006, **232**, 57–67.
182. Th. Heinze, T. Liebert and A. Koschella, *Esterification of Polysaccharides*, Springer Verlag, Heidelberg, 2006.
183. T. Liebert, St. Hornig, St. Hesse and Th. Heinze, Microscopic visualization of nanostructures of cellulose derivatives, *Macromol. Symp.*, 2005, **223**, 253–266.
184. T. Liebert and Th. Heinze, Tailored Cellulose Esters: Synthesis and Structure Determination, *Biomacromolecules*, 2005, **6**, 333–340.
185. Th. Heinze, K. Petzold and St. Hornig, Novel Nanoparticles based on xylan, *Cellulose Chem. Technol.*, 2007, in press.
186. S. Matsumura and S. Yoshikawa, Biodegradable poly(carboxylic acid) design. In *Agricultural and Synthetic Polymers Biodegradability and Utilization Development, ACS Symposium Series 433*, (Eds) J. E. Glass and G. Swift, American Chemical Society, Washington DC, 1990, pp. 124–135.
187. S. Matsumura, S. Maeda and S. Yoshikawa, Molecular design of biodegradable functional polymers, 2, Poly(carboxylic acid) containing xylopyranosediyl groups in the backbone., *Makromol. Chem.*, 1990, **191**, 1269–1274.
188. S. Matsumura, M. Nishioka and S. Yoshikawa, Enzymatically degradable poly(carboxylic acid) derived from polysaccharide, *Makromol. Chem. Rapid Commun.*, 1991, **12**, 89–94.

189. R. Andersson, J. Hoffman, N. Nahar and E. Scholander, An NMR study of the products of oxidation of cellulose and (1→4)-β-D-xylan with sodium nitrite in orthophosphoric acid, *Carbohydr. Res.*, 1990, **206**, 340–346.
190. T. J. Painter, Preparation and periodate oxidation of C-6-oxycellulose: conformational interpretation of hemiacetal stability, *Carbohydr. Res.*, 1977, **55**, 95–103.
191. Th. Heinze, D. Klemm, M. Schnabelrauch and I. Nehls, Properties and following reactions of homogeneously oxidized cellulose. In *Cellulosics: Chemical*, (Eds) J. F. Kennedy, G. O. Phillips and P. A. Williams, Biochemical and Material Aspects, Horwood, London, 1993, pp. 340–355.
192. J. R. Jerez, B. Matsuhiro and C. C. Urzúa, Chemical modifications of the xylan from, *Palmaria decipien*, *Carbohydr. Polym.*, 1997, **32**, 155–159.
193. N. P. Barroso, J. Costamagna, B. Matsuhiro and M. Villagran, El xilano de, *palmaria decipiens*: Modificacion quimica y formacion de un complejo de Cu(II), *Bol. Soc. Chil. Quim.*, 1997, **42**, 301–306.

CHAPTER 5

Starch and its Derived Products: Biotechnological and Biomedical Applications

JOHN F. KENNEDY,*[a] CHARLES J. KNILL,[a] LIU LIU[a] AND PARMJIT S. PANESAR[a,b]

[a] Chembiotech Laboratories, Institute of Advanced Science & Technology, Kyrewood House, Tenbury Wells, Worcestershire, WR15 8SG, UK;
[b] Department of Food Technology, Sant Longowal Institute of Engineering & Technology, Longowal 148 106, Punjab, India

5.1 Introduction

Starches are the principal food reserve polysaccharides in plants and are the major source of carbohydrates in the human diet, traditionally supplying ~70–80% of consumed calories.[1] Starch is present in all staple foods, *e.g.* wheat (*Triticum vulgare*), maize (*Zea mays*), rice (*Oryza sativa*), potato (*Solanum tuberosum*), sago (*Metroxylon* palm species), tapioca/cassava (*Manihot esculenta*), rye (*Secale cereale*), barley (*Hordeum vulgare*), oats (*Avena sativa*), *etc.*[2–12] Other, less common sources of starch include sorghum (*Sorghum bicolour*), yam/sweet potato (*Disoscorea* species), and arrowroot (*Maranta arundinacea*), and a host of other tubers/roots.[9–15] Starch, taken from grains, tubers and roots, has been consumed as food and feed for centuries, and is of great economic importance, being isolated on an industrial scale from many sources. Wheat is the leading cereal grain produced, consumed and traded in the world, followed closely by rice and maize.[16] The industrial uses of starch arise from its unique character since it can be used directly as intact granules, in the swollen granular

RSC Polymer Chemistry Series No. 1
Renewable Resources for Functional Polymers and Biomaterials
Edited by Peter A. Williams

Published by the Royal Society of Chemistry, www.rsc.org

state, in the dispersed form, as a film dried from a dispersion, as an extrudate powder, after controlled partial hydrolysis to a mixture of oligosaccharides, after hydrolysis and isomerisation to glucose and fructose syrups, or after chemical modification.[17]

5.1.1 Composition and Structure

Starch is a mixture of two D-glucan homopolymers, composed of α-D-glucopyranosyl (α-D-Glc*p*) units, namely amylose and amylopectin. Amylose is essentially a linear polymer consisting of (1→4)-linked α-D-Glc*p* units (Figure 5.1), although some (1→6)-linkages are thought to exist about every 180–400 units.[9,11,12,18–31] The abundance of hydroxyl groups makes amylose hydrophilic, and its uniform linear-chain nature permits crystallisation from solution as well as in a semi-solid state in films and coatings. Both the linearity and the hydroxyl groups cause amylose to orient in parallel thus permitting molecular re-association and hydrogen bonding between hydroxyl groups of adjacent chains forming a network of junction zones between molecules. With time this may result in the formation of a three-dimensional gel network held together by hydrogen bonding wherever close alignment has occurred, resulting in opacity/cloudiness in amylose solutions known as 'retrogradation' or 'set-back'.[4,9,21,22,28,29,32,33] Ultimately, under favourable conditions, a crystalline order appears. Thus, amylose solutions are unstable and firm resilient gels can result, and crystalline regions can thus develop, providing gel strength (rigidity) and stability, which may require significant heating to reverse.[34,35] The degree of polymerisation (DP) of amylose is ~200–20 000, giving a molecular weight of ~30–3200 kDa.[1,9,11,12,18,21,22,29,31,36,37] The molecular weight of rye amylose was determined as being ~218 kDa, which is somewhat lower than for maize starch

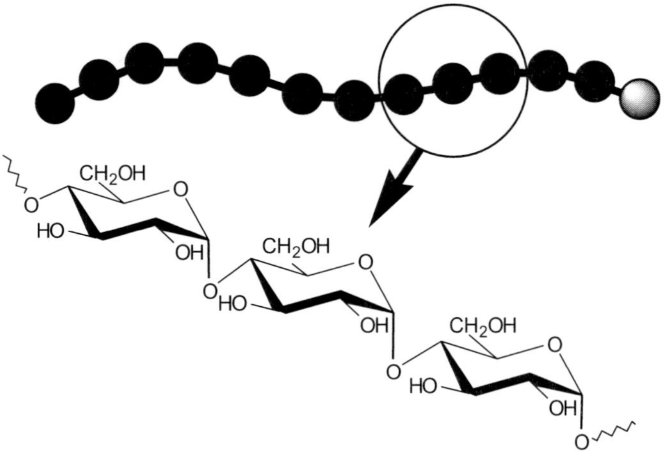

Figure 5.1 Basic structure of amylose.

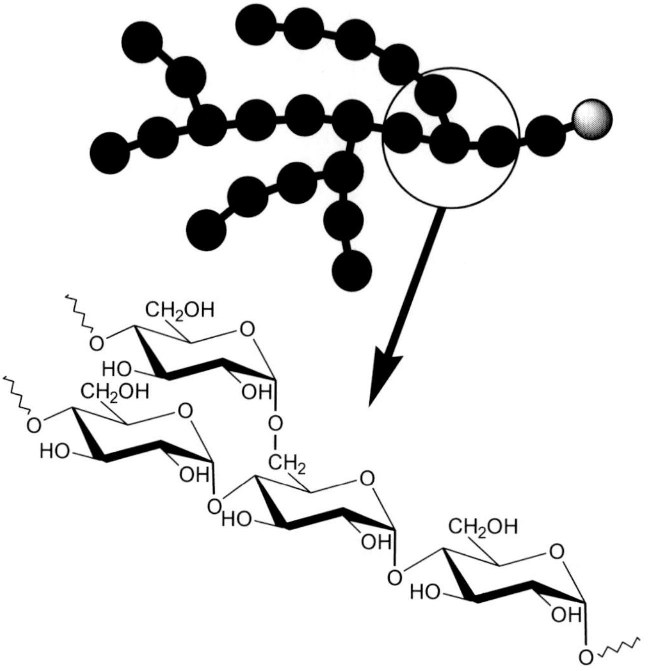

Figure 5.2 Basic structure of amylopectin.

amylose (~250 kDa) and wheat and triticale starch amylose (~260–285 kDa).[38] Some potato starch amylose fractions are of the order of 1000 kDa.[12,35]

Amylopectin is a highly branched form of amylose, consisting of (1→4)-linked α-D-Glc*p* units and a significant proportion of 1,4,6-tri-O-substituted residues acting as (1→6)-linkage branch points (Figure 5.2).[9,11,12,18,20,22,24,28–31,39] Several models have been proposed for amylopectin to account for its physicochemical properties. These include the laminated structure, the 'herringbone' model, the randomly branched structure, the tassel-on-a-string representation, and the cluster model (also known as the racemose or grape structure).[11,12,20,22,36,40] Amylopectin chains can be classified into three types, namely A, B and C chains. A chains are linked to the molecule only through their potential reducing end, while B chains also carry other chains, and C chains carry the single reducing end group (Figure 5.3).[9,22,30,36,41] The linear outer branches of amylopectin (A chains) can also participate in gel formation, since they are essentially like small amylose fragments. For relatively large molecular weights, each model has a characteristic ratio of A : B chains.[36] The initial determination of the A : B ratio favoured the randomly branched structure,[42] with most natural starches having an A : B chain ratio in the range of 1.1–1.5.[43]

The degree of branching and chain lengths between branch points in amylopectin varies not only with the source of the sample but also within each sample.[21] Rye amylopectin has been shown to have branch points about every 19–21 glucose units,[38,44] which is similar to wheat, triticale and other

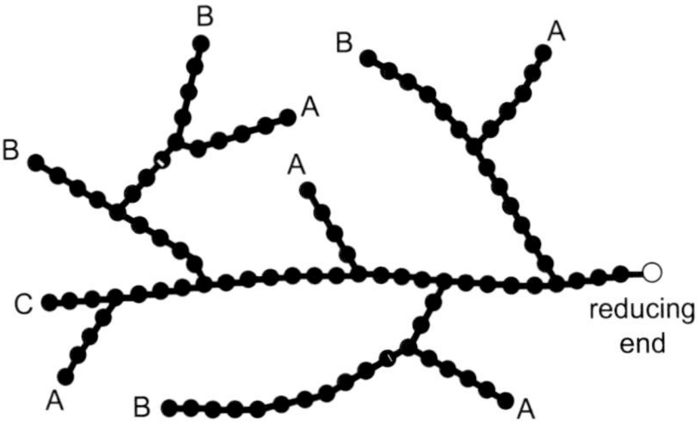

Figure 5.3 Representation of A, B and C chains in amylopectin.

amylopectins (every ~20–25 glucose units).[25,38] Hydrolysis of starch with a debranching enzyme (such as isoamylase) and measurement of the molecular weight of the resulting linear chains by gel permeation chromatography (GPC) can give an excellent profile of branch lengths. Amylopectin molecules are thus very large, with molecular weights ranging from 10 000 to over 500 000 kDa.[1,9,28,35–37]

The amylose : amylopectin ratio varies according to the source of the starch and its maturity, but is ~1 : 3 for most general starches. One unusual genetic variety of maize arose in China; its starch granules contained essentially only amylopectin (<2% amylose). When the maize kernel was cut with a knife the cut surface appeared shiny as though it contained wax, and the maize was referred to as 'waxy maize'. This was developed into a high-yielding hybrid in the US.[30,45] Waxy sorghum and glutinous rice are similar. Oppositely, high-amylose maize (amylomaize) was also genetically developed in the US with starch amylose contents of 60–80% produced commercially.[30,42] High amylose content mutant cultivars of pea and barley also exist. The majority of starches contain ~15–35% amylose, *e.g.* the amylose content of potato, rice and rye starches are ~18–23%, ~16–30% and ~24–26%, respectively.[38,46] Most natural starches contain both amylose and amylopectin; however, in some starches significant amounts of an intermediate that seems to be either a less branched amylopectin or a slightly branched amylose are also present. This intermediate is typical for many cereal starches but is not found in potato starch.[36,47,48] Besides water, amylose and amylopectin, natural starches also contain some minor constituents (*e.g.* protein, lipids, *etc.*). In potato starch esterified phosphate groups are present in the amylopectin fraction, attached to ~1 in every 215–560 anhydroglucose units, with most (~60–70%) being attached *via* the C6 hydroxyl group and the other third *via* the C3 hydroxyl, and ~90% of them attached to B chains (Figure 5.3).[1,42]

5.1.2 Physicochemical Characteristics

Nature has chosen the starch granule as an almost universal form for packaging and storing carbohydrates in green plants. Starch granules are quasi-crystalline and cold water-insoluble. The shape of the granules is somewhat characteristic

Table 5.1 Physicochemical characteristics of starch granules of various botanical origin.[a]

Starch	*Type*	*Diameter/ μm*	*Morphology*	*Gelatinisation temp./°C*	*Pasting temp./°C*	*Amylose (%)*
Wheat	cereal	1–55	round, lenticular	52–85	77	25–28
Maize	cereal	2–30	round, polygonal	62–72	80	25–28
Waxy maize	cereal	2–30	round, polygonal	63–72	74	<2
Amylo-maize	cereal	2–30	polygonal, irregular, elongated	63–170	>90	50–90
Rice	cereal	1–9	polygonal, spherical	68–78	81	19
Potato	tuber	5–100	oval, spherical	58–68	64	20–21
Sago	pith	15–65	oval, truncated	69–74	74	26
Tapioca	root	4–35	oval, truncated	52–73	63	17

[a]Data obtained from ref. 1, 4, 6, 9, 11, 14, 20, 28, 37 and 41.

of the source of the starch, and they range in size from sub-micron elongated granules to oval granules well over 100 μm.[4,12,29] A summary of starch granule dimensions is provided in Table 5.1.[1,4,6,9,11,14,20,28,31,41] The shapes of the granules include nearly perfect spheres, typical for small wheat starch granules; discs, typical for the large granules of wheat and rye; polyhedral granules, as in rice and maize; 'oyster shell' (oval, egg-shaped, ellipsoidal) irregular granules as often found in potato starch; and highly elongated irregular filamentous granules as in high-amylose maize starch.[49] In the case of potato starch there is a gradual change from an oval towards a spherical shape with decreasing granule diameter. As in wheat and barley, rye starch consists of populations of large and small granules.[44] The large granules increase in size during maturation, and at maturity vary in size from 10 to >35 μm in diameter. The distribution curve is broader and the maximum diameter larger than in wheat starch. The small starch granules accumulate relatively late in development and have diameters <10 μm.[50] Wheat starch granules are lenticular and have a bimodal or trimodal size distribution (1–5, 5–14 and >14 μm).[1]

Amylose is found naturally in three crystalline modifications designated A (cereal), B (tuber) and C (smooth pea and various beans).[24,30,36] Cereal starches (and also small starch granules of some tropical tubers) give the A-crystalline pattern (monoclinic lattice) and thus possess densely packed crystallites.[51] Potato starch and certain other tropical tuber starches, which are morphologically similar with respect to their granule shapes and sizes, as well as some amylose-rich starch granules (amylomaize, barley and wrinkled pea) have a B-crystalline pattern (hexagonal lattice). They, therefore, contain much more water than type A starches.[52,53] Starch granules from other tropical tubers and seeds as well as most legume starches possess the C pattern.[53,54] Precipitated amylose complexes (with iodine, long-chain alcohols and fatty acids), adopt the so-called V-structure (Verkleisterung) (often associated with the A, B or C patterns), which can appear after gelatinisation, although such patterns are also reported to exist in native starches.[30,53] The so-called amylose linearity is further complicated by a twisting of the chain into a helix, and it is the different degrees of hydration of the helix that give rise to the A, B and C forms.[37] B-amylose helices contain 6 α-D-Glc*p* units per unit cell (helical turn) and 3–4 molecules of water of hydration, with left-handed helices being slightly more energetically favorable.[42] Starch granules are made up of amylose and/or amylopectin molecules arranged radially. They contain both crystalline and amorphous (non-crystalline) regions in alternating layers. The clustered branches of amylopectin occur as packed double helices. It is the packing together of these double-helical structures that forms the many small crystalline areas comprising the dense layers in starch granules that alternate with less dense amorphous layers. Amylose molecules occur among the amylopectin molecules. The origin of the growth of the granule is called the hilum.[1] A range of single- and double-stranded helices have been identified and characterised from native starches.[48] Starch chains usually show a high proportion of local glycosidic conformations characteristic of V-type structures[55] that can be stabilised by complexation with iodine, fatty acids, monoglycerides, *etc.* An example of the arrangement/interactions of amylose/amylopectin in starch is provided (Figure 5.4).

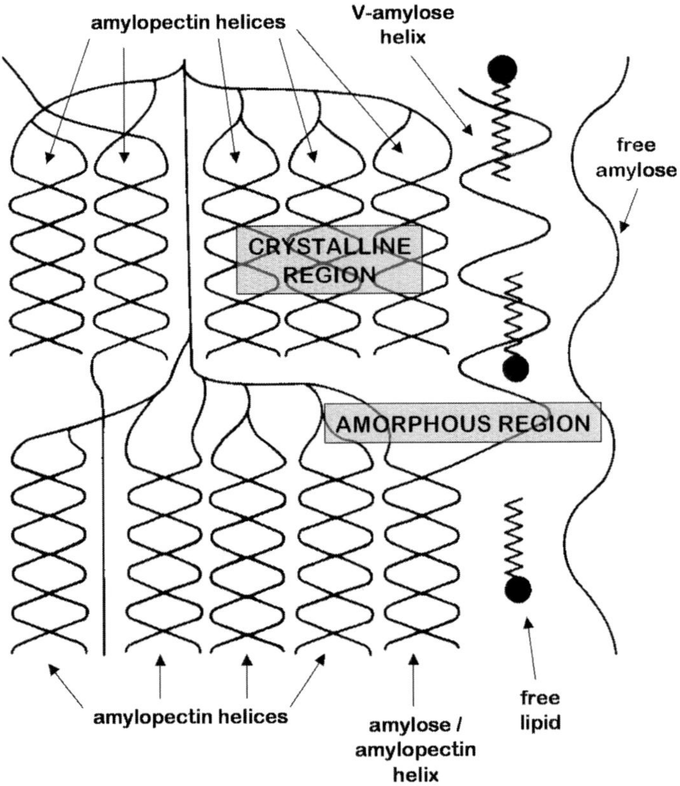

Figure 5.4 Example of amylose/amylopectin interactions in a starch granule.

Double helical structures are found for part of the amylopectin component within granules (and are formed from both amylose and amylopectin after gelatinisation). A minimum of 10 glucose residues is required for double helix formation,[56] although chains as short as 6 residues can co-crystallise.[57] The outer branch length of amylopectin is the major determinant of which double helical polymorph is found in native starch granules. Relatively short branches lead to A-type with longer branches giving B-type order.[36] A- and B-type polymorphs have very similar individual helical structures but differ in packing arrangements.[56] Amylose can be leached from many starches leaving the amylopectin as well as the granule crystallinity largely intact, which has led to the conclusion that amylose is mainly in the amorphous phase.[20,43,48] Conversely, amylopectin is the main component of the crystalline fraction. However, some amylopectin is homogeneously mixed with the amylose in the amorphous phase.[12,58] There is no sharp demarcation between the crystalline and amorphous phases of starch granules, and it is generally believed that some or all of the starch molecular chains run continuously from one phase to another.[59]

Native starch granules have a crystallinity varying from 15 to 45%,[58] thus, crystallinity is not the principal mode of organisation of the starch granule polymers. Starch granules can therefore be described as being made up of alternating semi-crystalline and crystalline shells that are between 120 and 400 nm thick.[59] The level of helical order in starch granules is often significantly greater than the extent of crystalline order. Consequently, it appears that much of the amylopectin in semi-crystalline shells is in the double helical form, although it is not crystalline.[54,60] Considerable evidence exists to indicate that the crystalline shells consist of alternating amorphous and crystalline 'lamellae', which are approximately 9–10 nm thick.[61] These lamellae are believed to represent the crystalline (side chain clusters) and amorphous regions (branching regions) of the amylopectin molecules.[59]

Undamaged starch granules are insoluble in cold water, but can imbibe water reversibly; that is, they can swell slightly, and then return to their original size on drying. As the temperature is increased the process becomes irreversible and eventually the granule bursts to form a starch paste, and a rapid onset in the development of viscosity is seen. In the initial phase of this heating process, swelling begins in the least organised amorphous, inter-crystalline regions of the granule.[9,11,20,33,41,59] This results in porous amylopectin-based granules suspended in a hot amylose solution.[24] Further heating leads to uncoiling or dissociation of double helical regions and disappearance of the amylopectin crystallite structure, and an increase in viscosity occurs. Upon continued heating and hydration the granule weakens to the point where it can no longer resist mechanical or thermal shearing, and a starch paste results.[29] Pasting is defined as the phenomenon following gelatinisation in the dissolution of starch. It involves granular swelling, exudation of molecular components from the granule, and eventually, total disruption of the granules.[11,19] A starch paste consists of a continuous phase of solubilised amylose and/or amylopectin and a discontinuous phase of granule remnants (granule ghosts and fragments). A granule ghost consists of the outer insoluble envelope of the granule (but is not a membrane).[62] Complete molecular dispersion can only be accomplished under conditions of high temperature, high shear, and excess water, which are seldom, if ever, encountered in most applications.[1] However, in many instances of processing, what is required is the breakdown of the granule structure *via* gelatinisation.[63,64]

Not all starch granules in a sample burst at the same temperature, but the range of temperature of gelatinisation is characteristic of starch from a particular source. Starch gelatinisation can be defined as the collapse (disruption) of molecular orders within the starch granule manifested in irreversible changes in properties such as granular swelling, native crystalline melting, and starch solubilisation. The point of initial gelatinisation and the range over which it occurs is governed by starch concentration, method of observation, granular type, and heterogeneities within the granule population under observation.[19–21] The length of double helices is proposed to be a determinant of gelatinisation temperature. Amylose double helices probably occur over a length scale of ~40–80 residues and melt at ~150 °C.[65,66] Typical amylopectin-based double helices occur over ~15–20 residues and melt at ~60–80 °C. Gelatinised starches/starch pastes can undergo retrogradation on cooling. A typical rheological profile for the gelatinisation/pasting/retrogradation of a potato starch is displayed in Figure 5.5, and shows the previously detailed changes in viscosity during initial rapid heating, maintenance at a constant temperature, and controlled cooling. Rheological measurements are routinely performed on starches of different botanical origin and their derivatives to compare and contrast their pasting/gelling characteristics and assess their suitability for specific applications.[4,13,28,67–77]

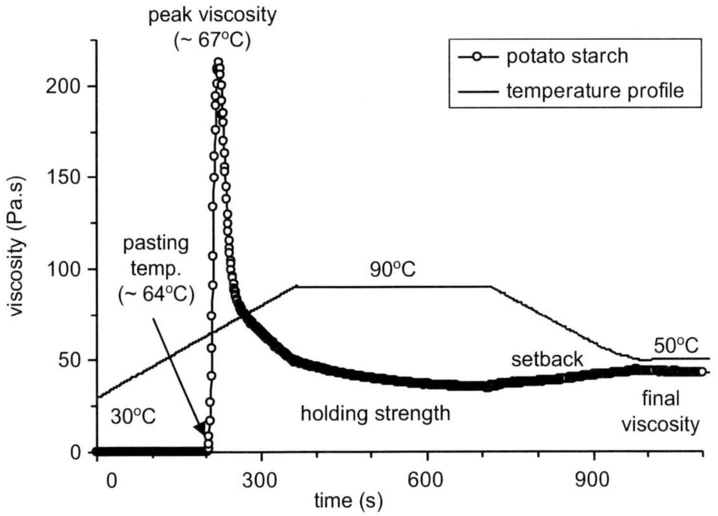

Figure 5.5 Typical rheological profile for potato starch gelatinisation (2.5% w/w aqueous dispersion analysed using a TA Instruments AR1000 rheometer).

5.2 Biotechnological Production of Starch Hydrolysis Products

The most diverse and numerous enzymes for carbohydrate hydrolysis and modification are those that act on starch (Figure 5.6), which can be utilised to produce a vast range of hydrolysis products.[78–81] Enzymes that are capable of catalysing the hydrolysis of the (1→4)-linkages in amylose/amylopectin are called amylases and are widely produced by plants, bacteria, fungi and animals.[82] In mammals amylases are mainly produced by the salivary glands and the pancreas.[83] α-Amylase (1,4-α-D-glucan glucanohydrolase, EC 3.2.1.1) is an endoglycosidase, attacking glucans away from the chain ends at an internal glycosidic bond and producing a rapid drop in viscosity (Figure 5.6).[1,36,41,82,84–88] Varying types of oligosaccharides are produced, characteristic of the type of α-amylase. Traditionally, α-amylase has been obtained from *Aspergillus oryzae*, but the enzymes obtained from various thermophilic *Bacillus* species (*e.g. Bacillus amyloliquefaciens* and *Bacillus licheniformis*) have the advantage of temperature stability.[12,84,87,89,90]

Dextrins produced by α-amylase action can be further processed by a variety of enzymes including β-amylases for the production of maltose syrups and glucoamylases for the production of glucose syrups. β-Amylase (1,4-α-D-glucan maltohydrolase, EC 3.2.1.2) attacks amylose/amylopectin in an *exo* fashion from the non-reducing ends, releasing β-maltose and a high molecular weight limit dextrin when the enzyme reaches a (1→6)-linkage in amylopectin (Figure 5.6).[12,36,41,83–85,87,88] β-amylases occur widely in many plants, with barley, wheat, sweet potatoes and soybeans being common sources.[86]

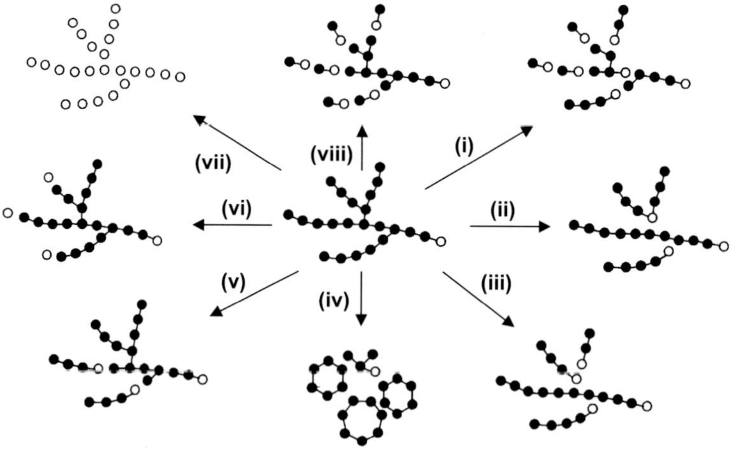

Figure 5.6 Visual overview of the major starch hydrolysing enzymes: (i) α-amylase; (ii) pullulanase; (iii) isoamylase; (iv) cyclodextrin D-glucotransferase; (v) exo-(1,4)-α-D-glucanase; (vi) α-D-glucosidase; (vii) glucoamylase; (viii) β-amylase.

Glucoamylase (γ-amylase; amyloglucosidase; glucan 1,4-α-glucosidase; 1,4-α-D-glucan glucohydrolase, EC 3.2.1.3) occurs almost exclusively in fungi and attacks amylose/amylopectin in an *exo* fashion from the non-reducing ends, releasing β-D-glucopyranose (Figure 5.6). It does not produce limit dextrins as it can catalyse hydrolysis of both (1→4)- and (1→6)-linkages (although at very different rates) and can therefore convert starch to glucose.[12,36,41,83–88] Glucoamylases are produced by a variety of organisms, although commercial products are obtained principally from *Aspergillus* or *Rhizopus*.[91]

α-D-Glucosidase (α-D-glucoside glucohydrolase; EC 3.2.1.20) is produced as an intracellular, cell-bound or extracellular enzyme by many fungi (*Aspergillus niger*), yeasts (*Saccharomyces cerevisiae*) and bacteria (*Bacillus* species).[85,86,88] α-D-Glucosidases act by exo-hydrolysis on the terminal (1→4)- and (1→6)-linkages of disaccharides to liberate α-D-glucose (Figure 5.6). Polysaccharides are attacked very slowly, if at all; however, the enzyme can work on starch in conjunction with other enzymes that liberate oligosaccharides.[86,88]

Isoamylases or debranching enzymes hydrolyse the (1→6)-linkages in amylopectin (Figure 5.6). These include amylo-1,6-glucosidase (dextrin 6-α-D-glucosidase; EC 3.2.1.33), pullulanase (amylopectin 6-glucanohydrolase; α-dextrin endo-1,6-α-glucosidase; EC 3.2.1.41), and isoamylase (glycogen 6-glucanohydrolase; EC 3.2.1.68.[12,36,41,84–88] The precise location of the (1→6)-linkages in the substrate is of the utmost importance since it affects the ability of the various strains of the enzymes to act on different substrates. For example, pullulanase from *Bacillus macerans* has very little action on amylopectin but is able to degrade its β-limit dextrin, whilst the pullulanase from *Aerobacter aerogenes* almost totally debranches amylopectin and its β-limit dextrin.[86] Amylo-1,6-glucosidase hydrolyses terminal (1→6)-linkages in limit dextrins.[86]

5.2.1 Maltodextrins

Commercial starch hydrolysates are classified in terms of dextrose equivalence (DE). Maltodextrins are defined as non-sweet starch hydrolysates that consist of α-D-glucose units linked primarily by (1→4) glycosidic linkages with a DE of 3–20.[1,84] DE is defined as the percentage of reducing sugar calculated as dextrose (glucose) on a dry weight basis. Therefore, in terms of molecular size, maltodextrins bridge the gap between starch and mono-/di-saccharides.[92] Starch hydrolysates with a DE greater than 20 are designated as various kinds of syrups depending on their source and composition. In general maltodextrins are fully water-soluble (some are even cold water soluble) carbohydrates of low bulk density, and are metabolised similarly to starch. They have very little or no sweetness and a bland, not starchy flavour that does not mask other flavours.[93] Due to these properties maltodextrins have considerable applications in the food industry, particularly in convenience and processed foods.

The production of maltodextrins is, by definition, achieved by the hydrolysis of starch down to glucose polymers with an average chain length of 5–10 glucose units per molecule. The properties of maltodextrins are controlled by their DE and DP, which change with the degree of hydrolysis.[84,92,94] Theoretically they can be produced by either enzymic or acidic hydrolysis; however, in practice acid hydrolysis produces too much free glucose (and large fragments) and maltodextrins thus produced have a strong tendency to retrograde. Linear starch fragments in low DE acid hydrolysates large enough to re-associate form insoluble aggregates causing hazy solutions, which are undesirable for some applications.[95,96] Thus commercially, they are invariably prepared from starch by controlled enzymic hydrolysis.

A starch slurry is initially liquefied by heating (70–90 °C) at neutral pH in the presence of a bacterial α-amylase to a DE of 2–15. The liquid starch hydrolysate is then autoclaved (110–115 °C) to gelatinise any remaining insoluble starch and, on cooling, subjected to further enzymic treatment to reach the desired DE. Some hydrolysis schemes employ both acid- and enzyme-catalysed hydrolysis. Initial acid-catalysed hydrolysis of a starch slurry to a DE of 5–15 is followed by neutralisation and further hydrolysis with a bacterial α-amylase (*e.g.* from *Bacillus subtilis* or *Bacillus mesentericus*) to produce maltodextrins that are haze-free and exhibit no retrogradation upon storage.[93,97] Such systems overcome the problem of hazy solutions and result in maltodextrins with low hygroscopicity and high water solubility. The source and variety of the starch is relatively unimportant for these processes. The final product is concentrated in vacuum evaporators to give finished syrups containing about 75% solids or, more regularly, is spray dried to a white powder containing ~3–5% moisture. The dextrinisation reaction is allowed to progress until the required DE product is obtained. Low DE value maltodextrins will tend to retrograde in solutions whereas those with higher DE values will form less viscous solutions and will exhibit increasing sweetness. Two types of maltodextrin are in commercial use: those ranging from about 10–14 DE and those ranging from about 15–19 DE. The compositions of these products depend not only on DE, but also on the method of hydrolysis employed in their manufacture. The saccharide component profiles of maltodextrins obtained by acid-catalysed hydrolysis are somewhat different from those obtained by enzyme- or acid/enzyme- catalysed hydrolysis, even though they can possess the same DE. Acid hydrolysates tend to contain greater proportions of higher molecular weight dextrins and glucose. The former makes them less water-soluble and they easily retrograde.[93]

5.2.2 Glucose and Fructose Syrups

Glucose syrup is a refined, concentrated aqueous solution of D-glucose, maltose and other polymers of D-glucose obtained by controlled partial hydrolysis of edible starch, with a DE of 20–100.[21,79,84,93,98] Acid conversion is reproducible though random in its attack on starch and, though little or no influence can be made on the individual sugar composition/spectrum, it is possible by controlling temperature, time and acid level to make products of reasonably constant composition for a given degree of hydrolysis. In products where lower degrees of hydrolysis are required to give thickening, viscosity, body, *etc.* without appreciable sweetness it is not possible to use acid hydrolysis to produce materials in which every starch molecule has been reduced in molecular weight sufficiently to give a non-retrograding, clear, stable solution. Enzymic hydrolysis is therefore utilized, particularly α-amylase, glucoamylase, and often pullulanase.[79,84,93,99]

Judicious use of fungal α-amylase and glucoamylase can give rise to syrups with various ratios of glucose, maltose and higher oligosaccharides. These high conversion syrups generally contain about 35–43% glucose, 30–40% maltose and 8–15% maltotriose. It is important that these syrups have a high DE and yet be sufficiently stable not to crystallise at temperatures down to 4 °C at ~80–82% dry substance.[89] Glucoamylases are also used to convert acid liquefied starch hydrolysates of a moderate degree of hydrolysis under mild conditions into products which have DE levels >90. Likewise, maltose syrups can be produced using β-amylase.[100]

A significant development in the production of syrups was the introduction of advanced enzyme engineering to convert high-purity D-glucose (derived from starch) using glucose isomerase (EC 5.3.1.18) into a mixture of D-glucose and D-fructose equivalent to invert sugar from sucrose, thereby opening to the starch industry the previously unavailable but enormous sweeteners market.[93,98,101,102] High fructose syrups (HFSs) are produced commercially in many countries, due to their extensive use in the processed food and beverage industries. HFSs are the best alternative to sucrose where starchy resources are abundant and cheap. A range of sweeteners can be derived from starch using commercial-scale biotechnological processes (Figure 5.7).[103]

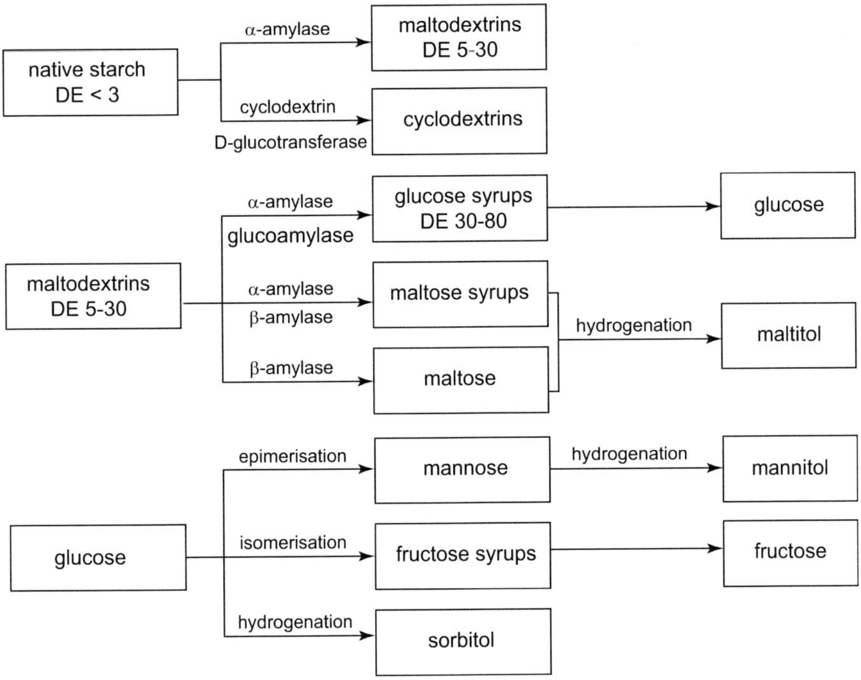

Figure 5.7 Commercial sweeteners biotechnologically derived from starch.

5.2.3 Cyclodextrins

Cyclodextrin D-glucotransferase (1,4-α-D-glucan 4-α-D-[1,4-α-d-glucano]-transferase cyclising; EC 2.4.1.19), commonly isolated from *Bacillus macerans* and *Bacillus megaterium*, is capable of hydrolysing starch to a series of non-reducing cyclomalto-oligosaccharides called cyclodextrins (previously referred to as Schardinger dextrins or cycloamyloses) (Figure 5.6).[36,41,86,104] The structure of α-cyclodextrin, and the molecular dimensions of α-, β- and γ-cyclodextrins, composed of 6, 7 and 8 α-D-Glc*p* units respectively are presented (Figure 5.8).[1,84,105,106]

As a consequence of cyclodextrin conformation, all secondary hydroxyl groups are located on one side of the torus-like cyclodextrin molecule, while all primary hydroxyl groups are on the other side. The 'lining' of the internal cavity is formed by hydrogen atoms and glycosidic oxygen-bridge atoms, making this surface slightly apolar. Molecules, or functional groups of molecules, having molecular dimensions that correspond to the cyclodextrin cavity, being less hydrophilic than water, can be included in the cyclodextrin cavity if both components are dissolved in water. In solution the slightly apolar cyclodextrin cavity is occupied by water molecules, which is energetically unfavourable (polar–apolar interactions) and they are therefore readily substituted by appropriate 'guest molecules' which are less polar than water. This often advantageously modifies the various physical and chemical properties of the included/encapsulated molecules, is simpler and cheaper than many other methods of encapsulation, and powder-like crystalline inclusion complexes can be isolated.[93,105,107–110.]

In the pharmaceutical industry cyclodextrins are utilised as an auxiliary substance for improving the stability and the absorption qualities of active ingredients in tablets.[111–114] This makes the production and introduction of numerous active substances possible whose stability, compatibility or

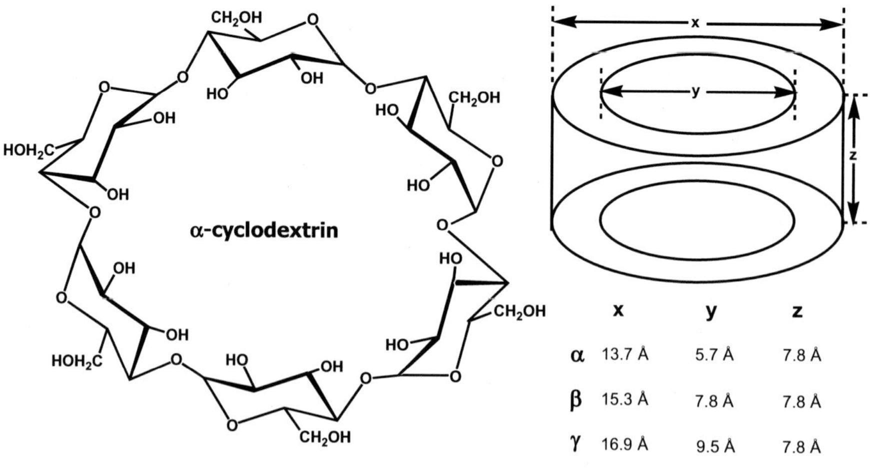

	x	y	z
α	13.7 Å	5.7 Å	7.8 Å
β	15.3 Å	7.8 Å	7.8 Å
γ	16.9 Å	9.5 Å	7.8 Å

Figure 5.8 Cyclodextrin structure and cavity dimensions.

absorption features prevented their use. Volatile compounds can be stabilised without losses through evaporation; liquid compounds can be transformed into crystalline, compressible forms; solubility in water as well as the rate of dissolution of poorly soluble substances can be increased; bad taste and smell can be masked by complex formation; and easily oxidisable substances can be protected against atmospheric oxidation.[93]

5.3 Chemical Modification

The manufacture of starch employs a variety of extraction processes that isolate purified starch from the other constituents of the raw material, the objective being to recover the insoluble starch granules as undamaged or intact as possible. Unprocessed native starches are generally too weak and functionally too restricted for application in advanced technologies. Processing/modification is therefore generally necessary to engender a range of desired functionalities, and polysaccharides (particularly starch, cellulose and chitin/chitosan) represent a diverse group of renewable bioresources that provide a complex and essentially inexhaustible substrate library for chemical derivatisation. Chemical modification of polysaccharides is based upon hydroxyl group chemistry and a modified polysaccharide can be defined as one whose hydroxyl groups have been altered by chemical reaction (*e.g.* by oxidation, esterification, etherification, cross-linking, *etc*).[115–117] The physicochemical properties of chemically modified polysaccharides are largely dependent upon the degree of substitution (DS), which is defined as the average number of substituted hydroxyl groups per anhydroglucose unit. In the case of amylose, there are three hydroxyl groups available for substitution, attached to the C2, C3 and C6 carbon atoms, the C1 and C4 hydroxyl groups being involved in the glycosidic linkages that form the polysaccharide backbone. Thus, the maximum theoretical average DS value, and maximum DS value for a single anhydroglucose monomer unit, is three.[117] Obviously, this value is reduced in the case of amylopectin, since a branched anhydroglucose unit has a maximum theoretical DS value of two (on the C2 and C3 hydroxyl groups, since the C6 hydroxyl group is part of the glycosidic linkage of the branch), thus the higher the degree of branching the lower the maximum theoretical average DS value. With higher degrees of branching steric factors are also significant as is the size of the derivatising agents and substituent group (making high DS values essentially unachievable).

Such chemical modifications affect hydrogen bonding, charge interactions and hydrophobic character thereby altering the nature of the interactions between the polysaccharide chains. Since starch is inherently water-soluble after disruption of the granular structure, a relatively high DS is not required to impart solubility or dispersibility, therefore the majority of industrial applications require only a low DS value (<0.2) to dramatically change the physicochemical properties of the polysaccharide.[117] An overview of the effect of modifications on starch properties is provided in Table 5.2.

Table 5.2 Effect of modifications on starch properties.

Modification	*Objective*	*Benefits*
Pre-gelatinisation	Cold-water thickening properties	Eliminates need to cook – convenience and energy saving
Thermal treatment	Strengthen granules and retard swelling/viscosity increase	Functional native starch with improved tolerance to heat, acid and shear
Acid thinning	Lower viscosity and increase gel strength	Enhance textural properties at higher concentrations
Enzyme conversion	Varied viscosity and gel strength, thermo-reversibility and sweetness	Contributes texture and rheology
Oxidation	Carbonyl and carboxyl groups, better clarity, less retrogradation	Improves adhesion of coatings, soft stable gels at higher dosage
Stabilisation	Prevent granule shrinkage, lower gelatinisation temperature	Chill and freeze/thaw stability extends shelf-life
Cross-linking	Strengthen granules and retard swelling/viscosity increase	Improved tolerance to heat, acid and shear

Low DS derivatives are generally manufactured by reacting starch in an aqueous suspension/slurry of ~30–45% solids, usually at pH 7–12. Sodium hydroxide is commonly employed to produce the alkaline pH. Conditions are adjusted to prevent gelatinisation of granular starch and to allow recovery of the starch derivative in granular form by filtration or centrifugation and drying. The derivative may be washed to remove unreacted reagents, by-products, salts and other solubles before drying. To prevent the swelling of the starch under strongly alkaline conditions, sodium chloride may be added to a concentration of ~10–30%. There is a limit to the level of substitution that can be made in aqueous slurry while maintaining the starch in granular form, because the granule gelatinisation temperature is lowered as the DS increases. If the level of hydrophilic substituents becomes high enough, the starch derivative gelatinises, becomes dispersible at room temperature, and is said to be cold-water-dispersible. A non-swelling solvent, such as acetone, is used to prepare higher DS, cold-water-dispersible derivatives in granule form.[117]

5.3.1 Oxidation

Although many reagents oxidise starch, alkaline hypochlorite is the most commonly utilised commercial reagent.[9] Starch is oxidised to obtain low-viscosity, high-solids dispersions with resistance to viscosity increases or gelling in aqueous dispersion. The oxidation causes some depolymerisation, reducing viscosity, and introduces relatively bulky carboxylic acid (–COOH) and

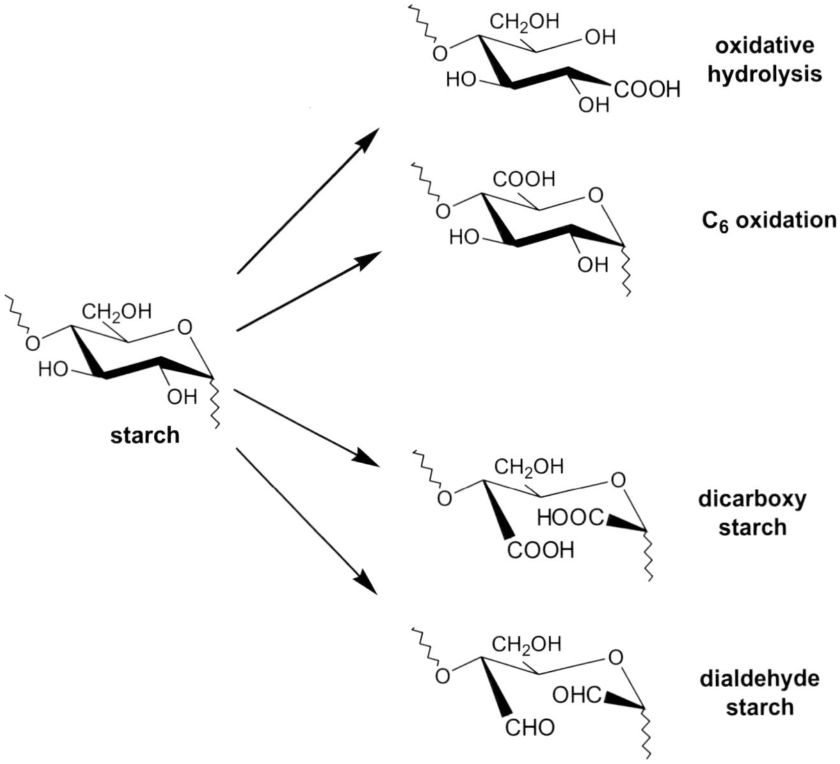

Figure 5.9 Examples of starch oxidation chemistry.

aldehyde (–CHO) groups. The steric hindrance of the bulky groups disrupts any retrograding tendency, thus reducing gel strength and providing viscosity stability, which is advantageous in many applications.[19,37,118] An overview of oxidation chemistry is provided in Figure 5.9; more detailed information on oxidation procedures and mechanisms is available.[115,117,119] Since hypochlorite oxidation results in scission of some glycosidic linkages, some of the starch is solubilised and removed during the separation and washing process. In general, oxidised starches are sensitive to heat, tending to yellow when exposed to heat and on prolonged storage. They gelatinise at a lower temperature than native starches, and produce aqueous dispersions of greater clarity and lower viscosity, with less tendency to retrograde, thus the pastes are more fluid.[117]

5.3.2 Stabilisation

Esterification or etherification with monofunctional reagents can be used to substitute bulky groups onto starch in order to take up space and sterically hinder any tendency for dispersed (cooked), linear fragments to realign and retrograde. The effectiveness of stabilisation depends upon the number and nature of the substituted group. This type of modification is therefore often called stabilisation, and the products called stabilised starches.[1] The primary objective of stabilisation is to prevent retrogradation and thereby enhance shelf life through tolerance to temperature fluctuations such as freeze-thaw cycles. A relatively low DS is adequate to achieve this stabilisation effect, thus many commercial starch acetates are generally granular with a DS of <0.2.[22,115] Polysaccharide esters are generally prepared in two ways. Aqueous reactions under controlled pH generally produce low DS value esters (<2), whereas non-aqueous processes can produce higher DS value esters (up to ~ 3).[120,121] The maximum DS attainable without gelatinisation varies with the particular starch, but the upper limit is ~ 0.5. Microscopic examination of the granule with a DS up to 0.2 reveals no discernible difference from native granules.

Stabilisation is also achieved by starch etherification. Hydroxyalkyl derivatives are produced by reacting starch with alkyl epoxides, the most common being hydroxypropylstarch (Figure 5.10) produced using propylene oxide and carboxymethylstarch produced using sodium chloroacetate.[19,21,115,119,122] Hydroxypropyl starches have a stronger decreasing effect on gelatinisation temperature than acetate groups, and have a remarkable stabilising effect over a wide pH range, especially at low temperature, making them useful in cold storage applications.[9,115,117] Hydroxyethyl starches are also utilised in a variety of commercial and biomedical applications.[123,124]

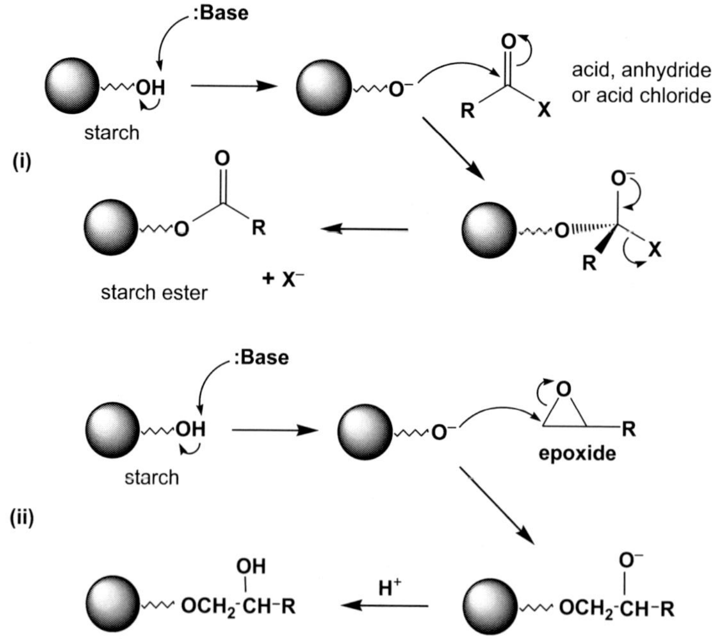

Figure 5.10 Starch stabilisation chemistry: (i) esterification; (ii) hydroxypropylation.

5.3.3 Cross-linking

When starch is treated with multifunctional reagents, cross-linking occurs. The reagent introduces intermolecular bridges or cross-links between molecules, thereby markedly increasing the average molecular weight. Since starch contains many hydroxyl groups, some intramolecular reaction also takes place. This is not significant in the usual granular reactions because the close packing of starch molecules favours intermolecular cross-linking. The reaction of starch with multifunctional agents can also be used to bind starch to another substrate. Cross-linking granular starch reinforces hydrogen bonds holding the granule together. This produces considerable changes in the gelatinisation and swelling properties of the starch granule with cross-linking as low as one cross-link per 100–3000 anhydroglucose units.[9,37,115] This toughening of the granule leads to a restriction in the swelling of the granule during gelatinisation, the degree of which is related to the amount of cross-linking. Thus, cross-linked granules are less fragile and more resistant to fragmentation by shear, high temperature, and low pH, and thus will maintain higher working viscosities and show less viscosity breakdown than untreated starches. As the amount of cross-linking increases, viscosity breakdown becomes less and the high peak viscosity stabilises, and the rate of gelatinisation and swelling of the granules decreases.[19,115] At high cross-linking levels, granules no longer gelatinise in boiling water.

A wide range of derivatising agents can be used for cross-linking, such as bifunctional etherifying and/or esterifying agents, *e.g.* epichlorohydrin (Figure 5.11), bisepoxides, dibasic organic acids, anhydrides or acid chlorides, *e.g.* adipic acid, adipic anhydride or adipoyl chloride.[22,125] Phosphorous oxychloride, sodium trimetaphosphate and sodium tripolyphosphate are used to produce phosphate cross-linked starch (Figure 5.12).[19,22,126] A wide range of

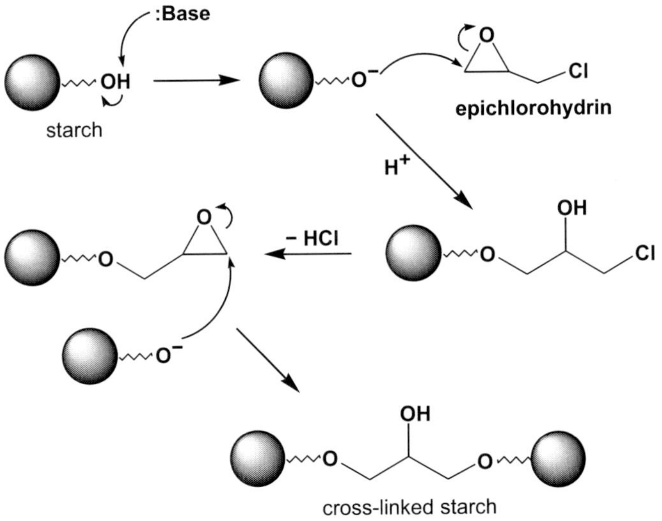

Figure 5.11 Cross-linking of starch with epichlorohydrin.

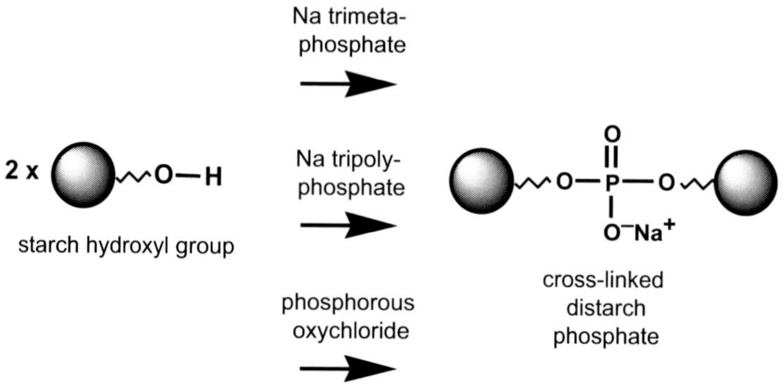

Figure 5.12 Phosphate cross-linking of starch.

other cross-linking reagents has been investigated.[117,118] The cross-linking of starch with epichlorohydrin has been studied extensively.[127,128] The need for three-dimensional matrices with controlled release properties for pharmaceutical and agrochemical uses has increased interest in unmodified and modified starch gelation.[129–132] Cross-linking is often used in combination with other derivatisation treatments to maintain dispersion viscosity upon exposure to high temperature cooking, high shear, or acid.

5.4 Specific Biomedical Applications

The most important requirement for the use of biodegradable polymers in medical applications is their biocompatibility in a specific environment, together with the non-cytotoxicity of their degradation products, and favourable physicochemical and mechanical properties.[133] The performance of a medical device is controlled by two sets of characteristics: those which determine the ability of a device to perform the appropriate and specific function, and those which determine the compatibility of the material within the body, namely biofunctionality and biocompatibility. As such, the approach in the assessment of material biocompatibility encompasses the evaluation of the effects of physiological environments on materials and of the materials effects on the environment.[134] It is generally accepted that biocompatibility means not only absence of a cytotoxic effect but also positive effects in the sense of biofunctionality, *i.e.* promotion of biological processes which further the intended aim of the application of a biomaterial.[135]

5.4.1 Orthopaedic Implants

One of the most prominent application areas for biomaterials is for orthopaedic implant devices. Both osteoarthritis and rheumatoid arthritis affect the structure of freely movable (synovial) joints, such as the hip, knee, shoulder, ankle, and elbow. The pain in such joints, particularly weight-bearing joints, can be considerable, and the effects on ambulatory function quite devastating. It has been possible to replace these joints with prostheses, and the relief of pain along with restoration of mobility is well known to many patients. The use of degradable polymeric materials for orthopaedic implant devices as bone fillers or in fixation of some fractures is a relatively common procedure.[136,137] A material to be used in such applications must exhibit adequate mechanical properties coupled with controlled degradation rates and an appropriate biological behaviour in terms of interaction with living tissues.[138] The scarcity of polymers that meet these demanding requirements has motivated the search for novel biodegradable materials with improved mechanical properties.

Starch-based polymers and composites have been introduced as promising biomaterials for orthopaedic applications.[139–142] These materials are biodegradable and, when adequately reinforced and processed by non-conventional injection moulding routes, possess a stiffness matching that of bone. To evaluate the *in vitro* biocompatibility of these materials, several *in vitro* culture assays are available. Most are based upon the assessment of cell death, adhesion, proliferation, morphology and biosynthetic activity.[143,144] The evaluation of biomaterials proposed for orthopaedic applications has been performed using cultured osteoblasts or osteoblast-like cells.[145,146] Enhanced cell adhesion and proliferation are required, and depend not only on the cell type but also on the physical and chemical properties of the material surface,[147–149] since surface topography and chemistry play important roles in cell orientation.[150–152]

Hydroxyapatite is a bioactive material known to promote the differentiation of osteoblastic cells *in vitro*, and therefore starch-based blends with different synthetic components have been reinforced with increasing percentages of hydroxyapatite in order to evaluate the effect on osteoblast-like cells in terms of cell adhesion/morphology and proliferation.[153,154] Biodegradable starch-based polymers/hydroxyapetite composites with mechanical properties matching those of human bone, and suitable three-dimensional structures (generated by solid blowing agents), have been produced and their biocompatibility and potential suitability for orthopaedic and tissue engineering applications evaluated.[155,156]

A starch/ethylene vinyl alcohol (SEVA) blend and a composite of SEVA reinforced with hydroxyapatite have been evaluated in both *in vitro* and *in vivo* biocompatibility assays for orthopaedic application as temporary bone replacement/fixation implants.[157] Cell culture methods have been used for *in vitro* analysis, and *in vivo* tissue reactions have been evaluated in an intramuscular and intracortical bone implantation model using goats. In both models SEVA-based materials did not induce adverse reactions, which in addition to their bone-matching mechanical properties made them promising materials for bone fixation and/or replacement. Atomic force microscopy (AFM) has been utilised for high-resolution *in situ* imaging of calcium phosphate layer formation on the surface of SEVA/hydroxyapatite composites.[158] The induction period necessary for nucleation was <24 h in simulated body fluid solution and growth of the calcium phosphate layer was continuous for up to 128 h. Calcium phosphate growth on the composite surface displayed a strong *in vitro* bioactivity demonstrating the potential of such composites for bonding to bone when implanted *in vivo*.

The cytotoxicity and cell adhesion properties of SEVA, and a blend of corn starch and cellulose acetate (SCA), and their respective composites with hydroxyapatite, have also been assessed.[159,160] SEVA blends were found to be less cytotoxic than SCA, although cells adhered better to SCA surfaces. Considering the overall behaviour, such materials have good potential in applications such as bone replacement/fixation and/or tissue engineering scaffolding. SEVA, SCA, and corn starch/polycaprolactone (SPCL) blends and their composites with hydroxyapatite have also been assessed with respect to their suitability for use in orthopedic applications.[153] The cell morphology on the surface of the materials was found to be appropriate in almost every case. Thus, starch-based biomaterials can be seen as good substrates for osteoblast-adhesion and proliferation, which demonstrates their potential for use in orthopaedic applications.

5.4.2 Bone Cements

Bone cements are materials employed in orthopaedic surgery and dental applications for the fixation of joint prostheses to act as a load distributor between the artificial implant and the bone, as well as filling self-curing materials for bone and dental cavities. Since the 1960's, polymethylmethacrylate (PMMA) has been used in this field, due to its biostability and good mechanical properties.[161] However, one of the major problems of these materials is the fixation into bone structure, so that the most frequent long-term complication is loosening of the prosthetic component, caused by both mechanical and biological factors.[162] Regarding the use of self-curing formulations for the sealing and filling of dental and bone cavities, one of the most important problems is the bonding of the filling agent or sealant to the wall or the surface of the cavities. In order to solve these problems, researchers investigated the incorporation of ceramic compounds,[163] modifying the liquid phase with different acrylic polymers,[164] use of adhesive additives,[165] and different mixing techniques.[166]

The incorporation of a biodegradable component into acrylic bone cement formulations, such as carboxymethyl cellulose, in order to create porous cements that could facilitate bone regeneration was previously proposed.[167] Starch-based polymers are known biodegradable materials and exhibit a range of properties that may permit their use as bone replacement biomaterials,[141,154] and bone cement formulations.[168] Partially biodegradable acrylic bone cements based on corn starch/cellulose acetate blends (SCA), with free radical polymerisation of methylmethacrylate (MMA) and acrylic acid (AA) at low temperature were developed.[169] The heterogeneous morphology of the cured cements could be positively applied for the formation of a relatively porous material. The developed systems showed a range of properties that could permit their application as self-curing bone cements, exhibiting several advantages with respect to other commercially available bone cements. SCA and poly(2-hydroxyethyl methacrylate) have been investigated more recently.[170]

5.4.3 Tissue Engineering Scaffolds

Tissue engineering emerged in the last decade of the 20th Century as an alternative approach to circumvent the existing limitations in the current therapies for organ failure or replacement, which were mainly related to the difficulty in obtaining tissues or organs for transplantation. Tissue engineering can be defined as the design and construction in the laboratory of living, functional components that can be used for regeneration of malfunctioning tissues. It is an interdisciplinary field that brings together the principles of the life sciences and medicine with those of engineering and has three basic components: cells, scaffolds and signals. The tissue engineering approach to repair and regenerate damaged tissues is based on the use of scaffolds, which act as supports for the regenerating tissue. The advent of tissue engineering has been motivated by the challenge of producing tissue substitutes that can restore the structural features and physiological functions of natural tissues *in vivo*.[171] There is a continuous search for materials that exhibit suitable properties that can be tailored to several tissue systems.[172]

In most cases, biocompatible, degradable polymers are utilised to induce surrounding tissue ingrowth or to serve as temporary scaffolds for transplanted cells to attach, grow and maintain differentiated functions.[173,174] However, the requirements for such type of materials are complex. In addition to being biocompatible both in as-implanted and degraded form, these scaffolds have to exhibit appropriate mechanical properties to provide the correct stress environment for the new tissue.[175] The material must be designed with a degradation rate that assures the strength of the scaffold is retained until the newly grown tissue takes over the synthetic support.[171,175] Also, the scaffolds should be porous and permeable to permit the ingress of cells and nutrients and should exhibit the appropriate surface chemistry for cell attachment and proliferation.[171,174] The methods of manufacturing such scaffolds in a reproducible manner are crucial to their success, and must allow for the necessary scale-up of the developed tissue engineering technology. The techniques used to manufacture scaffolds for tissue engineering are dependent upon the properties of the polymer and its intended application. It must allow the preparation of scaffolds with complex 3D geometries with controlled porosity and pore size, since these factors are associated with the supply of nutrients to transplanted and regenerated cells.[176] Various processing techniques have been developed to fabricate these scaffolds, such as solvent casting, particulate leaching, membrane lamination, fibre bonding, high-pressure-based methods, melt-based technologies, *etc.*[173,176,177] The major problem associated with the scaffolds produced by such methods is their mechanical weakness, which does not allow for their use in hard-tissue regeneration where high-strength scaffolds are required.[175] This has led to the search for better ways of producing porous scaffolds so as to simultaneously optimise their physical and chemical properties.

Studies on the development of an injection moulding method to produce biodegradable scaffolds from a range of corn-starch-based polymers were performed.[178,179] In some cases, hydroxyapatite was also used as a reinforcement of the biodegradable polymers. The developed methodology consisted of a standard conventional injection moulding process, on which a solid blowing agent based on carboxylic acids was used to generate the foaming of the bulk of the moulded part. The developed route allows the production of scaffolds having a compact skin and a porous core, with promising mechanical properties. Therefore, it is possible to manufacture biodegradable polymer scaffolds in an easy (melt-based processing) and reproducible manner. The scaffolds can be moulded into complex shapes, and the blowing additives do not affect the non-cytotoxic behaviour of the starch-based materials. 3D scaffolds with complex shapes and an appropriate morphology, without significant deterioration of mechanical properties, can be obtained. By simply varying the processing parameters, materials formulation and quantity of blowing agent, the porosity, pore size and pore structure can be controlled without decreasing significantly the mechanical properties of starch blends.[178] Materials manufactured from different starch-based blends, exhibit significant differences in their mechanical and degradation properties, which could make them suitable for different particular applications, such as high-strength scaffolds for hard-tissue regeneration.

A cartilage tissue engineering approach combining SPCL fibre mesh scaffolds with bovine articular chondrocytes has been investigated.[180] SPCL microparticles incorporating bioactive agents have been developed for drug delivery and tissue engineering applications,[181] and osteoblasts and endothelial cells have been co-cultured on SPCL scaffold for the *in vitro* development of vascularisation.[182] The use of synthetic and or natural polymeric materials for bone and cartilage repair has been recently reviewed.[183]

5.4.4 Drug Delivery Systems

One of the fastest growing areas for implant application is in devices for the controlled and targeted delivery of drugs. Many attempts have been made to incorporate drug reservoirs into implantable devices for a sustained and controlled release. Some of these technologies use polymeric materials as vehicles for drug delivery. Many new drugs exhibit poor wetting behaviour and low aqueous solubility. This is particularly an issue for pre-clinical studies like toxicological trials, in which considerably higher doses and volumes are being administered compared to clinical studies. Pre-clinical vehicles typically contain high levels of surfactants that can exert biological effects. However, the biological inertness of vehicles is pivotal for application in pre-clinical studies, stressing the need to find new excipients to overcome formulation problems associated with the discovery of new drugs.

Biodegradable polymers such as poly(lactic acid) (PLA), poly(glycolic acid) (PGA) and their respective copolymers have been applied in several drug delivery systems.[184,185] Cyclodextrins have been extensively investigated in such drug delivery systems in a variety of biomedical applications.[114,186–193] Attempts have been reported in trying to use starch-based polymers in these types of applications,[129,168,194] due to their biodegradability.[195] Maize starch acetates have been evaluated with respect to their controlled release properties,[196] as have acetate, aminoethyl and carboxymethyl-derivatives of epichlorohydrin cross-linked starches with good mucoadhesion properties.[197,198] Starch-based microspheres produced by emulsion crosslinking with potential media dependent responsive behaviour for drug delivery have been produced.[199] Enzymatically-mediated drug delivery carriers for bone tissue engineering applications, which combine biodegradable starch-based microparticles and differentiation agents have also been developed.[200]

Starch-based films are useful in pharmaceutical formulations, particularly in drug delivery applications.[201] Amylose, one of the major fractions of starch, possesses the ability to form films when prepared under appropriate conditions. Colon-specific drug delivery may be possible by application of dried amylose films to pharmaceutical formulations. The microstructure of the film can be resistant to the action of pancreatic α-amylase, but is digestible by amylases of the colonic microflora. However, under simulated gastro-intestinal conditions, coatings made solely of amylose swell, become porous and allow drug release. Incorporation of insoluble polymers into the amylose film, to control amylose swelling, provides a solution to this problem. Among the range of cellulose and acrylate based copolymers assessed, a commercially available ethylcellulose (Ethocel®) was found to control the swelling most effectively.[202] The *in vitro* dissolution of various coated pellets under simulated gastric and small intestinal conditions, using commercially available pepsin and pancreatin, demonstrated the resistance of the amylose-Ethocel® coat. *In vitro* drug release under simulated gastric and small intestinal conditions can be prevented with additional thermal treatment. Physical blends of starch graft copolymers have more recently been evaluated as matrices for colon targeting drug delivery systems.[203]

The technical feasibility of surfactant-free drug suspensions with starch derivatives indicated that octenyl succinate-modified starches, adequately wetted drugs and homogenous tasteless suspensions were obtained.[204] The modified starches exhibited only little influence on the viscosity as well as on the yield point in contrast to the rheological effects of xanthan gum. This gelling agent was the main stabilising excipient as the modified starches hindered to a lesser extent sedimentation. A commercially available extended release matrix material, Kollidon® SR, composed of polyvinylacetate (PVA) and polyvinyl pyrrolidone (PVP), was evaluated with respect to its ability to modulate the *in vitro* release of a weakly basic drug.[205] Addition of the highly swellable maize starch and the water-soluble lactose accelerated the drug release in a more pronounced manner compared to water-insoluble calcium phosphate. Compound release from matrix tablets prepared by wet granulation was faster than the drug release from tablets prepared by direct compression.

The possibility of obtaining porous SPLA scaffolds by means of an immersion precipitation technique (phase inversion) combined with supercritical fluid technology has been recently developed.[206,207] Dexamethasone-loaded SPLA matrices were produced by such supercritical phase inversion techniques, and *in vitro* sustained release of dexamethasone was achieved over 21 days and the kinetics found to be governed by both the drug diffusion and the swelling of the matrix.[208] Retrograded waxy maize starch gels have been evaluated as a tablet matrix for the controlled release of theophylline.[209]

Novel excipients for controlled-release matrix tablets have been produced by grafting ethylmethacrylate onto tapioca starch and hydroxypropyl tapioca starch.[210] Pregelatinised glutinous rice starch has been used as a sustained release agent for tablet preparations.[211] A novel starch-based interpolyelectrolyte complex (with κ-carrageenan) has been produced as a matrix for controlled drug release.[212]

5.4.4 Starch-containing Hydrogels

Hydrogels are hydrophilic three-dimensional networks, held together by covalent bond cross-links and weak cohesive forces in the form of hydrogen or ionic bonds. If enough interstitial space exists within the network, water molecules can become trapped and immobilised, filling the available free volume. Hydrophilic hydrogels are of great interest for biotechnological and biomedical applications.[213] Their utility as biomaterials is well-known; it is due to their permeability of small molecules, soft consistency, low interfacial tension, facility for purification and mainly high equilibrium water content which make them similar to the physical properties of living tissues.[214] Hydrogels can be applied as an interface between bone and an implant,[215] as artificial skin,[216] and in-controlled release applications.[217]

Hydrophilic hydrogels based on native starch, pure starch components and their derivatives are important due to their swellability in water, biocompatibility and biodegradability. Several approaches for the preparation of biodegradable starch hydrogels have been used.[218] These starch-based hydrogels have potential biomedical applications, especially as drug delivery carriers. The synthesis strategies include free radical graft polymerisation of hydrophilic vinyl monomers onto the starch substrate or free radical copolymerisation of hydrophilic vinyl monomers with the derivatised starch in the presence of polymerisable cross-linker, cross-linking by chemical reactions with complementary groups, radiation-induced polymerisation and cross-linking and novel self-assembly processes.[219,220] Seidel *et al.* synthesised stable starch-based hydrogels by cross-linking carboxymethyl starch (CMS) ($DS = 0.45$) using polyfunctional carboxylic acids such as malic, tartaric, citric, malonic, succinic, glutaric and adipic acids.[221]

Chemically cross-linked starch-based hydrogels have been prepared via a two-stage process, where the starch was first functionalised with reactive double bonds and then cross-linked by free radical polymerisation in water. Maize starch was first modified using allyl chloride.[222] The allyl-modified starch obtained was then copolymerised with methacrylic acid and a combination of methacrylic acid and acrylamide with potassium persulfate (KPS) as an initiator, resulting in highly cross-linked, biodegradable, starch-based hydrogel products. A novel hydrogel membrane was prepared by cross-linking polyvinyl alcohol with a starch suspension using glutaraldehyde as the cross-linking agent.[223] The obtained membrane showed sufficient strength, and the absence of free aldehydic groups from uncross-linked glutaraldehyde, for potential utilisation as an artificial skin. pH-Sensitive hydrogels for controlled release applications have been produced *via* free radical graft copolymerisation of polymethacrylic acid onto starch,[224] and carboxymethyl starch,[225] and arylamide onto starch.[226] New cross-linked gelatinised starch–xanthan gum hydrogel systems have also been developed for controlled release applications.[227] Hydrogels have also been produced from carboxymethyl cassava starch cross-linked with di- or polyfunctional carboxylic acids.[228] Starch-based aerogels have also been evaluated as drug-carriers.[229]

5.5 Concluding Remarks

This review of starch sources, physicochemical characteristics and the vast range of materials that can be derived from starch and its hydrolysis products clearly demonstrates the untapped potential of this highly versatile renewable raw material. As fossil-fuel-derived materials become exhausted, scientists will have to turn to such biomaterials to replace existing plastic materials with functionally acceptable naturally derived substitutes. Also, in the biomedical field, starch-based bioproducts with unique bioproperties (controlled and/or designed biological activity, biocompatibility and biodegradability) are continuing to be developed and applied in a range of application areas. The future for starch utilisation, beyond simple sustenance, is therefore a bright one.

References

1. J. N. BeMiller and R. L. Whistler, in *Food Chemistry*, (Ed.) O. R. Fenema, Marcel Dekker, New York, 1996, p. 157.
2. E. L. Mitch, in *Starch, Chemistry and Technology*, (Eds) R. L. Whistler, J. N. Bemiller and E. F. Paschall, Academic Press, New York, 1984, p. 479.
3. W. Bushuk, in *Rye: Production, Chemistry, and Technology*, (Ed.) W. Bushuk, AACC, St. Paul, 2001, p. 1.
4. N. Singh, J. Singh, L. Kaur, N. S. Sodhi and B. S. Gill, *Food Chem.*, 2003, **81**, 219.
5. B. L. D'Appolonia and P. Rayas-Duarte, in *Wheat: Production, Properties and Quality*, (Eds) W. Bushuk and V. F. Rasper, Blackie Academic, Glasgow, 1994, p. 107.

6. S. Z. Dziedzic and M. W. Kearsley, in *Handbook of Starch Hydrolysis Products and their Derivatives*, (Eds) S. Z. Dziedzic and M. W. Kearsley, Blackie Academic, Glasgow, 1995, p. 1.
7. B. O. Juliano, in *Starch, Chemistry and Technology*, (Eds) R. L. Whistler, J. N. BeMiller and E. F. Paschall, Academic Press, New York, 1984, p. 507.
8. J. W. Knight and R. M. Olson, in *Starch, Chemistry and Technology*, (Eds) R. L. Whistler, J. N. BeMiller and E. F. Paschall, Academic Press, New York, 1984, p. 491.
9. J. N. BeMiller, in *Industrial Gums: Polysaccharides and Their Derivatives*, (Eds) R. L. Whistler and J. N. BeMiller, Academic Press, San Diego, 1993, p. 579.
10. D. A. Corbishley and W. Miller, in *Starch, Chemistry and Technology*, (Eds) R. L. Whistler, J. N. BeMiller and E. F. Paschall, Academic Press, New York, 1984, p. 469.
11. J. J. M. Swinkels, in *Starch Conversion Technology*, (Eds) G. M. A. van Beynum and J. A. Roels, Marcel Dekker, New York, 1985, p. 15.
12. H. F. Zobel and A. M. Stephen, in *Food Polysaccharides and their Applications*, (Eds) A. M. Stephen, G. O. Phillips and P. A. Williams, Marcel Dekker, New York, 2006, p. 25.
13. R. Hoover, *Carbohydr. Polym.*, 2001, **45**, 253.
14. N. Lindeboom, P. R. Chang and R. T. Tyler, *Starch/Staerke*, 2004, **56**, 89.
15. S. A. Watson, in *Starch, Chemistry and Technology*, (Eds) R. L. Whistler, J. N. BeMiller and E. F. Paschall, Academic Press, New York, 1984, p. 417.
16. G. S. Ranhotra, in *Wheat: Production, Properties and Quality*, (Eds) W. Bushuk and V. P. Rasper, Blackie Academic, Glasgow, 1994, p. 12.
17. J. O. B. Carioca, H. L. Arora, P. V. Pannir Selvam, F. C. A. Tavares, J. F. Kennedy and C. J. Knill, *Starch/Staerke*, 1996, **48**, 322.
18. W. A. Atwell, *Wheat Flour*, Eagan Press, St. Paul, 2001.
19. W. A. Atwell, L. F. Hood, D. R. Lineback, E. Varriano-Marston and H. F. Zobel, *Cereal Foods World*, 1988, **33**, 306.
20. J. M. V. Blanshard, in *Starch: Properties and Potential*, ed. T. Galliard, John Wiley and Sons, Chichester, 1987, p. 16.
21. H. J. Cornell and and A. W. Hoveling, *Wheat: Chemistry and Utilization*, Technomic Publishing, Lancaster, 1998.
22. T. P. Coultate, *Food: The Chemistry of its Components*, Royal Society of Chemistry, Cambridge, 2002.
23. J. A. Curá, P.-E. Jansson and C. R. Krisman, *Starch/Staerke*, 1995, **47**, 207.
24. V. J. Morris, in *Functional Properties of Food Macromolecules*, (Eds) S. E. Hill, D. A. Ledward and J. R. Mitchell, Aspen Publishers, Gaithersburg, 1998, p. 143.
25. S. G. Ring, T. R. Noel and V. J. Bull, in *Seed Storage Compounds: Biosynthesis, Interactions, and Manipulation*, (Eds) P. R. Shewry and K. Stobart, Oxford University Press, Oxford, 1993, p. 25.

26. Y. Takeda, T. Shitaozono and S. Hizukuri, *Carbohydr. Res.*, 1990, **199**, 207.
27. Y. Takeda, S. Tomooka and S. Hizukuri, *Carbohydr. Res.*, 1993, **246**, 267.
28. D. J. Thomas and W. A. Atwell, *Starches*, Eagan Press, St. Paul, 1999.
29. P. Walstra, *Physical Chemistry of Foods*, Marcel Dekker, New York, 2003.
30. V. P. Yuryev, L. A. Wasserman, N. R. Andreev and V. B. Tolstoguzov, in *Starch and Starch Containing Origins*, (Eds) V. P. Yuryev, A. Cesàro and W. J. Bergthaller, Nova Science Publishers, New York, 2002, p. 23.
31. T. Galliard and P. Bowler, *Crit. Rep. Appl. Chem.*, 1987, **13**, 55.
32. S. Kitamura, K. Hakozaki and T. Kuge, in *Food Hydrocolloids: Structures, Properties, and Functions*, (Eds) K. Nishinari and E. Doi, Plenum Press, New York, 1993, p. 179.
33. L. Slade and H. Levine, in *Frontiers in Carbohydrate Research–1: Food Applications*, (Eds) R. P. Millane, J. N. BeMiller and R. Chandrasekaran, Elsevier Applied Science, London, 1989, p. 215.
34. C. J. A. M. Keetels, T. van Vliet and H. Luyten, in *Food Macromolecules and Colloids*, (Eds) E. Dickinson and D. Lorient, Royal Society of Chemistry, Cambridge, 1995, p. 472.
35. H. F. Zobel, in *Starch, Chemistry and Technology*, (Eds) R. L. Whistler, J. N. BeMiller and E. F. Paschall, Academic Press, New York, 1984, p. 285.
36. S. Hizukuri, J.-I. Abe and I. Hanashiro, in *Carbohydrates in Food*, (Ed.) A.-C. Eliasson, CRC Press, Boca Raton, 2006, p. 305.
37. P. Murphy, in *Handbook of Hydrocolloids*, (Eds) G. O. Phillips and P. A. Williams, Woodhead Publishing, Cambridge, 2000, p. 41.
38. C. P. Berry, B. L. D'Appolonia and K. A. Giles, *Cereal Chem.*, 1971, **48**, 415.
39. S. G. Ring, *Chem. Br.*, 1995, **31**, 303.
40. H. F. Zobel, in *Developments in Carbohydrate Chemistry*, (Eds) R. J. Alexander and H. F. Zobel, AACC, St. Paul, 1992, p. 1.
41. A. Guilbot and C. Mercier, in *The Polysaccharides-Vol 3*, (Ed.) G. O. Aspinall, Academic Press, Orlando, 1985, p. 209.
42. R. L. Whistler and J. R. Daniel, in *Starch, Chemistry and Technology*, (Eds) R. L. Whistler, J. N. BeMiller and E. F. Paschall, Academic Press, New York, 1984, p. 153.
43. D. J. Manners, *Carbohydr. Polym.*, 1989, **11**, 87.
44. P. R. Shewry and D. B. Bechtel, in *Rye: Production, Chemistry, and Technology*, (Ed.) W. Bushuk, AACC, St Paul, 2001, p. 69.
45. R. L. Whistler, in *Starch, Chemistry and Technology*, (Eds) R. L. Whistler, J. N. BeMiller and E. F. Paschall, Academic Press, New York, 1984, p. 1.
46. B. O. Juliano and C. P. Villareal, *Grain Quality Evaluation of World Rices*, International Rice Research Institute (IRRI), Manila, 1993.
47. D. J. Manners, in *New Approaches to Research on Cereal Carbohydrates*, (Eds) R. D. Hill and L. Munck, Elsevier Science Publishers, Amsterdam, 1985, p. 45.

48. M. J. Gidley, in *Starch: Advances in Structure and Function*, (Eds) T. L. Barsby, A. M. Donald and P. J. Frazier, Royal Society of Chemistry, Cambridge, 2001, p. 1.
49. L. E. Fitt and E. M. Snyder, in *Starch, Chemistry and Technology*, (Eds) R. L. Whistler, J. N. BeMiller and E. F. Paschall, Academic Press, New York, 1984, p. 675.
50. R. Karlsson, R. Olered and A.-C. Eliasson, *Starch/Staerke*, 1983, **35**, 335.
51. A. Imberty, H. Chanzy, S. Perez, A. Bulion and V. Tran, *Macromolecules*, 1987, **20**, 2634.
52. A. Imberty and S. Perez, *Biopolymers*, 1988, **27**, 1205.
53. D. J. Gallant, B. Bouchet, A. Bulion and S. Perez, *Eur. J. Clin. Nutr.*, 1992, **46**, S3.
54. D. J. Gallant, B. Bouchet and P. M. Baldwin, *Carbohydr. Polym.*, 1997, **32**, 177.
55. M. J. Gidley and S. M. Bociek, *J. Am. Chem. Soc.*, 1988, **110**, 3820.
56. B. Pfannemüller, *Int. J. Biol. Macromol.*, 1987, **9**, 105.
57. M. J. Gidley and P. V. Bulpin, *Carbohydr. Res.*, 1987, **161**, 291.
58. H. F. Zobel H F, *Starch/Staerke*, 1988, **40**, 44.
59. D. French, in *Starch, Chemistry and Technology*, (Eds) R. L. Whistler, J. N. BeMiller and E. F. Paschall, Academic Press, New York, 1984, p. 183.
60. M. J. Gidley and S. M. Bociek, *J. Am. Chem. Soc.*, 1985, **107**, 7040.
61. P. J. Jenkins, R. E. Cameron and A. M. Donald, *Starch/Staerke*, 1993, **45**, 417.
62. N. J. Atkin, R. M. Abeysekera and A. W. Robards, *Carbohydr. Polym.*, 1998, **36**, 193.
63. A. M. Donald, P. A. Perry and T. A. Waigh, in *Starch: Advances in Structure and Function*, (Eds) T. L. Barsby, A. M. Donald and P. J. Frazier, Royal Society of Chemistry, Cambridge, 2001, p. 45.
64. P. J. Jenkins and A. M. Donald, *Carbohydr. Res.*, 1998, **308**, 133.
65. M. J. Gidley and P. V. Bulpin, *Macromolecules*, 1989, **22**, 341.
66. J.-L. Jane and J. F. Robyt, *Carbohydr. Res.*, 1984, **132**, 105.
67. I. L. Batey, in *The RVA Handbook*, ed. G. B. Crosbie and A. S. Ross, AACC, St. Paul, 2007, p. 19.
68. P. N. Bhandari, R. S. Singhal and D. D. Kale, *Carbohydr. Polym.*, 2002, **47**, 365.
69. C. R. Chen and H. S. Ramaswamy, *Food Res. Int.*, 1999, **32**, 319.
70. B. Jauregui, M. E. Muñoz and A. Santamaria, *Int. J. Biol. Macromol.*, 1995, **17**, 49.
71. L. Kaur, N. Singh and N. S. Sodhi, *Food Chem.*, 2002, **79**, 183.
72. L. Kaur, N. Singh, N. S. Sodhi and H. S. Gujral, *Food Chem.*, 2002, **79**, 177.
73. V. Kislenko, L. Oliynyk and A. Golachowski, *J. Colloid Interface Sci.*, 2006, **294**, 79.
74. N. Singh, N. Inouchi and K. Nishinari, *Food Hydrocolloids*, 2006, **20**, 923.
75. N. Singh, L. Kaur, K. S. Sandhu, J. Kaur and K. Nishinari, *Food Hydrocolloids*, 2006, **20**, 532.

76. P. A. Sopade and K. Kiaka, *J. Food Eng.*, 2001, **50**, 47.
77. J.-Y. Thebaudin, A.-C. Lefebvre and J.-L. Doublier, *Lebensm.-Wiss. Technol.*, 1998, **31**, 354.
78. B. L. Dasinger, D. M. Fenton, R. P. Nelson, F. F. Roberts and S. J. Truesdell, in *Starch Conversion Technology*, (Eds) G. M. A. van Beynum and J. A. Roels, Marcel Dekker, New York, 1985, p. 237.
79. D. Howling, in *Glucose Syrups: Science and Technology*, ed. S. Z. Dziedzic and M. W. Kearsley, Elsevier Applied Science, Barking, 1984, p. 1.
80. F. H. Otey and W. M. Doane, in *Starch, Chemistry and Technology*, (Eds) R. L. Whistler, J. N. BeMiller and E. F. Paschall, Academic Press, New York, 1984, p. 389.
81. K. Ravi-Kumar and S. Umesh-Kumar, in *Food Biotechnology*, (Eds) K. Shetty, G. Paliyath, A. Pometto and R. E. Levin, CRC Press, Boca Raton, 2006, p. 709.
82. R. E. Hebeda and W. M. Teague, in *Developments in Carbohydrate Chemistry*, (Eds) R. J. Alexander and H. F. Zobel, AACC, St. Paul, 1992, p. 65.
83. J. R. Robyt, in *Starch, Chemistry and Technology*, (Eds) R. L. Whistler, J. N. BeMiller and E. F. Paschall, Academic Press, New York, 1984, p. 87.
84. P. H. Blanchard and F. R. Katz, in *Food Polysaccharides and their Applications*, (Eds) A. M. Stephen, G. O. Phillips and P. A. Williams, Marcel Dekker, New York, 2006, p. 119.
85. D. Howling, *Int. Biodeterior.*, 1989, **25**, 15.
86. J. F. Kennedy, J. M. S. Cabral, I. Sá-Correia and C. A. White, in *Starch: Properties and Potential*, (Ed.) T. Galliard, John Wiley and Sons, Chichester, 1987, p. 115.
87. P. J. Reilly, in *Starch Conversion Technology*, (Eds) G. M. A. van Beynum and J. A. Roels, Marcel Dekker, New York, 1985, p. 101.
88. C. A. White and J. F. Kennedy, in *Carbohydrate Chemistry*, (Ed.) J. F. Kennedy, Oxford University Press, Oxford, 1988, p. 342.
89. F. G. Priest and J. R. Stark, *in Biotechnology of Amylodextrin Oligosaccharides*, American Chemical Society, Washington, 1991, p. 72.
90. P. Gacesa and J. Hubble, in *Bioconversion of Waste Materials to Industrial Products*, (Ed.) A. M. Martin, Elsevier Applied Science, London, 1991, p. 39.
91. G. M. Frost and D. A. Moss, in *Biotechnology Vol. 7a*, (Ed.) J. F. Kennedy, VCH, Weinheim, 1987, p. 65.
92. C. E. Morris, *Food Eng.*, 1984, **56**, 48.
93. J. F. Kennedy, C. J. Knill and D. W. Taylor, in *Handbook of Starch Hydrolysis Products and their Derivatives*, (Eds) M. W. Kearsley and S. Z. Dziedzic, Blackie Academic, Glasgow, 1995, p. 65.
94. A. M. Frye and C. S. Setser, in *Low-Calorie Foods Handbook*, (Ed.) A. M. Altschul, Marcel Dekker, New York, 1993, p. 211.
95. G. G. Birch, M. N. Azudin and J. M. Grigor, in *Biotechnology of Amylodextrin Oligosaccharides*, (Ed.) R. B. Friedman, American Chemical Society, Washington, 1991, p. 261.

96. S. G. Ring and M. A. Whittam, in *Biotechnology of Amylodextrin Oligosaccharides*, (Ed.) R. B. Friedman, American Chemical Society, Washington, 1991, p. 273.
97. I. Jasutiene and V. Kolodzeiskis, in *Starch and Starch Containing Origins*, (Eds) V. P. Yuryev, A. Cesàro and W. J. Bergthaller, Nova Science Publishers, New York, 2002, p. 211.
98. N. E. Lloyd and W. J. Nelson, in *Starch, Chemistry and Technology*, (Eds) R. L. Whistler, J. N. BeMiller and E. F. Paschall, Academic Press, New York, 1984, p. 611.
99. B. E. Norman, *Starch/Staerke*, 1982, **34**, 340.
100. J. Wilms, in *Starch and Starch Containing Origins*, (Eds) V. P. Yuryev, A. Cesàro and W. J. Bergthaller, Nova Science Publishers, New York, 2002, p. 199.
101. A. P. G. Kieboom and H. van Bekkum, in *Starch Conversion Technology*, (Eds) G. M. A. van Beynum and J. A. Roels, Marcel Dekker, New York, 1985, p. 263.
102. R. van Tilburg, in *Starch Conversion Technology*, (Eds) G. M. A. van Beynum and J. A. Roels, Marcel Dekker, New York, 1985, p. 175.
103. A. Krutošíková and and M. Uher, *Natural and Synthetic Sweet Substances*, Ellis Horwood, Chichester, 1992.
104. S. Kobayashi, in *Enzymes for Carbohydrate Engineering*, (Eds) K.-H. Park, J. F. Robyt and Y.-D. Choi, Elsevier Science, Amsterdam, 1996, p. 23.
105. H. Dodziuk, in *Cyclodextrins and their Complexes*, (Ed.) H. Dodziuk, Wiley-VCH, Weinheim, 2006, p. 1.
106. S. Li and W. C. Purdy, *Chem Rev.*, 1992, **92**, 1457.
107. A. N. Kalinkevich, I. I. Chemeris, P. A. Pavlenko, V. V. Pilipenko, Y. V. Rogulsky, A. N. Bugai, S. N. Danilchenko and and L. F. Sukhodub, in *Starch and Starch Containing Origins*, (Eds) V. P. Yuryev, A. Cesàro and W. J. Bergthaller, Nova Science Publishers, New York, 2002, p. 154.
108. B. W. Müller and U. Brauns, *Int. J. Pharm.*, 1985, **26**, 77.
109. L. M. Tasic, M. D. Jovanovic and Z. R. Djuric, *J. Pharm. Pharmacol.*, 1992, **44**, 52.
110. H. van Dorne, *Eur. J. Pharm. Biopharm.*, 1993, **39**, 133.
111. T. Loftsson and M. E. Brewster, *J. Pharm. Sci.*, 1996, **85**, 1017.
112. T. Loftsson and D. Duchêne, *Int. J. Pharm.*, 2007, **329**, 1.
113. E. M. Martin Del Valle, *Process Biochem.*, 2004, **39**, 1033.
114. K. Uekama, F. Hirayama and and H. Arima, in *Cyclodextrins and their Complexes*, (Ed.) H. Dodziuk, Wiley-VCH, Weinheim, 2006, p. 381.
115. G. Fleche, in *Starch Conversion Technology*, (Eds) G. M. A. van Beynum and J. A. Roels, Marcel Dekker, New York, 1985, p. 73.
116. C. J. Knill and and J. F. Kennedy, in *Polysaccharides Structural: Diversity and Functional Versatility*, (Ed.) S. Dumitrui, Marcel Dekker, New York, 2005, p. 605.
117. M. W. Rutenberg and and D. Solarek, in *Starch, Chemistry and Technology*, (Eds) R. L. Whistler, J. N. BeMiller and E. F. Paschall, Academic Press, New York, 1984, p. 311.

118. O. B. Wurzburg, in *Food Polysaccharides and their Applications*, (Eds) A. M. Stephen, G. O. Phillips and P. A. Williams, Marcel Dekker, New York, 2006, p. 87.
119. M. Yalpani, *Polysaccharides: Syntheses, Modifications and Structure/Property Relations*, Elsevier Science Publishers, Amsterdam, 1988.
120. M. I. Khalil, A. Hashem and A. Hebeish, *Starch/Staerke*, 1995, **47**, 394.
121. M. M. Tessler and R. L. Billmers, *J. Environ. Polym. Degrad.*, 1996, **4**, 85.
122. J. F. Kennedy, *Adv. Carbohydr. Chem. Biochem.*, 1974, **29**, 305.
123. S. A. Kozek-Langenecker, *Anesthesiology*, 2005, **103**, 654.
124. J. Treib, J. F. Baron, M. T. Grauer and R. G. Strauss, *Intensive Care Med.*, 1999, **25**, 258.
125. I. Šimkovic, M. Hricovíni, R. Mendichi and J. J. G. van Soest, *Carbohydr. Polym.*, 2004, **55**, 299.
126. N. Atichokudomchai and S. Varavinit, *Carbohydr. Polym.*, 2003, **53**, 263.
127. K. P. R. Kartha and H. Srivastava, *Starch/Staerke*, 1985, **37**, 297.
128. Đ. Ačkar, J. Babić, D. Šubarić, M. Kopjar and B. Miličević, *Carbohydr. Polym.*, 2010, **81**, 76.
129. J. Heller, S. H. Pangburn and K. V. Roskos, *Biomaterials*, 1990, **11**, 345.
130. M. C. Levy and M. C. Andry, *Int. J. Pharm.*, 1990, **62**, 27.
131. D. Trimnell and B. S. Shasha, *J. Controlled Release*, 1988, **7**, 25.
132. R. E. Wing, S. Maiti and W. M. Doane, *Starch/Staerke*, 1987, **39**, 422.
133. J. M. Pachence and and J. Kohn, in *Principles of Tissue Engineering*, (Eds) R. Lanza, R. Langer and W. Chick, Academic Press, New York, 1997, p. 263.
134. D. F. Williams, *Mater. Res. Soc. Symp. Proc.*, 1986, **55**, 117.
135. C. J. Kirkpatrick, F. Bittinger, M. Wagner, H. Kohler, T. G. van Kooten, C. L. Klein and M. Otto, *Proc. Inst. Mech. Eng. H.*, 1998, **212**, 75.
136. N. Ashammakhi and P. Rokkanen, *Biomaterials*, 1997, **18**, 3.
137. P. Mainil-Varlet, B. Rahn and S. Gogolewski, *Biomaterials*, 1997, **18**, 257.
138. J. E. Bergsma, R. R. M. Bos, F. R. Rozema, W. Jong and G. de Boering, *J. Mater. Sci.: Mater. Med.*, 1996, **7**, 1.
139. R. L. Reis and A. M. Cunha, *J. Mater. Sci.: Mater. Med.*, 1995, **6**, 786.
140. R. L. Reis, A. M. Cunha, P. S. Allan and M. J. Bevis, *Polym. Advan. Technol.*, 1996, **7**, 784.
141. R. L. Reis, A. M. Cunha, P. S. Allan and M. J. Bevis, *Polym. Adv. Technol.*, 1997, **16**, 263.
142. A. J. Salgado, O. P. Coutinho, R. L. Reis and J. E. Davies, *J. Biomed. Mater. Res. A*, 2007, **80A**, 983.
143. A. Dekker, C. Panfil, M. Valdor, H. Richter and C. J. Kirkpatrick, *Cells Mater.*, 1994, **4**, 101.
144. C. Morrison, R. Macnair, C. MacDonald, A. Wykman, I. Goldie and M. H. Grant, *Biomaterials*, 1995, **16**, 987.
145. L. Bacakova, V. Stary, O. Kofronova and V. Lisa, *J. Biomed. Mater. Res.*, 2001, **54**, 567.
146. H. Zreiqat, C. R. Howlett, A. Zannettino, P. Evans, G. Schulze-Tanzil, C. Knabe and M. Shakibaei, *J. Biomed. Mater. Res.*, 2002, **62**, 175.

147. L. Calandrelli, B. Immirzi, M. Malinconico, G. Orsello, M. G. Volpe, F. Della Ragione, V. Zappia and A. Oliva, *J. Biomed. Mater. Res.*, 2002, **59**, 611.
148. M. J. Dalby, L. Di Silvio, N. Gurav, B. Annaz, M. V. Kayser and W. Bonfield, *Tissue Eng.*, 2002, **8**, 453.
149. P. Filippini, G. Rainaldi, A. Ferrante, B. Mecheri, G. Gabrielli, M. Bombace, P. L. Indovina and M. T. Santini, *J. Biomed. Mater. Res.*, 2001, **55**, 338.
150. K. Anselme, P. Linez, M. Bigerelle, D. Le Maguer, A. Le Maguer, P. Hardouin, H. F. Hildebrand, A. Iost and J. M. Leroy, *Biomaterials*, 2000, **21**, 1567.
151. M. J. Dalby, L. Di Silvio, G. W. Davies and W. Bonfield, *J. Mater. Sci.: Mater. Med.*, 2000, **12**, 805.
152. H. Zreiqat, P. Evans and C. R. Howlett, *J. Biomed. Mater. Res.*, 1999, **44**, 389.
153. A. P. Marques and R. L. Reis, *Mater. Sci. Eng., C*, 2005, **25**, 215.
154. R. L. Reis, A. M. Cunha, P. S. Allan and M. J. Bevis, *Med. Plast. Biomater.*, 1997, **4**, 46.
155. M. E. Gomes, A. S. Ribeiro, P. B. Malafaya, R. L. Reis and A. M. Cunha, *Biomaterials*, 2001, **22**, 883.
156. M. E. Gomes, R. L. Reis, A. M. Cunha, C. A. Blitterswijk and J. D. de Bruijn, *Biomaterials*, 2001, **22**, 1911.
157. S. C. Mendes, R. L. Reis, Y. P. Bovell, A. M. Cunha, C. A. van Blitterswijk and J. D. de Bruijn, *Biomaterials*, 2001, **22**, 2057.
158. I. B. Leonor, A. Ito, K. Onuma, N. Kanzaki and R. L. Reis, *Biomaterials*, 2003, **24**, 579.
159. A. P. Marques, R. L. Reis and J. A. Hunt, *Biomaterials*, 2002, **23**, 1471.
160. A. J. Salgado, M. E. Gomes, A. Chou, O. P. Coutinho, R. L. Reis and D. W. Hutmacher, *Mater. Sci. Eng., C*, 2002, **20**, 27.
161. H. U. Cameron, R. H. Mills, R. W. Jackson and I. Macnab, *Clin. Orthop. Relat. Res.*, 1974, **100**, 287.
162. L. D. T. Topoleski, P. Ducheyne and J. M. Cuckler, *J. Biomed. Mater. Res.*, 1990, **24**, 135.
163. C. I. Vallo, P. E. Montemartini, M. A. Fanovich, J. M. Porto and T. R. Cuadrado, *J. Biomed. Mater. Res.*, 1999, **48**, 150.
164. C. Elvira, B. Vlazquez, J. San Romlan, B. Levenfeld, P. Ginebra, X. Gil and J. A. Planell, *J. Mater. Sci.: Mater. Med.*, 1998, **9**, 679.
165. S. Morita, K. Furuya, K. Ishihara and N. Nakabayashi, *Biomaterials*, 1999, **19**, 1601.
166. G. Lewis, J. S. Nyman and H. H. Triew, *J. Biomed. Mater. Res., B.*, 1997, **38**, 221.
167. P. J. van Mullen, J. R. de Wijn and J. M. Vaandrager, *Ann. Plast. Surg.*, 1988, **21**, 576.
168. C. S. Pereira, A. M. Cunha, R. L. Reis, B. Vlazquez and J. San Romlan, *J. Mater. Sci.: Mater. Med.*, 1998, **9**, 825.

169. I. Espigares, C. Elvira, J. F. Mano, B. Vlazquez, J. S. Romlan and R. L. Reis, *Biomaterials*, 2002, **23**, 1883.
170. L. F. Boesel, S. C. P. Cachinho, M. H. V. Fernandes and R. L. Reis, *Acta Biomater.*, 2007, **3**, 175.
171. R. C. Thomson, M. C. Wake, M. Yaszemski and A. G. Mikos, *Adv. Polym. Sci.*, 1995, **122**, 247.
172. S. V. Madihally and H. W. T. Matthew, *Biomaterials*, 1999, **20**, 1133.
173. L. Lu and A. Mikos, *Mater. Res. Soc. Bull.*, 1996, **21**, 28.
174. W. W. Minuth, M. Sittinger and S. Kloth, *Cell Tissue Res.*, 1998, **291**, 1.
175. N. Rotter, J. Aigner, A. Nauman, H. Planck, C. Hammer, G. Burmester and M. Sittinger, *J. Biomed. Mater. Res.*, 1998, **42**, 347.
176. R. Thomson, M. Yaszemski and and A. Mikos, in *Principles of Tissue Engineering*, (Eds) R. Lanza, R. Langer and W. Chick, Academic Press, New York, 1997, p. 263.
177. D. J. Mooney, D. F. Baldwin, N. P. Suh and J. P. Vacanti, *Biomaterials*, 1996, **17**, 1417.
178. M. E. Gomes, J. S. Godinho, D. Tchalamov, A. M. Cunha and R. L. Reis, *Mater. Sci. Eng., C*, 2002, **20**, 19.
179. N. M. Neves, A. Kouyumdzhiev and R. L. Reis, *Mater. Sci. Eng., C*, 2005, **25**, 195.
180. J. Oliveira, A. Crawford, J. Mundy, A. Moreira, M. Gomes and P. Hatton, *J. Mater. Sci.: Mater. Med.*, 2007, **18**, 295.
181. E. R. Balmayor, K. Tuzlakoglu, H. S. Azevedo and R. L. Reis, *Acta Biomater.*, 2009, **5**, 1035.
182. M. I. Santos, R. E. Unger, R. A. Sousa, R. L. Reis and C. J. Kirkpatrick, *Biomaterials*, 2009, **30**, 4407.
183. D. Puppi, F. Chiellini, A. M. Piras and E. Chiellini, *Prog. Polym. Sci.*, 2010, **35**, 403.
184. L. Youxin and T. Kisse, *J. Controlled Release*, 1993, **27**, 247.
185. K. J. Zhu, L. Xiangzhou and Y. Shilin, *J. Appl. Polym. Sci.*, 1990, **39**, 1.
186. K. Uekama, F. Hirayama and T. Irie, *Chem. Rev.*, 1998, **98**, 2045.
187. M. E. Brewster, W. R. Anderson, K. S. Estes and N. Bodor, *J. Pharm. Sci.*, 1991, **80**, 380.
188. R. Challa, A. Ahuja, J. Ali and R. K. Khar, *AAPS PharmSciTech.*, 2005, **6**, E329.
189. F. Hirayama and K. Uekama, *Adv. Drug Delivery Rev.*, 1999, **36**, 125.
190. T. Loftsson and E. Stefánsson, *Drug Dev. Ind. Pharm.*, 1997, **23**, 473.
191. T. Loftsson, N. Leeves, B. Bjornsdottir, L. Duffy and M. Masson, *J. Pharm. Sci.*, 1999, **88**, 1254.
192. R. A. Rajewski and V. J. Stella, *J. Pharm. Sci.*, 1996, **85**, 1142.
193. V. J. Stella and R. A. Rajewski, *J. Pharm. Res.*, 1997, **14**, 556.
194. C. Elvira, J. F. Mano, J. San Roman and R. L. Reis, *Biomaterials*, 2002, **23**, 1955.
195. C. Bastioli, in *Degradable Polymers: Principles and Applications*, (Eds) G. Scott and D. Gilead, Chapman and Hall, London, 1995, p. 112.
196. L. Chen, X. Li, L. Li and S. Guo, *Curr. Appl. Phys.*, 2007, **7S1**, e90.

197. J. Mulhbacher, P. Ispas-Szabo, M. Ouellet, S. Alex and M. A. Mateescu, *Int. J. Biol. Macromol.*, 2006, **40**, 9.
198. F. Onofre, Y.-J. Wang and A. Mauromoustakos, *Carbohydr. Polym.*, 2009, **76**, 541.
199. P. B. Malafaya, F. Stappers and R. L. Reis, *J. Mater. Sci.: Mater. Med.*, 2006, **17**, 371.
200. E. R. Balmayor, K. Tuzlakoglu, A. P. Marques, H. S. Azevedo and R. L. Reis, *J. Mater. Sci.: Mater. Med.*, 2008, **19**, 1617.
201. S. Milojevic, J. M. Newton, J. H. Cummings, G. R. Gibson, R. L. Botham, S. G. Ring, M. Stockham and M. C. Allwood, *J. Controlled Release*, 1996, **38**, 75.
202. E. L. McConnell, J. Tutas, M. A. M. Mohamed, D. Manning and A. W. Basit, *Cellulose*, 2007, **14**, 25.
203. I. Silva, M. Gurruchaga and I. Goñi, *Carbohydr. Polym.*, 2009, **76**, 593.
204. M. Kuentz, P. Egloff and D. Röthlisberger, *Eur. J. Pharm. Biopharm.*, 2006, **63**, 37.
205. H. Kranz and T. Wagner, *Eur. J. Pharm. Biopharm.*, 2006, **62**, 70.
206. A. R. C. Duarte, J. F. Mano and R. L. Reis, *J. Supercrit. Fluids*, 2009, **49**, 279.
207. A. R. C. Duarte, S. G. Caridade, J. F. Mano and R. L. Reis, *Mater. Sci. Eng., C*, 2009, **29**, 2110.
208. A. R. C. Duarte, J. F. Mano and R. L. Reis, *Acta Biomater.*, 2009, **5**, 2054.
209. H.-S. Yoon, J. H. Lee and S.-T. Lim, *Carbohydr. Polym.*, 2009, **76**, 449.
210. M. Casas, C. Ferrero and R. Jiménez-Castellanos, *Carbohydr. Polym.*, 2010, **80**, 71.
211. J. Peerapattana, P. Phuvarit, V. Srijesdaruk, D. Preechagoon and A. Tattawasart, *Carbohydr. Polym.*, 2010, **80**, 453.
212. H. J. Prado, M. C. Matulewicz, P. R. Bonelli and A. L. Cukierman, *Carbohydr. Res.*, 2009, **344**, 1325.
213. A. S. Hoffman, *Adv. Drug Delivery Rev.*, 2002, **43**, 3.
214. E. Karadag, D. Saraydin, S. Centinkaya and O. Guven, *Biomaterials*, 1996, **17**, 67.
215. P. A. Netti, J. C. Shelton, P. A. Revell, C. Pirie, S. Smith, L. Ambrosio, L. Nicolais and W. Boneld, *Biomaterials*, 1993, **14**, 1098.
216. C. D. Young, J. R. Wu and T. L. Tsou, *J. Membr. Sci.*, 1998, **146**, 83.
217. A. Abusafieh, S. Siegler and S. R. Kalidindi, *J. Biomed. Mater. Res.*, 1997, **38**, 314.
218. L.-M. Zhang, C. Yang and L. Yan, *J. Bioact. Compat. Pol.*, 2005, **20**, 297.
219. V. D. Athawale and V. Lele, *Starch/Staerke*, 2000, **53**, 7.
220. V. D. Athawale and L. Vidyagauri, *Carbohydr. Polym.*, 1998, **35**, 21.
221. C. Seidel, W. M. Kulicke, C. Heb, B. Hartmann, M. D. Lechner and W. Lazik, *Starch/Staerke*, 2001, **53**, 305.
222. S. P. Bhuniya, S. Rahman, A. J. Satyanand, M. M. Gharia and A. M. Dave, *J. Polym. Sci., A: Polym. Chem.*, 2003, **41**, 1650.

223. K. Pal, A. K. Banthia and D. K. Majumdar, *Afr. J. Biomed. Res.*, 2006, **9**, 23.
224. A. E.-H. Ali and A. Al Arifi, *Carbohydr. Polym.*, 2009, **78**, 725.
225. M. R. Saboktakin, A. Maharramov and M. A. Ramazanov, *Carbohydr. Polym.*, 2009, **77**, 634.
226. A. J. M. Al-Karawi and A. H. R. Al-Daraji, *Carbohydr. Polym.*, 2010, **79**, 769.
227. A. Shalviri, Q. Liu, M. J. Abdekhodaie and X. Y. Wu, *Carbohydr. Polym.*, 2010, **79**, 898.
228. O. S. Lawal, J. Storz, H. Storz, D. Lohmann, D. Lechner and W.-M. Kulicke, *Eur. Polym. J.*, 2009, **45**, 3399.
229. T. Mehling, I. Smirnova, U. Guenther and R. H. H. Neubert, *J. Non-Cryst. Solids*, 2009, **355**, 2472.

CHAPTER 6

Gum Arabic and other Exudate Gums

GLYN O. PHILLIPS[a] AND ALED O. PHILLIPS[b]

[a] Phillips Hydrocolloids Research Ltd, 45 Old Bond Street, London W1S 4AQ, UK; [b] Institute of Nephrology, University of Wales College of Medicine, Cardiff CF14 4XN, Wales, UK

6.1 Introduction

Breaking the bark of trees in a tropical environment, either by insect attack or by delibreate wounding, can yield a sticky polysaccharide gum containing a small amount of protein. The oldest used in medicine is gum arabic. This is an exudate from acacia trees. Throughout the world there are at least 1000 species of acacia, but only two species (*Acacia senegal* and *Acacia seyal*) are permitted for use in food and medicines by the world's regulatory bodies. Thus their safe use now requires careful characterisation, in a way that was not even contemplated previously. This article will describe such characterisation and show how the structure relates to the various functionalities required of this gum in a variety of applications. It is used in food and in medicine; the experience in both areas is complementary, since they draw on the same properties, and for this reason both are discussed in this article. The other tree exudate gums, used in food and medicine, gum tragacanth and gum karaya, have not be studied in such detail as gum arabic but have excellent properties which could be more extensively harnessed in future.

The article will deal first with gum arabic, its structure and new methods of characterisation. Examples of its functionality will be given and the role of the main protein component, the arabinogalactan protein, identified.

RSC Polymer Chemistry Series No. 1
Renewable Resources for Functional Polymers and Biomaterials
Edited by Peter A. Williams

Published by the Royal Society of Chemistry, www.rsc.org

The regulatory definition of the species acceptable in food and pharmaceuticals is given. How structure controls the functional application is described particularly with regard to emulsification, which is the main characteristic which determines its effectiveness in oil beverages, food dressings, ointments and oily pharmaceutical preparations. The other exudate gums, tragacanth and karaya, are also described.

The most recent and exciting application of these old gums is their new use as dietary fibre, which allow them to perform both in a physical role and also to ferment through colonic microflora to give short-chain fatty acids, which have a very beneficial effect on colon health.

6.2 Gum Arabic

6.2.1 Origin

Gum arabic or gum *Acacia* is a tree gum exudate and has been an important article of commerce since Ancient times. It was used by the Egyptians for embalming mummies and also for paints for hieroglyphic inscriptions. Traditionally the gum has been obtained mainly from the *Acacia senegal* species. The trees grow widely across the Sahelian belt of Africa, situated north of the equator up to the Sahara desert and from Senegal in the west to Somalia in the east. The gum oozes from the stems and branches of trees (usually after five years of age or more) when subjected to stress conditions such as drought, poor soil or wounding. Production is stimulated by 'tapping', which involves removing sections of the bark with an axe, taking care not to damage the tree. The sticky gummy substance dries on the branches to form hard nodules which are picked by hand and are sorted according to colour and size. Commercial samples commonly contain *Acacia* species other than *Acacia senegal*, notably *Acacia seyal*. In Sudan the gums from *Acacia senegal* and *seyal* are referred to as hashab and talha respectively. The former is a pale to orange–brown solid which breaks with a glassy fracture and the latter is darker, more friable and is rarely found in lumps in export consignments. Hashab is undoubtedly the premier product but the lower priced talha has found recent uses which have boosted its value. It is not possible to identify precisely the exact balance between these two products in the market place since it is continually changing. Table 6.1 gives the botanical classification of the acacia trees from which the regulatory approved species *Acacia senegal* and *Acacia seyal* are derived.

More recently there has been new work on the generic status of *Acacia* (*leguminosae*: *mimosoideae*).[1]

6.2.2 Regulatory Requirements

There has been constant debate over the past 20 years about whether *Acacia seyal* is an acceptable gum for use in food and medicine.[2] The situation appears now to have been resolved and the FAO WHO Joint Committee for Food Additives (JECFA) Specification approved by the main international

Table 6.1 Botanical classification of gum acacia trees according to Bentham's & Vassal's main division of the genus.

Australian species – subgen. Heterophyllum Vas

Series **PHYLLODINEAE BENTH.**
A. tetragonophylla, A. calamifolia, A. uncinata, A. montana, A. pruinocarpa, A. victoriae, A. aestivalis, A. bancroftii, A. mabellae, A. microbotrya, A. murrayana, A. penninervis, A. podalyriifolia, A. prainii, A. pycnantha, A. retinodes, A. rostellifera, A. rubida, A. salicina, A. saligna (syn. A.cyanophylla), A. goorginae, A. harpophylla, A. cyclops, A. latescens.

Series **JULIFLORAE Benth.**
A. pubifolia, A. acradenia, A. aneura, A. kempeana, A. coolgardiensis, A. microneura, A. auriculiformis, A. leptostachya, A. stereophylla, A. torulosa, A. holosericea, A. mangium, A. beauverdiana.

Series **BOTRYCEPHALAE Benth.**
A. dealbata, A. deanei, A. decurrens, A. alata, (syn. A. terminalis), A. filicifolia, A. leucoclada, A. mearnsii, (syn. A. mollissima), A. parramattensis, A. parvipinnula, A. silvestris, A. trachyphloia.

Indian, African & American species

Series **GUMMIFERAE Benth. = Subgen. ACACIA Vas.** also called **"Acacia seyal complex"**
A. abyssinica subsp.calophylla, A. nilotica (syn. A. adansonii and A. arabica), A. drepanolobium, A. farnesiana, A. gerrardii, A. giraffae, A. hebeclada, A. karroo, A. kirkii, A. leucophloea, A. nebrownii, A. nubica, A. reficiens, A. rigidula, A. seyal, A. sieberana (var.villosa, var. woodii), A. seyal (var. fistula, var. seyal), A. tortilis (subsp. heteracantha).

Series **VULGARES Benth. = Subgen. ACULEIFERUM Vas.** also called **"Acacia senegal complex"**
A. berlandieri, A. polyacantha subsp. campylacantha, A. catechu, A. erubescens, A. fleckii, A. goetzii subsp. goetzii, A. laeta, A. mellifera, subsp detinens, A. senegal, A. sundra.

regulatory body is given in Table 6.2, which if followed permits the use of gum arabic throughout the world in food and medicine. Nevertheless, it must be conceded that there are geographical variations, none of which, however, contradict the JECFA Specification of gum arabic as a food additive.

Up until 2008 gum arabic was subjected to the constraints of its regulatory definition as a food additive (E, 414 and INS 414). This described its function as: "thickener/emulsifier, stabiliser, glazing agent, bulking agent". This status did not recognise its use for decades as an ingredient in gum confectioneries (gum balls and pastilles) *etc*. For the same reason there was no regulatory basis for its designation as a dietary fibre, since this could not be accommodated within E414. There was clearly a problem in dealing with materials such as gum arabic which had more than one functionality. Various Working Parties of the EC could not find a way of dealing with this anomaly. To be allowed an additional functionality would need consideration under the Novel Foods Legislation (EC 258/97).

However, since 2008 there has been a groundbreaking transformation in the regulatory status of gum arabic. The hitherto restrictive status as a food

Table 6.2 Specification approved by the FAO WHO Joint Committee for Food Additives (JECFA) and accepted by Codex Alimentarius.[a]

SYNONYMS	Gum arabic (*Acacia senegal*), gum arabic (*Acacia seyal*), Acacia gum; arabic gum; INS No. 414
DEFINITION	Gum arabic is a dried exudate obtained from the stems and branches of *Acacia senegal* (L.) Willdenow or *Acacia seyal* (fam. *Leguminosae*) Gum arabic consists mainly of high-molecular weight polysaccharides and their calcium, magnesium and potassium salts, which on hydrolysis yield arabinose, galactose, rhamnose and glucuronic acid. Items of commerce may contain extraneous materials such as sand and pieces of bark, which must be removed before use in food.
C.A.S. number	9000-01-5
DESCRIPTION	Gum arabic (*A. senegal*) is a pale white to orange-brown solid, which breaks with a glassy fracture. The best grades are in the form of whole, spheroidal tears of varying size with a matt surface texture. When ground, the pieces are paler and have a glassy appearance. Gum arabic (*A. seyal*) is more brittle than the hard tears of gum arabic (*A. senegal*). Gum arabic is also available commercially in the form of white to yellowish-white flakes, granules, powder, roller dried, or spray-dried material. An aqueous solution of 1 g in 2 ml flows readily and is acid to litmus.
FUNCTIONAL USES	Emulsifier, stabiliser, and thickener
CHARACTERISTICS	
IDENTIFICATION	
Solubility	One gram dissolves in 2 ml of water; insoluble in ethanol
Gum constituents	Proceed as directed under Gum Constituents Identification (FNP 5) using the following as reference standards: arabinose, galactose, mannose, rhamnose, galacturonic acid, glucuronic acid and xylose. Arabinose, galactose, rhamnose and glucuronic acid should be present. Additional spots corresponding to mannose, xylose and galacturonic acid should be absent.
Optical rotation	Gum from *A. senegal*: aqueous solutions are levorotatory Gum from *A. seyal*: aqueous solutions are dextrorotatory Test a solution of 10 g of sample (dry basis) in 100 ml of water (if necessary, previously filtered through a No. 42 paper or a 0.8 μm millipore filter), using a 200-mm tube.
PURITY	
Loss on drying	Not more than 15% (105°, 5 h) for granular and not more than 10% (105°, 4 h) for spray dried material Unground samples should be powdered to pass through a No. 40 sieve and mixed well before weighing
Total ash	Not more than 4 %
Acid-insoluble ash	Not more than 0.5 %
Acid-insoluble matter	Not more than 1 %
Starch or dextrin	Boil a 1 in 50 solution of the sample, cool and add a few drops of Iodine T.S. No bluish or reddish colour should be produced.

Table 6.2 (*Continued*)

Tannin-bearing gums	To 10 ml of a 1 in 50 solution of the sample, add about 0.1 ml of ferric chloride TS. No blackish colouration or blackish precipitate should be formed.
Microbiological criteria	*Salmonella* Spp.: Negative per test *E. coli*: Negative in 1 g
Lead	Not more than 2 mg/kg Prepare a sample solution as directed for organic compounds in the Limit Test and determine by *atomic absorption spectroscopy*

[a]Prepared at the 51st JECFA (1998) and published in FNP 52 Add 6 (1998); republished in FNP 52 Add 7 (1999) to include editorial changes. Supersedes specifications prepared at the 49th JECFA (1997), published in FNP 52 Add 5 (1997). ADI "not specified", established at the 35th JECFA in 1989.

"additive" no longer remains a constraint. It is now an acknowledged "ingredient" in view of its long use in this function. In considering this matter the United Kingdom Food Safety Agency first examined the evidence, and in view of its acceptance the EC has similarly confirmed that gum arabic no longer requires examination under the Novel Foods legislation for those established uses outside the "additive" category. Codex and the EC in 2009 also finally accepted a legal definition of "dietary fibre" to which gum arabic conforms scientifically. Gum arabic is thus a flexible ingredient and additive poised to find multiple new applications.

6.2.3 The Chemical Components of Gum Arabic[3–5]

All acacia gums in chemical terms are arabinogalactan proteins (AGPs). AGPs are found in most higher plants and in many of their secretions. They are a group of macromolecules characterised by a high proportion of carbohydrate in which D-galactose and L-arabinose are the predominant monosaccharides. There is also a low proportion of protein, typically containing high levels of hydroxyproline. AGPs and AGs (arabinogalactan without protein) are found in flowering plants from every taxonomic group tested. In higher plants AGPs occur in leaves, stems, rods, floral parts, seeds and in large quantities in the trunks of some angio- and gymno- sperms. The acacia gums, therefore, belong to a catholic family of structurally related AGPs.

The component sugars are D-galactose, L-arabinose, L-rhamnose, D-glucuronic acid and 4-0-methyl glucuronic acid and the proportions vary in *A. seyal* and *A. senegal* (Table 6.3). The amino acids which make up the protein present in both species are shown in Table 6.4. Approximately 50% of the proteinaceous matter found in gum arabic consists of hydroxyproline, serine and proline. There are variations in protein content for various gums of the acacia species ranging from 0.13 to 10.4%.

The structural linkages of the sugars in the carbohydrate (AG) moieties are illustrated in Figure 6.1.[6] In the polysaccharide chain there are uniform blocks of (1→3)-linked D-galactopyranosyl residues; these blocks are comparable in size to those postulated for many AGs of simpler structure, so confirming the relationship of the acacia gums within the wider AGP family. The point of attachment of the AG to the protein chain is probably *via* hydroxyproline.

Table 6.3 Sugar composition of *Acacia senegal* and *Acacia seyal*.

	Acacia senegal (%)	*Acacia seyal (%)*
D-galactose	44	38
L-arabinose	27	46
L-rhamnose	13	4
D-glucuronic acid	14.5	6.5
4-O-methyl D-glucuronic acid	1.5	5.5
Nitrogen	0.36	0.15

Table 6.4 Amino acid composition of *Acacia senegal* and *Acacia seyal* gums (residues/1000 residues).

	Acacia senegal	*Acacia seyal*
Hyp	256	240
Asp	91	65
Thr	72	62
Ser	144	170
Glu	36	38
Pro	64	73
Gly	53	51
Ala	28	38
Cys	3	
Val	35	42
Met	2	
Ile	11	16
Leu	70	85
Tyr	13	13
Phe	30	24
His	52	51
Lys	27	18
Arg	15	11

6.2.4 The Molecular Architecture of Gum Arabic[7–9]

Both *A. seyal* and *A. senegal* are polydisperse in character. These gums are not a discrete chemical species but are complex polysaccharides consisting of several sugars as previously noted. As already noted both gums also contain a small amount (1–3%) of protein as an integral part of the structure. The carbohydrate structure has shown that it consists of a core of β-(1→3)-linked galactose units with extensive branching at the C6 position (Figure 6.1). The branches consist of galactose and arabinose and terminate with rhamnose and glucuronic acid.

The gum consists of three broad molecular fractions, which differ principally in their size and protein contents. Most of the gum (~90%) contains very little protein and has a molecular mass of 1×10^5–1×10^6 (average ~4.0×10^5). This is designated arabinogalactan (AG). A second fraction, (~10% of the total)

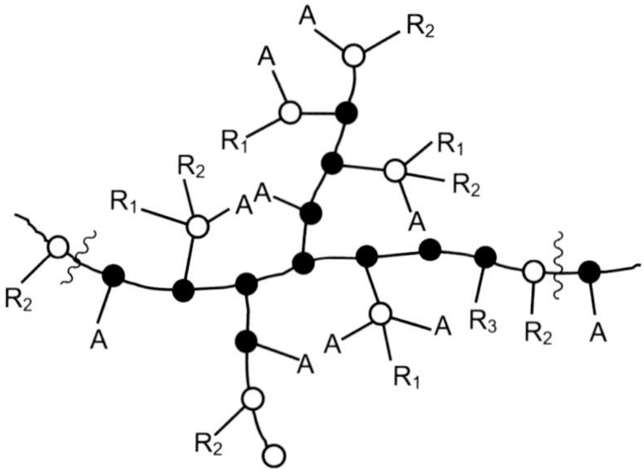

Figure 6.1 Molecular structure for *A. senegal* gum.[6] A = arabinose; filled circle = 3-linked galactose (galactose attached); open circle = 6-linked galactose (galactose or glucuronic acid attached or end group); R_1 = rhamnose-glucuronic acid; R_2 = galactose-3-arabinose; R_3 = arabinose-3-arabinose-3-arabinose.

contains ~10% protein and has a molecular mass of $1–3\times10^6$ and has been shown to have a 'wattle-blossom'-type structure where blocks of carbohydrate of molecular mass $\sim3.5\times10^5$ are connected to a common polypeptide chain. This is designated arabinogalactan protein (AGP). The third fraction (~2% of the total) contains up to 50% protein and has a molecular mass of $\sim1\times10^5$. This is designated glycoprotein (GP). The high degree of branching gives rise to a very compact molecular structure for all of the fractions and results in solutions of very low viscosity. Fraction 2 (AGP) has been shown to be responsible for the gum's excellent ability to stabilise oil-in-water emulsions.[7–9]

Size exclusion chromatography (SEC; also called gel permeation chromatography, GPC), coupled to an on-line absolute molecular weight determining device (such as a light scattering photometer) and a concentration sensitive detector (such as refractive index or ultraviolet) is currently the best available technique for the characterisation of gum arabic and quantifying the amounts of the three components AGP, AG and GP. The light scattering detector utilises the principle that the intensity of light scattered elastically by a molecule (Raleigh scattering) is directly proportional to the molecular weight (mass detector). By using the refractive index detector connected directly after the light scattering it is possible to measure the molecular weight of each fraction as it elutes from the GPC column. In addition to these two detectors, it is also possible to use an ultraviolet (UV) detector at 214 nm, which specifically shows the amount of protein in the fractionated gum. The typical elution behaviour of gum arabic after being separated by GPC is as follows:

1. The light scattering response shows two distinctive peaks. The first peak has a high response since it corresponds to the high molecular weight material (AGP) content. The second peak (AG) is broader with a lower response and it accounts for the rest of the gum (~90%).

2. The refractive index (RI) response also shows two peaks but the response is opposite to that in light scattering. This is because it is a concentration detector and since the AGP is only 10% of the total gum its peak is smaller than that of the AG and GP which consists ~90% of the total mass.
3. The UV response shows three peaks. The first peak is for the AGP, which has the protein core and the carbohydrate attached to it. The second peak appears as a shoulder immediately after the AGP and corresponds to the AG. Finally the third peak elutes before the total volume and it corresponds to the glycoprotein (GP). The GP peak is not detected on the light scattering (mass detector) since it has low molecular weight. Also it cannot be seen on the refractive index (concentration detector).

The fractionation possible using this new and powerful technique is shown in Figure 6.2. From this data it is possible to measure the weight average molecular weight of the gum arabic (at 6.22×10^5) and quantify the amount of AGP present (10.6%). Table 6.5 shows how a series of commercial gum arabic samples can be characterised. These samples are drawn from a primary producer who presented them as identical. Even so, it is evident that there is a large natural variation between them. They can also be expected to behave differently in particular applications and this is found to be the case.

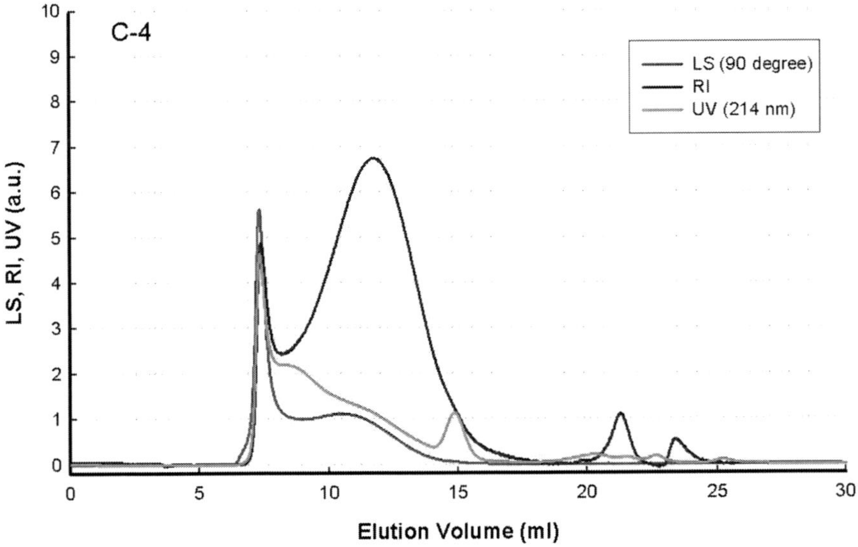

Figure 6.2 GPC chromatograms of gum arabic showing the light scattering (LS), refractive index (RI) and UV at 214 nm. The data was normalised as injected mass of 0.4 mg to display on the same y-axis scale.

Table 6.5 Molecular weight parameters of commercial gum arabic samples determined by GPC-MALLS.

M_{wt} (Average)	Rg	M_{wt} 1. AGP 2. AG	% AGP	Rg
$5.77 \times 10^5 \pm 0.21$	25.2	$2.21 \times 10^6 \pm 0.04$	13.5	31.5
		$3.47 \times 10^5 \pm 0.12$		17.6
$6.13 \times 10^5 \pm 0.24$	26.4	$2.47 \times 10^6 \pm 0.08$	13.0	32.7
		$3.64 \times 10^5 \pm 0.15$		18.9
$8.28 \times 10^5 \pm 0.30$	31.7	$3.54 \times 10^6 \pm 0.14$	14.0	36.7
		$4.65 \times 10^5 \pm 0.14$		25.3
$5.37 \times 10^5 \pm 0.20$	24.7	$1.85 \times 10^6 \pm 0.06$	14.9	29.4
		$3.30 \times 10^5 \pm 0.13$		19.7
$5.30 \times 10^5 \pm 0.21$	22.5	$1.99 \times 10^6 \pm 0.07$	12.3	27.2
		$3.41 \times 10^5 \pm 0.14$		18.2
$8.88 \times 10^5 \pm 0.32$	33.8	$3.75 \times 10^6 \pm 0.15$	16.0	39.0
		$4.57 \times 10^5 \pm 0.14$		26.0
$8.34 \pm 0.32 \times 10^5$	21.6	$3.06 \pm 0.10 \times 10^6$	14.9	29.6
		$4.73 \pm 0.19 \times 10^5$	91.9	–
$5.83 \pm 0.26 \times 10^5$	–	$1.99 \pm 0.08 \times 10^6$	12.1	23.7
		$3.85 \pm 0.17 \times 10^5$	88.1	–

Notes

- Weight average molecular weight (Mw) for the whole gum is given in column 2.
- Rg is the RMS-radius of gyration – a measure of overall size.
- The molecular weights of AGP and AG can be measured from Figure 2.
- The % of AGP is given in column 3.

6.2.5 How Structure Affects Functional Performance

In pharmaceutical formulations and in food beverage emulsions, it is the ability of gum arabic to form stable emulsions which is the necessary property. This functionality is achieved by bridging oil–water interfaces, so coating small droplets of the oil or fat to prevent their coalescence. In this respect it is the AGP component which coats the oil or fat droplet to form the emulsion. The hydrophobic moiety of the gum arabic bridges the hydrophilic barrier to coat the oil droplet, with the driving force being directed by the entropic energy of the hydrophilic carbohydrate groups residing in the water layer. This leads to the ability to stabilise smaller droplet sizes and to provide considerably greater long-term stability for the emulsion.[8,9] This behaviour is shown in Figure 6.3.

The molecular parameters shown in Table 6.5 can provide an accurate indicator of the emulsification effectiveness of a gum arabic sample. A standard method can be used to measure both the initial droplet size and the stability of gum arabic emulsions.

6.2.6 To Establish an "Emulsification Index" for Variable Gum Arabic Samples

1. Gum arabic (1 kg) should be dissolved in ion exchanged water (4 kg), centrifuged to remove insolubles, and prepared into 20 wt% aqueous solutions.

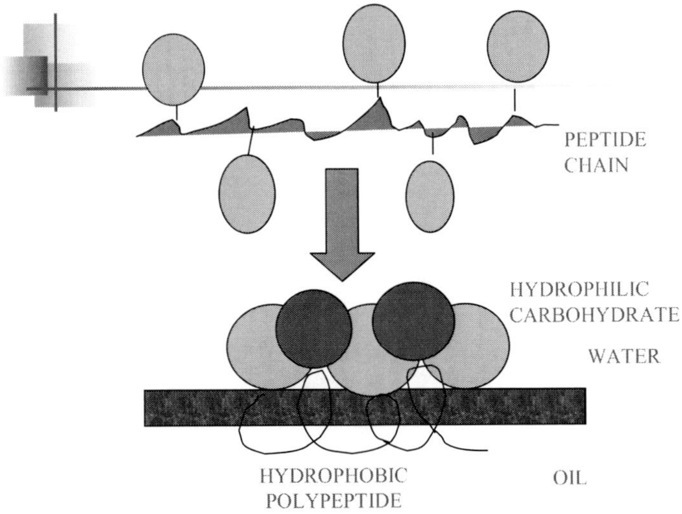

Figure 6.3 The figure shows the AG carbohydrate moieties strung along the peptide chain to form AGP in what is termed the "wattle blossom model". The hydrophobic protein coats the oil droplet and the AG hydrophilic groups align themselves into the water layer.

2. These 20 wt% aqueous gum arabic solutions (850 g) to be added to medium-chain triglycerides (octanoic/decanoic acid triglyceride O.D.O.) during agitation.
3. Each of the mixtures must be emulsified 4 times using a homogeniser at a pressure of 44 MPa (450 kg cm^{-2}).
4. The particle size diameter of the emulsion just after preparation (initial) can be measured using a particle size distribution analyzer. The emulsification effectiveness can be evaluated by the initial particle size of emulsion.
5. The emulsion must then be subjected to acceleration testing (7 days storage at 60 °C). The particle size diameter of emulsion after the acceleration test is measured using a particle size distribution analyzer. The emulsification stability is evaluated by the change in particle size of emulsion after the acceleration test.
6. The change in particle size after the acceleration test (7 days storage at 60 °C) provides a parameter which designates the category of the gum sample. Therefore, the gum samples which showed a change of 0.1 μm or less were given category 1 (good emulsifier). A change >0.1 μm–1.0 μm put the sample in category 2. The less stable emulsions which showed a change >1.0 μm were given category 3 (poor emulsifier).

The emulsification effectiveness was measured for the two samples for which molecular parameters are given in Table 6.6 and the results shown in Table 6.7. To produce a good emulsion the gum must produce a small droplet size and enable this to be stable over a long period. In our work we have measured the

Table 6.6 Molecular weight parameters of gum arabic determined by GPC-MALLS. Weight average molecular weight (Mw) for the whole gum is given in column 2. % mass means the recovered mass which is calculated using a dn/dc value of 0.141, Rg is the RMS-radius of gyration. M_{wt} processed as two peaks means that the high molecular weight fraction (peak 1) was processed separately and the remainder of the gum processed as the second peak.

Sample No.	*M_{wt} processed as one peak*	*% mass*	*P*	*Rg*	*M_{wt} processed as two peaks*	*% mass*	*P*	*Rg*	*State and species of the sample*
1	$5.83 \pm 0.26 \times 10^5$	100	1.91	–	$1.99 \pm 0.08 \times 10^6$	12.09	1.19	23.7	kibbled/senegal
					$3.85 \pm 0.17 \times 10^5$	88.08	1.43	–	
2	$8.34 \pm 0.32 \times 10^5$	106	2.23	21.6	$3.06 \pm 0.10 \times 10^6$	14.85	1.29	29.6	kibbled/senegal
					$4.73 \pm 0.19 \times 10^5$	91.9	1.43	–	

Table 6.7 Particle droplet size before and after the acceleration test for the determination of the emulsification index for selected samples given in Table 6.6.

Sample No.	*Initial particle size/μm*	*Particle size/μm after acc. test*	*Change/ μm*	*State of gum: emulsification index*
1	0.74	2.57	1.83	kibbled Category 3
2	0.63	0.71	0.08	kibbled/Category 1

Table 6.8 Molecular weight of the fraction at the peak height of the refractive index (Mp RI) and at the UV (Mp UV) for GA samples 1 and 2 in Table 6.5.

Sample No.	*Mp (RI)*	*Mp (UV)*	*Comments*
1	$2.67 \pm 0.09 \times 10^6$ $3.06 \pm 0.15 \times 10^5$	$1.44 \pm 0.05 \times 10^6$	(Poor emulsifier) category 3
2	$4.04 \pm 0.13 \times 10^6$ $3.55 \pm 0to.15 \times 10^5$	$4.23 \pm 0.13 \times 10^6$	(Good emulsifier) category 1

droplet size and used an accelerated stability test to compare the effectiveness of each of the gums.

From this type of evaluation it can be concluded that:

- The best emulsifiers would have the highest average molecular weight, having a higher molecular weight for peak 1 and the related protein (AGP). These measurements are reflected in the light scattering R_g values of the overall behaviour and of the high molecular weight protein.
- A point to note is that the effective emulsifiers also have the higher molecular weights and a greater proportion of the protein at peak 3 (GP).

These principles, when used to evaluate specially produced samples, indicate that on the basis of the parameters quoted sample 1 should be a poor emulsifier and 2 a good emulsifier. This has been confirmed by experimental measurements. The data in Table 6.8 show that the good emulsifier has a molecular weight of protein almost 3 times higher than the poor emulsifier. Figure 6.4 shows the distribution of protein throughout the various components which constitute the polydisperse system that makes up gum arabic. The best emulsifier has the protein within the highest molecular weight region.

6.2.7 Functionalities which are Related to Molecular Structure

The applications in the food sector depend on the action of gum arabic as a protective colloid, stabiliser, or adhesive when in contact with water, and its ability to thicken and provide a dietary food fibre with low calorific value. It can impart its desirable qualities though its influence over the viscosity, body and texture of the food. It is also a good emulsifier and a foam stabiliser in beer

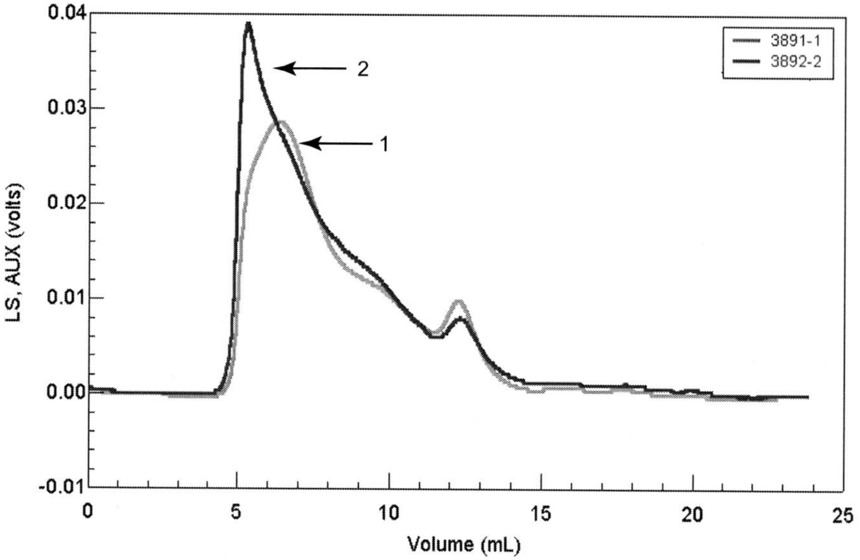

Figure 6.4 GPC chromatogram showing protein distribution (at 214 nm) for Samples 1 and 2.

and when used in spray–dried flavours, since it is capable of film-forming and giving an impenetrable film around the flavour particle. Confectionery applications depend on its ability to interact and bind water, thicken as a gel and so prevent sugar crystallisation. Here also its emulsification quality is important to enable fat to be distributed throughout the product and not move to the surface and make the food appear greasy. In this application, as in beverages, it needs to be able to coat the fat or oil droplet using its high molecular protein structure.

As a stabiliser in dairy products it needs to prevent water crystallisation and the formation of ice crystals and is thus dependent on its water absorbing properties.

In the baking industry it is used because of its adhesive property in glazes and toppings.

Similar functionalities are necessary in other application sectors, such as pharmaceuticals where gum arabic is used as a suspending agent, in syrups, emulsions, antiseptic preparations, medicines, wound dressings, cosmetics and adhesives, paints, inks, lithography and textiles. Table 6.9 illustrates the necessary functionality in each of the applications.[10]

6.3 Gum Tragacanth

6.3.1 Definition

It is defined by the Food Chemical Codex as: "the dried gummy exudation obtained from *Astralagus gummifer* Labillardiere or other Asiatic species of *Astralagus* (Fam. *Leguminosae*) a shrub originally located in the Middle East".

In the European Pharmacopoeia, gum tragacanth is defined as "the air-hardened gummy exudates, flowing naturally or obtained by incision from the

Table 6.9 Functionality of gum arabic necessary for particular applications.

Application	*Necessary functionality*
Emulsification in beverages and confectionary	Coating of an oil or a fat droplet by protein
Encapsulation of flavour essential oils; vitamins aromatic compositions, plant essences *etc.*	Forming a hard film to avoid penetration of oxidising agents
Bakery for toppings and glazes and to prevent sugar crystallisation and paper to paper adhesion (stamps and cigarette papers)	Adhesion
Texture and flavour modification in confectionary	Interact and bind water, to thicken as a gel
Chewing gum and confectionary ju-jus, pastilles and gum drops	High water absorption and tenacity
Lace curtain effect on beer, marsh mallows and whipping creams	Foam stabilisation – structure forming and interfacial rheology
Salad dressings	Fat distribution and fat replacement
Dietary Fibre in dairy products, processed fruits, bakery items, frozen desserts and food for diabetics	Traverse through intestine to ferment in the colon to provide short chain fatty acids. Low calorific value

trunk and branches of *Astralagus gummifer* Labillardiere and certain other species of *Astralagus* from Western Asia.

Although the *Astralagus* genus comprises more than 2000 species, most commercially traded gum tragacanth is obtained from two species: *Astralagus gummifer* Labill. and *A. microcephalus* Willd. The plants are small, low bushy perennial shrubs having a large taproot along with branches and grow wildly in the dry deserts and mountainous regions of South West Asia, from Pakistan to Greece and in particular in Iran and Turkey. Plants develop a mass of gum in the centre of the root, which swells in the summer heat. If the stem is slit, soft gum is exuded.

6.3.2 Properties

Table 6.10

Solubility in cold water	Two components: one which swells in water and a water-soluble component
Functionality	• acid stable and resistant • bifunctional emulsifier • creamy texturepseudoplastic • film former • bodying agent • adhesive • suspending agent • thickener
pH	5.0–6.0

6.3.3 Typical Product Specification of a Commercial Gum Tragacanth

Table 6.11

Appearance	Off-white to creamy coloured fine powder
Loss on drying	12% maximum
Ash	3% maximum
Acid insoluble ash	0.3% maximum
Viscosity 1% in water	800 ± 150 cps[2]
Particle size	90% minimum pass 150 mesh BSS
Heavy metals: Pb	10% maximun
Heavy metal: As	3% maximum
Microbiology	*Salmonella*/*Escherichia coli* absent

6.3.4 Applications

- Oil and flavour emulsions
- Soft drinks
- Confectionery and icings
- Bakery emulsions and fillings
- Ice creams, ices and sherbets
- Pourable salad dressings and sauces
- Pharmaceuticals and cosmetics

6.3.5 Composition

Chemically, gum tragacanth consists of two fractions. One fraction is termed tragacanthic acid or bassorin, which represents 60–70% of the total gum. It is insoluble in water but has the capacity to swell and form a gel. Another smaller fraction termed tragacanthin is soluble in water to give a colloidal hydrosol solution.

After acid hydrolysis the major components are: D-galacturonic acid and D-galactose, L-fucose, D-xylose, L-arabinose and L-rhamnose. D-Galacturonic acid is also a major component of tragacanthic acid. For more details see ref. 11.

6.3.6 Regulatory Status

Gum tragacanth is classified as "generally regarded as safe" (GRAS) within the USA. It is also classified as "acceptable daily intake (ADI) not specified", which is the highest category of safety evaluation by the Joint WHO/FAO Expert Committee for Food Additives (JECFA) and has the number E 413 in the list of food additives approved by the Scientific Committee for Food of the European Community.

6.3.7 Current Position

The demand for gum tragacanth fell during the 1980s from several thousand tons to several hundred tons per year at this time. The main reason is competition from xanthan gum, which was finally approved for food use in the 1980s. To make matters worse there was intervention from the Iranian government at origin, which led to a sharp rise in export price, which was already increasing due to the higher labour costs because of more attractive employment opportunities within the village communities. However, due to certain unique properties there are certain applications for which gum tragacanth cannot be replaced satisfactorily by xanthan or any other gum. Indeed, it is now due for a revival.

6.4 Karaya Gum

Gum karaya, also known as *Sterculia* gum, is defined by the Food Chemical Codex as the dried gummy exudation from the Indian *Sterculia urens* tree or the African *Sterculia setifera* tree. The bark of the tree is stripped in small sections to induce the gum flow. The pieces vary in size and colour, with predominantly yellow, pinkish, and brown translucent crystalline pieces. Pieces of tree bark and other admixtures are typically found in the gum; this darkens the colour and increases the necessity of sorting the gum based upon cleanliness. Gum karaya tends to have an acetic acid smell and taste based upon the acetyl groups within its structure. The bulk of commercial gum karaya is obtained from *S. urens* trees from India. The trees grow to a height of about ten metres on the dry, rocky hills and plateaux of central and northern India. The average tree can be tapped about five times during its lifetime, with a total yield of up to 5 kg per season. The exudate is allowed to solidify on the tree. The gum "tears" are broken into fragments less than 25 mm in diameter, then cleaned, sorted and graded according to colour and purity before selling to importers in Western countries.

Gum karaya was for many years a major Indian export. However, the world demand has fallen from around 6000 tons in the early 1980s to less than 3000 tons recently. The demand has slackened as other materials replaced gum karaya in some of its major applications. Also the Indian government introduced export regulations in order to maintain high prices, which interfered with the supply chain. It appears that most of the tonnage lost by India has been acquired by African exporters.

6.4.1 Structure

Gum karaya is a complex, branched, partially acetylated polysaccharide with a reported molecular weight of 9–10 million Daltons. On average new gum karaya contains about 10–14% acetyl groups from which the acetic acid is split off on ageing. On hydrolysis, gum karaya produces D-galacturonic acid, L-rhamnose, D-galactose and small amounts of D-glucuronic acid. The total uronic acid content can be up to 35–40%. About 1% proteinaceous components are also bound into the structure, but the amino acid content varies widely with the different species. Structurally, the main chain is made up of L-rhamnose, linked (1→4) to D-galacturonic acid, which is linked (1→2) to L-rhamnose. There are side chains where D-glucuronic acid is linked (1→3) to galacturonic acid and D-galactose linked (1→4) to L-rhamnose and (1→2) to D-galacturonic acid. Calcium and magnesium are the major cations linked with the uronic acids in the gum structure.

6.4.2 Uses and Applications

Applications are mainly based on its stable viscosity in acidic conditions, (although less so than tragacanth gum), and excellent water-binding and adhesive properties. It can be used as a thickener and emulsifier in low pH products due to its acid resistance, as binder in meat products, paper and textile industries, as an adhesive when partially wet with water and as a gelling agent when used with alkali such as sodium borate to form soft gels. It can also be used as a bulking agent when it absorbs water and swells. Since karaya gum swells in water to form a sticky, hypo-allergenic gel it has been the gum of choice for colostomy rings and denture adhesives. In Europe and Japan, it is widely used as an active ingredient in laxatives, and there are still some remnants of the once considerable use of gum karaya in food products, such as ice cream stabilisers. Gum karaya quality is assessed based upon its thickening power (viscosity) and purity (lack of BFOM: bark and foreign organic matter).

6.4.3 Regulatory Status

Gum karaya is classified as "generally recognised as safe" (GRAS) within the USA. It is classed as "ADI not specified" by the Joint WHO/FAO Expert Committee for Food Additives and has the number E-416 in the list of food additives approved in the European Union. However, in Europe, the human intake is restricted to an upper limit of 12 mg per kg body weight per day based on toxicological studies. Recent registration in the EU has limited gum karaya use in the following food products: cereal and potato-based snacks, nut coatings, fillings, toppings and coatings for fine bakery wares, desserts, emulsified sauces, egg-based liqueurs, chewing gum and dietary food supplements. For more details see ref. 11.

6.5 Concluding Remarks

Quality control of natural gum arabic should first ensure that the minimum requirements of the applicable regulatory system are met. The identification criteria must be fulfilled: solubility, confirmation of the presence of the core sugars (but excluding mannose, xylose and galacturonic acid), optical rotation, purity, total ash, acid-insoluble ash, acid-insoluble matter, absence of starch, dextrin and tannin-bearing gums, lead content not more than 2mg kg^{-1} and negative tests for *Salmonella* spp and *E.coli*.[10] This basic information should be supplemented with rheological data such as intrinsic viscosity, which would inexpensively indicate the average viscosity average molecular weight. For specific applications a study of the structure–function relationship should be undertaken, supported by a molecular profile of the gum as described in this paper.

There is need for further structure–function studies on gums tragacanth and karaya to enable the same degree of process and formulation control to be possible as with gum arabic. The role of the international community in protecting and developing the supplies and quality of these gums also needs to be improved.

The most exciting developments in health-related areas are the possibilities of the increased use of these natural gums as dietary fibres, to combat obesity and Western-style dietary deficiencies.[13–37]

References

1. G. Dondain and G. O. Phillips, *The Regulatory Journey of Gum Arabic Food and Food Ingredients Journal of Japan*, 1999, No 17939–17956.
2. B. R. Maslin, J. T. Miller and D. S. Seigler, Overview of the generic status of *Acacia* (Leguminosae: Mimosoideae), *Australian Systematic Botany*, 2003, **16**, 1–18.
3. P. A. Williams, O. H. M. Idris and G. O. Phillips, Structural Analysis of Gum from *Acacia senegal* (Gum arabic). In *Cell and Developmentat Biology of Arabinogalactan-Proteins*, Chapter 21, (Ed.) Nathnagel et el, Kluwer Academic/Plenum Publishers, 2000, pp. 241–251.
4. P. A. Williams and G. O. Phillips, Gum Arabic, In *Handbook of Hydrocolloids*, Chapter 9, (Eds) G. O. Phillips and P. A. Williams, Woodhead Publishing Ltd, 2000, pp. 155–168.
5. B. Biswas, S. Biswas and G. O. Phillips, The relationship of specific optical rotation to structural composition for Acacia and related gums, *Food Hydrocolloids*, 2000, **14**(2000), 601–608.
6. A. M. Stephen and S. C. Churms, in *Food polysaccharides and their applications*, (Ed.) A. M. Stephen, Marcel Dekker Inc., New York, 1995, pp. 377–440.
7. M. E. Osman, A. R. Menzies, P. A. Williams, G. O. Phillips and T. C. Baldwin, The molecular characterisation of the polysaccharide gum from *Acacia senegal*, *Carbohydr. Res.*, 1993, **246**, 303–318.
8. R. C. Randall, G. O. Phillips and P. A. Williams, The role of the proteinaceous component on the emulsifying properties of gum arabic, *Food Hydrocolloids*., 1988, **2**, 131–140.
9. R. C. Randall, G. O. Phillips and P. A. Williams, Fractionation and characterisation of gum from *Acacia-senegal*, *Food Hydrocolloids*, 1989, **3**, 65–75.
10. FAO/WHO Compendium of Food Additives, 1999, 52, Add. 7. 49–50.

11. W. Weipeng, Tragacanth and karaya, In *Handbook of Hydrocolloids*, Chapter 13, (Ed) G. O. Phillips and P. A. Williams, Woodhead Publishing Ltd., 2000, pp. 232–247.
12. AOAC Official Method: Cereal Foods, Chapter 32 (2000), p.7-12 "AOAC Official Methods Of Analysis", AOAC Official Method 991. 43.
13. S. A. Bingham, Mechanism and experimental and epidemiological evidence relating dietary fibre and starch to protection against large bowel cancer, *Proc Nutr Soc*, 1999, **49**, 153–171.
14. G. D'Argenio, V. Cosenza, M. Della Cave and P. Iovino, *et al.*, Butyrate enemas in experimental colitis and protection against large bowel cancer in a rat model, *Gasteroenterology*, 1996, **110**, 1727–1734.
15. A. Hague, A. M. Manning, K. A. Hanlon, L. I. Huschtscha, D. Hart and C. Paraskeva, Sodium butyrate induces apoptosis in human colonic tumor cell lines in a p53-independent pathway: implications for the possible role of dietary fibre in the prevention of large-bowel cancer, *Int J Cancer*, 1993, **55**, 498–505.
16. S. Siavoshiam, J. P. Segain, M. Kornprobst, C. Bonnet, C. Cherbut, J. P. Galmiche and H. M. Blottiere, Butyrate and trichostatin A effects on the proliferation/differentiation of human intestinal epithelial cells: induction of cyclin D3 and p21 expression, *Gut*, 2000, **46**, 507–514.
17. C. Pellizzaro, D. Coradini and G. G. Daidone, Modulation of angiogenesis-related protein synthesis by sodium butyrate in colon cancer cell line HT29, *Carcinogenesis*, 2002, **23**, 735–740.
18. M. D. Saemann, O. Parolini, G. A. Bohmig, P. Kelemen and P. M. Krieger, *et al.*, Bacterial metabolite interference with maturation of human monocyte-derived dendritic cells, *J Leukocyte Biol*, **71**, 238–246.
19. M. D. Saemann, G. A. Bohmig, C. H. Osterreicher, H. Burtscher, O. Parolini, C. Diakos and J. Stockl, *et al.*, Anti-inflammatory effects of sodium butyrate on human monocytes: potent inhibition of IL-12 and up-regulation of IL-10 production, *FASEB J, 2000*, 2002, **14**, 2380–2382.
20. E. A. Williams, J. M. Coxhead and J. C. Mathers, Anti-cancer effects of butyrate: use of micro-array technology to investigate mechanisms, *Proc Nutr Soc*, 2003, **62**, 107–115.
21. A. Wachtershaouse and J. Stein, Rationale for the luminal provision of butyrate in intestinal diseases, *Eur J Nutr*, (Aug) 2000, **39**(4), 164–171.
22. T. May, R. I. Mackie, G. C. Fahey, J. C. Cremin and K. A. Garleb, Effect of fibre source on short-chain fatty acid production and on the growth and toxin production by Clostridium difficlie, *Scand J Gastroenterology*, Oct 1994, **29**(10), 916–922.
23. J. G. Smith, W. H. Yokoyama and J. B. German, Butyrate from the diet: Actions at the level of gene expression, *Critical Reviews in Food Science and Nutrition*, 1998, **38**, 259–295.
24. T. M. S. Wolever, K. B. Schrade, E. Tshihlias and M. I. McBurney, Do colonic short-chain fatty acids contribute to the long-term adaptation of blood lipids in, subjects with type 2 diabetes consuming a high-fibre diet?, *Am J Clin Nutr*, 2002, **75**, 1023–1039.

25. D. Kromhout, *et al.* (D. Kromhout, E. B. Bosschieter, C. De Lezenne Coulander), Dietary fibre and 10 year mortality from coronary heart disease, cancer and all causes., *Lancet*, 1982, **ii**, 518–522.
26. E. B. Rim, A. Ascherio, E. Giovannucci, D. Spiegelman, M. F. Stampfer and W. C Willett, Vegetable, fruit and cereal fibre intake and risk of coronary heart disease among men, *JAMA*, 1996, **275**, 447–451.
27. J. Stamler, L. Elliot, Q. Appel, M. Chan, B. Buzzard, A. R. Dennis, P. Dyer, P. Elmer, D. Greenland, H. Jones and J. Kesteloot, *et al.*, Hum Higher-blood pressure in middle-aged American adults with less-education-role of multiple dietary factors: the INTERMAP study, *Hypertens*, 2003, **17**(9), 655–775.
28. Y. Jang, J. H. Lee, O. Y. Kim, H. Y. Park and S. Y. Lee, Consumption of whole grain and legume powder reduces insulin demand, peroxidation, and plasma homocysteine concentrations in patients with coronary artery disease: randomized controlled clinical trial. Arteriosclerosis Thrombosis and Vascular, *Biology*, 2001, **21**, 2065–2071.
29. Y. S. Diniz, A. C. Ciognam, C. R. Padavani, M. D. Silva, L. A. Faine, C. M. Galhardi, H. G. Rodrigues and E. L. Novelli, Dietary restriction and fibre supplementation: oxidative stress and metablolic shifting for cardiac health, *Can J Physiol Pharmacol*, 2003, **81**(11), 1042–1048.
30. J. J. Pins, D. Geleva, K. Leeman, C. Fraze, P. J. O'Connor and L. M. Cherney, Do whole-grain, cereal fibre oat cereals reduce the need for antihypertensive medications and improve blood pressure control?, *Journal of Family Practice*, 2002, **51**, 353–359.
31. M. A. Martaugh, D. R. Jacobs, B. Jacob, L. M. Steffen and L. Maruart, Epidemiological support for the protection of whole grains agains diabetes, *Proc Nutr Soc*, 2003, **62**, 143–149.
32. H. Yatzidis, Preliminary studies with locust bean gum a new sorbent with great potential, *Kidney Int*, 1977, **12**(Suppl 8), 152A.
33. H. Yatzidis, D. Koutsicos and P. Digenis, Newer oral sorbents in uremia, *Clin Nephrol*, 1979, **11**(2), 105.
34. D. S. Rampton, S. L. Cohen, V. B. Crammond, J. Goibbons, M. F. Lilburn, J. Y. Rabet, A. J. Vince, J. D. Wager and O. M. Wrong, Treatment of chronic renal failure with dietary fibre, *Clin Nephrol*, 1984, **21**(3), 15.
35. D. Z. Bliss, T. P. Stein, C. R. Scleifer and R. G. Settle, Supplementation with gum arabic fibre increases fecal nitrogen excretion and lowers serum urea nitrogen concentration in chronic renal failure patients consuming a low-protein diet, *American Journal of Clinical Nutrition*, 1996, **63**, 392–398.
36. B. H. Ali, A. A. Al-Qarawi, E. M. Haroun and H. M. Mousam, The effect of treatment with gum Arabic on gentamicin nephrotoxicity in rats: a preliminary study, *Renal Failure*, 2003, **25**(1), 15–20.
37. D. Lairon, S. Bertrais, S. Vincent, N. Arnault, P. Galan, M. C. Boutron and S. Hercberg, Dietary fibre intake and clinical indices in the French Supplementation en Vitamines et Mineraux AntioXydants (SU.VI.MAX) adult cohort, *Proc Nutr Soc*, 2003, **62**, 11–15.

CHAPTER 7

Alginates: Existing and Potential Biotechnological and Medical Applications

KURT I. DRAGET AND GUDMUND SKJÅK-BRÆK

Norwegian Biopolymer Laboratory (NOBIPOL), Department of Biotechnology, Norwegian University of Science and Technology (NTNU), N-7491 Trondheim, Norway

7.1 Introduction

Alginates occur both as a structural component in marine brown algae (*Phaeophyceae*) and as capsular polysaccharides in some bacteria. All commercial alginates are at present still extracted from algal sources, although present research points towards a possible production by microbial fermentation and also post-polymerization modification of the alginate molecule. The latter is most likely to be of utmost importance in high-tech specialty applications within pharmacy and biotechnology. Industrial applications of alginates, such as their ability to retain water, their gelling, viscosifying and stabilizing properties, accounts for the most important quantitative uses of alginates. Forthcoming qualitative applications within biotechnology and medicine are, on the other hand, either based on specific biological effects of the alginate molecule itself, or its unique, gentle and almost temperature-independent sol–gel transition in the presence of multivalent cations (*e.g.* Ca^{2+}), which makes alginate highly suitable as an immobilization matrix for living cells. Traditional industrial exploitation of alginates has to a large extent been based on empirical knowledge.

RSC Polymer Chemistry Series No. 1
Renewable Resources for Functional Polymers and Biomaterials
Edited by Peter A. Williams

Published by the Royal Society of Chemistry, www.rsc.org

Alginates have, however, now entered into more knowledge-demanding areas such as pharmacy and biotechnology. Advanced research within these areas now functions as a locomotive for a further detailed understanding of structure–function relationships, which also will be the main focus of the present chapter.

7.2 Chemical Composition and Conformation

Alginates, being a family of unbranched binary copolymers, consist of (1→4) linked β-D-mannuronic acid (M) and α-L-guluronic acid (G) residues (see Figure 7.1a and b) of widely varying composition and sequence. In the 1960s, alginate was separated into three fractions by partial acid hydrolysis.[1–5] Two of these contained almost homopolymeric molecules of G and M, respectively, while a third fraction consisted of nearly equal proportions of both monomers, and was shown to contain a large number of MG dimer residues. It was hence concluded that alginate was to be regarded as a true block copolymer composed of homopolymeric regions of M and G, termed M- and G-blocks respectively, interspersed with regions of alternating structure (MG-blocks; see Figure 7.1c). Furthermore, it was also shown[6–8] that alginates have no regular repeating unit, and that the distribution of the monomers along the polymer chain cannot be described by Bernoullian statistics. Knowledge of the monomeric composition is hence not sufficient to determine the sequential structure of alginates. It was suggested[7] that a second order Markov model was required for a general approximate description of the monomer sequence in alginates. Further understanding of the mode of action of the different mannuronan C5 epimerases (see below) is likely to increase the understanding of the monomer distribution in the near future. The main difference at the molecular level between algal and bacterial alginates is the presence of *O*-acetyl groups at position C2 and/or C3 in the bacterial alginates.[9]

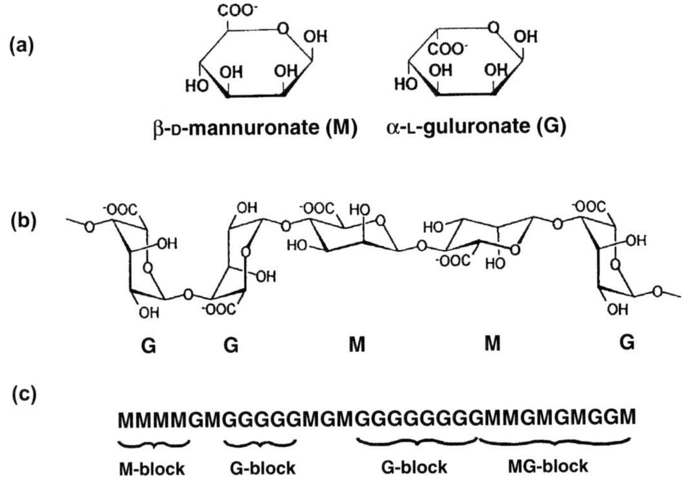

Figure 7.1 Structural characteristics of alginates: alginate monomers, (b) chain conformation, (c) block distribution.

Knowledge of the monomer ring conformations is necessary to understand the polymer properties of alginates. X-Ray diffraction studies of mannuronate-rich and guluronate-rich alginates showed that the guluronate residues in homopolymeric blocks were in the $^{1}C_{4}$ conformation,[10] while the mannuronate residues had the $^{4}C_{1}$ conformation[11,12] (see Figure 7.1a). Hence, alginate contains all four possible glycosidic linkages: diequatorial (MM), diaxial (GG), equatorial–axial (MG) and axial–equatorial (GM); see Figure 7.1b.

The diaxial linkage in G-blocks results in a large hindered rotation around the glycosidic linkage, which may account for the stiff and extended nature of the alginate chain.[11] Additionally, taking the polyelectrolyte nature of alginate into consideration, the electrostatic repulsion between the charged groups on the polymer chain will increase the chain extension and hence also the intrinsic viscosity. Relative dimensions for the neutral, unperturbed alginate chain much higher than for amylose derivatives and even slightly higher than for some cellulose derivatives have been reported.[13]

7.3 Sources and Source Dependence

Commercial alginates are mainly produced from *Laminaria hyperborea, Macrocystis pyrifera, Laminaria digitata, Ascophyllum nodosum, Laminaria japonica, Eclonia maxima, Lessonia nigrescens, Durvillea antarctica* and *Sargassum* spp. Table 7.1 gives some sequential parameters (determined by high-field NMR spectroscopy) for samples of these types of alginates. The composition and sequential structure may, however, vary according to seasonal and growth conditions.[1,14] Alginates with more extreme compositions containing up to 100% mannuronate can be isolated from bacteria.[15] Alginate with a very high content of guluronic acid, being of importance for the mechanical properties of the alginate gel (see below), can be prepared from special algal tissues such as the outer cortex of old stipes of *L. hyperborea* (see Table 7.1)

Table 7.1 Composition and sequence parameters of algal alginates (adapted from ref. 66).

Source	F_G	F_M	F_{GG}	F_{MM}	$F_{GM,MG}$
Laminaria japonica	0.35	0.65	0.18	0.48	0.17
Laminaria digitata	0.41	0.59	0.25	0.43	0.16
Laminaria hyperborea, blade	0.55	0.45	0.38	0.28	0.17
Laminaria hyperborea, stipe	0.68	0.32	0.56	0.20	0.12
Laminaria hyperborea, outer cortex	0.75	0.25	0.66	0.16	0.09
Lessonia nigrescens[a]	0.38	0.62	0.19	0.43	0.19
Ecklonia maxima	0.45	0.55	0.22	0.32	0.32
Macrocystis pyrifera	0.39	0.61	0.16	0.38	0.23
Durvillea antarctica	0.29	0.71	0.15	0.57	0.14
Ascophyllum nodosum, fruiting body	0.10	0.90	0.04	0.84	0.06
Ascophyllum nodosum, old tissue	0.36	0.64	0.16	0.44	0.20

[a]Data provided by Bjørn Larsen.

or by enzymatic modification *in vitro* using mannuronan C-5 epimerases from *Azotobacter vinelandii*[15] (see below). This family of enzymes is able to epimerize M-units into G-units in highly different patterns from almost strictly alternating to very long G-blocks. The epimerases from *A. vinelandii* have been cloned and expressed, and they represent at present a powerful new tool for the tailoring of alginates.

7.4 Properties

Compared to other gelling polysaccharides, the most striking feature of alginate's physical properties is the selective binding of multivalent cations, being the basis for gel formation, and the fact that the sol–gel transition of alginates is not particularly influenced by temperature.

7.4.1 Selective Ion Binding

The basis for the gelling properties of alginates is their specific ion-binding characteristics.[1,16–19] It has been shown that the selective binding of certain alkaline earth metal ions (*e.g.* the strong and cooperative binding of Ca^{2+} relative to Mg^{2+}) increases markedly with increasing content of L-guluronate residues in the chains. Polymannuronate blocks and alternating blocks are almost without selectivity. This is illustrated in Figure 7.2, where a marked hysteresis in the binding of Ca^{2+} ions to G-blocks is also seen.

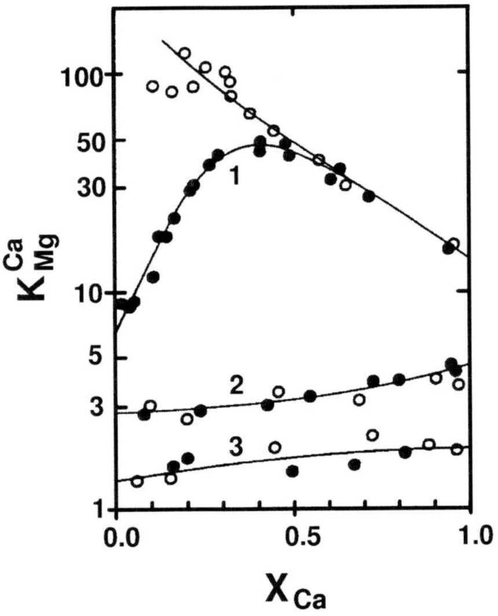

Figure 7.2 Selectivity coefficients, K_{Mg}^{Ca}, as a function of ionic composition (X_{Ca}) for different alginate fragments. Curve 1: fragments with 90% guluronate residues. Curve 2: alternating fragment with 38% guluronate residues. Curve 3: fragments with 90% mannuronate residues. Filled circles: dialysis of the fragments in their Na^{+} form. Open circles: dialysis first against 0.2 M $CaCl_2$, then against mixtures of $CaCl_2$ and $MgCl_2$.

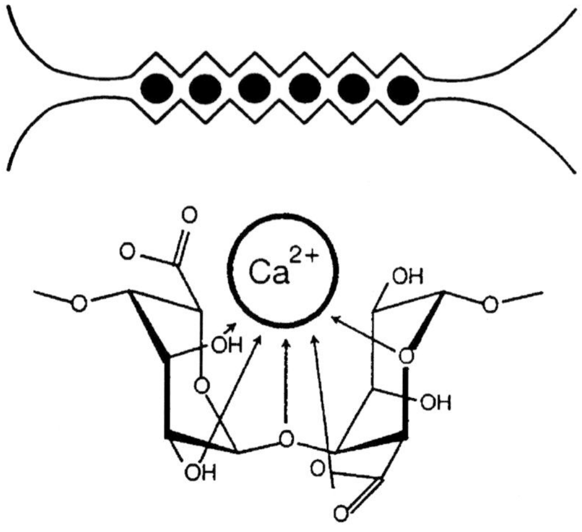

Figure 7.3 The "egg-box" model for the binding of divalent cations between homopolymeric blocks of α-L-guluronate residues, and a probable binding site in a GG-sequence.

The high selectivity between similar ions such as those from the alkaline earth metals indicates that some chelation takes place, caused by structural features in the G-blocks. Grant *et al.* attempted to explain this phenomenon by the so-called "egg-box" model.[20] Although more accurate steric arrangements have been suggested, as supported by X-ray diffraction[21] and NMR spectroscopy,[22] the simple "egg-box" model still persists (Figure 7.3), being principally correct and giving an intuitive understanding of the ion-binding properties of alginates. The original dimerization of alginate molecules in the "egg-box" model is at present questionable, as data from small-angle X-ray scattering on alginate gels suggest lateral association far beyond a pure dimerization with increasing [Ca^{2+}] and G-content of the alginate.[23] Also the fact that isolated and purified G-blocks (totally lacking elastic segments; typically with a degree of polymerization, DP = 20) are able to act as gelling modulators when mixed with a gelling alginate betokens higher order junction zones.[24]

7.4.2 Ionic and Acid Gel Formation

A very rapid and irreversible binding reaction of multivalent cations is typical for alginates; a direct mixing of these two components therefore rarely produces homogeneous gels. A controlled introduction of cross-linking ions is made possible by the two fundamental methods for preparing an alginate gel: the diffusion method and the internal setting method. The diffusion method is characterized by allowing a cross-linking ion (*e.g.* Ca^{2+}) to diffuse from a large outer reservoir into an alginate solution (Figure 7.4a). Diffusion setting is characterized by rapid gelling kinetics, and is indeed utilized for immobilization purposes (see below) where each droplet of alginate solution makes one single gel bead containing the entrapped (bio-)active agent.[25]

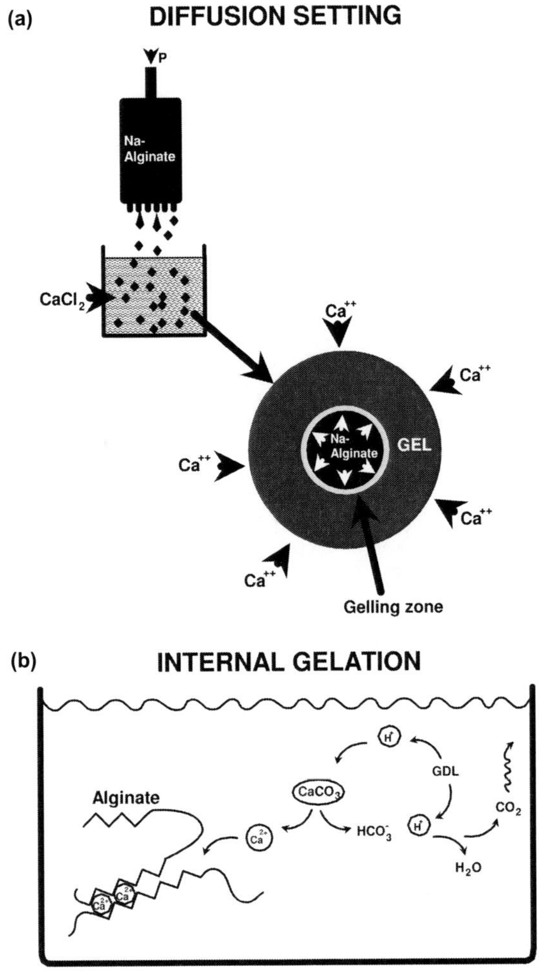

Figure 7.4 Principal differences between the diffusion method exemplified by the immobilization technique and the internal gelation method exemplified by the $CaCO_3$/GDL technique.

Internal setting differs from the former in that the Ca^{2+} ions are released in a controlled fashion from an inert calcium source within the alginate solution (Figure 7.4b). Controlled release is usually obtained by a change in pH, by a limited solubility of the calcium salt source and/or the presence of chelating agents. The main difference between internal and diffusion setting is the gelling kinetics, which is not diffusion-controlled in the former case. With internal setting, the tailor-making of an alginate gelling systems for a given manufacturing process is possible due to the controlled, internal release of cross-linking ions.[26]

It is also well known that alginates may form acid gels at pH values below the pK_a values of the uronic residues, but these alginic acid gels have traditionally not been as extensively studied as their ionically cross-linked counterparts. With the exception of some pharmaceutical uses (see below), the number of applications so far is also rather limited. The preparation of an alginic acid gel has to be performed with care. Direct addition of acid to *e.g.* a Na-alginate solution leads to an instantaneous precipitation rather than a gel. The pH must therefore be lowered in a controlled fashion.

7.4.3 Gel Properties

In contrast to most gelling polysaccharides, alginate gels are cold setting, implying that they set more or less independently of temperature. The kinetics of the gelling process can, however, be strongly modified by a change in temperature, but a sol–gel transition will always occur if gelling is favored (by *e.g.* the presence of cross-linking ions). It is also important to realize that the properties of the final gel will most likely change if gelling occurs at different temperatures due to the non-equilibrium nature of alginate gels.[18]

As the selective binding of ions is a prerequisite for alginate gel formation, the alginate monomer composition and sequence also has a profound impact on the final properties of calcium alginate gels. Figure 7.5 shows gel strength as a function of the average length of G-blocks larger than one unit ($N_{G>1}$). This empirical correlation shows that there is a profound effect on gel strength when $N_{G>1}$ changes from 5 to 15. These values coincide with the range of G-block lengths found in commercial alginates and are important with respect to *e.g.* the mechanical properties of alginate beads for immobilization purposes.

An important feature of gels made by the diffusion setting method is that the final gel often exhibits an inhomogeneous distribution of alginate, the highest concentration being at the surface and gradually decreasing towards the center of the gel. Extreme alginate distributions have been reported[27] with a fivefold increase at the surface (as calculated from the concentration in the original

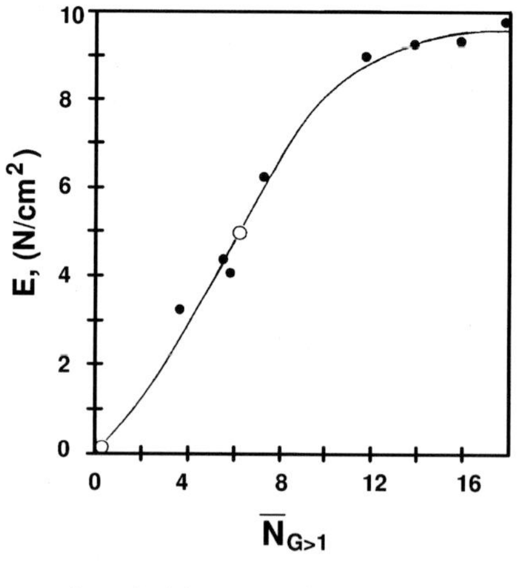

Figure 7.5 Elastic properties of alginate gels as function of average G-block length.

alginate solution), and virtually zero concentration in the center. This result has been explained by the fact that the diffusion of gelling ions will create a sharp gelling zone that moves from the surface towards the center of the gel. The activity of alginate (and of the gelling ion) will equal zero in this zone, and alginate molecules will diffuse from the internal, non-gelled part of the gelling body towards the zero activity region.[27,28] An inhomogeneous alginate distribution may or may not be beneficial in the final product. It is therefore important to know that the degree of homogeneity can be controlled, and which parameters govern the final alginate distribution. Maximum inhomogeneity is reached by placing a high-G, low molecular weight alginate gel within a solution containing a low concentration of the gelling ion and in the absence of non-gelling ions. Maximum homogeneity is reached by a high molecular weight alginate gelled with high concentrations of both gelling and non-gelling ions.[27]

The presence of non-gelling ions in alginate gelling systems also affects the stability of the gels. It has been shown that alginate gels start to swell markedly when the ratio between non-gelling and gelling ions becomes too high, and that the observed destabilization increases with decreasing F_G (fraction of G residues).[29]

As in the case of the alginic acid gel, it has been shown[30] that their gel strength becomes independent of pH below approximately 2.5. The modulus of these gels seems to be rather independent of the history of formation, suggesting that the acid gels (unlike the ionic cross-linked gels) are closer to equilibrium. Figure 7.6 shows the observed elastic moduli of acid gels made

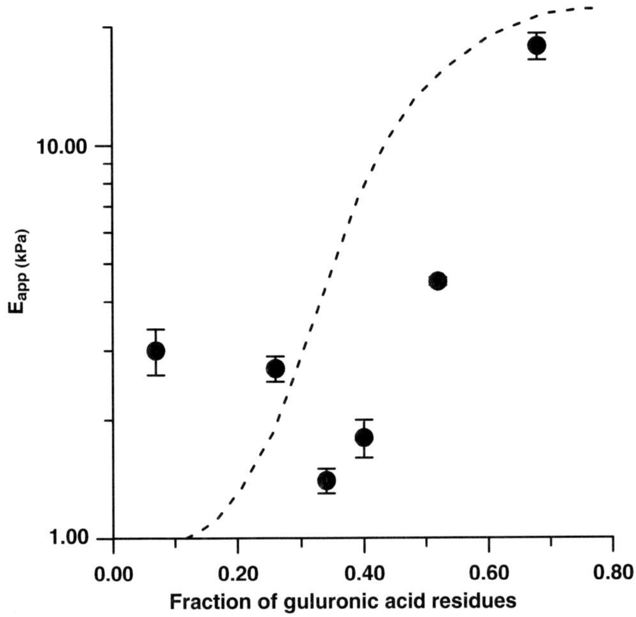

Figure 7.6 Young's modulus (E_{app}) of alginic acid gels at apparent equilibrium as function of guluronic acid content. The dashed line refers to expected results for Ca^{2+} cross-linked alginate gels.

from alginates with different chemical composition, together with expected values for ionically crosslinked gels. From these data, it can be concluded that acid gels resemble ionic gels in the sense that high contents of guluronate (long stretches of G-blocks) give the strongest gels. It is, however, also seen that polymannuronate sequences support alginic acid gel formation, whereas poly-alternating sequences seem to perturb this transition. The obvious demand for homopolymeric sequences in acid gel formation suggests that cooperative processes may be involved just as in the case of ionic gels.[30]

7.4.4 Biological Properties of the Alginate Molecule

Through a series of papers, it has lately been established that the alginate molecule itself has different effects on biological systems. This is more or less to be expected due to the wide variety of possible chemical compositions and molecular weights of alginate preparations. A biological effect of alginate was initially hinted at in the first animal transplantation trials of encapsulated Langerhans islets for diabetes control (see below). The overgrowth of alginate capsules by phagocytes and fibroblasts, resembling a foreign body/inflammatory reaction, was reported.[31] In bioassays, the induction of the tumor necrosis factor (TNF) and interleukin 1 (IL-1) showed that the inducibility depended upon the content of mannuronate in the alginate sample.[32] This result directly explains the observed capsule overgrowth; mannuronate-rich fragments, which do not take part in the gel network, will leach out of the capsules and directly trigger an immune response.[33] This observed immunologic response can, at least partly, be linked to $(1 \rightarrow 4)$ glycosidic linkages since other homopolymeric di-equatorial polyuronates, like D-glucuronic acid (C6-oxidized cellulose), also exhibit this feature.[34] The immunologic potential of polymannuronate has now been observed in *in vivo* animal models (see below).

7.5 Tailoring Alginates by *in vitro* Modification

One element of development, which furthers suggests a rapid movement of alginate into the specialty polymers rather than staying a commodity, is the discovery of the mannuronan C5 epimerases. Alginate with a high content of guluronic acid can be prepared from special algal tissues, by chemical fractionation or by *in vitro* enzymatic modification of the alginate using these mannuronan C-5 epimerases from *A. vinelandii.*[35–38] These epimerases, which convert M to G in the polymer chain, have recently allowed for the production of highly programmed alginates with respect to chemical composition and sequence. *A. vinelandii* encodes a family of 7 exo-cellular iso-enzymes with the capacity to epimerize all sorts of alginates and other mannuronate-containing polymers as shown in Figure 7.7, where the mode of action of AlgE4 (giving alternating introduction of G) is presented.

Although the genes have a high degree of homology, the enzymes they encode exhibit different specificities. Different epimerases may give alginates

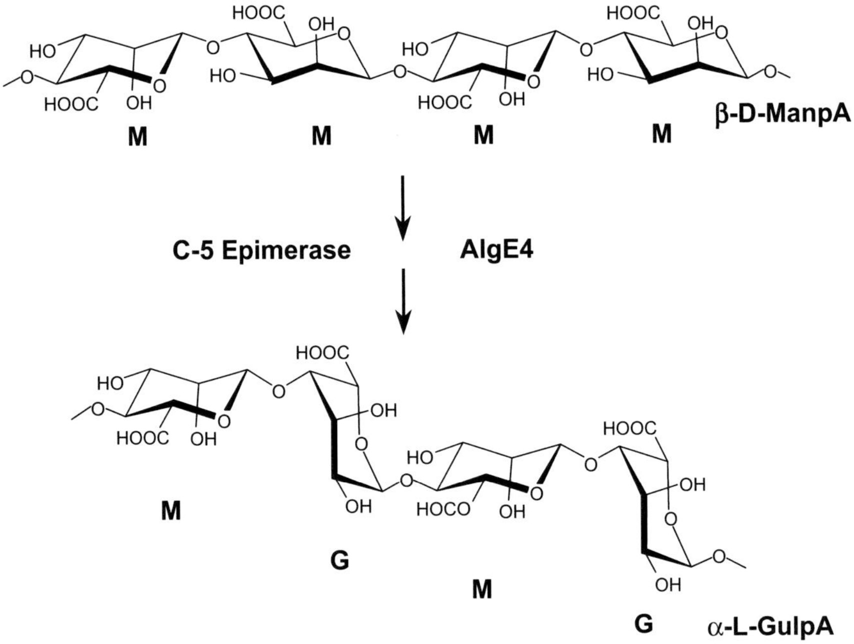

Figure 7.7 Mode of action for the mannuronan C5-epimerase AlgE4.

Table 7.2 The seven AlgE epimerases from *A. vinelandii.*

Type	[kDa]	Modular structure	Products
AlgE1	147.2	A1 | R1 | R2 | R3 | A2 | R4	Bi-functional G-blocks + MG-blocks
AlgE2	103.1	A1 | R1 | R2 | R3 | R4	G-blocks (short)
AlgE3	191	A1 | R1 | R2 | R3 | A2 | R4 | R5 | R6 | R7	Bi-functional G-blocks + MG-blocks
AlgE4	57.7	A1 | R1	MG-blocks
AlgE5	103.7	A1 | R1 | R2 | R3 | R4	G-blocks (medium)
AlgE6	90.2	A1 | R1 | R2 | R3	G-blocks (long)
AlgE7	90.4	A1 | R1 | R2 | R3	Lyase activity + G-blocks + MG-blocks

A - 385 amino acids, R - 155 amino acids

with different distributions of M and G and, thus, alginates with tailored physical and chemical properties can be made. None of the enzymatically modified polymers are, however, commercially available at present. Table 7.2

lists the modular structure of the mannuronan C-5 epimerase family and their specific actions.

7.6 Applications of Alginates in Medicine and Biotechnology

Before treating these special and scientifically most important applications, it should be stated that it becomes rather unnatural to distinguish between medical and biotechnological uses of alginates. This is due to the fact that there is a huge degree of overlap between these two areas. We have therefore chosen to present the material in terms of traditional medical uses of alginates, and in new and potential uses of alginates within biotechnology and medicine. It should also be made clear that, at present, there are no pharmaceutical products claiming clinical effects of the alginate molecule itself.

7.6.1 Traditional Uses of Alginate in Medicine and Pharmacy

Alginates have for decades been used as devices in various human health applications. Examples here are drug delivery systems (DDS), traditional wound dressings, as a dental impression material and in some formulations preventing gastric reflux.

For a recent summary of the past, present and future use of alginates in oral DDS, please see Tønnesen and Karlsen.[39] The main advantage of using alginates in DDS is their property of preserving a solid-like attribute (gel) under two different conditions (with the aid of multivalent ions and the alginic acid gel). This property opens up the possibility of the protection of delicate compounds against the acid influence of gastric juice, both by preventing convection flow and by acting as a buffering agent in the stomach when the DDS is manufactured using Na- or Ca-alginate. The latter property is linked to the pK_a-value of the uronic acid residues. Oral administration of colon-targeted drugs based on alginates as the excipient has been successfully obtained in tablets containing viable lactic acid bacteria[40] and colon-specific delivery of bee venom peptide by applying alginate gel beads containing liposomes.[41]

Alginate/chitosan microcapsules have also been suggested as a possible system for a controlled delayed release of orally administered drugs. Positive results have been reported on *e.g.* acid-labile water-soluble mistletoe lectins for potential cancer therapy[42] and on controlled release of Tramadol-HCl.[43] Focus has also been placed on the use of DDS for controlled release of drugs for the treatment of systemic hypertension in order to obtain a reduced administration frequency. Here, alginates have been used as an excipient for verapamil.[44,45]

Alginates have also been used as a dental impression material (*e.g.* Jeltrate) for several decades.[46] It is based on a dry mix of alginate and gelling agents which will set within minutes upon the addition of water. These types of materials have also been suggested for use in devices for different applications such as taking an impression of the nose prior to rhinoplasty,[47] as molds for measuring wound volumes[48] and as molds for the manufacturing of polycarbonate face masks for the treatment of hypertrophic scars.[49]

For more than 30 years, alginate-based raft-forming formulations have been used for the treatment of heartburn and gastric reflux.[50] They are marketed under various brand names, the most famous being Gaviscon, and they are based on alginate being administered in a soluble form. Upon contact with the gastric content, it will build a physical raft on top of the stomach preventing gastric reflux into the esophagus. These formulations have gained a high popularity as they provide symptom relief within minutes.[51] Alginate-based anti-reflux formulations like Gaviscon are also considered to be reasonably safe with little or no adverse effects as they are also prescribed to infants[52] and against heartburn during pregnancy.[53]

Alginates have been used successfully for decades in dressings for the management of epidermal and dermal wounds (*e.g.* Sorbsan, Seasorb and Kaltostat). Out of recently reported positive effects of alginate-based wound dressing, one can mention that applying Ca-alginate dressings appears to be an appropriate topical treatment of diabetic foot lesions with respect to both healing and tolerance[54] and that Ca-alginate dressings can preferably be used in post-surgical wounds being allowed to heal by second intention.[55] In addition to the immunological properties of the alginate molecule itself (see below), it has been pointed out that Ca^{2+} ions, playing an important role in the normal homeostasis of mammalian skin serving as a modulator in keratinocyte proliferation and differentiation, could be released form Ca-alginate fibers promoting early stage wound healing.[56]

With the newly discovered immunological properties of alginates rich in mannuronic acid residues (see below), some scientific projects have recently undertaken studies on such dressing to distinguish between effects of alginates and the effects of endotoxins on immunological responses.[57,58] There is no doubt that most of the wound dressings made from natural biomaterials do contain endotoxins, and in substantial amounts giving cytotoxic effects on proliferation of fibroblasts, but that immunological effects from the alginate itself are present in some products. It is, however, doubtful if the classical alginate-based wound dressings do contain immunologically active alginate in reasonable quantities since these would be of special quality, not readily available. In the years to come, it is therefore likely that we will see new and improved wound dressings based on alginate.[59,60]

As a sort of intermediate between traditional (being a spin-off of wound healing) and new applications of alginate within medicine, we find the use of hydrogels as polymer scaffolds in tissue engineering.[61,62] Polymer scaffolds are used as space filling agents, as delivery agents for bioactive molecules and as three-dimensional structures that can organize cells to direct the formation of a desired tissue. Gels made of natural biopolymers are an appealing scaffold material due to the fact that they are structurally similar to the extracellular matrix of many tissues, that they can often be processed under mild conditions and that they may be delivered in a minimally invasive manner.[62]

7.6.2 New and Potential Uses of Alginates in Biotechnology and Medicine

The entrapment of cells within Ca-alginate spheres has become the most widely used technique for the immobilization of living cells.[25] This immobilization procedure can be carried out in a single-step process under very mild conditions and therefore becomes compatible with most cells. The cell suspension is mixed with a sodium alginate solution and the mixture dripped into a solution containing multivalent cations (usually Ca^{2+}). The droplets then instantaneously form gel-spheres entrapping the cells in a three-dimensional lattice of ionically cross-linked alginate. Figure 7.8 shows one example of viable cells immobilized in an alginate bead. The possible uses for such systems in industry, medicine, and agriculture are numerous, ranging from the production of ethanol by yeast, and of monoclonal antibodies by hybridoma cells, to mass production of artificial seed by entrapment of plant embryos.[25]

Perhaps the most exciting prospect for alginate gel-immobilized cells is their potential use in cell transplantation. Here, the main purpose of the gel is to act as a barrier between the transplant and the immune system of the host. Different cells have been suggested for gel immobilization including parathyroid cells for the treatment of hypocalcemia and dopamine-producing adrenal chromaffin cells for treatment of Parkinson's disease.[63] However, the major interest has been focused on insulin-producing cells for the treatment of Type I diabetes. Alginate/poly-L-lysine capsules containing pancreatic Langerhans islets have been shown to reverse diabetes in large animals and have also been

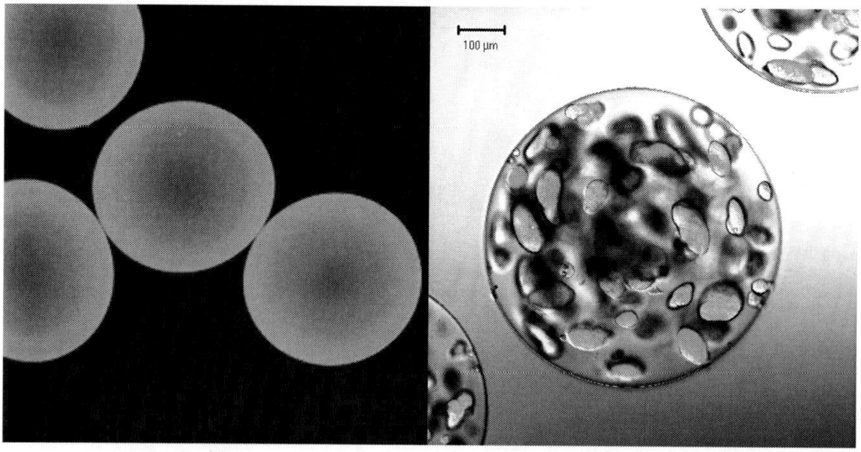

Figure 7.8 Confocal image of empty alginate capsules made from fluorescenamine labelled alginate (left), and (right) alginate encapsulated human embryonic kidney cells (293 EBN) transfected with the gene for endostatin.

Table 7.3 Some potential biomedical application of alginate encapsulated cells. For details, please see ref. 63–65.

Cell type	*Treatment of*
Adrenal chromaffin cells	Parkinson's disease
Hepatocytes	Liver failure
Paratyroid cells	Hypocalcemi
Langerhans islets (β-cells)	Diabetes
Genetically altered cells	Cancer

clinically tested in humans.[64] Table 7.3 lists some potential biomedical applications of alginate encapsulated cells. As a result of this development, one alginate producer (Novamatrix, a subsiduary of FMC Biopolymer) now commercially manufactures an ultrapure alginate quality highly compatible with mammalian biological systems. These are low in pyrogens and facilitate the sterilization of the alginate solution by filtration due to a low content of aggregates.

7.6.2.1 *Controlling the Essential Properties of the Alginate Matrix Used for Immobilization*

The alginate bead matrix to be used for immobilization and transplantation for highly different cell systems will have to be optimized in each case. For a general recent review on the challenges and perspectives of cell microencapsulation, please see Orive *et al.*[65] In general, however, alginate gel beads should ideally be characterized by: high mechanical and chemical stability; controllable swelling properties; low content of toxic, pyrogenic and immunogenic contaminants; defined pore size; and a narrow pore size distribution. This may be obtained, at least in part, by:

- Selection and purification of alginates,
- Selection of gelling ions and control of the gelling kinetics,
- Combination with other polymers, and
- Chemical and enzymatic modification of alginates.

The mechanical and swelling properties of the gel beads depend strongly upon the monomeric composition, block-structure, and molecular weight of the alginate.[29] Beads with the highest mechanical strength are generally made from alginate with a high content of α-L-guluronic acid (>70%) and an average length of G-blocks ($N_{G>1}$) of about 15. Low viscous (low M_w) alginates are generally preferred because they are easier to sterilize by membrane filtration. One should, however, note that below a certain critical molecular weight the gel-forming properties of alginates are reduced.[66] The critical molecular weight depends on the concentration of alginate, as should be expected from the theory of polymer coil overlap.

The major limitation of the use of calcium alginate as a cell immobilization matrix is its sensitivity towards chelating compounds, such as phosphate and citrate, and towards non-gelling cations such as sodium or magnesium ions. Various ways to overcome this have been suggested, the simplest being to keep the gel beads in a medium containing a few millimoles per liter of free calcium and to keep the sodium calcium ratio less than 25 : 1 for high-G and 3 : 1 for low-G alginates. However, replacing calcium ions with other divalent cations having a higher affinity for alginate can also stabilize alginate gels. The affinity series for various divalent cations, also reflecting the resulting mechanical properties of a gel, is:[1]

$$Pb > Cu > Cd > Ba > Sr > Ca > Co, Ni, Zn > Mn$$

Due to a high toxicity, the use of most of the ions (in particular Pb, Cu and Cd) is strictly limited. Only stabilization with strontium and barium can be used for the entrapment of living cells.

Alginate forms strong complexes with polycations such as polysaccharides (*e.g.* chitosan,[67,68] polypeptides (*e.g.* polylysine[69,70]) or synthetic polymers (*e.g.* polyethylenimine[71,72]). These complexes do not dissolve in the presence of calcium chelators or non-gelling cations, and can thus be used both to stabilize the gel and to reduce/control the gel porosity.

Diffusion characteristics, being determined by pore size and size distribution, are essential for the use of alginate gels as an immobilization matrix. Self-diffusion of small molecules seems to be very little affected by the alginate gel matrix[73] whereas the gel network restricts the transport of molecules by convection. For larger molecules such as proteins, diffusional resistance occurs, although even large proteins with molecular weights $>3 \times 10^5$ Da will leak out of the gel beads with a rate depending on their molecular size.[29,73] The highest diffusion rates of proteins, indicating the most open pore structure, are found in beads made from high-G alginates.[29]

Several of the procedures for stabilizing the alginate gels mentioned above will also affect the porosity of the gel. The formation of polyanion–polycation membranes with polypeptides or chitosan has been used to prevent diffusion of antibodies through the capsule membrane.[74,75] Molecules as small as insulin have been retained after using poly-D-lysine as a polycation.[74]

Although alginates do fulfil the requirements for being additives in food and pharmacy, some alginates do contain small amounts of poly-phenols that might be harmful to sensitive cells. In connection with transplantation, the alginate must also be free from pyrogens and immunogenic materials such as proteins and complex carbohydrates. Some alginates show immune-stimulating activity,[34] but fibrotic reactions to alginate–polycation capsules are mainly caused by the polycation.[76]

7.6.3 Alginate as an Immune-stimulating Agent

One of the main initial problems with the injection of alginate capsules for treating diabetes was the overgrowth of the capsules with phagocytes and fibroblasts.[31] When alginates were tested for cytokine induction on human

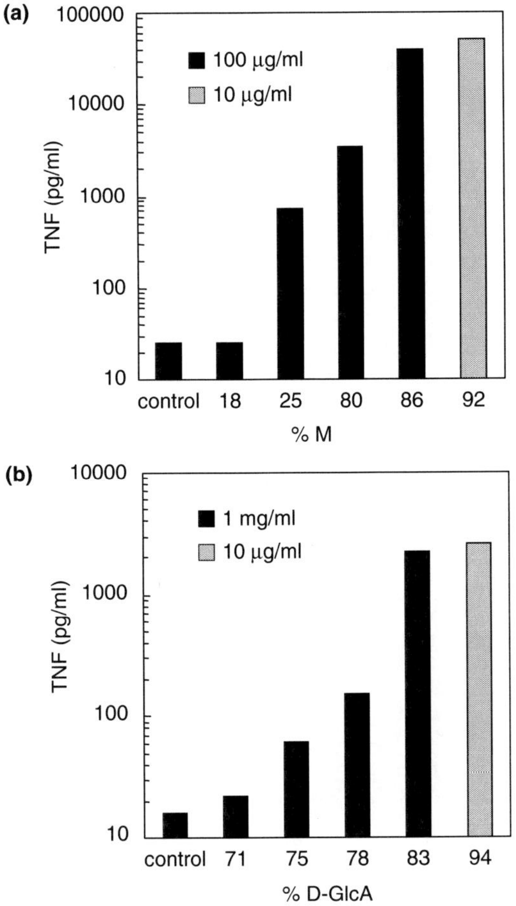

Figure 7.9 Stimulation of tumour necrosis factor (TNF) production as an effect of (a) M-content in alginate, and (b) GlcA-content in C6OXY.

monocytes, it became evident that the ability of alginates to induce TNF, IL-1 and IL-6 correlated with the ManA(mannuronate)-content of the alginate, as well as on the molecular size.[77] This is illustrated in Figure 7.9a, where alginates with different contents of mannuronic acid were tested for the induction of TNF from human monocytes. The highest potency is found for polymers containing more than 95% mannuronic acid residues and with a molecular size above 50 000 Da. This material was isolated from *Pseudomonas aeruginosa* and is designated poly-M. An analogue structure, a di-equatorially $\beta(1\rightarrow4)$ linked D-glucuronic acid, prepared by a selective oxidation at C-6 in cellulose (C6OXY), also stimulates monocytes to produce TNF although with less potency compared to poly-M (Figure 7.9b).

The 3D structure of C6OXY (94% D-GlcA) is similar to that of poly-M except that the consecutive uronic acid residues in C6OXY are broken up with D-Glc.

The stimulatory effect of C6OXY on TNF production depends on the amount of D-GlcA in the polymer, suggesting that the TNF induction from monocytes may occur with different types of β(1→4)-linked uronic polymers. The TNF-inducing capacity increases with molecular weight of the poly-M up to 50 000 Dalton. By degrading it into oligomers with a DP≈20, either with a controlled acid hydrolysis or by treatment with an M-specific lyase, the TNF-inducing capacity is lost. Immune stimulation can, however, be regained and even strongly potentiated by linking the oligomers to microparticles.[78]

7.6.3.1 Effects in vivo

The potent cytokine-inducing ability of β(1→4)-linked uronic acid polymers on monocytes *in vitro* points towards possible effects also in different *in vivo* models. Until recently, only limited knowledge has been available on the immune-stimulating effects of alginates in animal models. Table 7.4 summarizes the most important data on the biological effects of poly-M and C6OXY.

These data demonstrate that β(1→4)-linked uronic acid polymers are also active inducers of cytokine *in vivo*. Of particular interest is that poly-M can protect mice against lethal infection with *Escherichia coli* or *Staphylococcus aureus*.[34] In line with these observations, poly-M also gives a marked protection of mice against lethal irradiation.[80] Subjecting animals to irradiation leads to loss in the ability of the bone marrow to generate white blood cells. Irradiated animals thus may die from infections caused by bacteria which are normally well tolerated (Figure 7.10). In combination with sub-optimal concentration of colony stimulating factors, poly-M enhances the formation of GM-CSF colonies suggesting that poly-M can increase the production of myeloid blood cells.[80] This could be one mechanism behind the radio-protective effect of poly-M. The stimulating effect of poly-M on hematopoietic cells may be clinically important.

Alginate rich in mannuronic acid also has immune-stimulating properties in fish. Juvenile turbots fed with an alginate rich in mannuronic acid obtained an increased protection against one type of pathogenic bacteria.[81] Evidently, these data show that (1→4)-linked uronic acid polymers have potent biological effects in several biological systems and that these effects are most probably caused by stimulation of the monocytes/macrophages.

Table 7.4 Summary of the biological effects of poly-uronic acids on immune cell functions. For a complete review, please see ref. 79.

- Induction of TNF, IL-1, IL-6, GM-CSF, and IL-12 p40 in human monocytes
- Induction of TNF and IL-6 in mice
- Protection against lethal gram positive and gram negative infections
- Protection against lethal effects of irradiation
- Increases the generation of myeloid progenitor cells
- Increases the amount of antibody producing cells
- Increases non-specific immunity in turbots

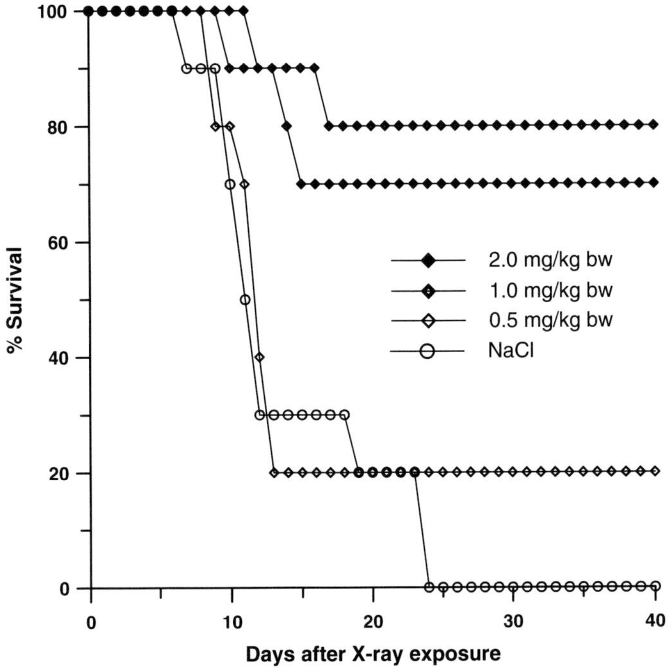

Figure 7.10 Effect of prophylactic (24 h; intraperitoneal) administration of mannuronan on the survival of lethally irradiated (7.3 Gy) C57Bl/6 mice.

7.6.3.2 Molecular Mechanisms

The immune response of both lipopolysaccharides (LPS) from gram negative bacteria and poly-M involve CD14 on the monocyte membrane,[82] and poly-M binds to CD14 on monocytes in the presence of serum.[83] The binding of both poly-M and LPS to monocytes can be inhibited by addition of G-blocks, suggesting a common binding site for these apparently different polysaccharides.[77] Several reports have later implicated a role for CD14 in response to a variety of different compounds such as soluble peptidoglycan fragments and protein-free phenol extracts from *S. aureus*,[84,85] rhamnose–glucose polymers from *Streptococcus mutans*,[86] chitosan from arthropods,[87] and insoluble cell walls from different Gram-positive bacteria.[88] These data imply that CD14 has a broad specificity for compounds containing different types of sugar residues.

LPS can also stimulate cells that do not express membrane CD14 indicating a wide variety of different cell types affected by LPS. In contrast, poly-M is not able to stimulate cell types lacking membrane CD14.[83] This suggests that LPS can interact with cells *via* several (different) modes of action. This broad stimulatory pattern of LPS is likely to be important for its lethal effect *in vivo*.

The apparent specific effect of uronic acid polymers on CD14-positive cells may result in low systemic toxicity and suggests potential applications as an immuno-modulator. Hence, poly-M may activate the non-specific immune system, resulting in increased protection against various types of infections.

7.7 Concluding Remarks

Alginates have for decades been used within medicine and biotechnology on an empirical and semi-empirical basis. With the emerging knowledge on the biological properties of the alginate molecule itself, on important parameters for controlling alginate capsules for immobilization and transplantation as well as the possibility of tailoring alginate molecules with the mannuronan C5 epimerases, it is reasonable to anticipate a flourishing increase in optimized and tailored alginate products within these areas.

References

1. A. Haug, *Composition and Properties of Alginates*, Thesis, Norwegian Institute of Technology, Trondheim, 1964.
2. A. Haug, B. Larsen and O. Smidsrød, A study of the constitution of alginic acid by partial hydrolysis, *Acta Chem. Scand.*, 1966, **20**, 183–190.
3. A. Haug and B. Larsen, A study on the constitution of alginic acid by partial acid hydrolysis, *Proc. Int. Seaweed Symp.*, 1966, **5**, 271–277.
4. A. Haug, B. Larsen and O. Smidsrød, Studies on the sequence of uronic acid residues in alginic acid, *Acta Chem. Scand.*, 1967a, **21**, 691–704.
5. A. Haug and O. Smidsrød, Fractionation of alginates by precipitation with calcium and magnesium ions, *Acta Chem. Scand.*, 1965, **19**, 1221–1226.
6. T. J. Painter, O. Smidsrød and A. Haug, A computer study of the changes in composition-distribution occurring during random depolymerisation of a binary linear heteropolysachharide, *Acta Chem. Scand.*, 1968, **22**, 1637–1648.
7. B. Larsen, O. Smidsrød, T. J. Painter and A. Haug, Calculation of the nearest-neighbour frequencies in fragments of alginate from the yields of free monomers after partial hydrolysis, *Acta Chem. Scand.*, 1970, **24**, 726–728.
8. O. Smidsrød and S. G. Whittington, Monte Carlo investigation of chemical inhomogeneity in copolymers, *Macromolecules*, 1969, **2**, 42–44.
9. G. Skjåk-Bræk, B. Larsen and H. Grasdalen, Monomer sequence and acetylation pattern in some bacterial alginates, *Carbohydr. Res.*, 1986, **154**, 239–250.
10. E. D. T. Atkins, W. Mackie and E. E. Smolko, Crystalline structures of alginic acids, *Nature*, 1970, **225**, 626–628.
11. O. Smidsrød, R. M. Glover and S. G. Whittington, The relative extension of alginates having different chemical composition, *Carbohydr. Res.*, 1973, **27**, 107–118.
12. H. Grasdalen, B. Larsen and O. Smidsrød, 13C-NMR studies of alginate, *Carbohydr. Res.*, 1977, **56**, C11–C15.
13. O. Smidsrød, Solution properties of alginate, *Carbohydr. Res.*, 1970, **13**, 359–372.

14. M. Indergaard and G. Skjåk-Bræk, Characteristics of alginate from Laminaria digitata cultivated in a high phosphate environment, *Hydrobiologia*, 1987, **151/152**, 541–549.
15. S. Valla, H. Ertesvåg and G. Skjåk-Bræk, Genetics and biosynthesis of alginates, *Carbohydr. Eur.*, 1996, **14**, 14–18.
16. O. Smidsrød and A. Haug, Dependence upon uronic acid composition of some ion-exchange properties of alginates, *Acta Chem. Scand.*, 1968, **22**, 1989–1997.
17. A. Haug and O. Smidsrød, Selectivity of some anionic polymers for divalent metal ions, *Acta Chem. Scand.*, 1970, **24**, 843–854.
18. O. Smidsrød, *Some Physical Properties of Alginates in Solution and in the Gel State*, Thesis, Norwegian Institute of Technology, Trondheim, 1973.
19. O. Smidsrød, Molecular basis for some physical properties of alginates in the gel state, *J. Chem. Soc. Farad. Trans*, 1974, **57**, 263–274.
20. G. T. Grant, E. R. Morris, D. A. Rees, P. J. C. Smith and D. Thom, Biological interactions between polysaccharides and divalent cations: The egg-box model, *FEBS letters*, 1973, **32**, 195–198.
21. W. Mackie, S. Perez, R. Rizzo, F. Taravel and M. Vignon, Aspects of the conformation of polyguluronate in the solid state and in solution, *Int. J. Biol. Macromol.*, 1983, **5**, 329–341.
22. C. A. Steginsky, J. M. Beale, H. G. Floss and R. M. Mayer, Structural determination of alginic acid and the effects of calcium binding as determined by high-field n.m.r., *Carbohydr. Res.*, 1992, **225**, 11–26.
23. B. T. Stokke, K. I. Draget, Y. Yuguchi, H. Urakawa and K. Kajiwara, Small angle X-ray scattering and rheological characterization of alginate gels. 1 Ca-alginate gels, *Macromolecules*, 2000, **33**, 1853–1863.
24. K. I. Draget, E. Onsøyen, T. Fjæreide, M. K. Simensen and O. Smidsrød, Use of G-block polysaccharides', *Intl. Pat. Appl.* #PCT/NO97/00176, 1997.
25. O. Smidsrød and G. Skjåk-Bræk, Alginate as immobilization matrix for cells, *Trends in Biotechnology*, 1990, **8**, 71–78.
26. K. I. Draget, K. Østgaard and O. Smidsrød, Homogeneous alginate gels; a technical approach, *Carbohydr. Polym.*, 1991, **14**, 159–178.
27. G. Skjåk-Bræk, H. Grasdalen and O. Smidsrød, Inhomogeneous polysaccharide ionic gels, *Carbohydr. Polym.*, 1989, **10**, 31–54.
28. A. Mikkelsen and A. Elgsæter, Density distribution of calcium-induced alginate gels, *Biopolymers*, 1995, **36**, 17–41.
29. A. Martinsen, G. Skjåk-Bræk and O. Smidsrød, Alginate as immobilization material: I. Correlation between chemical and physical properties of alginate gel beads, *Biotechnol. Bioeng.*, 1989, **33**, 79–89.
30. K. I. Draget, G. Skjåk-Bræk and O. Smidsrød, Alginic acid gels: the effect of alginate chemical composition and molecular weight, *Carbohydr. Polym.*, 1994, **25**, 31–38.
31. P. Soon-Shiong, M. Otterlei, G. Skjåk-Bræk, O. Smidsrød, R. Heintz, R. P. Lanza and T. Espevik, An immunologic basis for the fibrotic reaction to implanted microcapsules, *Transplant Proc.*, 1991, **23**, 758–759.

32. P. Soon-Shiong, E. Feldman, R. Nelson, R. Heints, Q. Yao, T. Yao, N. Zheng, G. Merideth, G. Skjåk-Bræk, T. Espevik, O. Smidsrød and P. Sandford, Long-term reversal of diabetes by the injection of immunoprotected islets, *Proc. Natl. Acad. Sci.*, 1993, **90**, 5843–5847.
33. B. T. Stokke, O. Smidsrød, F. Zanetti, W. Strand and G. Skjåk-Bræk, Distribution of uronate residues in alginate chains in relation to gelling properties 2:Enrichment of β-D-mannuronic acid and depletion of α-L-guluronic acid in the sol fraction, *Carbohydr. Polym.*, 1993, **21**, 39–46.
34. T. Espevik and G. Skjåk-Bræk, Application of alginate gels in biotechnology and biomedicine, *Carbohydr. Eur.*, 1996, **14**, 19–25.
35. H. Ertesvåg, B. Doseth, B. Larsen, G. Skjåk-Bræk and S. Valla, Cloning and Expression of an Azotobacter vinelandii Mannuronan C-5-epimerase, *Gene. J. Bacteriol*, 1994, **176**, 2846–2853.
36. H. Ertesvåg, H. K. Høidal, I. K. Hals, A. Rian, B. Doseth, S. Valla, A family of modular type mannuronan C-5-epimerase genes controls alginate structure in *Azotobacter vinelandii. Mol. Microbiol.* 1995, **16**, 719–731.
37. H. Ertesvåg, H. K. Høidal, G. Skjåk-Bræk and S. Valla, The *Azotobacter vinelandii* mannuronan C-5-epimerase AlgE1 consists of two separate catalytic domains, *J. Biol. Chem.*, 1998, **273**, 30927–30932.
38. H. K. Høidal, H. Ertesvåg, G. Skjåk-Bræk, B. T. Stokke and S. Valla, The recombinant *Azotobacter vinelandii* mannuronan C-5-Epimerase AlgE4 epimerizes alginate by a nonrandom attack mechanism, *J. Biol. Chem.*, 1999, **274**, 12316–12322.
39. H. H. Tønnesen and J. Karlsen, Alginate in drug delivery systems, *Drug development and industrial pharmacy*, 2002, **28**, 621–630.
40. M. Stadler and H. Viernstein, Optimization of a formulation containing viable lactic acid bacteria, *Int. J. Pharmaceutics*, 2003, **256**, 117–122.
41. X. Liu, D. W. Chen, L. P. Xie and R. Q. Zhang, Oral colon-specific drug delivery for bee venom peptide; development of a coated calcium alginate gel beads entrapped liposome, *J. Contr. Rel.*, 2003, **93**, 293–300.
42. S. Y. Lyu, Y. J. Kwon, H. J. Joo and W. B. Park, Preparation of alginate/chitosan microcapsules and enteric coated granules of mistletoe lectin, *Arch. Pharm. Res.*, 2004, 118–126.
43. N. Acosta, I. Aranaz, C. Peniche and A. Heras, Tramadol release from a delivery system based on alginate-chitosan microcapsules, *Macromol. Biosci.*, 2003, **3**, 546–551.
44. L. M. Prisant, B. Bottini, J. T. Dipiro and A. A. Carr, Novel drug-delivery systems for hypertension, *Am. J. Med.*, 1992, **93**(Suppl. 2A), S45–S55.
45. L. M. Prisant and W. J. Elliot, Drug delivery systems for treatment of hypertension, *Clinical Pharmacokinetics*, 2003, **42**, 931–940.
46. Anon, Council adapts American Dental Association specification no 18 (alginate impression material), *J. Am. Dent. Ass.*, 1968, **77**, 1354.
47. D. L. Wood, Dental alginate for taking impressions of the nose prior to rhinoplasty, *Plastic and Reconstructive Surgery*, 1997, **99**, 2609–2670.
48. N. A. Stotts, M. J. Salazar, D. WipkeTevis and E. McAdoo, Accuracy of alginate molds for measuring wound volumes when prepared and stored

under varying conditions, *Wounds-A Compendium of Clinical Research and Practice*, 1996, **8**, 158–164.
49. S. Locke, S. Smith, B. Szeliski-Scott and E. D. Lemaire, A clear polycarbonate fase mask for the treatment of hypertrophic scars, *J. Prostetics and Orthotiscs*, 1991, **3**, 182–190.
50. K. G. Mandel, B. P. Daggy, D. A. Brodie and H. I. Jacoby, Alginate-raft formulations in the treatment of heartburn and acid reflux, *Alimentary Pharmacology & Therapeutics*, 2000, **14**, 669–690.
51. S. Chatfield, A comparison of the efficacy of the alginate preparation, Gaviscon Advance, with placebo treatment of gastro-oesophageal reflux disease, *Current Medical Research and Opinion*, 1999, **15**, 152–159.
52. J. A. J. M. Taminiau, Gastro-oesophageal reflux in children, *Scand. J. Gastroenterology*, 1997, **32**(Suppl. 223), 18–20.
53. S. W. Lindow, P. Regnell, J. Sykes and S. Little, An open-label, multicentre study to assess the safety and efficacy of a novel reflux suppressant (Gaviscon Advance (R)) in the treatment of heartburn during pregnancy, *Int. J. Clin. Pract.*, 2003, **57**, 175–179.
54. J. D. Lalau, R. Bresson, P. Charpentier, V. Coliche, S. Erlher, G. H. Van, G. Magalon, J. Martini, Y. Moreau, S. Pradines, F. Rigal, J. L. Wermeau and J. L. Richards, Efficacy and tolerance of calcium alginate versus vaseline gauze dressings in the treatment of diabetic foot lesions, *Diabetes & Metabolism*, 2002, **28**, 223–229.
55. A. A. Marghoob, N. N. Artman and D. M. Siegel, Calcium alginate dressings with second intention healing of surgical wounds: Our experience, *Wounds – A Compendium of Clinical Research and Practice*, 1997, **9**, 50–55.
56. A. B. G. Lansdown, Calcium: a potent central regulator in wound healing in the skin, *Wound Repair and Regeneration*, 2002, **10**, 271–285.
57. A. Thomas, K. G. Harding and K. Moore, Alginates from wound dressings activate human macrophages to secrete tumour necrosis factor-alpha, *Biomaterials*, 2000, **21**, 1797–1802.
58. Y. Nagawa, T. Murai, C. Hasegawa, M. Hirata, T. Tsuchiya, T. Yagami and Y. Haishima, Endotoxin contamination in wound dressings made of natural biomaterials, *J. Biomed. Mat. Res. Part B – Appl. Biomat.*, 2003, **66B**, 347–355.
59. Y. Suzuki, M. Tanihara, Y. Nishimura, K. Suzuki, Y. Yamawaki, H. Kudo, Y. Kakimaru and Y. Shimizu, In vivo evaluation of a novel alinate dressing, *J. Biomed. Mat. Res.*, 1999, **48**, 522–527.
60. L. H. Wang, E. Khor, A. Wee and L. Y. Lim, Chitosan-alginate PEC membrane as a wound dressing: Assessment of incisional wound healing, *J. Biomed. Mat. Res.*, 2002, **63**, 610–618.
61. B. K. Mann, Biologic gels in tissue engineering, *Clinics in Plastic Surgery*, 2003, **30**, 601–609.
62. J. L. Drury and D. J. Mooney, Hydrogels for tissue engineering: scaffold design variables and applications, *Biomaterials*, 2003, **24**, 4337–4351.

63. W. M. Kühtreiber, R. P. Lanz, W. L. Chick, (Eds), Part III: Applications of cell encapsulated systems. In *Cell Encapsulation Technology and Therapeutics*, Birkhäuser, Boston, 1999, pp. 217–379.
64. P. Soon-Shiong, R. E. Heintz, N. Merideth, Q. X. Yao, Z. Yao, T. Zheng, M. Murphy, M. K. Moloney, M. Schmehl, M. Harris, R. Mendez and P. A. Sandford, Insulin independence in a type 1 diabetic patient after encapsulated islet transplantation, *The Lancet*, 1994, **343**, 950–951.
65. G. Orive, R. M. Hernandez, A. R. Gascon, R. Calafiore, T. M. S. Chang, P. de Vos, G. Hortelano, D. Hunkeler, I. Lacik and J. L. Pedraz, History challenges and perspectives of cell microencapsulation, *Trends in Biotechnology*, 2004, **22**, 87–92.
66. S. Moe, K. I. Draget, G. Skjåk-Bræk, O. Smidsrød, in Alginates, (Ed.) A. Stephen, *Food Polysaccharides*, Marcel Dekker, New York, 1995, pp. 245–287.
67. C. K. Rha, Chitosan as biomaterial, In *Biotechnology in the Marine Sciences*, (Eds) R. R. Colwell, E. R. Pariser, A. J. Sinskey, Wiley, New York, 1984, pp. 177–189.
68. O. Gåserød, O. Smidsrød and G. Skjåk-Bræk, Microcapsules of alginate-chitosan – I. A quantitative study of the interaction between alginate and chitosan, *Biomaterials*, 1998, **19**, 1815–1825.
69. F. Lim and A. M. Sun, Microencapsulated islets as bioartificial endocrine pancreas, *Science*, 1980, **210**, 908–910.
70. B. Thu, P. Bruheim, T. Espevik, O. Smidsrød and G. Skjåk-Bræk, Alginate polycation microcapsules. I. Interaction between alginate and polycation, *Biomaterials*, 1996, **17**, 1031–1040.
71. I. A. Veliky and R. E. Williams, The production of ethanol by *Saccharomyces cerevisiae* immobilized in polycation stabilized calcium alginate gels, *Biotechnol. Lett.*, 1981, **33**, 275–280.
72. H. Tanaka, H. Kurosawa, E. Kokufuta and I. A. Veliky, Preparation of immobilized glucoamylase using Ca-alginate gel coated with partially quaternized poly(ethyleneimine), *Biotechnol. Bioeng.*, 1984, **26**, 1393–1394.
73. H. Tanaka, M. Matsumura and I. A. Veliky, Diffusion characteristics of substrates in Ca-alginate gel beads, *Biotechnol. Bioeng.*, 1984, **26**, 53–58.
74. B. Kulseng, B. Thu, T. Espevik and G. Skjåk-Bræk, Alginate polylysine capsules as immune barrier: Permeability of cytokines and immunoglobulins over the capsule membrane, *Cell Transplant.*, 1997, **6**, 387–394.
75. O. Gåserød, A. Sannes and G. Skjåk-Bræk, Microcapsules of alginate-chitosan. II. A study of capsule stability and permeability, *Biomaterials*, 1999, **20**, 773–783.
76. B. L. Strand, L. Ryan, P. In't Veld, B. Kulseng, A. M. Rokstad, G. Skjåk-Bræk and T. Espevik, Poly-L-lysine induces fibrosis on alginate microcapsules via the induction of cytokines, *Cell Transplant.*, 2001, **10**, 263–275.
77. M. Otterlei, A. Sundan, G. Skjåk-Bræk, L. Ryan, O. Smidsrød and T. Espevik, Similar mechanisms of action of defined polysaccharides and

liposacccharides: Characterization of binding and tumor necrosis factor alpha induction, *Infect. Immun.*, 1993, **61**, 1917–1926.

78. G. Berntzen, T. Flo, A. Medvedev, L. Kilaas, G. Skjåk-Bræk, A. Sundan and T. Espevik, The tumor necrosis factor-inducing potency of lipopolysaccharide and uronic acid polymers is increased when they are covalently linked to particles, *Clin. Diagn. Lab. Immunol.*, 1998, **5**, 355–361.
79. G. Skjåk-Bræk, T. Flo, Ø. Halaas, T. Espevik, Immune stimulating properties of di-equatorially β(1→4) linked poly-uronides. In *Bioactive Carbohydrate Polymers*, Kluwer Academic Publishers, Dordracht (NL), 2000. pp. 85-93.
80. Ø. Halaas, W. M. Olsen, O. P. Veiby, D. Løvhaug, G. Skjåk-Bræk, R. Vik and T. Espevik, Mannuronan enhances survival of lethally irradiated mice and stimulates murine haematopoiesis *in vitro*, *Scand. J. Immunol.*, 1997, **46**, 358–365.
81. J. Skjermo, T. Defoort, M. Dehasque, T. Espevik, Y. Olsen, G. Skjåk-Bræk, P. Sorgeloos and O. Vadstein, Immonustimulation of juvenile turbot (*Scophthalmus maximus* L.) using an alginate with high mannuronic acid content administrated via the live food organism *Artemia*, *Fish Shellfish Immunol.*, 1995, **5**, 531–534.
82. M. Otterlei, Ø. K. Stgaard, G. Skjåk-Bræk, O. Smidsrød, P. Soon-Shiong and T. Espevik, Induction of cytokine production from human monocytes stimulated with alginate, *J. Immunother.*, 1991, **10**, 286–291.
83. T. Espevik, M. Otterlei, G. Skjåk-Bræk, L. Ryan, S. D. Wright and A. Sundan, The involvement of CD14 in stimulation of cytokine production by uronic acid polymers, *Eur. J. Immunol.*, 1993, **23**, 255–261.
84. B. Weidemann, H. Brade, E. T. Rietschel, R. Dziarski, V. Bazil, S. Kusomoto, H. D. Flad and A. J. Ulmer, Soluble peptidoglycan-induced monokine production can be blocked by anti-CD14 monoclonal antibodies and by lipid-A partial structures, *Infect. Immun.*, 1994, **62**, 4709–4715.
85. T. Kusunoki, E. Hailman, T. S.-C. Juan, H. S. Lichenstein and S. D. Wright, Moleculesfrom Staphylococcus aureus that bind CD14 and stimulate innate immune-responses, *J. Exp. Med.*, 1995, **182**, 1673–1682.
86. M. Soell, E. Lett, F. Holveck, M. Scholler, D. Wachsmann and J. Klein, Activation of human monocytes by streptococcal rhamnose glucose polymers is mediated by CD14 antigen, and mannan-binding protein inhibits TNF-alpha release, *J. Immunol.*, 1995, **154**, 851–860.
87. M. Otterlei, K. M. Vårum, L. Ryan and T. Espevik, Characterization of binding and TNF-α-inducing ability of chitosans on monocytes: the involvement of CD14, *Vaccine*, 1994, **12**, 825–832.
88. J. Pugin, D. Haumann, A. Tomasz, V. V. Kravchenko, M. P. Glauser, P. S. Tobias and R. J. Ulevich, CD14 is a pattern-recognition receptor, *Immunity*, 1994, **1**, 509–516.

CHAPTER 8

Pectins: Production, Properties and Applications

H.U. ENDRESS

Herbstreith & Fox KG, Turnstrasse 37, 75305, Neuenbürg, Germany

8.1 Introduction

Heteropolysaccharides rich in galacturonic acid are usually referred to as pectins but, according to its formal definitions (see Table 8.1), the term pectin may only be used for polysaccharides with a minimum galacturonic acid content of 65% for pectins used as a food additive[1–3] or 74% if used as USP grade pectin according to the United States Pharmacopeia.[4]

Pectins are part of the plant cell wall in all higher land plants and some aquatic ones, and are located in the highest concentration in tissue-connecting cells of the middle lamella. Pectin content decreases in the direction of the primary cell walls and is almost absent in lignified secondary walls.

Pectins are an inhomogeneous group of heteropolysaccharides consisting of zigzag-shaped chains of axial–axial poly(α-1,4-D-galacturonic acid) backbone (see Figure 8.1) with inserted α-1,2-L-rhamnopyranose units, resulting in kinks of the linear main chain as shown in Figure 8.2. The distribution of rhamnose in the chain is uneven; segments with low rhamnose content are called homogalacturonans or smooth regions, and segments with higher rhamnose content are referred to as rhamnogalacturonan-1 or hairy regions.[5] Albersheim's group also describes a minor and complex class of heteropolysaccharides referred to as rhamnogalacturonan-2.[6]

RSC Polymer Chemistry Series No. 1
Renewable Resources for Functional Polymers and Biomaterials
Edited by Peter A. Williams

Published by the Royal Society of Chemistry, www.rsc.org

Table 8.1 Regulations for purity requirements of pectins.

	EU/GER				
International specifications	*E 440 (i)*	*E 440 (ii)*	*Codex JECFA*	*FDA/FCC*	*USP*
1. Loss on drying	max. 12%	max. 12%	max. 12%	max. 12%	max. 10%
2. Acid-insoluble ash (3n HCl)	max. 1%	max. 1%	max. 1%	max. 1%	—
3. Total insolubles	—	—	max. 3%	max. 3%	—
4. Sodium methylsulfate	—	—	—	max. 0.1%	—
5. Free methyl-, ethyl- or isopropyl alcohol (in dry matter)	max. 1%	max. 1%	max. 1%	max. 1%	max. 1%
6. Sulphurdioxide (in dry matter)	max. 50 ppm	max. 50 ppm	max. 50 ppm	max. 50 ppm	max. 50%
7. Nitrogen content (pectins) E 440 (i)[a]	max. 1%	—	max. 2.5%	—	—
8. Nitrogen content (amidated pectins) E 440 (ii)[a]	—	max. 2.5%	max. 2.5%	—	—
9. Galacturonic acid[a]	min. 65%	min. 65%	min. 65%	min. 65%	min. 74%
10. Degree of amidation (E 440 (ii)	—	max. 25%	max. 25%	max. 25%	—
11. Sugar and organic acids	—	—	—	—	max. 160 mg kg^{-1}
12. Arsenic	max. 3 ppm	max. 3 ppm	—	—	max. 3 ppm
13. Lead	max. 5 ppm	max. 5 ppm	max. 5 ppm	max. 5 ppm	max. 5 ppm
14. Cadmium	max. 1 ppm	max. 1 ppm	—	—	—
15. Mercury	max. 1 ppm	max. 1 ppm	—	—	—
16. Heavy metals (as lead)	max. 20 ppm	max. 20 ppm	—	—	—
17. Pesticides	according to general food regulations				
18. Pathogenic germs	according to general food regulations				salmonella absent
19. Organic volatile impurities	according to general food regulations				defined limits

[a]Ash-free and dried basis

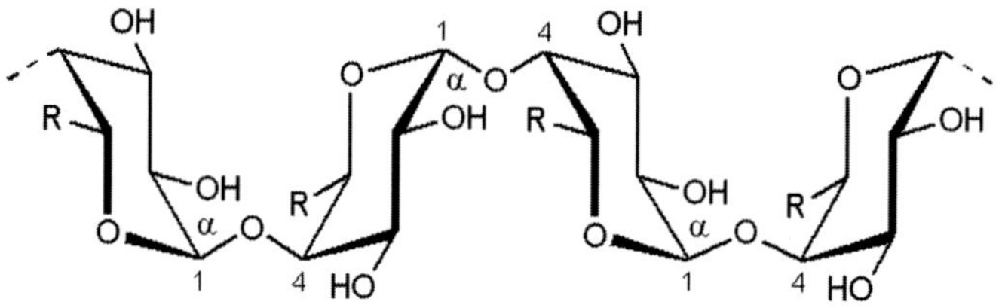

Figure 8.1 Zigzag-shaped pectin chain.

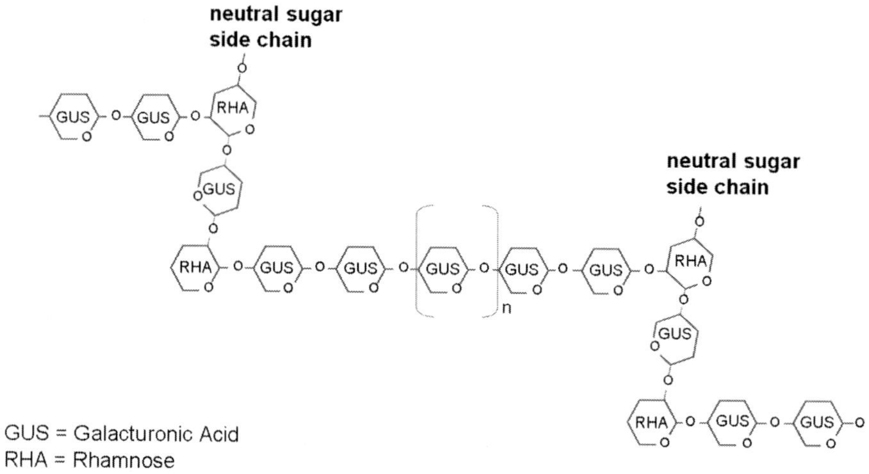

Figure 8.2 Pectin chain with inserted rhamnose units.

Rhamnose residues of pectins also carry sugar side chains consisting mainly of L-arabinose or D-galactose, and L-fucose is found at the terminal end of these side chains. D-Xylose, D-glucose, D-mannose, D-apiose and other rare sugars are also found as minor components. These minor sugars can be found as single unit side chains (*e.g.* D-xylose), short side chains, or complex structures in the cases of L-arabinose and D-galactose. Pectins with attached arabinans can also be isolated from many fruits and vegetables like apples (the proposed structure of apple pectin is shown in Figure 8.3), apricots, cabbage, carrots, onions, pears and sugar beet. Arabinans are branched polysaccharides with a backbone of α-1,5-linked arabinofuranosyl units with α-1,2- and α-1,3-linked arabinofuranosyl side chains.

Apples, citrus fruits, grapes, onions, potatoes, soy beans, tomatoes and others also contain arabinogalactans with two structurally different forms. Type 1 consists of an α-1,4-linked linear chain of D-galactopyranosyl residues with short chains of linear α-1,5-arabinans connected to C-3, whereas type 2 is a highly branched polysaccharide with ramified chains of α-1,3-and α-1,6-linked D-galactopyranosyl residues terminated by L-arabinofuranosyl or, to a smaller extent, L-arabinopyranosyl residues.[5]

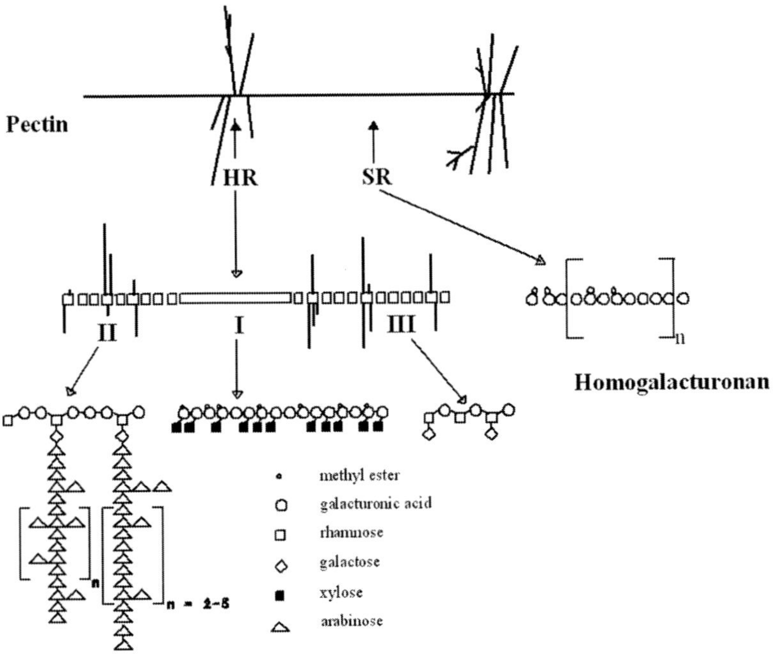

Figure 8.3 Postulated structure of apple pectin.[5]

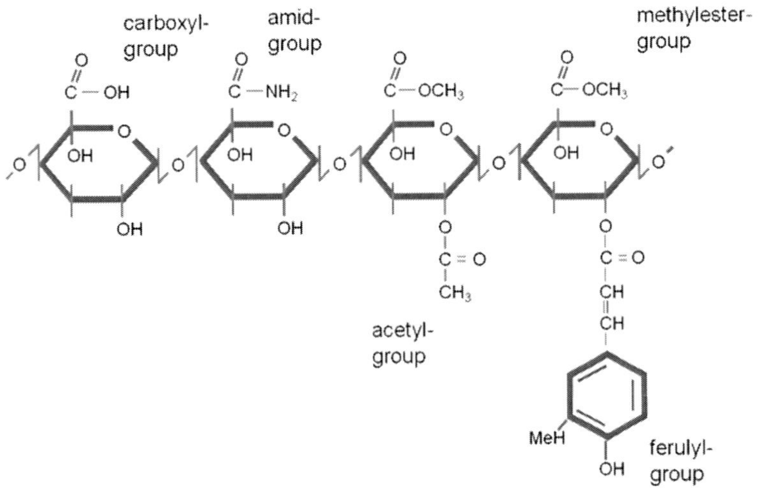

Figure 8.4 Pectin chain with substituted OH groups.

Pectins can be derivatized on both their carboxyl and hydroxy groups, as exemplified in Figure 8.4. The galacturonic acid units of pectins may be esterified with methanol (and rarely with ethanol) forming poly(galacturonic acid) methyl esters with different degrees of esterification (DE) or degree of methylation (DM).

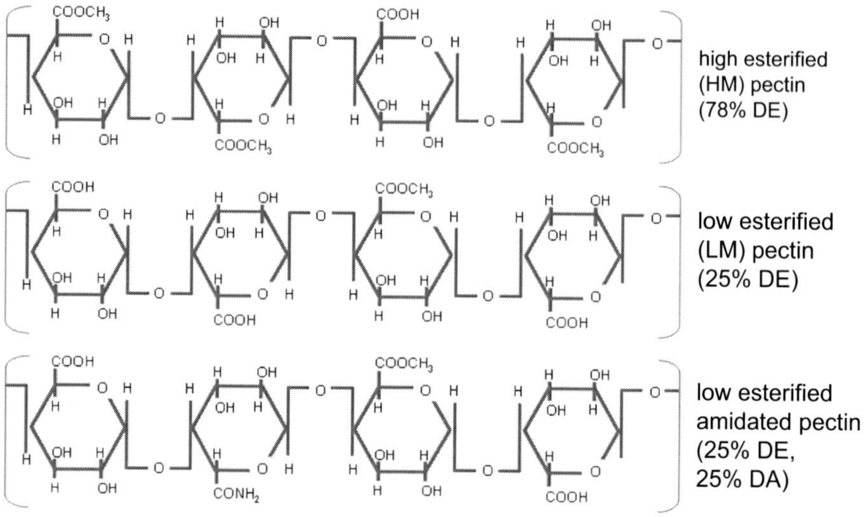

Figure 8.5 High and low ester pectins.

A DE of 50% is defined as the boundary between high methyl ester (HM) and low methyl ester (LM) pectins (see Figure 8.5). LM pectins can be extracted, whereas HM pectins are found in most plants. For instance, pectins isolated from apples and citrus fruits have DEs of up to 60–80%, sugar beet 60%, lemon and lime 70%, and apple 80%. The distribution of methyl esters, and accordingly the distribution of unesterified acid residues, also varies from plant to plant and changes during maturation by enzymic deesterification. Plant pectin esterases act blockwise resulting in deesterified blocks with free carboxyl groups which exhibit increased reactivity against divalent cations and positively charged proteins.

The galacturonic acid units may also be esterified with acetic acid (and other acyl compounds), mainly on C-2 and C-3, but potentially to all their hydroxy groups, giving a degree of acetylation (DAc) higher than 100%. Apple and citrus pectins show low acetyl contents, but sugar beet pectin exhibits high DAcs, which is the reason why extracted sugar beet pectins do not jellify.

This chapter provides a systematic review of pectins, dealing mainly with commercially produced pectins used in the food and pharmaceutical (nutraceutical) industries. It covers information on the sources of raw materials, industrial production, and details of different health food and related medical, biomedical and skin care applications. In-depth information on other pectins with a detailed structure–function relationship of different sources, mainly from medicinal herbs of traditional oriental medicine, can be found in papers by Yamada and colleagues.[7–10]

8.2 Industrial Sources and Production of Pectins

Commercial pectins are produced mainly from citrus peels, apple pomace and also from sugar beet pulp in lesser quantities. The technology of commercial production means that all of the raw materials need to be available in good storable quality all-year round. Thus, fresh citrus peels, apple pomace and sugar beet pulp are immediately dried after processing the fruits for juice or sugar production. Citrus peels are additionally washed before drying to avoid the peels turning brown. Sometimes citrus peels are even pasteurized to inactivate pectic enzymes which are responsible for blockwise deesterification of the pectin chains as described above. In Central and South America and Sicily, pectin is also extracted from fresh, wet citrus peels of local plants during the citrus harvest season.

The pectin contained in the cell walls is insoluble and is referred to as "protopectin". Therefore the raw material is treated with inorganic acids at elevated temperatures to solubilize the so-called protopectin. Under these conditions the galacturonosyl bonds are stable, but the bonds between sugars of the side chains, especially the arabinofuranosyl bonds, are split, thus leading to release of pectin from the cell wall network into the aqueous solution.

The resulting pectin solution is then separated from the depectinized raw material, clarified mechanically and concentrated by evaporation or membrane filtration. As pectin is insoluble in alcohol, it can be isolated from the aqueous solution by alcohol precipitation. Previously pectins were also isolated from the aqueous solution by precipitation with aluminium salts. But this technique is considered environmentally unfriendly and should not be used today. Furthermore, the presence of aluminium ions in Al-precipitated pectins is often associated with Alzheimer's disease; this is a strong argument against aluminium precipitation.

In the case of alcohol precipitation, the precipitated pectin is separated from the alcohol–water mixture, dried, ground and sieved to produce a powder with a defined particle size. When no other modification is introduced in the alcohol process, high methyl ester pectins are obtained. Deesterification with sodium or potassium hydroxide is too rapid and not controllable, but acidic deesterification can be employed to reduce the degree of esterification in a controllable manner in the following three ways.

1. Extraction conditions that produce lower degree of esterification.
2. Deesterification in aqueous solution which produces the most homogeneous distribution of DE.
3. Deesterification in alcoholic suspension of the dried and ground pectin.

A similar technique is also used to obtain amidated pectins by deesterification under alkaline conditions in ammonia (see Figure 8.6). This way methyl ester groups are converted to amide groups (–$CONH_2$). Free carboxyl groups are not amidated under these conditions.

Pectin esterases (PEs), mainly from plant origin, can also be used to obtain highly reactive HM pectins due to their blockwise deesterification mechanism. PE from moulds acts randomly and its use is not economic compared to acidic deesterification. As shown in Figure 8.7 HM pectins with identical DE can be

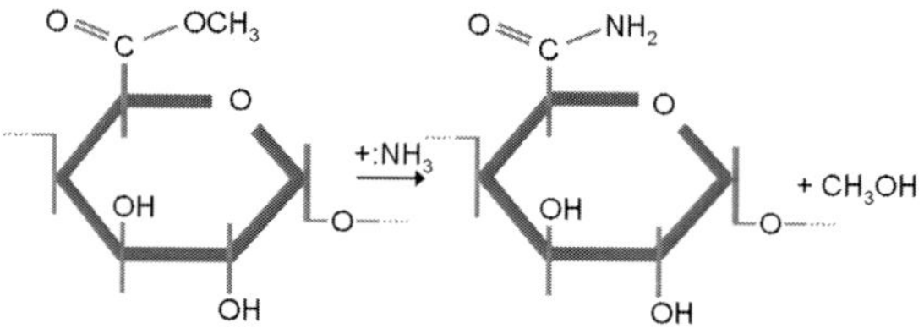

Figure 8.6 Ammonolysis of ester groups to amide groups.

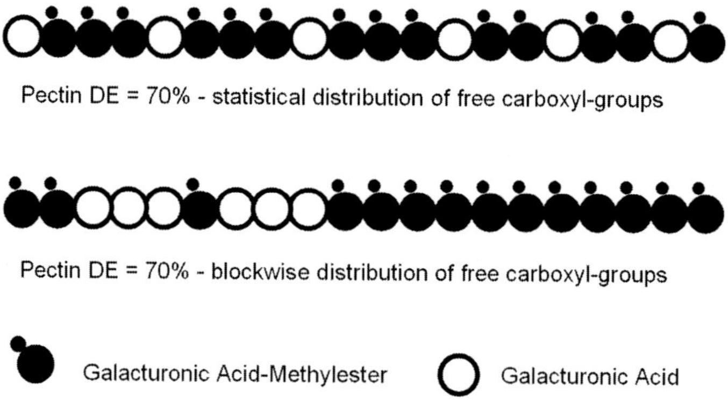

Figure 8.7 Statistical and block by block distribution.

distinguished by their degree of blockiness (statistical or blockwise distribution of free carboxyl groups).

Pectins can also be esterified up to almost 100% by diazomethane. (Caution: diazomethane is highly explosive on rough surfaces.) Pectins with such high DEs also gel relatively independently of pH as they contain almost no dissociated carboxyl groups to repel the chains from each other. Pectin esterification can also be achieved with methanol in sulfuric acid, but in this case sodium methyl sulfate is also formed, which should not exceed 0.1% in commercial pectins according to the Food Chemical Codex (FCC).

Enzymic pectin extraction has not been successful so far. Even the recently discovered rhamno-galacturonase has not changed this situation. Pectins could only be obtained in significant quantities with cellulases and hemicellulases in combination with pectinases, but this process leads to a deterioration of pectin quality by the action of the pectinases. This finding also amply confirms the complexity of the cell wall structure.

Low viscosity pectins can be obtained by enzyme treatment. Pectin lyases (PLs, EC 4.2.2.10) lead to no changes in DE, but produce a Δ-4,5-unsaturated galacturonic acid unit at the non-reducing end of the pectin chain. However, treatment with a combination of pectin esterase (PE, EC 3.1.1.11) and endo-polygalacturonase (endo-PG, EC 3.2.1.15) leads to low viscosity pectins with lower DEs. Enzymes involved in pectin degradation are shown in Figure 8.8.

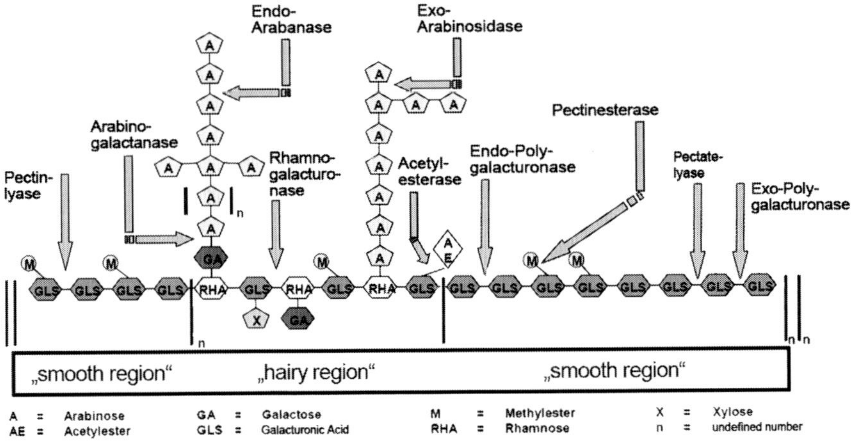

Figure 8.8 Enzymes involved in pectin degradation.

Sugar beet pectins are also esterified with acetic acid, mainly at C2 or C3. Commercial beet pectins produced by Herbstreith & Fox exhibit a DAc of up to about 40%. This acetylation is the reason why beet pectins do not exhibit gelling properties. Enzymic and acidic deacetylation processes produce beet pectins with still poorer gelling strengths due to their lower molecular weights (M_w).

The molecular weight of pectins can also be reduced by oxidizing agents or ball milling techniques which only slightly modify the degree of pectin esterification.

The balance of hydrophobic acetyl and methyl ester groups and the hydrophilic hydroxy groups in sugar beet pectin provide emulsifying properties that are superior to those of gum arabic.

8.3 Physical Properties and Chemical Stability of Pectins

8.3.1 Molecular Weight and Viscosity of Pectins

Pectins are mainly used in the food industry as thickening and gelling agents (INS No. 440, CAS number 9000-69-5), or for stabilizing acidified milk drinks and cloudy juices.[11,12] Depending on their molecular weight, pectins provide different viscosities and gel strengths. Molecular weight of commercial pectins is in the region of about 100 000, but much higher values can also be found in the literature, depending on the analytical method used. To create low and extra low viscosity pectins for dietetic applications the molecular weight may be reduced by enzymic or oxidative methods. Pectins are soluble in water but insoluble in organic solvents.

Dilute pectin solutions and solutions of pectins with very low molecular weight, used as low viscosity soluble dietary fiber, show an almost Newtonian viscosity. Higher concentrations of gelling pectins give pseudo-plastic solutions. The viscosity of pectin solution increases on the addition of divalent metal salts which crosslink the pectin chains. This viscosity increase is higher for pectins with a blockwise distribution of free carboxyl groups and with a decreasing degree of esterification. However, the addition of too much metal salt leads to precipitation of the corresponding pectinate salt. The addition of soluble solids, *e.g.* sugars, also increases the viscosity of pectin solution.

8.3.2 Chemical Stability of Pectins

Pectins are very stable under acidic conditions in a pH range from 2–4.5. Very low pHs (<2) and elevated temperatures result in deesterification of pectins; this leads to lower gelling temperatures for HM pectins, higher gelling temperatures for medium esterified pectins (due to the influence of calcium gelling mechanism), and higher calcium sensitivity for LM pectins.

At higher pH values, between 4.5 and 10, pectin chains are split by β-elimination. This splitting reaction occurs only next to a methylated carboxyl group, and consequently HM pectins are degraded faster than LM pectins. However, pectic acid and pectates are stable in this pH range because they lack any methyl ester groups. β-Elimination is activated by pectin lyase (PL) and produces a Δ-4,5 double bond at the non-reducing end of the galacturonic acid chain, which can be detected at 235 nm. Still higher pHs lead to the saponification of pectin. Thus deesterification becomes more effective than β-elimination as saponified pectins are less sensitive to β-elimination.

Gelation (setting) is the most interesting property of pectins for food uses. Gels are formed under defined conditions by cooling down hot pectin solutions (pectin sol). HM pectins gel with a high soluble solids percentage ($>55\%$ SS) and low pHs (about 2.8–3.4) by the so-called sugar–acid gelling mechanism. Pectin chains form bonding zones by hydrophobic interaction of the methyl ester groups and by hydrogen bonding of the hydroxy groups. Hydrophobic interactions are formed first at higher temperatures, followed by hydrogen binding during cooling. Addition of an acid is necessary to reduce the dissociation of the acidic groups, and thus preventing pectin chains from repelling each other.

Soluble solids in pectin sols or gels reduce the water activity and its hydrate cover, but, as shown in Figure 8.9, a certain relation must be maintained between soluble solids and pH. At too high pH values and too low soluble solids no gel is formed. On the other hand, at too low pH values and too high soluble solids the gelling temperature is too high and gelation starts in the cooking vessel at the cooking temperature, resulting in broken gel structures. This behavior is called pregelation.

Pectins are classified according to their gelling temperature into rapid set, medium rapid set and slow set pectins, provided the cooling rate is constant

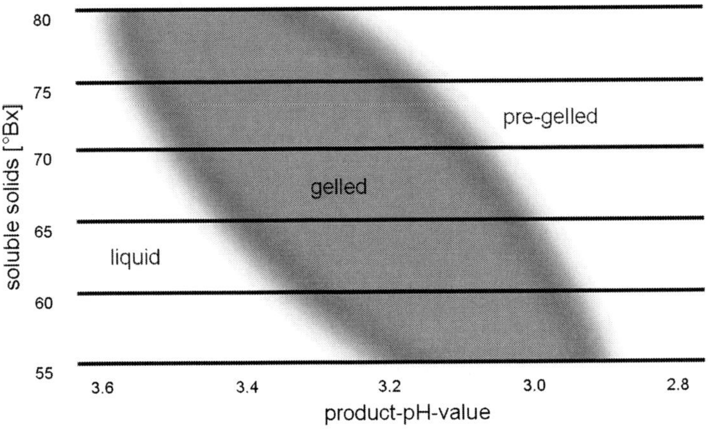

Figure 8.9 Setting range of high esterified pectins.

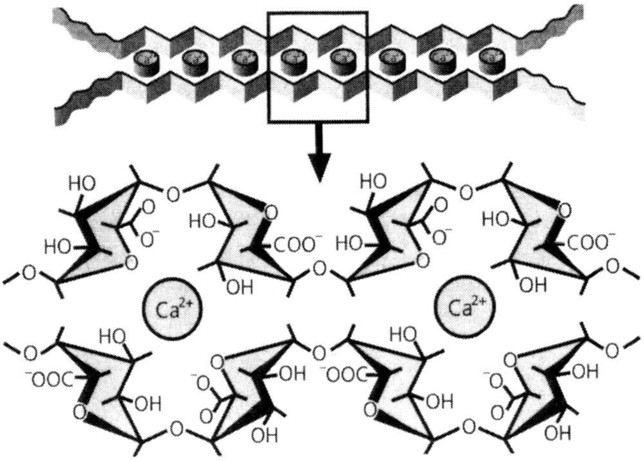

Figure 8.10 Gelation of LM pectin by calcium ions (egg box model).

so that time to gelation correlates with setting temperature. The setting temperature increases with increasing pectin concentration, increasing soluble solids and decreasing pH. On the other hand, the setting temperature decreases with decreasing degree of esterification to a minimum of about 58–62 DE; then it increases again due to the gelling mechanism of carboxyl groups with calcium ions.

LM pectins gel by the sugar–acid gelling mechanism and also with divalent alkali earth metals relatively independently of pH and soluble solids. In practice, calcium ions are normally used to influence pectin setting. Figure 8.10 shows two pectin chains forming a zigzag network, with cage spaces formed by four galacturonic acid units with two units from each pectin chain. Inside the cages calcium ions are bound as metal complexes like eggs in an egg box, giving the name "egg box model" for this gelling mechanism (as discussed more fully in ref. 11).

The calcium concentration necessary for maximum gel strength generally depends on the calcium sensitivity of the pectin, mainly depending on the degree of esterification but also on the distribution of free carboxyl groups along the pectin chains. This calcium requirement increases with increasing pH, increasing ionic strength and decreasing soluble solids. However, higher calcium ion concentrations are required on changing soluble solids from sucrose to fructose, and to sugar alcohols like sorbitol.

On adding calcium ions to a pectin sol, the viscosity increases gradually due to the association of pectin chains by calcium bridges, leading to optimum gel formation and then to the formation of insoluble calcium pectinates which are relatively soft gels depending on the gelling temperature. The reaction of LM pectins with calcium ions can also be used to obtain pectin beads or pectin stripes containing sugars, colors, flavors or other active ingredients by dropping or extruding the pectin solution into a bath of calcium chloride.

8.3.3 Enzymic Determination of Pectin, Poly(galacturonic acid) and Galacturonic Acid

Pectin is deesterified by a pure pectin esterase to poly(galacturonic acid), which is degraded to galacturonic acid (mono-galacturonate) by a pure poly-galacturonase (see Chapter 2). Pectin and pectate lyases have to be absent as these enzymes yield unsaturated galacturonates which cannot be utilized for the subsequent enzymetric determination, which would result in values which are too low for the analyzed pectin. In the following enzymetric steps mono-galacturonate is transformed to D-tagaturonate (5-oxo-L-galactonate) by D-glucuronate isomerase, and D-tagaturonate is then reduced by NADH-specific D-tagaturonate reductase. With this enzyme system, both D-galacturonate and D-tagaturonate can be determined by measuring the concentration (absorption) of NADH at 339 nm, Hg 334 nm or Hg 365 nm, which is proportional to substrate concentration.[13] Both D-glucuronate isomerase and D-tagaturonate reductase can be obtained from *Bacillus polymyxa*, and their separation is not necessary.

8.4 Medical Applications of Pectins

8.4.1 Effect of Pectin on Cholesterol and Lipid Metabolism

Pectins control the level of available cholesterol in the body mainly by the following two mechanisms.

- Binding of steroids like cholesterol ingested with food, thus reducing their absorption.
- Binding of bile acids, and thus reducing cholesterol synthesis.

Pectin increases the viscosity of the chymus and also forms complexes with low density lipoproteins (LDL) in the gut, thus enhancing their excretion and reducing their resorption. Work by Baig and Cerda indicates that the interaction between pectin and lipoproteins is of an electrostatic nature, with possible hydrogen bonding.[14] Falk and Nagyvary report that one part HM pectin binds 4 parts LDL, but LM pectins bind less LDL, meaning that the binding mechanism depends on the DE of the pectin.[15]

Pectins also bind bile acids which are synthesized in the liver and secreted into the gut in the small intestine. This means that pectins interact with bile acids in their enterohepatic pathway and reduce their reabsorption rate from the gut back into the liver. Thus, bile acid synthesis in the liver is increased by a feedback mechanism. The substrate for this synthesis is serum cholesterol which is mainly bound to serum LDL. This reduces the LDL level, the lipoprotein with the highest aterogenic potential. Pectins have almost no influence on high density lipoproteins (HDL), but their binding to LDL reduces the LDL/HDL ratio.

As early as 1961 it was reported that serum cholesterol levels can be lowered by the intake of food containing a certain quantity of pectin.[16] Early studies up to 1985 have been reviewed by Behall, Reiser and Endress, and are summarized in Table 8.2.[17–19] Apart from two reports which did not prove significant cholesterol reductions, all of the studies on a wide variety of subjects and experimental conditions indicate significant cholesterol reductions if 6–15 g day^{-1} pectin was given,[16,20–35] but cholesterol reduction is insignificant when less than 6 g day^{-1} is consumed.

Table 8.2 Summary of reported results on the cholesterol (Chol) lowering effect of pectin.

Diet form (n)	*Duration/weeks*	*Pectin consumed/g day^{-1}*	*% Cholesterol reduction*	*Ref.*
Controlled (24)	3	15	5	16
— (23)^b	7–9	9–12	0	20
Self served (6)	4	2–10	6^a	21
Self served (12)	4	36	12	22
Self served (12)	3	12	8	23
Controlled (9)	3	15	13	24
Controlled (6)	4	2	0	25
Controlled (10)	6	6	0	26
Controlled (5)	3	30	13	27
Self served (21)	6	15	9	28
Self served (11)	6	15	19	28
Controlled (62)	5	15	10	29
— (12)^b	2	9	10	30
Self served (10)	3	15	16	31
Self served (10)	3	15	18	31
Controlled (6)	3	36	10	32

^{a}On 6 g day^{-1} or more pectin.
^{b}No information given in reference on way of serving.

Table 8.3 Summary of reported results on the influence of pectin on lipid metabolism.[a]

Subjects/ diet[b]	*Duration*	*Pectin*[c]*/g day*$^{-1}$	*% Changes in lipid levels* *Total*	*LDL*	*HDL*	*TG*	*Ref.*
30/C	3	20	–17	–21	+4	nd	40
15/S	3	20	–12	–14	+12	nd	40
27	4	15	–15	nd	nd	nd	41
54	90 d	15+a	–34	nd	+25	–25	42
55			–34	nd	+34	–26	42
47	90 d	15+b	–36	–23	+36	–19	43
10	4	15	–112	–11	+21	nd	44, 45
40	4	10+c	–44	–45	+14	–55	46

[a]Decimals rounded to the nearest unit.
[b]C, controlled; S, self served; nd, not determined.
[c]a, pectin with sorbitol 1 : 1; b, pectin with sorbitol 2 : 1; c, with 1.5 g day^{-1} ω-3 fatty acid.

The influence of DE of pectins on their cholesterol lowering ability is controversial. Ershoff and Wells showed that pectins with approximately 10% methyl ester counteracted the increment of rat liver cholesterol induced by cholesterol feeding, whereas pectins with approximately 5% methyl ester were ineffective.[36] However, these results could not be confirmed by Judd and Truswell who obtained similar results in reducing serum cholesterol by using 15 g high ester (DE 71%), low ester (DE 37%) or amidated pectin.[31]

Further studies published since 1985 are summarized in Table 8.3. When 15 g pectin was given with 20 g day^{-1} fish oil, the cholesterol ester fraction of the plasma lipids was reduced by 44%. Another beneficial effect observed was a 30% decline in the fatty acid fraction by the end of the treatment period.[37] Plasma cholesterol and triglycerides were more effectively lowered in rats when apple pectin and apple polyphenols were fed to them together, suggesting possible interactions between apple pectins and polyphenols.[38]

In 1999 Brown *et al.* published a meta-analysis of 67 studies carried out between 1966 and 1996 on the effects of pectin (see Figure 8.11), psyllium, oat fiber and guar on cholesterol, triglycerides and lipoprotein levels.[39] The analysis indicates that these dietary fibers reduce total cholesterol and LDL levels, but did not alter HDL level significantly. Pectin is the most effective in lowering total cholesterol, LDL and triglyceride levels (see Table 8.4), but increasing the daily dosage to more than 10 g pectin did not improve its effect further.

Some studies indicate that the beneficial effect of pectin is proportional to the serum cholesterol level of the subject, and hence higher in patients with a high cholesterol level.[47] In one study 15 g day^{-1} pectin and 450 mg day^{-1} ascorbic acid were given to 11 patients (group 1) with high cholesterol levels and 21 healthy volunteers (group 2); the total cholesterol level was reduced by 18.7% in the first group and 8.6% in the second.[48]

Pfeiffer has also analyzed the influence of molecular weight, neutral sugar content and acetylation of pectins on serum cholesterol and lipoproteins in rats.[49] In the first trial, apple pectin, beet pectin, beet fibers and an arabinan

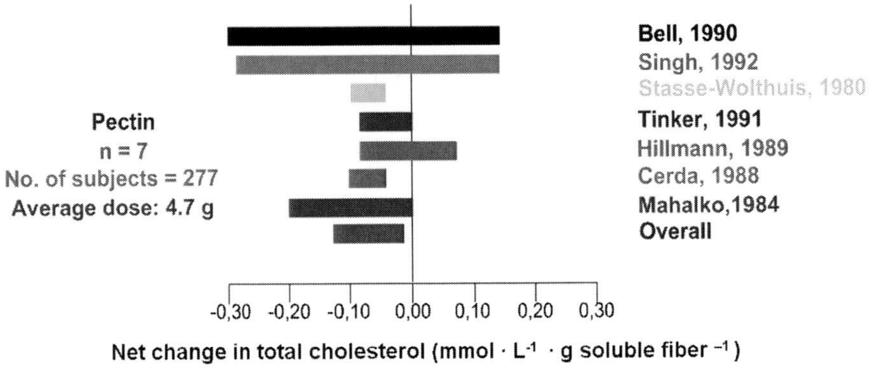

Figure 8.11 Changes of total cholesterol contents.[39]

Table 8.4 Meta analysis of 67 reported studies on the reduction of lipid levels by soluble dietary fibers.[39]

Lipid	*Soluble dietary fiber*	*Number of studies*	*Number of subjects*	*mg dl^{-1} change per g fiber*
Cholesterol	Oat products	26	1600	−1.43
	Psyllium	17	757	−1.08
	Pectin	7	277	−2.71
	Guar	17	341	−1.00
LDL cholesterol	Oat products	22	1439	−1.23
	Psyllium	17	757	−1.12
	Pectin	4	117	−2.13
	Guar	12	218	−1.28
HDL cholesterol	Oat products	24	1542	−0.07
	Psyllium	17	757	−0.07
	Pectin	7	277	−0.14
	Guar	15	302	−0.11
Triglycerides	Oat products	20	1374	+0.7
	Psyllium	16	720	+0.3
	Pectin	6	247	−1.8
	Guar	17	338	−0.9

were compared. Serum cholesterol was lowest in the apple pectin group; apple pectin and beet pectin showed similar effects on total lipid reduction, followed by beet fibers; and arabinan had no effect. In a second trial several beet pectins were compared: beet pectin with lower M_w and deacetylated beet pectin had the highest cholesterol lowering effect, and beet pectin with reduced neutral sugar content had the lowest effect. This shows that neither high molecular weight nor high viscosity is essential for the cholesterol lowering effect of pectins, but apple pectin with a high methyl ester content and high viscosity had the strongest serum cholesterol lowering effect. Thus, hydrophobic interactions of the methyl ester residues must also be taken into consideration.

Yamaguchi *et al.* have also compared three pectins with different molecular weights, M_w 750 000, 185 000 and 66 000 in rats.[50] The lowest M_w pectin had no hypocholesterolemic activity, but the medium M_w pectin showed low viscosity and retained its cholesterol lowering property. Although the molecular weights reported by Yamaguchi *et al.* may not be realistic for pectins (some methods give very high values), some viscosity effect on cholesterol reduction is indicated. Pectic substances are also involved in bile and metabolism.[5] An addition of 15 g pectin to the meal of ileostomy patients increased bile acid excretion by 35% and cholesterol excretion by 14%.[52] The increased bile acid excretion seemed to be independent of the DE of the pectins. Pectic substances not only accelerated the bile acid excretion, but also changed the bile acid profile.[53] The greatest interaction with bile acids was found with a very highly esterified pectin at pH 6 *in vitro*, but the interaction diminished with decreasing DE. This decrease was more marked for pectins with blockwise arrangement of free carboxyl groups formed by the action of a plant pectin esterase. Amidated and acetylated pectins had lower interactions with bile acids.[54]

Thus, the different mechanisms of pectin action on lipids can be summarized as follows.

1. Increasing viscosity of the chymus.
2. Increasing unperturbed (unstirred) water layer, *i.e.* a higher resorption barrier.
3. Binding of lipids and the formation of a matrix, thus hindering enzyme–substrate interactions.
4. Binding of enzymes like pancreatic lipase, and hence less triglyceride split.[55]
5. It is postulated that by higher production of SCFA from pectin, pH is reduced, thus changing microbial cholesterol synthesis, and consequently all parameters lead to a higher excretion of cholesterol and bile acids.

However, the influence of pectins on the excretion of triglycerides is still controversial. For instance, several studies show no effect beyond the non-specific increased excretion observed with all nutrients due to higher viscosity of the chymus. In one study, the combination of 10 g day^{-1} pectin with 1.5 g day^{-1} ω-3 fatty acids reduced the level of triglycerides by 55% (from 377 to 170 mg dl^{-1}), of total cholesterol by 44% on average (from 270 to 151mg dl^{-1}) or 30% if the values were slightly higher, or 50% when values were very high, and of LDL by 45.5% (from 255 to 139 mg dl^{-1}). But the level of HDL was increased on average by 14%.[46] In other studies a 15 g day^{-1} pectin diet increased the fecal excretion of neutral steroids in human by 17%, bile acids by 35%,[24] and acidic steroids by 11%.[56]

8.4.2 Effect of Pectin on Glucose and Insulin Concentrations

Following a carbohydrate-rich meal, the blood glucose level rises quickly, which may cause problems in diabetics. By the same mechanisms as discussed above for the effect of pectin on lipids, pectin also delays and reduces glucose absorption by gel formation and increased viscosity of the chymus. Several studies with insulin dependent and insulin independent diabetics have shown

Table 8.5 Summary of reported results on the influence of pectins on serum glucose and serum insulin.[a]

		Time (min) to significant decrease in		
Subjects (n)	*Pectin added/g*	*Serum glucose*	*Serum insulin*	*Ref.*
Diabetics (8)	10	30–90	30–120	57
Normal (13)	10	15, 30–90 (ns)	15–45	58
Gastric surgery/ dumping syndrome (5)	10.5	30, improved retention of load in stomach	nd	59
Diabetics (6)	14.5	ns	ns	60
Diabetics (6)	9 per m^2 body surface	30–60	ns	61
Normal (6)	14.5	30–45	(—)	62
Gastric surgery (23)	10–20	30	nd	63
Hypoglycaemic (3)	5	Hypoglycaemia averted	nd	63
Diabetics id (8)	15	15–90	nd	64
Diabetics id (7)	7	60–90	>180	65
Normal (6)	10	ns	ns	66
Normal (6)	10	60–90	ns	66
Diabetics (13)	10	60	ns	67
Normal (5) Obese (6) Diabetics (5)	10 + guar	Significant decrease in all, but most in obese and diabetic subjects		68
Normal (13)	20	ns	(—)	69

[a]ns, not significant; nd, not determined; (—), no change

that pectin lowers blood glucose and insulin levels after a carbohydrate meal (see Table 8.5).[57–72]

Pectin, when not standardized with sugar, has a glycemic index (GI) of almost zero. Sugars added for standardization increase the GI proportionally to the added amount of sugar. Pectins reduce the GI of all sugars, therefore they are often used in food designed for diabetics. According to Jenkins and Jenkins the addition of purified pectin to carbohydrate test meals flattens the glycemic response in both normal and diabetic volunteers, reduces the insulin requirement in patients on an artificial pancreas, and in the longer term reduces urinary glucose loss and improves diabetes control.[73]

In the context of high-fiber, high-carbohydrate diets, these findings have had a major impact in influencing the recommendations for the dietary management of diabetes. The mechanism of action of pectin on sugars appears in part to be the slowing down of absorption, rather than increasing the colonic loss of carbohydrates. Consequently postprandial gastro inhibitory peptide (GIP) and insulin levels are reduced, and the more viscous purified fibers like pectins appear most effective.

Effects of pectin on fatty acid and glucose absorption have also been studied in rats and humans by perfusing the intestine with linoleic acid and glucose solutions with and without pectin. Linoleic acid and glucose absorption decreased on increasing the concentration of pectin. The reduction in linoleic acid absorption was not caused by pectin binding or impaired micelle formation due to pectin–bile acid binding, but the unstirred water layer expanded with increasing concentrations of pectin. These results suggest that the expansion of the unstirred water layer is closely associated with the reduced absorption of fatty acids and glucose ingested with pectin.[74] This confirms the results previously reported by Gerencser *et al.* in rabbits,[75] and by Flourie *et al.* in humans.[76]

Siddhu *et al.* examined the effect of corn oil on postprandial glycaemia and insulinaemia when ingested with glucose, casein, cellulose and pectin in various combinations. They found that glucose–corn oil–pectin and glucose-corn oil–casein–pectin had the lowest glycemic and insulinaemic responses, with the response to the latter being lower. Corn oil by itself had only a modest effect on postprandial metabolic response to glucose, but the addition of protein and fiber, especially pectin, led to a significant attenuation of the glycemic and insulinaemic responses.[77]

8.4.3 Effect of Pectin on Digestive Enzymes and Hormones

In addition to increased viscosity, pectins are reported to reduce amylase activity by 10–40%, lipase activity by 40–80%, and trypsin activity by 15–80%.[78–82] The activity of pancreatic enzymes is reduced by complex formation between pectin and the enzymes, and by the increased viscosity of the digestive fluids, leading to reduced contacts between the enzymes and their substrates. In a study on the inhibition of lipase activity by citrus pectins, Tsujita *et al.* suggest that pectin may interact with emulsified substrates, and thus inhibit their surface contact with lipase.[83] Subsequent to these findings they filed a patent for health food containing pectins which inhibit lipase activity to prevent diseases caused by human lifestyles, especially obesity, hyperlipidemia and arteriosclerosis.[84] Increased cholesterol-7-α-hydroxylase activity has also been reported after pectin supplementation in rats.[85] This enzyme catalyzes bile acid synthesis from cholesterol and its increased activity may also reduce the LDL level.

Morgan *et al.*[86] and Levitt *et al.*[87] reported that pectin supplements reduce the production of GIP. This hormone reduces gastric motility, which may also have an influence on the gastric emptying half time. It also reduces insulin secretion, which could result in lower activity of the enzyme α-hydroxy-α-methylglutaryl (HMG)–CoA reductase, because its activity depends on insulin concentration. This could cause reduced cholesterol level, because HMG–CoA reductase is involved in an early step of the endogenous cholesterol synthesis.

8.4.4 Effect of Pectin on Atherosclerosis

Fibrinogen is an important risk factor in atherosclerosis, stroke and cardiovascular diseases (CHD). This risk increases with high serum cholesterol levels. It is also believed that not only fibrinogen concentration, but also the type of fibrin network may be an important risk factor in the development of CHD. Atherosclerosis-associated CHD and stroke, plus the related problems of hyperinsulinaemia, hyperlipidaemia and hypertension, are strongly related to the diet.[44] Veldman *et al.* have studied the possible effects of pectin on fibrinogen levels and fibrin network architecture.[44] In this study two groups of 10 male hyperlipidaemic volunteers each received a 15 g day^{-1} pectin supplement or a placebo for 4 weeks. Pectin supplementation significantly decreased the total cholesterol, LDL and apolipoprotein A and B. Significant changes in the characteristics of fibrin network in the plasma of the pectin-supplemented group included a higher permeability and a lower tensile strength, *i.e.* network structures which are believed to be less atherogenic. It is suspected that pectin modifies network characteristics by a combination of its effects on metabolism and altered fibrin conversion.

In a subsequent study the same group examined the possibility that pectin may also influence the fibrin network architecture *in vivo* through acetate.[45] Thus, the earlier study[44] was repeated with 6.8 g day^{-1} of acetate instead of a placebo. Fibrinogen levels remained almost unchanged, but significant changes were found in the characteristics of fibrin networks developed in plasma, as in the pectin group. The fibrin networks were more permeable, had lower tensile strength and were more lysable. From these results it seems highly likely that acetate may be responsible in part for the beneficial effects of pectin supplementation *in vivo*.

In the Los Angeles Atherosclerosis Study, the intima-media thickness (IMT) of the common carotid arteries was measured by ultrasonography in humans aged 40–60 ($n = 573$, 47% women).[88] A significant inverse association was observed between IMT progression and the intake of viscous fiber ($p < 0.05$) and pectin ($p < 0.01$). The ratio of total lipid to HDL cholesterol was inversely related to the intake of total fiber ($p < 0.01$), viscous fiber ($p < 0.05$) and pectin ($p < 0.01$). Thus, the intake of viscous fiber, especially pectin, appears to protect against IMT progression, and serum lipids may act as mediators between dietary fiber intake and IMT progression.

Frequent and long lasting high insulin concentrations promote vascular lesions, the primary locus of atherosclerosis. The suppression of postprandial insulin level by pectins may therefore have an anti-atherogenetic effect.[89] Dietary fibers like pectin may also increase peripheral insulin sensitivity in young and old adults.[90]

8.4.5 Pectins in Weight Management

Weight reduction by pectins involves a variety of non-specific mechanisms and is proportional to the gel forming and viscosifying properties of the pectin used. Low viscosifying properties can be compensated for by increasing the pectin dosage. In addition to the ability of pectins to act as soluble dietary fibers which can replace sugar and fat in certain recipes, the following properties of pectins may also play a role in weight reduction.

1. Delaying of gastric emptying half time.[62,69,91,92]
2. Longer mouth caecum transit time.
3. High water-retaining capacity in a gel matrix.

4. Prolonged feeling of satiety.
5. Reduced food consumption.
6. Reduced absorption of food components from the stomach.[76]
7. Immobilization of nutrients.
8. Hindrance of enzyme–nutrient complexes and actions.[93–96]
9. Reduced degradation and digestion of macromolecular nutrients.[96]
10. Expansion of the unstirred water layer.[75]
11. Reduced and delayed resorption of nutrients.
12. Higher non-specific excretion of nutrients.

For instance, a significantly high (80%) excretion of fatty acids is reported when healthy volunteers consumed 36 g day^{-1} pectin.[97] Pectins also reduce bile acid concentration in the small intestine, thus reducing the proportion of emulsified fat. Also when bile acid concentration is not sufficient, fat absorption is significantly reduced. Thus, DiLorenzo recommends pectins as a useful adjuvant in the treatment of disorders related to overeating.[98] Tiwary *et al.* have also studied different dosages of HM apple pectin in healthy US Army adults and concluded that pectin may have an important adjunct role in human nutrition in general and in obese persons in particular.[99]

Lipodystrophy has been described with increasing frequency in patients infected with HIV. In one study differences in the history of dietary intake between men with HIV who developed fat deposition and those who did not indicated that HIV-positive patients without fat deposition had a greater intake of overall energy (kcal kg^{-1}), total protein, total dietary fiber, soluble dietary fiber, insoluble dietary fiber and pectin than did HIV-positive patients with fat deposition.[100]

8.4.6 Effect of Pectin on Dumping, Short Bowel and Short Gut Syndromes

Dumping syndrome refers to early postprandial abdominal and vasomotor symptoms resulting from osmotic fluid shifts and the release of vasoactive neurotransmitters, and to late symptoms related to reactive hypoglycaemia. Effective relief from the symptoms of dumping can be achieved by dietary modifications that minimize the ingestion of simple carbohydrates and by avoiding fluid intake during ingestion of the solid portion of the meal. More severely affected individuals may respond to agents such as pectins, which increase the viscosity of intraluminal contents.[59,101–105]

Short bowel syndrome is characterized by weight loss, diarrhea and malabsorption. Pectin improves small and large bowel mucosal structure, prolongs intestinal transit and decreases diarrhea in rats. For instance, Roth *et al.* report that pectin significantly increases stool solidity and improves colonic water absorption following resection without significantly altering mucosal structure.[106] Sales *et al.* have described a pectin-supported diet program instead of total parenteral nutrition for patients with reduced length of remaining small bowel after bowel surgery due to mesenteric thrombosis or Crohn's disease; patients responded well to this approach and returned to their previous professional activities.[107] A pectin-supplemented diet evaluated in a 3-year old boy with short gut syndrome indicated that nitrogen absorption was higher and stomach-to-anus transit time was prolonged during pectin supplementation of the enteral feed.[108]

8.4.7 Effect of Pectin on Acute Intestinal Infections

The effect of different pectins on shigella, salmonella, klebsiella, enterobacter, proteus and citrobacter has been studied under clinical observations. This study revealed that pectin, irrespective of the raw material used for its production, had inhibitory effects on these microorganisms; the effect was most pronounced with the use of 5% pectin solution, and a rapid suppression of diarrhea and other manifestations of the infections were observed in patients.[109]

In another study, children with acute non-complicated diarrhea received either a preparation containing apple pectin and chamomile extract (Diarrhoesan) or a placebo. At the end of three days of treatment, the diarrhea ended in 33 out of 39 in the pectin–chamomile group, in 23 out of 40 in the placebo group, and pectin–chamomile reduced the duration of diarrhea by at least 5.2 h.[110] The therapeutic effects of cooked green banana and pectin in children with persistent diarrhea were also observed. By day 3 significantly more children receiving pectin or banana than controls recovered from diarrhea, *i.e.* 59%, 55% and 15% respectively. By day 4, these proportions increased to 82%, 78% and 23% respectively. Green banana and pectin also significantly reduced the amounts of stool, oral rehydration solution, intravenous fluid, frequency of vomiting and diarrhea duration.[111] A rice-based diet with green banana or pectin also reduced diarrhea in infants better than a rice-alone diet.[112]

Diarrhea is one of the most common complications in patients on tube feeding formulas. Supplementing these formulas with pectin significantly reduces the incidence of liquid stools and helps the normalization of colonic fluid composition.[113] Schultz *et al.* have also reported reduced rates of diarrhea when the tube feed for patients receiving antibiotics was supplemented with pectin.[114] Pectin and a combination of pectin with other substances like agar, tannic substances, iodine, kaolin, betonies, alkyl polyalcohols, aluminium phosphate, activated charcoal, sweet whey and nickel pectinate have also been claimed for the treatment of diarrhea.[19]

8.5 Pectins as Antidote in Metal Poisoning

As early as 1825 Braconnot, the discoverer of pectins, suggested that pectic substances are good antidotes for heavy metal poisoning because of the insolubility of the compounds they form. This and other early work is summarized in a 1951 book by Kertesz,[115] but what Braconnot did not know was that this is only half the story. Pectins also exchange cations according to the stability of their respective complexes.[116–122] This means that toxic metal ions can be complexed by pectins to enhance their excretion.[123–129] For instance, pectin has been evaluated as a prophylactic against Pb toxicosis.[130] The absorption of strontium into the bones of rats could also be suppressed by pectin.[131–133] *In vivo*, pectic substances form insoluble pectinates with heavy metals that are excreted with the stool and urine.[118,129,134–136] In another study typical symptoms of Pb poisoning disappeared when some factory workers ate 8–9 g pectin per day.[137]

Paskins-Hurlburt *et al.* have studied the affinity of sodium pectinate with a series of different metal ions and listed their results as follows:[118]

Stability of metal pectinates

$$Pb > Ba > Cd > Sr > Zn > Cu > Co > Ni > Fe > Hg > Cr > Mn > Mg$$

The complexation of pectin with divalent metal ions can be both intramolecular and intermolecular between two or more pectin chains (see the egg box model); the affinity is influenced by the DE of pectin[119,121,122] and by the distribution of free carboxyl groups, *i.e.* random or blockwise.[138] Complexation is also influenced by pH, ionic strength and the concentration and affinity of the other cations present.[139–141]

The significance of pH on the binding of essential metal ions (Ca, Cu, Fe, Zn) to pectin is demonstrated by Schlemmer in *in vitro* studies.[139–141] The highest binding constants are in the range of pH 4–6, except for iron which occurs at pH 7.0–7.5. In the jejunum, metal-ion absorption occurs at pH 6–7. Therefore no adverse effects on the bioavailability of Ca, Cu or Zn are observed, and only iron availability may be reduced.[129,139,140] However, this is also controversial in the literature, and the findings range from a significant reduction[142,143] to a slight reduction in bioavailability,[144,145] and to no change at all.[146–148]

For instance, in some studies supplemented pectin had no negative effects on Fe, Cu and Zn balances in short-term studies in humans.[146,149–151] Similar findings are reported for Ca and Mg in 5–6 week studies in humans.[29,97] In another study involving workers exposed to high levels of Pb no significant changes in Cu, Fe, Mg or Zn were found with 8 g day^{-1} pectin over a period of 6 weeks.[123]

The DE and molecular weight of pectin are also reported to influence iron absorption. In a study in rats, for instance, pectins with a high DE of 75 and low molecular weight of 89 000 improved iron absorption from 48% in the control group to 57% in the pectin fed group. Serum iron, transferring saturation, hematocrit and liver and spleen irons also increased compared to the control group, and also compared to the groups fed with high M_w or low DE and low M_w pectin.[152–154]

Interestingly when pectin or another negatively charged dietary fiber is taken alongside a high fiber intake in the form of fruits, vegetables and cereal products, a negative effect on the availability of minerals is sometimes reported. This may be due to other food components like phytate or lignin,[129] and may also be dependent on the duration of such studies. Harmuth-Hoene and Schelenz found that rats fed 10% carrageenan or agar-agar in the diet had a reduced resorption of Ca, Fe, Zn, Cu, Cr, and Co; Na-alginate decreased Fe, Cr, and Co absorption; and carob and guar gum interfered with the absorption of Zn, Cr, Cu, and Co. However, after 21 weeks of feeding on guar gum and agar-agar no deficit of these minerals in the body tissues, nor any difference in the control group, was found, suggesting that the rat is able to compensate for the increased fecal losses, presumably by reduced urinary losses.[155] Furthermore, it is also considered that gut mucosa may undergo an adoption process under the influence of fiber rich diets, which could be demonstrated in rats within 1–2 weeks. Thus, rats were fed a diet containing 15% fiber (pectin, cellulose, microcrystalline cellulose (MCC) and bran); the size and form of jejunal villi were assessed at 2 and 5 weeks, and were found to be more uniform at week 5 compared to week 2. In general it was found that the number of villi, and thus the resorption area, increases on increasing the dietary fiber content of the food and the duration of dietary supplementation. Therefore the reported short-term studies demonstrating negative effects of dietary fibers on the availability of minerals should be considered with caution and may be due to morphological changes.[129,156]

The ability of pectin to form insoluble salts has been used to quantify pectins as copper salts, known as the cuprizon method, and also by adding aluminium chloride to aqueous pectin extract to precipitate pectins industrially. However, as mentioned above this technology is no longer seen as appropriate because of the negative impact of aluminium ions on the environment and concern about an association with Alzheimer's disease.

In the case of Pb, pectin significantly increases excretion in both animals[118,124–127] and humans,[123,129,136,137] except for results in one publication.[145] For instance, Walzel *et al.* gave 8 g day^{-1} pectin with different DEs over a period of 6 weeks to 10 workers who were exposed to high Pb levels in an accumulator factory, and observed that their renal Pb excretion increased significantly from 55 to 127 ng Pb per ml urine. The results also showed significant decreases in blood Pb levels from 760 to 530 ng Pb per ml.[123,129] Blood cadmium also decreased in this study, but Cu, Fe, Mg, and Zn remained unchanged. This is attributed by Walzel *et al.*[129] to the high affinity[118] and high binding selectivity[121] of Pb towards pectin, which is almost independent of DE below 50%. This is in contrast to other cations like Cd,[120] Zn[157] and Ca.[158] Even with high DE pectins, Pb ions form relatively stable complexes.[121] This selective and strong binding mechanism exists under higher pH conditions present in the small intestines, where complications also take place, but only weak binding occurs in the acidic stomach.

A product called Medetopekt, a tablet composed of special LM apple pectin, apple fiber and apple powder, was first studied in rats and later in humans,[159] and was found effective for enhancing the excretion of lead, cadmium and strontium. In subsequent human studies in Kiev and Minsk, Medetopekt was tested against crude fiber tablets as the placebo in a double-blind study. Volunteers were given 3 × 3 tablets (550 mg each) from day 1 to 3, 3 × 7 tablets from day 4 to 6, and 3 × 10 from day 7.

In the first study the blood Pb level was measured before the treatment and at day 21. Medetopekt reduced Pb levels by 23% from 480 ng per ml blood to 370 ng per ml within 21 days, but crude fiber did not show significant changes. Renal excretion of Pb was also significantly increased compared to that of the crude fiber group. As the volunteers left the Pb poisoning environment during this study, Pb excretion in the crude fiber group dropped significantly, and the Medetopekt group also showed lower excretion at day 12 but higher at day 21. These results suggest an adoption phase to pectin utilization and degradation in the gut flora. Apparently after some time the microorganisms adapt to the pectin-enriched diet and produce more pectin-degrading enzymes. Consequently more oligogalacturonides are formed that could contribute to increasing renal Pb excretion.

In the case of strontium, ^{90}Sr was used, and the radioactivity of ^{90}Sr was measured in the urine. The excretion radioactivity was not significantly changed by crude fiber, but increased by 30 Medetopekt tablets (16.5 g) from 0.65×10^{-11} Curie per litre urine to 1.15×10^{-11} Curie per liter ($p < 0.05$) within 21 days.

Irrespective of how Pb entered the body, by eating or by breathing in, its renal excretion is always observed upon pectin supplementation. As pectin is not reabsorbed as a macromolecule, it is suggested that the active substances in this detoxification process are galacturonic acid oligomers which are formed by microorganisms in the colon and can be reabsorbed in the colon. This mechanism was demonstrated by Kohn *et al.* (cited in ref. 129) and by Walzel *et al.* both *in vitro* and *in vivo*.[160]

The binding affinity of saturated galacturonic acid oligomers with DPs of 1–9 towards Ca, Sr, Zn, Cd, Cu and Pb increases with increasing chain length; however, all the oligomers bind with different cations with different strengths. Ca, Sr and Zn bind very weakly by intramolecular electrostatic interactions, but Cd, Cu and Pb form stronger complexes. The strongest binding with Pb occurs for DP 5, with almost the same strength as for the polymeric chain. While the binding of a galacturonic acid monomer with Ca, Sr and Zn is negligible, it shows significant association with Cu and Pb.

No increased Pb excretion is observed by giving a galacturonic acid dose of 10 mg per kg body weight to Pb-exposed rats,[129] but a mixture of galacturonic acids with DPs of 1–3 given orally or intravenously significantly increased Pb excretion. Cu, Fe, Mg and Zn remained unchanged.[129] Walzel *et al.* report that an even more effective Pb excretion could be observed by intravenous injection of mixtures of Δ-4,5-unsaturated galacturonic acid oligomers (10 mg per kg body weight per day).[160] Here the excretion increases by 400% by the injection of a mixture of (mainly) unsaturated tri- and tetra-galacturonates, and by 600% by a mixture containing mainly unsaturated di-galacturonates, with less unsaturated trimers and almost no unsaturated tetramers. It is stated that the double bond is responsible for this higher binding affinity.

The essential mineral salts of pectins, and especially of oligo(galacturonic acid)s, can be used as a supplement to enhance the bioavailability of these elements. Lakatos and Meisel have also described complexes of saturated oligo(galacturonic acid) with Fe, Zn and Mg.[161]

8.6 Pectins as Soluble Dietary Fibers

Pectins are chemically almost totally stable under the conditions of the stomach and the small intestines; they are also not degraded by human or animal digestive enzymes, and hence barely resorbed in the small intestines before reaching the colon. It is on this basis that pectins are regarded as (soluble) dietary fibers. However, pectins are degraded by several types of microorganisms like *Bacteroides*, *Escherichia coli*, *Lactobacillus*, *Bifidobacterium*, *etc*., and the gut flora provide a complete spectrum of pectin-degrading enzymes including polygalacturonases, pectin esterases, pectin and pectate lyases. Thus, the human gut flora degrade pectins by 90–95%.[162,163] However, pectin degradation is reportedly less effective in rats,[164,165] and increases with decreasing DE,[166] and with the duration of the adoption period to a pectin-containing diet.[167] The fermentation of pectins occurs mainly in the caecum, the colon ascendens and the colon transversum colonized by these microorganisms.[163]

Human studies indicate that pectin degradation occurs not only in the colon, but also to some extent in the ileum.[168,169] For instance in patients with ileum syrinx, the applied pectins were recovered to a level of only 70%. Sandberg also suggests that microbial degradation of pectins occurs even in the very first sections of the gut, albeit at much lower extents compared to that in the colon.[168] Even though most authors deny this fermentation,[129] a certain alkaline degradation in this section of the intestines cannot be completely excluded.[170] This could explain the increased level of uronic acids in blood, liver and urine when pectins are applied.[171,172]

The degradation products of pectins in humans are mainly the short chain fatty acids (SCFAs), acetate, propionate and butyrate with molar ratios of 84 : 14 : 2 (ref. 163 and 173) or 81 : 10 : 9 at low pectin concentrations (2.5 mg ml^{-1}), 74 : 7 : 20 at high pectin concentrations (30 mg ml^{-1}),[174] and also some gases like methane, carbon dioxide and hydrogen. The SCFAs are almost completely utilized by the host microorganisms, as nutrients by the colon cells.

Incomplete pectin degradation yields oligo(galacturonic acid)s of different DPs. The non-reducing ends of these oligomers also may bear Δ-4,5-double bonds due to β-elimination under alkaline conditions or by action of pectin and pectate lyases. These products were also detected by Matsuura by incubating pectic acid with human feces.[175] Galacturonic acid as well as Δ-4,5-unsaturated di- and tri-galacturonic acids were generated with di-galacturonic acid as the main product. The corresponding enzymes were also found in animal feces.[176–178] Further products of galacturonic acid metabolism are furan-2,5-dicarbonic acid, and galactaric acid which may be transformed to acetoacetic acid.

In a European interlaboratory study on the fermentability of pectins in fresh human feces as inoculum, pectin degradation was 97.4%.[179] Some differences occurred between the laboratories with respect to the absolute values, but the donors were not the influential factors, and differences could be reduced by adding less substrate during incubation or using less dilute inocula. *In vitro* fermentation with inocula made from human feces, and from rat caecal contents, gave similar results. The digestible energy of HM pectins was analyzed in these laboratories with white Wistar rats at between 9.9 and 10.4 kJ per g dry weight at a fermentability of 92–95%.[180]

In comparing the fermentability of different substrates, at least for dietary fibers, the time lag in the adaptive response of the caecum and colon should also be taken into account. This time lag is suggested to be related to the time required for the microflora to adapt to the dietary polysaccharide.[181] When homogenates of human feces were incubated anaerobically with pectin, the yield of short chain fatty acids increased by 6.5 mmol per g pectin, or 1.05 mol per mol hexose equivalent.[182] *Bacteroides ovatus* preferentially utilized starch and pectin when grown on a mixture of polysaccharides in batch cultures, indicating that these carbohydrates are important substrates for the bacterium in the human large intestine.[183] LM pectins were also depolymerized and fermented faster than HM pectins by human fecal flora *in vitro*.[184]

The complete range of human fecal flora and cultures of defined species obtained from fecal flora were investigated *in vitro* to determine their ability to ferment pectin. The spectrum and the amount of unsaturated oligo-(galacturonic acid)s formed as intermediate products of pectin fermentation changed permanently in the culture media during incubation with complete fecal flora. After 24 h, no oligo(galacturonic acid)s were detected. The pectin-degrading activity of pure cultures of *Bacteroides thetaiotamicron* was lower than the pectin-degrading activity of the complete fecal flora. Cocultures of *B. thetaiotamicron* and *E. coli* exhibited intermediate levels of degradation activity. In pure cultures of *E. coli* no pectin-degrading activity was found. Saturated oligo(galacturonic acid)s were not found during pectin fermentation. The disappearance of oligo(galacturonic acid)s in later stages of fermentation was accompanied by the increased formation of short chain fatty acids.[185]

Dongowski *et al.* have investigated the degradation, metabolism and some properties of pectin in the intestinal tract of rats.[186] Conventional and germ-free rats were fed a 3-week pectin-free diet, or diets containing 6.5% pectin (DE 34.5%, 70.8% and 92.6%). The M_w distribution of the pectins isolated from intestinal contents of the germ free rats was unaffected by the diet (pectin passes the small intestine as macromolecule). No or very little galacturonan was found in the caecum, colon or feces of most of the conventional rats. In the colon contents of some conventional rats, di- and tri-galacturonic acid were present. Total anaerobic and *Bacteroides* counts were greater in the pectin-fed groups. Concentration of SCFA was higher in the caecum and feces of all pectin-fed groups. With increasing degree of esterification, the formation rate of SCFA decreased in the caecum of conventional rats. During *in vitro* fermentation of pectin with faecal flora of rats, unsaturated oligo(galacturonic acid)s appeared as intermediate products. LM pectin was fermented faster than HM pectin *in vivo* and *in vitro*. Pectin-fed rats had greater ileum, caecum and colon weights[186] and the weights of caecal wall and caecal contents, as well as the total bacterial population of the caecum, had increased.[187]

8.7 Prebiotic Fermentation of Pectin and Galacturonic Acid Oligomers

Crociani *et al.* have reported that pectin was fermented by 10% of the strains of *Bifidobacteria* tested for their ability to ferment complex carbohydrates,[188] and Slovakova *et al.* have explained why some strains of *Bifidobacteria* can ferment pectin and some cannot.[189] They compared the fermentation of apple pectin by *B. pseudolongum* P6 from rabbit caecum and by a *Bifidobacterium sp.* not growing on pectin, and a *Streptococcus bovis* X4. *B. pseudolongum* P6 fermented pectin *via* a modified Entner–Doudoroff pathway by extracellular endo-polygalacturonase (EC 3.2.1.15),[188–191] as in pectin-utilizing bacteria of the rumen. Cell extracts also showed the activity of 2-keto-3-deoxy-6-phosphogluconate (KDPG) aldolase (EC 4.1.2.14).

KDPG aldolase activity has been reported in pectin-utilizing rumen bacteria, *Treponema saccharophilum*,[192] *Butyrivibrio fibrisolvens*, *Prevotella ruminicola*[190] and *Lachnospira multiparus*,[191] and also in various saprophytic bacteria and phytopathogens. However, KDPG aldolase activity was not found in *Streptococcus bovis* and in the *Bifidobacterium sp.* not growing on pectins. The metabolites of pectin degradation with *B. pseudolongum* P6 were formate, acetate, lactate, succinate and ethanol, 3.22 ± 0.23, 6.01 ± 0.64, 1.58 ± 0.25, 0.66 ± 0.10, 0.32 ± 0.10 mmol per g pectin, respectively, but no carbon dioxide formation was observed.

Olano-Martin *et al.* have also compared the bifidogenic properties of pectins with different DEs, and of oligosaccharides derived from them, in pure and mixed fecal cultures.[193] In general, a greater fermentation selectivity was observed with lower DEs. A size selectivity effect was also observed, with oligosaccharides being more size selective than the pectins they were derived from. Good growth on HM pectins was observed for *Bifidobacterium lactis* but *B. pseudolongum*, *B. bifidum* Bb12, *Lactobacillus plantarum* 0207, *L. casei* shirota and *L. acidophilus* preferred LM pectins. *Bifidobacterium angulatum* and *B. infantis* showed no growth on HM pectins, but grew on LM pectins (see Table 8.6).

8.8 Effect of Pectin on Mutagens and Pathogens

8.8.1 Does Pectin Reduce Cancer Risk?

Dietary fibers can bind mutagenic substances, and thus reduce the risk of cancer. To understand the effect of dietary fibers on carcinogenesis, in some studies mutagenic substances are applied to the colon to provoke the incidence of tumour formation. These studies examined the effect of dietary fiber on colon tumour formation in rats or mice, mostly after either intragastric or subcutaneous administration of 1,2-dimethylhydrazine and azoxymethane, or intrarectal instillation of methylnitrosourea.

However, in this kind of study design the number of carcinomas is dependent on the residence time of the mutagenic substance in the colon. Insoluble dietary

Table 8.6 Prebiotic nature of pectins.[a]

Microorganism	*Specific growth rate*[b]*/μh*$^{-1}$ *HM Pectin*	*LM Pectin*	*Oligo HM Pectin*	*Oligo LM Pectin*
Bifidobacterium angulatum	NG ±	2.11 ± 0.19	1.89 ± 0.4	1.56 ± 0.21
Bifidobacterium lactis	7.95 ± 0.41	640 ± 0.83	5 ± 1.7	7.2 ± 1.1
Bifidobacterium infantis	NG ±	2.25 ± 0.29	0.55 ± 0.2	2.2 ± 0.79
Bifidobacterium pseudolongum	4.82 ± 0.16	5.13 ± 0.82	2.02 ± 0.1	14.16 ± 1.01
Bifidobacterium adolesceutis	NG ±	3.25 ± 0.69	2.04 ± 0.2	4.24 ± 0.35
Bifidobacterium bifidum Bb 12	3.2 ± 0.19	10.88 ± 1.52	NG ±	8.88 ± 0.57
Lactobacillus casei shirota	3.47 ± 0.28	23.03 ± 6.76	8.29 ± 0.3	9.11 ± 2.48
Lactobacillus acidophilus	3.36 ± 0.28	13.56 ± 0.58	6.54 ± 1.6	13.06 ± 1.62
Lactobacillus pentosus	0.79 ± 0.06	3.51 ± 0.31	NG ±	0.76 ± 0.02
Lactobacillus plantarum 0207	10.4 ± 1.82	58.23 ± 6.94	NG ±	44.93 ± 3.78
Lactobacillus casei subsp cremoris LC5	6.61 ± 0.94	8.79 ± 0.13	8.1 ± 2	6.52 ± 0.41
Bacteroides distasonis	0.47 ± 0	14.43 ± 0.39	1.75 ± 0.9	10.54 ± 1.07
Bacteroides ovatus	6.15 ± 0.243	8.39 ± 0.006	5.28 ± 0.2	10.19 ± 1.774
Bacteroides fragilis	0.4 ± 0.03	2.65 ± 0.12	0.72 ± 0.1	1.58 ± 0.08
Bacteroides thethaiotaomicron	0.27 ± 0.04	4.88 ± 0.14	NG ±	10.14 ± 1.11
Clostridium perfringens	0.33 ± 0.07	5.42 ± 0.08	0.38 ± 0.3	3.36 ± 0.04
Clostridium ramosum	0.53 ± 0.09	2.15 ± 0.09	NG ±	2.23 ± 0.44
Clostridium inocuum	1.34 ± 0.27	13.92 ± 3.12	9.76 ± 1.5	NG ±
Enterococcus faecalis	9.32 ± 1.32	25.23 ± 0.33	9.39 ± 0.3	16.24 ± 0.27
Escherichia coli	21 ± 1.2	31.77 ± 1.72	25.61 ± 1.1	32.42 ± 0.3

[a]Values of specific growth rates are mean values of triplicate determinations ± standard deviation. NG indicates no growth.
[b]Specific growth rates of elected gut bacteria on 1% high methylated pectin (HMP), 1% low methylated pectin (LMP), 1% pectic-oligosaccharides derived from high methylated pectin (Oligo HMP) and 1% pectic-oligosaccharides derived from low methylated pectin (Oligo LMP).

fibers usually accelerate the excretion of mutagens, but soluble dietary fibers sometimes prolong the clearance of the gut, especially when too little liquid is given. In these cases an increased incidence of tumor cells may be reported, but reports on the reduction of carcinogenesis by pectins prevail.[183–198]

To understand the potential role of colonic bacteria in colon carcinogenesis, the effect of different pectin types on fecal bacterial enzymes (β-glucuronidase, β-glucosidase and tryptophanase) during azoxymethane (AOM)-induced colon carcinogenesis has been investigated. A diet supplemented with 20% apple or citrus pectin decreased the number of colon tumors. Apple pectin feeding decreased fecal β-glucuronidase and tryptophanase levels. Furthermore, a significant decrease in the activity of β-glucuronidase was observed in the apple pectin group during the initiation phase. These findings suggest that the protective effect of pectin on colon carcinogenesis may be dependent on the type of pectin, and also dependent on β-glucuronidase activity in the initiation stage of carcinogenesis.[199]

In another study, Tazawa *et al.* observed a significant decrease in the number of tumors and the incidence of 1,2-dimethylhydrazine (DMH)-induced colon tumors by a diet supplemented with 20% apple pectin. The prostaglandin E2 (PGE2) level in distal colonic mucosa in the apple pectin fed rats was lower than that in basal diet fed rats.[200] Fecal β-glucuronidase activity in the apple pectin fed group, which is considered a key enzyme for the final activation of 1,2-dimethylhydrazine metabolism to carcinogenesis in the colon lumen, was significantly lower in the apple pectin fed group than that in the control group at the initiation stage of carcinogenesis. This effect may partially depend on the decrease of PGE2 concentration in colon mucosa and on the type of pectin, and may also be related to fecal enzyme activity.

Tazawa *et al.* further reported that a diet supplemented with 20% apple pectin also significantly decreased the number and the incidence of AOM-induced colon tumors in rats.[201] Again the PGE2 level in distal colon mucosa and blood of portal vein was lower in rats fed pectin than those fed the basal diet, and the decrease of PGE2 was dose-dependent. The results suggest an anti-inflammatory effect in the bowel. Rats fed with apple pectin showed a significantly lower incidence of hepatic metastasis than those fed the basal diet. The number and tumor score of colorectal tumors induced by DMH in transgenic mice carrying human c-Ha-ras genes were also significantly reduced by ingestion of apple pectin compared to a control diet.[202]

The colonic crypt contains highly proliferative cells in its base, and differentiated cells on its luminal surface. Carcinogenesis affects this orderly cellular distribution. Avivi-Green *et al.* have studied the effect of a pectin-supplemented diet on the expression of apoptosis-related proteins along the crypt–lumen axis during DMH-induced carcinogenesis.[203] The pectin-enriched diet induced up-regulation of active caspase-1 (20 kDa) and caspase-3 precursor in DMH-treated rats.

Pectin also enhanced caspase-3 activity in all colonocyte populations, in both non-DMH and DMH-treated rats. Luminal colonocytes exhibited higher caspase-3 activity than proliferative colonocytes of rats fed a standard diet in untreated and DMH-treated rats, whereas in pectin-fed untreated rats, an equal activity was measured among all colonocyte populations. In DMH-treated rats, the cleaved poly(ADP-ribose) polymerase subunit (89 kDa) was detected in the luminal colonocytes of pectin fed rats, and its level was higher than in rats fed the standard diet. The cell death regulatory protein bak was equally expressed in isolated colonocytes of rats of both dietary groups treated with DMH, and in normal rats fed with pectin, whereas in the untreated rats fed a standard diet a higher expression was observed in differentiated colonocytes.

In the DMH-treated rats, Bcl-2 expression was lower in all colonocytes harvested from rats fed with pectin compared to rats on the standard diet. The apoptotic index in DMH-treated groups was higher in rats receiving the pectin diet in both types of colonocytes. The average tumor number and volume per rat were lower in rats fed with pectin. Avivi-Green *et al.* assume that the butyrate produced by fermentation of pectin may play a key role in these effects.[203]

According to Olano-Martin *et al.*, who analyzed the anti-proliferative effects of pectin and pectic oligosaccharides on human colonic adenocarcinoma cell line HT 29, the induction of apoptosis in tumor cells is an important protective mechanism against colorectal cancer.[204] Olano-Martin *et al.* observed a significant reduction in attached cell numbers after three days' incubation. The increased apoptosis frequency after incubation with 1% (w/v) pectin or pectic oligosaccharides was demonstrated by caspase-3 activity and DNA laddering on agarose gel electrophoresis. It is concluded that pectins and their degradation products may contribute to the reported protective effects of fruits against colon cancer.

8.8.2 Effects of Pectin Hydrolyzates on Pathogens

Pathogenic organisms and cell destructive substances must bind to the cell surface to cause infection, and glycosidic structures play an important role in this adhesion. The ability of cells to metastasize also appears to be partly related to the cohesiveness of cells. In other words, for a tumor to spread, it may require a clump of cells rather than a single cell or a few cells together. Cellular interactions are mediated by a carbohydrate binding protein at the cell surface called galectin-3. In animal studies, modified citrus pectin inhibits the spontaneous pulmonary metastases.[205] In human studies there is a correlation between the level of galectin expression and the tumor stage in human colorectal, gastric and thyroid cancers. In agarose cell cultures, anti-galectin monoclonal antibodies inhibit the growth of tumor cells.[206]

Pienta *et al.* report that oral administration of modified citrus pectin (MCP) interferes with cell–cell interactions that are mediated by cell surface carbohydrate binding galectin-3. *In vitro* tests show a time-dependent and dose-dependent inhibition of cell adhesion by tumor cell lines, and *in vivo* tests in rats produce significantly fewer metastases.[207] Further work on MCP is published by Platt and Raz,[208] Inohara and Raz,[209] Naik *et al.*,[210] Hsieh and Wu,[211] Strum *et al.*[212] and Weiss *et al.*[213] Hayashi *et al.* also report that MCP reduces the growth of solid primary tumors.[214] Zhu *et al.* have demonstrated the enhancement of human natural killer cells by a rhamnogalacturonan comparable to MCP.[215]

Nangia-Makker *et al.* have studied galectin-3-mediated function during tumor angiogenesis *in vitro* by assessing the effect of MCP on capillary tube formation in human umbilical vein endothelial cells (HUVECs) in Matrigel.[216] Tumor growth, angiogenesis and spontaneous metastasis *in vivo* were significantly reduced in mice fed with MCP. *In vitro*, MCP inhibited HUVEC morphogenesis in a dose-dependent manner, and at concentrations of 0.1 and 0.25% it inhibited the binding of 10 pg ml^{-1} galectin-3 to HUVECs by 72.1% and 95.8% respectively, and at 0.25% and 1 pg ml^{-1} galectin-3 the inhibition was 100%. Nangia-Makker *et al.* conclude that orally given MCP inhibits carbohydrate-mediated tumor growth, angiogenesis and metastasis *in vivo*, presumably *via* its effect on galectin-3 function.[216] A phase 2 pilot study also showed that prostate specific antigen doubling time in 7 out of 10 men increased after taking MCP for 12 months compared to before taking it.[217]

However, it is necessary to state that MCP is not a clearly defined compound. Early work by Raz and Pienta *et al.* claimed MCP to be a galactosyl rich pectin fragment with a low molecular weight, a high pH value and a DE of less than 5%.[205,207] This author tried to produce MCP according to the above-mentioned patent, but failed. MCP became of public (commercial) interest on the internet with all its lights and shadows. If you want to order MCP on the web you can find all kinds of pectin products. This makes it difficult to come to a clear conclusion. There seems to be a potential in pectin fragments for preventing cancer. Gut flora may produce these fragments from native plant and vegetable pectins, and from isolated pectin, too. Further studies should hopefully clarify this confused situation.

According to Moro, carrot soup is described to reduce the adhesion of pathogenic germs like pathogenic *E. coli* to epithelial cells of the gastro-intestinal and urogenital tract by up to 90%.[218,219] The active components of this carrot soup are α-1,4-galacturonic acid units in pectin with DE of 20–80%. Guggenbichler *et al.* also report that the galacturonic acid monomer has no activity, but unesterified di-galacturonic acid and tri-galacturonic acid blocks show adhesions of 91.7% and 84.6%, respectively, and that activity decreases with increasing chain length.[219] Thus, it can be concluded that a di-galacturonide fitting into the receptor is necessary for blocking of cell adhesion, and the activity decreases with molecular size, because additional galacturonic acid units protrude from the receptor and are ineffective.

In contrast to the above argument, Stahl and Boehm claim that neither the DP nor DE affects the adhesion of microorganisms or viruses to the cell.[220] In fact carbohydrates with terminal Δ-4,5-unsaturated galacturonic acid at their non-reducing end are said to be the active substance. A mixture of carbohydrates with different DPs of 2–40, preferably DPs 2–10 is said to have high anti-adhesive activity against pathogenic substances and microorganisms.

Another patent claims that pectin hydrolyzates reduce or prevent the adhesion of pathogenic substances and organisms to eukaryotic cells, especially mammalian cells, or block galectin-3-induced cell–cell or cell–matrix interactions which could cause tumors and metastases.[221] Pectin hydrolyzates are produced by different pectinases starting with pectin lyase EC 4.2.2.10 or an endo-polygalacturonase EC 3.2.1.15, or the opposite way round, and not necessarily followed by deesterification with pectin esterase EC 3.1.1.11. Galacturonides bearing a double bond exhibit higher anti-adhesion activities, which increase with increasing DE and their galacturonic acid contents. The active carbohydrate preferably has a DE of DP 2–10 and unsaturated galacturonide content of 36.5–46%.

These preparations should be useful for treating prostate and kidney cancer, Kaposi's sarcoma, chronic leukemia, mammary carcinoma, mammary adenocarcinoma, sarcoma, rectum carcinoma, faucal carcinoma, bronchial carcinoma, melanoma and tumors of the stomach, small intestine, colon and lung in humans and other mammals.

8.8.3 Other Claimed Medical Effects of Pectins

A variety of other properties and effects have been attributed to pectins, as summarized below.

1. A diet containing 4.2% pectin can promote the regression of gallstones in Syrian hamsters by 52% after 50 days.[222]
2. A pectin-rich diet in combination with drugs is suggested for curing chronic destructive non-suppurating cholangitis.[223]
3. The anti-inflammatory role of pectin on sodium dextran sulfate-induced experimental colitis.[224]
4. The addition of pectin to an elemental diet improves the healing of experimental colonic anastomoses by the presence of SCFA produced from the fermentation of pectin.[225]
5. Changes in intraluminal pressure in a rat colon with aging without and with pectin were measured. The pressure increased with age and decreased in the latter half of the study, and the motility index was lowest in the pectin group. The results suggest that long-term ingestion of pectin might have a prophylactic effect on the development of diverticula.[226]
6. Indomethacin (IDM) injected into rats subcutaneously reduced the unstirred water layer dose-dependently, and subsequently many IDM-induced ulcers were observed. When pectin was given together with solid stock food, ulcer formation was significantly inhibited.[227]
7. Pectin supplementation (1% in intragastric infusion) significantly decreased body weight loss, organ water content and intestinal myeloperoxidase levels, and increased mucosal protein, DNA and RNA in enterocolitis rats. Intestinal permeability increased by administration of methotrexate (MTX), and pectin supplementation significantly reversed the increased permeability in the distal small bowel and colon. Pectin supplementation also lowered the magnitude of bacterial translocation, decreased plasma endotoxin levels and restored the bowel microecology.[228]

8. A significant reduction in peroxy radical-induced mucosal damage to the rat jejunum was noted when pectin was perfused before peroxy perfusion. Full protection against hydroxyl radical-induced mucosal damage was achieved when pectin was perfused before damage induction.[229]
9. Pectin was effective in binding dioxin isomers, and enhanced fecal excretion of 1,2,3,4,7,8-HxCDD in mice.[230]
10. Pectin was effective in the inhibition of a male's reproductive toxicity caused by environmental hormones including dioxins. Male reproductive toxicity is described as a reduction in the number of sperm cells, the reduction of seminal fluid, the reduced concentration of testosterone, the reduced motility of sperm cells, the increased number of modified sperm cells, genital malformation, prostate diseases and other tissue problems relating to reproduction.[231]
11. It is claimed that pectin reduces the resorption of exogenous substances, namely caffeine or creatine, to prevent short-term peaks in serum concentration, or maintain the serum concentration of the substance.[232]
12. A mixture of HM pectin and pectic acid (2 : 1) is recommended for patients with chronic renal failure to enhance the excretion of uremic toxicants.[233]

8.9 Biomedical Effects of Pectins on Cell Morphology and Proliferation

As previously discussed, dietary fibers influence the morphology and ultrastructure of the intestines. For instance, the effect of pectin on jejunal and ileal morphology and ultrastructure has been studied on adult male mice fed a semisynthetic diet containing 8% cellulose or pectin for 30 days. No significant differences in the jejunal villus height between the two groups were found, but the jejunal crypt depth and both the ileal villus height and crypt depth of the mice fed the pectin diet were significantly greater than those of the mice fed the cellulose diet. Numerous intercellular spaces were observed in the jejunal absorptive cells of the pectin group, but not in the cellulose group. Moreover, the ileal absorptive cells of the pectin group contained numerous peroxisomes, but only few in case of the cellulose group.[234] Similarly, male Wistar rats fed an elemental diet containing 2.5% pectin for 14 days showed significant increases in villus height and crypt depth in the small intestine with corresponding increases in plasma enteroglucagon levels.[224]

In another study pectin supplementation resulted in significant increases in the length, weight and number of Ki-67-positive cells in the ileum, caecum and colon.[235] In the case of the caecum of Sprague-Dawley rats, the concentration of SCFA was positively associated with the number of cells per crypt column, total cells per crypt and the proliferative zone. In contrast, in the distal colon there was no significant correlation between SCFA concentration and cell proliferation. The data suggest that pectin stimulates caecal cell proliferation through the production of SCFA.[236] The effect of pectin on cell proliferation in the proximal colon of rats is also highly dependent on the source of fat in the diet; it exerts a hyperproliferative effect when the fat is corn oil, but not when beef tallow or fish oil. This indicates that both pectin and fat modulate colon cell proliferation in an interactive site-specific manner.[237]

8.10 Pectins in Controlled and Targeted Drug Delivery to the Colon

Although oral delivery has become a widely acceptable route for administration of therapeutic drugs, the gastrointestinal tract (GIT) presents several formidable barriers to drug delivery. Colonic drug delivery has gained increasing importance not just for the delivery of drugs for treatment of local diseases associated with the colon, but also for the potential delivery of proteins and therapeutic peptides in general. It also appears that β-carotene is absorbed too early in the small intestine when applied as an isolated substance. Native β-carotene in intact plant cells is obviously absorbed in more distal regions of the gastrointestinal tract when liberated by microbial disintegration of the cells. Thus, the plant cell may be a good example for the design of a controlled drug delivery device. To achieve successful colonic delivery, the drug must be protected from absorption and the environment of the upper GIT, and then be released into the proximal colon which is considered the optimum site for colon targeted delivery of drugs. Colon targeting is naturally of value for the topical treatment of diseases of colon such as Crohn's disease, ulcerative colitis, colorectal cancer and amebiasis.

8.10.1 Different Modes of Pectin-based Drug Delivery Systems

Various strategies for targeting drugs for the colon include covalent linkage of the drug to a carrier, coating with pH sensitive polymers, formulation of timed released systems, the use of carriers that are degraded specifically by colonic bacteria, bioadhesive systems and osmotic-controlled drug delivery systems.[238] Apart from the last named system, all the other strategies can be realized with pectin and pectin–hydrocolloid mixtures.

Pectin and some other hydrocolloids remain intact in the physiological media of the stomach and small intestine, but once the dosage form enters the colon, they are acted upon by pectinases or other polysaccharidases, and degrade and release the drug in the vicinity of the bioenvironment of the colon. However, they should be protected while gaining entry into the stomach and small intestine.

Pectin-based delivery devices can be produced by chemical attachment or microencapsulation of the drug. A formulation coated with enteric polymers can release the drug when the pH moves towards the alkaline range, whereas a multicoated formulation may pass the stomach and release the drug after a lag time of 3–5 h, equivalent to the small intestinal transit time. A drug coated with bioadhesive polymers that selectively provide adhesion to the colonic mucosa may release the drug in the colon.[238] Soft capsules composed of gelatin and pectinate are also claimed to be suitable for colonic drug delivery.[239] Liu *et al.* have reviewed recent developments in pectin-based formulations.[240] However, if drug release is directly dependent on pectin degradation by colonic microflora it is important that the composition of this microflora is relatively consistent across a diverse human population.

Pressed tablets of calcium pectinate, an insoluble salt of pectin, have been evaluated as colon-specific drug delivery carriers by Rubinstein *et al.*[241] Wakerly *et al.* have also combined pectin and ethyl cellulose as a film coating for colonic drug delivery.[242] Increasing the proportion of ethyl cellulose, and increasing the coat weight, reduced drug release at pH 1 and pH 7.4. In the case of amidated calcium pectinate, drug release was faster for low amidation, but slower if too much calcium was used with a high degree of amidation.[243] Matrix tablets of calcium pectinate with pectin showed faster and more reproducible disintegration in the colon than tablets made from calcium pectinate with guar.[244]

Pectin is a polyanion and forms polymeric or polyion complexes (coacervates) with polycations like chitosan and gelatin.[245–247] Thus, coacervate complexes of pectin with chitosan and gelatin, and blends with hydroxypropylmethylcellulose (HPMC), have also been studied for drug delivery by Macleod *et al.*,[248–250] Ofori-Kwakye and Fell,[251–254] Saravanan *et al.*[255] and McMullen *et al.*[256] Ahrabi *et al.* have also investigated several pectins and ethylcellulose (EC) for their suitability as matrix tablets. Coating pellets of the model drug ropivacaine with Eudragit L 100, a polymethacrylic, reduced the ropivacaine release under simulated upper GI conditions without interference with subsequent pectinase activity.[257]

The drug release rate from theophylline pellets coated with HM pectin–Eudragit NE30D complexes was dependent on the HM pectin content. The lowest theophylline release was observed when the HM pectin content of the complexes was 20%, *i.e.* optimal complexation between HM pectin and Eudragit. The theophylline release from the coated pellets was slower in presence of the pectinolytic enzymes when the pectin content of complexes is higher than 20% w/w. On the other hand, the effect of the enzymes induced an increase of the theophylline release when the HM pectin content of the coatings ranged between 10 and 15%.[258] Poly(galacturonic acid) has also been used as an additive in Eudragit RS films to control the release of colonic delivery systems.[259]

Pectin crosslinked with epichlorohydrin to different degrees has also been described for drug delivery, and increasing crosslinking results in a reduction in its rate of degradation by pectinase.[260] Turkoglu and Ugurlu have reported that pectin alone is not sufficient to protect the core tablets, as judged by GI transit time; thus they used HPMC to overcome this problem.[261] The drug release rate from compressed matrix tablets could also be increased by increasing the pectin : HPMC ratio.[262] Colonic drug delivery by zinc pectinate gel micro particles has also been reported.[263]

Drug microencapsulation in pectin by various processes, including solvent evaporation,[264] suspension techniques,[265,266] fluidized bed,[267] spray drying[268] and formation-dried calcium pectinate gel beads,[269] as well as the formation of compressed tablets[270] and pectinate pellets by extrusion–spheronization,[271,272] have also been reported. Similarly, coating core tablets with calcium pectinate by interfacial complexation has also been described for sustained drug release.[273] A modulated drug release in a binary system of LM pectin and sodium alginate,[274,275] and in a ternary system of LM pectin, calcium alginate and cellulose acetate phthalate,[276,277] is described by Pillay and Fassihi *et al.* Other ternary matrix systems based on pectin, HPMC and gelatin,[278] and of alginate–pectin-poly-L-lysine particulates have been described by Liu and Krishnan.[279]

Guérin *et al.* encapsulated *Bifidobacteria* in pectin–protein gel beads by a transacylation reaction to protect the microorganisms against gastric juice and bile.[280] Intestinal patches for oral drug delivery that adhere to the intestinal wall have been produced from amidated pectin,[281] and by sandwiching crosslinked bovine serum albumin microspheres between ethylcellulose and Carbopol®–pectin films.[282] These patches reduce drug loss into the intestinal lumens, and thus improve trans-luminal flux compared to that from the drug solution. Oral mucosal bioadhesive tablets of pectin and HPMC for sublingual administration have also been described by Miyazaki *et al.*[283]

Extensive work is published in the Chinese literature on the use of pectin for the delivery of bismuth for protecting the stomach mucous membrane,[284] the treatment of heliobacter pylori positive duodenal ulcer[285] and heliobacter pylori infection.[286] The potential for localized prolonged delivery of triamcinolone acetonide to the skin by hydrocolloid patches based on pectin has also been reported.[287] A pulsatile drug delivery system based on erodible pectin plugs containing a pectinolytic enzyme has been developed in which drug release is controlled by enzymic degradation of pectin.[288]

8.10.2 Pectins in Related Biomedical and Medicinal Applications

Mn(II)-pectin is also discussed as a new intravenous and oral magnetic resonance imaging contrast agent[289] and, on a related topic, the thickness of a contrast film based on barium sulfate has been modified with pectin.[290] Pectin coated on polypropylene hollow fibers has also been evaluated as a composite membrane for selective extracorporeal removal of human LDL-cholesterol.[291]

A comprehensive summary of pectin used as medicaments and pharmaceuticals has been published by Endress in 1991.[19] More recent publications describe anti-reflux formulations of potential interest for the treatment of patients classified as having endoscopy negative gastro-oesophageal reflux disease (GORD),[292,293] *i.e.* patients with normal oesophageal mucosa at endoscopy, but with heartburn as their main complaint. The same pectin-based anti-reflux formulations reduce postprandial H^+ concentrations and the amount of radiolabeled food reaching the oesophagus, as measured by the mean time during which the oesophageal pH falls below 4 in comparison to placebo.[294] Prevention of gastro-oesophageal reflux by pectin gel in severely disabled patients has also been reported.[295]

8.11 Pectins in Skincare Products

Orabase was the first pectin-containing product launched in 1960 for treatment of stomatitis (oral ulcers). This wound dressing was a paste, which also contained gelatin and carboxymethylcellulose (CMC). It was applied to the treatment of inflamed ulceriferous lesions of the oral cavity due to stomatitis and related disorders.[296] A combination with added steroid hormones was also sold as Kenalog. The first bandage based on a combination of Orabase with a hydrophobic polymer, polyisobutylene, was developed in 1967. This product, which was also called a hydrocolloid dressing, is still the basis of many stoma seals.

More generally a wide variety of pectin-containing products is used for various skincare applications including, among others, skin barriers and general medical adhesives. The difference between barriers and general medical adhesives is that barriers are usually composed of hydrophilic and hydrophobic polymers, and contain at least one hydrophilic polymer. The hydrophilic polymer, or hydrocolloid, absorbs water released by the skin, and also some liquor released at the stoma. Typical polymer blends used as skin barriers are composed of 40% polyisobutylene, 20% pectin, 20% gelatin and 20% other hydrocolloids like CMC or karaya gum. Thus, when discussing the type and features of these compositions it is very difficult to appreciate which hydrophilic polymer is responsible for which physical properties, and what kind of skin-protecting properties each may have.

Numata *et al.* have compared several hydrocolloids and evaluated their effectiveness in maintaining the crista cutis, and describes karaya gum as best, followed by citrus pectin.[297] Gross and Irving describe a skincare preparation composed of karaya gum paste and a compressed wafer composed of gelatin, pectin and methyl cellulose, with drainable collection appliances.[298] This product protects the inflamed skin from fistula discharge and allows healing to proceed beneath the dressing. A number of other pectin-containing stoma seals have been described.[299–304]

Lund *et al.* describe a pectin-based barrier over an anchoring tape to minimize epidermal stripping.[305] This barrier provides effective adhesion and protects neonatal skin from damage that may be caused during tape removal. This type of barrier should be applied on tapes that are used to anchor medical equipment on premature infants.[306] Typical pectin-containing wound dressing compositions include a water-insoluble freeze-dried gel based on poly(vinyl alcohol) and 2% pectin,[307] pectin and polyurethane foam,[308] multilayered bandages for skin disorders[309–311] and microporous adhesives for surgical purposes.[312] A three-stage wound dressing composition based on 35–50% pectin in the first stage, and 5–10% pectin in the second stage, has also been reported by Alvarez.[313]

8.12 Pectin as Raw Material for L-Ascorbic Acid

As early as 1935 Reichstein and Gruessner[314] recognized that ascorbic acid occurring in plants usually contains large quantities of galacturonic acid; they concluded that ascorbic acid is produced from galacturonic acid, and hence their attempts to rearrange galacturonic acid to ascorbic acid. Although Reichstein and Gruessner were right in their conclusion, they failed in their attempted synthesis of ascorbic acid because the transformation is much more complex than they could have realized then. However, some 10 years later Isbell suggested the following transformation steps from galacturonic acid to ascorbic acid.[315]

Beet pulp → [hydrolysis] → galacturonic acid → [hydrogenate] → L-galactonic acid → [oxidize] → 2-keto-L-galactonic acid → [enolize and lactonize] → ascorbic acid

Since then a number of papers on the synthesis of ascorbic acid from galacturonic acid have been published. Popova and Krachanova suggested sunflower heads as a raw material, and tried to optimize the yield of galacturonic acid by enzymic degradation of isolated pectins.[316] Kulbe *et al.* proposed a reaction sequence for the enzymic synthesis from pectin by applying a concept called intrasequential cofactor regeneration.[317,318] Thus, pectin was hydrolyzed enzymically by a combination of PE and PG to galacturonic acid, which was purified by membrane separation. The resulting D-galacturonic acid was reduced to L-galactonic acid by a hexonate dehydrogenase and NADPH, followed by oxidation to 2-keto-L-galactonic acid by galactonate dehydrogenase and $NADP^+$, and finally 2-keto-L-galactonic acid was converted to ascorbic acid by a simple acid- or base-catalyzed cyclization.

A further improvement of this enzymic synthesis has been reported by using a different enzyme combination. Here D-galacturonic acid is reduced to L-galacturonic acid by an NADPH-dependent urinate reductase (UR), with the continuous regeneration of the native coenzyme NADPH by NAD(P)-dependent glucose dehydrogenase (from *Bacillus cereus*) in a charged ultrafiltration membrane reactor. After lactonization, L-galactono-γ-lactone is oxidized by L-gulono-γ-lactone oxidase (GulOx] to yield 2-keto-L-gulonic acid, which spontaneously rearranges to L-ascorbic acid.[319]

8.13 Concluding Remarks

Pectins are soluble dietary fibers. Depending on raw material and production, the viscosity varies. HM pectins bind mainly LDL and reduce the risk for atherosclerosis, reduce the glucose absorption. Risk factors for cardiovascular disease are: high blood lipids, high cholesterol, highly-oxidized cholesterol, high blood pressure, hyperglycemia, and insulin resistance. As several of these factors are positively influenced by pectins, the application of pectins to reduce high blood pressure and the risk of cardiovascular disease (CVD) may be interesting. Pectins are stable in the stomach and the small intestine and are degraded by enzymes of the colon bacteria. Pectins and their degradation products, the oligogalacturonides, may be considered as prebiotics. Oligogalacturonides (saturated or Δ-4,5-unsaturated) with different M_w (and degree of polymerization, DP), and degree of esterification (DE) may be used in infant formula to strengthen gut health and the immune system and may protect patients and healthy men from poisonous gut bacteria and toxic substances. Oligogalacturonides and LM pectins may be used as antidotes for heavy metal poisoning in polluted areas.

It may be beneficial that food supplements and drugs are resorbed in certain districts of the colon. This can be achieved by encapsulating these substances with pectins of different degrees of esterification, crosslinked with a crosslinking agent, if necessary.

8.14 Abbreviations and Symbols

AOM	azoxymethane
CAS	Chemical Abstract Service
CHD	cardiovascular heart disease
CMC	carboxymethylcellulose
DAc	degree of acetylation
DE	degree of esterification
DM	degree of methylation
DMH	dimethylhydrazine
DNA	deoxyribonucleic acid
DP	degree of polymerization
EC	enzyme commission number
EC	ethylcellulose
endo-PG	endo-polygalacturonase
EU	European Union
FAO	Food and Agricultural Organization
FCC	Food Chemical Codex
FDA	food and drug administration
GER	Germany
GI	glycemic index
GIP	gastro inhibitory peptide
GIT	gastro intestinal tract
GORD	gastro-oesophageal reflux disease
GulOx	gulono-γ-lactone oxidase
HDL	high density lipoproteins
HIV	human immunodeficiency virus
HM	high methyl ester
HMG	α-hydroxy-α-methylglutaryl
HPMC	hydroxypropylmethylcellulose
HUVECs	human umbilical vein endothelial cells
HxCDD	hexachlorodibenzo-p-dioxin
IDM	indomethacin
IMT	intima-media thickness
INS	international numbering system

JECFA	joint expert committee on food additives
LDL	low density lipoproteins
LM	low methyl ester
MCC	microcrystalline cellulose
MCP	modified citrus pectin
MTX	methotrexate
Mw	molecular weight
$NADP^+$	nicotinamide ademine dinucleotide
NADPH	nicotinamide ademine dinucleotide phosphate
PE	pectin esterase
PGE2	prostaglandin E2
PL	pectin lyase
RNA	ribonucleic acid
SCFA	short chain fatty acids
SS	soluble solids
UR	urinate reductase
USP	United States Pharmacopeia

References

1. European Commission Directive 96/77/EC.
2. FAO-Food and Nutrition Paper 52 Add 9 JECFA-Specifications for identity and purity of food additives, 2001.
3. FCC 7th edition, National Academy Press, Washington, DC, 2010.
4. United States Pharmacopeia (USP 34).
5. H. A. Schols and A. G. J. Voragen. In *Pectins and their Manipulation*, (Eds) G. B. Seymour and J. P. Knox, Blackwell Publishing, 2002, p. 1.
6. M. W. Spellmann, M. McNeil, A. G. Darvill and P. Albersheim, *Carbohydr. Res.*, 1983, **122**, 131.
7. H. Yamada. In *Pectins and Pectinases*, (Eds) J. Visser and A. G. J. Voragen, Elsevier Science BV, 1996, p. 173.
8. H. Yamada, H. Kiyohara and T. Matsumoto. In *Advances in Pectin and Pectinase Research*, (Eds) F. Voragen, H. Schols and R. Visser, Kluwer Academic Publishers, 2003, p. 481.
9. M. Hirano, H. Kiyohara and H. Yamada, *Planta Med.*, 1994, **60**(5), 450.
10. M. H. Sakurai, T. Matsumoto, H. Kiyohara and H. Yamada, *Planta Med.*, 1996, **62**(4), 341.
11. H. U. Endress, F. Mattes and K. Norz. In *Handbook of Food Technology and Engineering* (Ed.) Y. H. Hui, Vol. 3, CRC Taylor & Francis, Boca Raton, 2005, p. 140–1.
12. B. R. Thakur, R. K. Singh and A. K. Handa, *Critical Review in Food Science and Nutrition*, 1997, **37**(1), 47.
13. (a) K. Gierschner and H. U. Endress. In *Methods of Enzymic Analysis Vol 6 Metabolites 1: Carbohydrates*, (Ed.) H. U. Bergmeyer, Verlag Chemie GmbH, Weinheim, 1984, p. 313; (b) K. Gierschner and H. U. Endress.

In *Methods of Enzymic Analysis Vol 6 Metabolites 1: Carbohydrates*, (Ed.) H. U. Bergmeyer, Verlag Chemie GmbH, Weinheim, 1984, p. 71; (c) K. Gierschner and H. U. Endress. In *Methods of Enzymic Analysis Vol 6 Metabolites 1: Carbohydrates*, (Ed.) H. U. Bergmeyer, Verlag Chemie GmbH, Weinheim, 1984, p. 70.
14. M. M. Baig and J. J. Cerda, *Am. J. Clin. Nutr.*, 1981, **34**, 50.
15. J. D. Falk and J. J. Nagyvary, *J. Nutr.*, 1982, **112**, 182.
16. A. Keys, F. Grande and J. T. Anderson, *Proc. Soc. Exp. Biol. Med.*, 1961, **106**, 555.
17. K. Behall and S. Reiser. In *Chemistry and Function of Pectins*, (Ed.) M. L. Fishmen and J. J. Jen, ACS Symposium Series 310, Washington DC, 1986, p. 248.
18. S. Reiser, *Food Technol.*, 1987, **41**, 91.
19. H. U. Endress. In *The Chemistry and Technology of Pectin*, (Ed.) R. H. Walter, Academic Press, New York, 1991, p. 251.
20. M. J. Fahrenbach, B. A. Riccardi, J. C. Saunders, I. N. Lourie and J. G. Heider, *Circulation*, 1965, **31/32**(Suppl. 2), 1141.
21. G. H. Palmer and D. G. Dixon, *Am. J. Clin. Nutr.*, 1966, **18**, 437.
22. D. J. A. Jenkins, A. R. Leeds, A. Gassull, H. Houston, D. Goff and M. Hill, *Clin. Sci. Mol. Med.*, 1976a, **51**, 8.
23. P. N. Durrington, C. H. Bolton, A. P. Manning and M. Hartog, *Lancet*, 1976, **21**, 394.
24. R. M. Kay and A. S. Truswell, *Am. J. Clin. Nutr.*, 1977, **30**, 171.
25. T. L. Raymond, W. E. Connor, D. S. Lin, S. Warner, M. M. Fry and S. L. Connor, *J. Clin. Invest.*, 1977, **60**, 1429.
26. F. Delbarre, J. Rondier and A. deGéry, *Am. J. Clin. Nutr.*, 1977, **30**, 463.
27. D. J. A. Jenkins, D. Reynolds, A. R. Leed, A. L. Walker and J. H. Cummings, *Am. J. Clin. Nutr.*, 1979, **32**, 2430.
28. E. Ginter, E. J. Kubec, J. Vozar and P. Bobek, *Intl. J. Vit. Nutr. Res.*, 1979, **49**, 406.
29. M. Stasse-Wolthuis, H. F. F. Albers, J. G. G. van Jeveren, J. W. de Jong, J. G. A. J. Hautvast, R. J. J. Hermus, M. B. Katan, W. G. Brydon and M. A. Eastwood, *Am. J. Clin. Nutr.*, 1980, **33**, 1745.
30. H. Nakamura, T. Islukawa, N. Tada, A. Kagami, K. Koudo, E. Miyazaml and S. Takeyama, *Nutr. Rep. Int.*, 1982, **26**, 215.
31. P. A. Judd and A. S. Truswell, *Br. J. Nutr.*, 1982, **48**, 451.
32. A. D. Challen, W. J. Branch and J. H. Cummings, *Human Nutr. Clin. Nutr.*, 1983, **37**, 209.
33. T. A. Miettinen and S. Tarpila, *Clin. Chim. Acta.*, 1977, **79**, 471.
34. P. Schwandt, W. O. Richter, P. Weisweiler and G. Neureuther, *Atherosclerosis*, 1982, **44**, 379.
35. K. Hundhammer and M. Marshall, *Akt. Ernähr.*, 1983, **8**, 222.
36. B. H. Ershoff and A. F. Wells, *Exp. Med. Surg.*, 1962, **20**, 272.
37. J. P. Sheehan, I. W. Wei, M. Ulchaker and K. Y. Tserng, *Am. J. Clin. Nutr.*, 1997, **66**(5), 1183.

38. O. Aprikian, V. Duclos, S. Guyot, C. Besson, C. Manach, A. Bernalier, C. Morand, C. Remesy and C. Demigne, *J. Nutr.*, 2003, **133**(6), 1860.
39. L. Brown, B. Rosner, W. W. Willett and F. M. Sacks, *Am. J. Clin. Nutr.*, 1999, **69**, 30.
40. U. Schuderer, *Dissertation*, University of Giessen, Germany, 1986.
41. J. J. Cerda, *X. International Fruit Juice Convention*, Orlando, FL, USA, 1988.
42. J. Groudeva-Popova, *Z. Lebensm. Unters. Forsch.*, 1996, **A 204**, 374.
43. J. Groudeva-Popova, M. Krachanova, A. Djurdjev and C. Krachanov, *Folia Medica*, 1996, **39**, 39.
44. F. J. Veldman, C. H. Nair, H. H. Vorster, W. J. Vermaak, J. C. Jerling, W. Oosthuizen and C. S. Venter, *Throm. Res.*, 1997, **86**(3), 183.
45. F. J. Veldman, C. H. Nair, H. H. Vorster, W. J. Vermaak, J. C. Jerling, W. Oosthuizen and C. S. Venter, *Throm. Res.*, 1999, **93**(6), 253.
46. V. P. Bartz, *Ernährung & Medizin*, 2002, **17**, 149.
47. I. H. Ullrich, *J. Am. Coll. Nutr.*, 1987, **6**, 19.
48. M. G. Aroch, *Slovak Patent* CODON: CZXXA9 CS 228809 B 0101, 1986.
49. R. Pfeiffer, *Fachverlag Koehler*, Giessen, 2000.
50. F. Yamaguchi, S. Uchida, S. Watabe, H. Kojima, N. Shimizu and C. Hatanaka, *Biosci. Biotechnol. Biochem.*, 1995, **59**(11), 2130.
51. M. A. Eastwood, W. G. Brydon and K. Tadesse. In *Medical aspects of dietary fiber*, (Ed.) G. A. Spiller and R. M. Kay, Plenum Press, New York 1980, p. 1.
52. I. Bosaeus, N. G. Carlsson, A. S. Sandberg and H. Andersson, *Hum. Nutr. Clin. Nutr.*, 1986, **40**, 429.
53. P. E. Pfeffer, L. W. Doner, P. D. Hoagland and G. G. McDonald, *J. Agric. Food Chem.*, 1981, **29**, 455.
54. G. Dongowski, *Z. Lebensm. Unters. Forsch.*, 1995, **201**(4), 390.
55. R. A. Baker, *Food Technology*, 1994, **11**, 133.
56. J. K. Ross and J. E. Leklem, *Am. J. Clin. Nutr.*, 1981, **34**, 2068.
57. D. J. A. Jenkins, D. V. Goff, A. R. Leeds, K. G. M. M. Alberti, T. M. S. Wolever, M. A. Gassull and T. D. R. Hockaday, *Lancet*, 1976, **2**, 172.
58. D. J. A. Jenkins, A. R. Leeds, M. A. Gassull, B. Cochet and K. G. M. M. Alberti, *Ann. Intern. Med.*, 1977, **86**, 20.
59. A. R. Leeds, D. N. Ralphs, P. Boulos, F. Ebied, G. Metz, J. B. Dilawari, A. Elliott and D. J. A. Jenkins, *Proc. Nutr. Soc.*, 1978, **37**, 23A.
60. D. J. A. Jenkins, T. M. S. Wolever, A. R. Leeds, M. A. Gassull, P. Halsman, J. Dilawari, D. V. Goff, G. L. Metz and K. G. M. M. Alberti, *Brit. Med. J.*, 1978, **1**, 1392.
61. L. Monnier, T. C. Pham, L. Aguirre, A. Orsetti and J. Mirouze, *Diabetes Care*, 1978, **1**, 83.
62. S. Holt, R. C. Heading, D. C. Carter, L. F. Prescott and P. P. Tothill, *Lancet*, 1979, **24**, 637.
63. D. Labayle, J. C. Chaput, C. Buffet, C. Rousseau, P. Francois and J. P. Etienne, *Nouv. Presse Med.*, 1980, **9**, 223.

64. S. Vaaler, K. F. Hanssen and O. Aagenaes, *Acta Med. Scand.*, 1980, **208**, 389.
65. T. Poynard, G. Slama, A. Delage and G. Tchobroutsky, *Lancet*, 1980, **1**, 158.
66. L. A. Gold, J. P. McCourt and T. J. Merimee, *Diabetes Care*, 1980, **3**, 50.
67. D. R. R. Williams, W. P. T. James and I. E. Evans, *Diabetologia*, 1980, **18**, 379.
68. Y. Kanter, N. Eitan, G. Brook and D. Barzilai, *Israel J. Med. Sci.*, 1980, **16**, 1.
69. S. E. Schwartz, R. A. Levine, A. Singh, J. R. Scheidecker and N. S. Track, *Gastoenterology*, 1983, **83**, 817.
70. A. Sahi, R. L. Bijlani, M. G. Karmarkar and U. Nayar, *Nutr. Res.*, 1985, **5**, 1431.
71. K. S. Sandhu, M. M. el Samahi, I. Mena, C. P. Dooley and J. E. Valenzueala, *Gastroenterology*, 1987, **92**, 486.
72. G. Tunali, D. Stetten, U. Schuderer and H. Hofmann, Wiss. DGE-Kongress 1990, München Ernährungs-Umschau aus Forschung und Praxis, **37**, 141.
73. D. J. Jenkins and A. L. Jenkins, *Proc. Soc. Exp. Biol. Med.*, 1985, **180**(3), 422.
74. K. Fuse, T. Bamba and S. Hosoda, *Dig. Dis. Sci.*, 1989, **34**(7), 1109.
75. G. A. Gerencser, J. Cerda, C. Burgin, M. M. Baig and R. Guild, *Proc. Soc. Exp. Biol. Med.*, 1984, **176**(2), 183.
76. B. Flourie, N. Vidon, C. H. Florent and J. J. Bernier, *Gut.*, 1984, **25**(9), 936.
77. A. Siddhu, S. Sud, R. L. Bijlani, M. G. Karmarkar and U. Nayar, *Indian J. Physiol. Pharmacol.*, 1991, **35**(2), 99.
78. G. Isaksson, I. Lundquist, B. Akesson and I. Ihse, *Digestion*, 1982, **25**, 39.
79. G. Isaksson, I. Lundquist and I. Ihse, *Gastroenterology*, 1982, **82**, 918.
80. S. Dutta and J. Hlasko, *Am. J. Clin. Nutr.*, 1985, 517.
81. W. E. Hansen, *Pancreas.*, 1987, **2**(2), 195.
82. W. E. Hansen, *Int. J. Pancreatol.*, 1986, **1**(5–6), 341.
83. T. Tsujita, M. Sumiyosh, L. K. Han, T. Fujiwara, J. Tsujita and H. Okuda, *J. Nutr. Sci. Vitaminol.*, 2003, **49**(5), 340.
84. J. Tsujita, T. Tsujita, T. Fujiwara, A. Okamoto, S. Hamada, S. Kikuchi, Y. Fujii and A. Nomura, *JP Patent*, 2002, 199534.
85. H. B. Matheson, I. S. Colon and J. A. Story, *J. Nutr.*, 1995, **125**(3), 454.
86. L. M. Morgan, T. J. Goulder, D. Tsiolakis, V. Marks and K. Alberti, The effect of unabsorbable carbohydrates on gut hormones, *Diabetologia*, 1979, **17**, 85.
87. N. S. Levitt, A. I. Vinik, A. A. Sive, P. T. Child and W. P. U. Jackson, *Diabetes Care*, 1980, **3**, 515.
88. H. Wu, K. M. Dwyer, Z. Fan, A. Shircore, J. Fan and J. H. Dwyer, *Am. J. Clin. Nutr.*, 2003, **78**(6), 1085.
89. N. W. Flodin, *J. Am. Coll Nutr.*, 1986, **5**, 417.
90. N. K. Fukagawa, J. W. Anderson, G. Hageman, V. R. Young and K. L. Minaker, *Am. J. Clin. Nutr.*, 1990, **52**, 524.

91. S. E. Schwartz, R. A. Levine, R. S. Weinstock, R. S. Petokas, C. A. Mills and F. D. Thomas, *Am. J. Clin. Nutr.*, 1988, **48**, 1413.
92. K. S. Sandhu, M. M. el Samahi, I. Mena, C. P. Dooley and J. E. Valenzuela, *Gastroenterol.*, 1987, **92**(2), 486.
93. I. L. Gatfield and R. Stute, *FEBS Lett.*, 1972, **28**, 29.
94. F. Wilson and J. Dietschy, *Biochim. Biophys. Acta*, 1974, **363**, 112.
95. G. Dunaif and B. O. Schneeman, *Am. J. Clin. Nutr.*, 1981, **34**, 1034.
96. W. E. Hansen and G. Schulz, In *Loesliche und fixierte Inhibitoren der Amylase in Guar und anderen Ballaststoffen*, (Ed.) K. Huth, S. Karger, Basel, Switzerland, 1983, p. 144.
97. J. H. Cummings, D. A. T. Southgate, W. J. Branch, H. S. Wiggins, H. Houston, D. J. A. Jenkins, T. Jivraj and M. J. Hill, *Br. J. Nutr.*, 1979, **41**, 477.
98. C. DiLorenzo, C. M. Williams, F. Hajnal and J. E. Valenzuela, *Gastoenterology*, 1988, **95**, 1211.
99. C. M. Tiwary, J. A. Ward and B. A. Jackson, *J. Am. Coll. Nutr.*, 1997, **16**(5), 423.
100. K. M. Hendricks, K. R. Dong, A. M. Tang, B. Ding, D. Spiegelman, M. N. Woods and C. A. Wanke, *Am. J. Clin. Nutr.*, 2003, **78**(4), 790.
101. A. R. Leeds, D. N. Ralphs, F. Ebled, G. Metz and J. B. Dilawari, *Lancet.*, 1981, **1**(8229), 1075.
102. O. Lawaetz, A. M. Blackburn, S. R. Bloom, Y. Aritas and D. N. L. Ralphs, *Scand. J. Gastroenterol.*, 1983, **18**, 327.
103. E. Harju, *Int. Surg.*, 1990, **75**, 27.
104. I. Samuk, R. Afriat, T. Horne, T. Bistritzer, J. Barr and I. Vinograd, *J. Pediatr. Gastroenterol Nutr.*, 1996, **23**(3), 235.
105. W. L. Hasler, *Curr. Treat. Options Gastroenterol.*, 2002, **5**(2), 139.
106. J. A. Roth, W. L. Frankel, W. Zhang, D. M. Klurfeld and J. L. Rombeau, *J. Surg. Res.*, 1995, **58**(2), 240.
107. T. R. Sales, H. O. Torres, C. M. Couto and E. B. Carvalho, *Nutrition*, 1998, **14**(6), 508.
108. Y. Finkel, G. Brown, H. L. Smith, E. Buchanan and I. W. Booth, *Acta Paediatr. Scand.*, 1990, **79**(10), 983.
109. E. G. Potievskii, Sh. Sh. Shavakhabov, V. M. Bondarenko and Z. D. Ashubaeva, *Zh. Mikrobiol. Epidemiol. Immunobiol.*, 1994, (Suppl. 1), 106.
110. S. de la Motte, S. Bose-O'Reilly, M. Heinisch and F. Harrison, *Arzneimittelforschung*, 1997, **47**(11), 1247.
111. G. H. Rabbani, T. Teka, B. Zaman, N. Majid, M. Khatun and G. J. Fuchs, *Gastroenterol.*, 2001, **121**(3), 554.
112. C. Triplehorn and P. S. Millard, *ACP. J. Club.*, 2002, **136**(2), 67.
113. D. M. Zimmaro, R. H. Rolandelli, M. J. Koruda, R. G. Settle, T. P. Stein and J. L. Rombeau, *JPEN. J. Parenter Enteral Nutr.*, 1989, **13**(2), 117.
114. A. A. Schultz, B. Ashby-Hughes, R. Taylor, D. E. Gillis and M. Wilkins, *Am. J. Crit. Care*, 2000, **9**(6), 403.

115. Z. I. Kertesz (Ed.) *The Pectic Substances*, Interscience Publishers, New York, London, 1951.
116. B. F. Harland, *Nutr. Res. Rev.*, 1989, **2**, 133.
117. H. H. G. Jellinek and P. Y. A. Chen, *J. Polymer Sci*, 1972, **A–1**(10), 287.
118. A. J. Paskins-Hurlburt, Y. Tanaka, S. C. Skoryna, W. Moore, J. F. Stara and J. R. Stara, *Environ. Res.*, 1977, **14**, 128.
119. R. Kohn, A. Malovikova, W. Bock and G. Dongowski, *Nahrung/Food*, 1981, **25**, 853.
120. A. Malovikova and R. Kohn, *Collection Czechoslov. Chem. Commun.*, 1982, **47**, 702.
121. A. Malovikova and R. Kohn, *Collection Czechoslav. Chem. Commun.*, 1979, **44**, 2915.
122. R. Kohn, *Pure Appl. Chem.*, 1975, **42**, 371.
123. E. Walzel, W. Bock, M. Kujawa, R. Macholz, M. Raab and H. Woggon, In *Mengen- und Spurenelemete, Arbeitstagung Leipzig*, 1987, 149.
124. A. D. Bezzubov, O. G. Vasilieva and A. I. Khatina, *Gig. Truda Prof. Zabol.*, 1960, **4**(3), 32.
125. G. I. Bondarev, A. A. Anisova, T. E. Alekseeva and J. K. Syzrancev, *Vopr. Pitanija*, 1979, **2**, 65.
126. O. D. Livshic, *Vopr. Pitanija*, 1969, **4**, 76.
127. T. Nlculescu, E. Rafaila, R. Eremia and E. Balasa, *Igiena*, 1968, **17**, 421.
128. I. M. Trakhtenberg, V. P. Lukovenko, T. K. Korolenko, V. A. Ostroukhova, P. L. Demchenko, T. E. Rabotiaga and V. V. Krotenko, *Lik. Sprava.*, 1995, **1–2**, 132.
129. E. Walzel, In *Aktuelle Aspekte der Ballaststoffforschung*, (Eds) J. Schulze and W. Bock, Behr's Verlag, 1993, pp. 239–275.
130. G. I. Bondarev, A. A. Anisova, T. E. Alexeewa and J. K. Syzrancev, *Laboratorija profilakticeskogo pitanija, Instituta pitanija AMN SSSR*, Moskwa, 1978, Postupila I/III, 65.
131. N. S. McDonald, R. E. Nusbaum, F. Ezmirlian, R. C. Barbera, G. V. Alexander, P. Spain and D. E. Rounds, *J. Pharmacol. Exp. Ther.*, 1952, **104**, 348.
132. A. A. Rubanovskaya, *Postupila v redakciju*, 1960, **16**(XI), Institut gigieny truda i profzabolevanij AMN SSSR, 43.
133. G. Patrick, *Nature*, 1967, **216**, 815.
134. A. D. Bezzubov and A. I. Khatina, *Vsesojuzni naucno-issledowatelskij institut konditerskoj promislennosti, Postupila v. redakciju*, 1960, **19**(XII), 39.
135. M. Markova, G. Grescheva, K. Nikolova, K. Koen and R. Angelova, *Third Sci. Pract. Symp. HEI*, Varna 21.22.04.1976.
136. S. Stantschev, C. Kratschanov, M. Popova and N. Kirtschev, *Letopisi Chlg.-Epidemiol. Sofija*, 1980, **12**(7), 95.
137. S. Stantschev, C. Kratschanov, M. Popova, N. Kirtschev and M. Martschev, *Z. ges. Hyg.*, 1979, **25**, 585.
138. O. Markovic and R. Kohn, *Experienta*, 1984, **40**, 842.
139. U. Schlemmer, *Ernährungs-Umschau*, 1983, **30**, 232.

140. U. Schlemmer, *Jahresbericht der Bundesforschunganstalt für Ernährung*, 1982, A 6.
141. U. Schlemmer, *Food Chem.*, 1989, **32**, 223.
142. L. Monnier, C. Colette, L. Aguirre and J. Mirouze, *Am. J. Clin. Nutr.*, 1980, **33**, 1225.
143. A. S. Sandberg, R. Ahderinne, H. Andersson, B. Hallgren and L. Hulten, *Human Nutr.: Clin. Nutr.*, 1983, **37c**, 171.
144. B. F. Harland, S. A. Smith, M. P. Howard, R. Ellis and J. C. Smith, *J. Am. Diet. Assoc.*, 1988, **88**, 1562.
145. A. R. Wapnir, S. A. Moak and F. Lifshitz, *Am. J. Clin. Nutr.*, 1980, **33**, 2303.
146. K. Y. Lei, M. W. Davis, M. M. Fang and L. C. Young, *Nutr. Rep. Internat.*, 1980, **22**, 459.
147. G. A. Spiller, M. C. Chernoff and J. E. Gates, *Nutr. Rep. Internat.*, 1980, **22**, 353.
148. M. M. Baig, C. W. Burgin and J. J. Cerda, *J. Nutr.*, 1983, **119**, 2385.
149. A. Plant, C. Kies and H. M. Fox, *Fed. Proc.*, 1979, **38**, 549.
150. J. Grudeva-Popova and I. Sirakova, *Folia Med. (Plovdiv.)*, 1998, **40**, 41.
151. L. M. Drews, C. Kies and H. M. Fox, *Am. J. Clin. Nutr.*, 1979, **32**, 1893.
152. M. Kim, M. T. Atallah, C. Amarasiriwardena and R. Barnes, *J. Nutr.*, 1996, **126**(7), 1883.
153. M. Kim and M. T. Atallah, *J. Nutr.*, 1993, **123**, 117.
154. M. Kim and M. T. Atallah, *J. Nutr.*, 1992, **122**, 2298.
155. A. E. Harmuth-Hoene and R. Schelenz, *J. Nutr.*, 1980, **110**, 1774.
156. K. C. Mercurio and P. A. Behm, *J. Food Sci.*, 1981, **46**, 1462.
157. A. Malovikova and R. Kohn, *Collection Czechoslov. Chem. Commun.*, 1983, **48**, 3154.
158. R. Kohn and I. Furda, *Collection Czechoslov, Chem. Commun.*, 1967, **32**, 4470.
159. V. A. Ostapenko, A. I. Tepliakow, A. S. Prokopovich and T. I. Chegerova, *Med. Tr. Prom. Ekol.*, 2001(5), 44.
160. E. Walzel, H. Anger, D. Bleyl, W. Bock, r. Kohn, M. Kujawa, A. Malovikova and M. Raab, In *Mengen-und Spurenelemente*, (Eds) M. Anke, C. Brückner, B. Groppel, H. Guertler, M. Gruen, I. Lombeck and H. J. Schneider, Arbeitstagung Leipzig, Germany, 1990, 156.
161. B. Lakatos and J. Meisel, *US Patent* 4.225.592, Int. Cl. A61K31/70.
162. S. C. Werch and A. C. Ivy, *Am. J. Dig. Dis.*, 1941, **8**, 101.
163. J. H. Cummings and G. T. Macfarlane, *J. Appl. Bacteriol.*, 1991, **70**, 443.
164. M. Nyman, N. G. Asp, J. Cummings and H. Wiggins, *Brit. J. Nutr.*, 1986, **55**, 487.
165. P. J. Van Soest, J. Jeraci, T. Foose, K. Wrick and F. Ehle. In *Proceedings of fiber in human and animal nutrition symposium*, Wellington, New Zealand, 1983, 75.
166. M. Nyman and N. G. Asp, *Brit. J. Nutr.*, 1982, **47**, 357.
167. M. Nyman and N. G. Asp, *Brit. J. Nutr.*, 1985, **54**, 635.

168. A. S. Sandberg, *Naringsforskning*, 1983, **2**, 65.
169. W. D. Holloway, C. Tasman-Jones and K. Maher, *Am. J. Clin. Nutr.*, 1983, **37**, 253.
170. S. Viola, G. Zimmermann and S. Mokady, *Nutr. Rep. Int.*, 1970, **1**, 367.
171. W. Bock, G. Pose and S. Augustat, *Biochem. Z.*, 1964, **341**, 64.
172. N. Gilmore, *Ph. D. thesis* 1965, Michigan State University.
173. H. N. Englyst, S. Hay and G. T. Macfarlane, *FEMS Microbiology Ecology*, 1987, **95**, 163.
174. P. B. Mortensen, H. Hove, M. R. Clausen and K. Holtug, *Scand. J. Gastroenterol.*, 1991, **26**, 1285.
175. Y. Matsuura, *Agric. Biol. Chem.*, 1991, **55**(3), 885.
176. M. Wojciechovicz, *Acta Microbiol. Pol. Ser. A.*, 1971, 45.
177. M. Wojciechovicz and A. Ziolecki, *Appl. Environ. Microbiol.*, 1979, **37**, 136.
178. M. Wojciechowicz and A. Ziolecki, *J. Appl. Bact.*, 1984, **56**, 515.
179. J. L. Barry, C. Hoebler, G. T. Macfarlane, S. Macfarlane, J. C. Mathers, K. A. Reed, P. B. Mortensen, I. Nordgaard, I. R. Rowland and C. J. Rumney, *Br. J. Nutr.*, 1995, **74**, 303.
180. G. Livesey, T. Smith, B. O. Eggum, I. H. Tetens, M. Nyman, M. Roberfroid, N. Delzenne, T. F. Schweizer and J. Decombaz, *Br. J. Nutr.*, 1995, **74**(3), 289.
181. G. Brunsgaard, B. O. Eggum and B. Sandstrom, *Comp. Biochem. Physiol. A. Physiol.*, 1995, **111**(3), 369.
182. A. J. Vince, N. I. McNeil, J. D. Wager and O. M. Wrong, *Br. J. Nutr.*, 1990, **63**, 17.
183. B. A. Degnan, S. Macfarlane and G. T. Macfarlane, *J. Appl. Microbiol.*, 1997, **83**(3), 359.
184. G. Dongowski and A. Lorenz, *Carbohydr. Res.*, 1998, **314**(3–4), 237.
185. G. Dongowski, A. Lorenz and H. Anger, *Appl. Environ. Microbiol.*, 2000, **66**(4), 1321.
186. G. Dongowski, A. Lorenz and J. Proll, *J. Nutr.*, 2002, **132**(7), 1935.
187. A. K. Mallett, A. Wise and I. R. Rowland, *Food Chem. Toxicol.*, 1984, **22**(6), 415.
188. F. Crociani, A. Alessandrini, M. M. Mucci and B. Biavati, *Int. J. Food Microbiol.*, 1994, **24**, 199.
189. L. Slovakova, D. Duskova and M. Marounek, *Lett. Appl. Microbiol.*, 2002, **35**, 126.
190. M. Marounek and D. Duskova, *Lett. Appl. Microbiol.*, 1999, **29**, 429.
191. D. Duskova and M. Marounek, *Lett. Appl. Microbiol.*, 2001, **3**, 159.
192. B. J. Paster and E. Canale-Parola, *Appl. Environ. Microbiol.*, 1985, **50**, 212.
193. E. Olano-Martin, G. R. Gibson and R. A. Rastall, *J. Appl. Microbiol.*, 2002, **93**, 505.
194. D. W. Heitman, W. E. Hardman and I. L. Cameron, *Carcinogenesis*, 1992, **13**(5), 815.
195. Y. H. Jiang, J. R. Lupton and R. S. Chapkin, *J. Nutr.*, 1997, **127**, 1938.

196. M. Higashimoto, H. Yamato, T. Kinouchi and Y. Ohnishi, *Mutat. Res.*, 1998, **415**(3), 219.
197. C. V. Rao, D. Chou, B. Simi, H. Ku and B. S. Reddy, *Carcinogenesis*, 1998, **19**, 1815.
198. A. Hensel and K. Meier, *Planta Med.*, 1999, **65**(5), 395.
199. H. Ohkami, K. Tazawa, I. Yamashita, T. Shimizu, K. Murai, K. Kobashi and M. Fujimake, *Jpn. Cancer Res.*, 1995, **86**(6), 523.
200. K. Tazawa, H. Okami, I. Yamashita, Y. Ohnishi, K. Kobashi and M. Fujimaki, *J. Exp. Clin. Cancer Res.*, 1997, **16**, 33.
201. K. Tazawa, K. Yatuzuka, M. Yatuzuka, J. Koike, H. Ohkami, T. Salto, Y. Ohnishi and M. Salto, *Hum. Cell.*, 1999, **12**(4), 189.
202. K. Ohno, S. Narushima, S. Takeuchi, K. Itoh, T. Mitsuoka, H. Nakayama, T. Itoh, K. Hioki and T. Nomura, *Exp. Anim.*, 2000, **49**(4), 305.
203. C. Avivi-Green, Z. Madar and B. Schwartz, *Int. J. Mol. Med.*, 2000, **6**(6), 689.
204. E. Olano-Martin, G. H. Rimbach, G. R. Gibson and R. A. Rastall, *Anticancer Res.*, 2003, **23**(1A), 341.
205. A. Raz and K. J. Pienta, *US Patent* 5 834 442, 1998.
206. A. Raz and R. Lotan, *Cancer Metastasis Rev.*, 1987, **6**, 433.
207. K. J. Pienta, H. Naik, A. Akhtar, K. Yamazaki, T. S. Replogle, J. Lehr, T. L. Donat, L. Tait, V. Hogan and A. Raz, *J. Natl. Cancer Inst.*, 1995, **87**(5), 348.
208. D. Platt and A. Raz, *J. Natl. Cancer Inst.*, 1992, **18**, **84**(6), 438.
209. H. Inohara and A. Raz, *Glycoconj J.*, 1994, **11**(6), 527.
210. N. Naik, M. J. Pilat and T. Donat, *Proc. Am. Assoc. Cancer Res.*, 1995, **36**, A377.
211. T. Hsieh and J. M. Wu, *Biochem. Mol. Biol. Int.*, 1995, **37**, 833.
212. S. Strum, M. Scholz, J. McDermed, M. McCulloch and I. Eliaz, *International Conference on Diet and Prevention of Cancer*, 28.05.-02.06.1999, Tampere, Finland.
213. T. Weiss, M. McCulloch and I. Eliaz, *International Conference on Diet and Prevention of Cancer*, 28.05.-02.06.1999, Tampere, Finland.
214. A. Hayashi, A. C. Gillen and J. R. Lott, *Altern. Med. Rev.*, 2000 **5**(6), 546.
215. H. G. Zhu, T. M. Zoller, A. Klein-Franke and F. A. Anderer, *J. Cancer Res. Clin. Oncol.*, 1994, **120**(7), 383.
216. P. Nangia-Makker, V. Hogan, Y. Honjo, S. Baccarini, L. Tait, R. Bresalier and A. Raz, *J. Natl. Cancer Inst.*, 2002, **18**(94), 1854.
217. B. W. Guess, M. C. Scholz, S. B. Strum, R. Y. Lam, H. J. Johnson and R. I. Jennrich, *Prostate Cancer Prostatic Dis.*, 2003, **6**(4), 301.
218. J. P. Guggenbichler, A. De Bettignies-Dutz, P. Meissner, S. Schellmoser and J. Jurenitsch, *Pharmaceutical and Pharmacological Letters*, 1997 **7**, 35.
219. J. P. Guggenbichler, *European Patent* 1998, 0 716 605 B1.
220. B. Stahl and G. Boehm, *International Patent* 2001, WO 01/60378 A2.

221. M. Kunz, M. Munir and M. Vogel, *International Patent* 2000, WO 02/42484 A2.
222. D. Kritchevsky, S. A. Tepper and D. M. Klurfeld, *Experientia*, 1984, **40**(4), 350.
223. D. Muting, R. Fischer, J. F. Kalk and P. Kruck, *Fortschr. Med.*, 1982, **100**, 1179.
224. A. Andoh, T. Bamba and M. Sasaki, *J. Parenter Etneral Nutr.*, 1999, **23**(5 Suppl), 70.
225. R. H. Rolandelli, M. J. Koruda, R. G. Settle and J. L. Rombeau, *Surgery*, 1986, **99**(6), 703.
226. H. Murakami, S. Iwane, A. Munakata, S. Nakaji, K. Sugawara, S. Tsuchida and D. Sasaki, *Dig. Dis. Sci.*, 2001, **46**(6), 1247.
227. K. Sugiyama, T. Bamba, S. Nakajo and S. Hosoda, *Nippon Shokakibyo Gakkai Zasshi*, 1991, **88**(10), 2636.
228. Y. Mao, B. Kasravi, B. Nobaek, L. O. Wang, D. Adawi, G. Roos, U. Stenram, G. Molin, S. Bengmark and B. Jeppsson, *Scand. J. Gastroenterol.*, 1996, **31**(6), 558.
229. R. Kohen, V. Shadmi, A. Kakunda and A. Rubinstein, *Br. J. Nutr.*, 1993, **69**(3), 789.
230. O. Aozasa, S. Ohta, T. Nakao, H. Miyata and T. Nomura, *Chemosphere*, 2001, **45**, 195.
231. K. N. Kim, *WO Patent* 01/47531 A1, 2001.
232. J. Decombaz, *European Patent Application* 262 518 A1, 2003.
233. O. V. Piatchina, M. V. Odintsova and luS Khotimchenko, *Vopr. Pitan.*, 2003, **72**, 43.
234. M. Tamura and H. Suzuki, *Ann. Nutr. Metab.*, 1997, **41**(4), 255.
235. T. Fukunaga, M. Sasaki, Y. Araki, T. Okamoto, T. Yasuoka, T. Tsujikawa, Y. Fujlyama and T. Bamba, *Digestion*, 2003, **67**(1–2), 42.
236. J. Zhang and J. R. Lupton, *Nutr. Cancer*, 1994, **22**(3), 267.
237. D. Y. Lee, R. S. Chapkin and J. R. Lupton, *Nutr. Cancer*, 1993, **20**, 107.
238. M. K. Chourasia and S. K. Jain, *J. Pharm. Sci.*, 2003, **6**(1), 33.
239. S. Lee, S. La, C. Lim, S. Lee, B. Seo and C. Pai, *US Patent* 6 319 518, 2001.
240. L. Liu, M. L. Fishman, J. Kost and K. B. Hicks, *Biomaterials*, 2003, **24**(19), 3333.
241. A. Rubinstein, R. Radai, M. Ezra, S. Pathak and J. S. Rokem, *Pharm. Res.*, 1993, **10**, 258.
242. Z. Wakerly, J. T. Fell, D. Attwood and D. Parkins, *Pharm. Res.*, 1996, **13**(8), 1210.
243. Z. Wakerly, J. T. Fell, D. Attwood and D. Parkins, *J. Pharm. Pharmacol.*, 1997, **49**(6), 622.
244. D. A. Adkin, C. J. Kenyon, E. I. Lerner, I. Landau, E. Strauss, D. Caron, A. Penhasi, A. Rubinstein and I. R. Wilding, *Pharm. Res.*, 1997, **14**(1), 103.
245. A. Mitrevej, N. Sinchaipanid, Y. Rungvejhavuttivittaya and V. Kositchaiyong, *Pharm. Dev. Technol.*, 2001, **6**(3), 385.

246. T. H. Kim, Y. H. Park, K. J. Kim and C. S. Cho, *Int. J. Pharm.*, 2003, **250**, 371.
247. M. Hiorth, I. Tho and S. A. Sande, *Eur. J. Pharm. Biopharm.*, 2003, **56**, 175.
248. G. S. Macleod, J. H. Collett and J. T. Fell, *J. Control Release*, 1999, **58**(3), 303.
249. G. S. Macleod, J. T. Fell and J. H. Collett, *Int. J. Pharm.*, 1999, **188**(1), 11.
250. G. S. Macleod, J. T. Fell, J. H. Collett, H. L. Sharma and A. M. Smith, *Int. J. Pharm.*, 1999, **187**, 251.
251. K. Ofori-Kwakye and J. T. Fell, *Int. J. Pharm.*, 2003, **250**, 431.
252. K. Ofori-Kwakye and J. T. Fell, *Int. J. Pharm.*, 2001, **226**(1–2), 139.
253. K. Ofori-Kwakye and J. T. Fell, *Int. J. Pharm.*, 2003, **250**(1), 251.
254. K. Ofori-Kwakye, J. T. Fell, H. L. Sharma and A. M. Smith, *Int. J. Pharm.*, 2004, **270**(1–2), 307.
255. M. Saravanan, G. S. Kishore, S. Ramachandran, G. S. Rao and S. K. Sridhar, *Indian Drugs*, 2002, **39**(7), 368.
256. J. N. McMullen, D. W. Newton and C. H. Becker, *J. Pharm. Sci.*, 1984, **73**(12), 1799.
257. S. F. Ahrabi, G. Madsen, K. Dyrstad, S. A. Sande and C. Graffner, *Eur. J. Pharm. Sci.*, 2000, **10**(1), 43.
258. R. Semdé, K. Amighi, M. J. Devleeschouwer and A. J. Moes, *Int. J. Pharm.*, 2000, **197**(1–2), 181.
259. P. Sriamornsak, J. Nunthanid, S. Wanchana and M. Luangtana-Anan, *Pharm. Dev. Technol.*, 2003.
260. R. Semdé, A. J. Moes, M. J. Devleeschouwer and K. Amighi, *Drug Dev. Ind. Pharm.*, 2003, **29**, 203.
261. M. Turkoglu and T. Ugurlu, *Eur. J. Pharm. Biopharm.*, 2002, **53**(1), 65.
262. H. Kim and R. Fassihi, *Pharm. Res.*, 1997, **14**(10), 1415.
263. I. El-Gibaly, *Int. J. Pharm.*, 2002, **232**(1–2), 199.
264. S. Y. Lin, K. S. Chen and H. S. Teng, *J. Microencapsul.*, 1999, **16**(1), 39.
265. T. W. Wong, L. W. Chan, H. Y. Lee and P. W. Heng, *J. Microencapsul.*, 2002, **19**(4), 511.
266. T. Wong, H. Lee, L. Chan and P. Heng, *Int. J. Pharm.*, 2002, **242**(1–2), 233.
267. M. M. Meshali, E. Z. el-Dien, S. A. Omar and L. A. Luzzi, *J. Microencapsul.*, 1992, **9**(1), 67.
268. P. Giunchedi, U. Conte, P. Chetoni and M. F. Saettone, *Eur. J. Pharm. Sci.*, 1999, **9**(1), 1.
269. P. Sriamornsak, *Eur. J. Pharm. Sci.*, 1999, **8**(3), 221.
270. S. Sungthongjeen, T. Pitaksuteepong, A. Somsiri and P. Sriamornsak, *Drug Dev. Ind. Pharm.*, 1999, **25**(12), 1271.
271. I. Tho, E. Anderssen, K. Dryrstad, P. Kleinebudde and S. A. Sande, *Eur. J. Pharm. Sci.*, 2002, **16**(3), 143.
272. I. Tho, S. A. Sande and P. Kleinebudde, *Eur. J. Pharm. Biopharm.*, 2002, **54**(1), 95.

273. P. Sriamornsak, S. Prakongpan, S. Puttipipathkhachorn and R. A. Kennedy, *J. Contr. Release*, 1997, **47**(3), 221.
274. V. Pillay and R. Fassihi, *J. Control Release*, 1999, **20**(59), 229.
275. V. Pillay and R. Fassihi, *J. Control Release*, 1999, **20**(59), 243.
276. V. Pillay and M. P. Danckwerts, *J. Pharm. Sci.*, 2002, **91**(12), 2559.
277. V. Pillay, M. P. Danckwerts and R. Fassihi, *Drug Deliv.*, 2002, **9**, 77.
278. H. Kim and R. Fassihi, *J. Pharm. Sci.*, 1997, **86**(3), 316.
279. P. Liu and T. R. Krishnan, *J. Pharm. Pharmacol.*, 1999, **51**, 141.
280. D. Guérin, J.-C. Fuillemard and M. Subirade, *J. Food Protection*, 2003, **66**(11), 2076.
281. C. T. Musabayane, O. Munjeri and T. P. Matavire, *Ren. Fail.*, 2003, **25**(4), 525.
282. Z. Shen and S. Mitragotri, *Pharm. Res.*, 2002, **19**(4), 391.
283. S. Miyazaki, N. Kawasaki, T. Nakamura, M. lwatsu, W. M. Hou and D. Attvvood, *Int. J. Pharm.*, 2000, **204**(1–2), 127.
284. Y. R. Song, S. Z. Wang, D. F. Wang and L. P. Niu, *Guang Pu Xue Yu Guang Pu Fen Xi*, 2002, **22**(6), 1043.
285. Y. Nie, Y. Li, H. Wu, W. Sha, H. Du, S. Dai, H. Wang and Q. Li, *Helicobacter*, 1999, **4**, 128.
286. S. Zhang, S. Zhang, Z. Yu and Y. Wang, *Zhonghua Yi Xue Za Zhi*, 2002, **82**(13), 872.
287. G. P. Martin, D. Ladenheim, C. Marriott, D. A. Hollingsbee and M. B. Brown, *Drug Dev. Ind. Pharm.*, 2000, **26**(1), 35.
288. I. Krogel and R. Bodmeier, *Pharm. Res.*, 1999, **16**(9), 1424.
289. Y. Mino, H. Kitagaki, M. Sasaki, K. Ishii, T. Mori, K. Yamada and O. Nagasawa, *Biol. Pharm. Bull.*, 1998, **21**, 1385.
290. H. P. Ronneburg, *Radiol. Diagn. (Berlin)*, 1989, **30**, 215.
291. D. Lewinska, W. Piatkiewicz and S. Rosinski, *Int. J. Artif. Organs.*, 1997, **20**(11), 650.
292. T. Havelund, C. Aalykke and L. Rasmussen, *Eur. J. Gastroenterol. Hepatol.*, 1997, **9**(5), 509.
293. T. Havelund and C. Aalykke, *Scand. J. Gastroenterol.*, 1997, **32**(8), 773.
294. E. T. Waterhouse, C. Washington and N. Washington, *Int. J. Pharm.*, 2000, **209**(1–2), 79.
295. R. Kato and J. Kishibayashi, *No To Hattatsu*, 2002, **34**(5), 437.
296. S. Anazawa, Y. Ohmura and R. Yoshikawa, (Eds), *Skin Barriers for Stoma Care*, Alcare Co Ltd, Kinshi, Sumida-ku, Tokyo, Japan, 2001.
297. S. Numata, R. Yoshikawa and S. Anazawa, *Journal of Japanese Society of Stoma Rehabilitation*, 1994, **10**(1), 25.
298. E. Gross and M. Irving, *Br. J. Surg.*, 1977, **64**(4), 258.
299. R. D. Cilento, A. L. Lavia, J. L. Chen and J. A. Hill, *US Patent* 4 166 051, 1979.
300. J. L. Chen, R. D. Cilento, J. A. Hill and A. L. Lavia, *German Patent* DE 28 25 196A1, 1987.
301. J. L. Chen, R. D. Cilento, J. A. Hill and A. L. Lavia, *US Patent* 4 253 460, 1981.

302. J. L. Chen, R. D. Cilento, J. A. Hill and A. L. Lavia, *US Patent* 4 204 540, 1980, *US Patent* 4 192 785, 1980.
303. H. O. Larsen and E. L. Sorensen, *US Patent* 4 231 369, 1980.
304. D. F. Doehnert and A. S. Hill, *US Patent* 4 505 976, 1985.
305. C. Lund, J. M. Kuller, C. Tobin, L. Lefrak and L. S. Franck, *J. Obstet. Gynecol. Neonatal Nurs.*, 1986, **15**(1), 39.
306. E. J. Dollison and J. Beckstrand, *Neonatal Netw.*, 1995, **14**(4), 35.
307. M. Nambu, *US Patent* 4 524 064, 1985.
308. R. D. Cilento and F. M. Freeman, *US Patent* 4 773 408, 1988a, *US Patent* 4 773 409, 1988b.
309. H. Mathews and P. L. Steer, *US Patent* 4 341 207, 1980.
310. F. M. Freeman and J. M. Pawelchak, *US Patent* 4 538 603, 1985.
311. F. M. Freeman and J. M. Pawelchak, *US Patent* 4 728 642, 1988.
312. R. D. Cilento, A. L. Lavia and C. Riffkin, *US Patent* 4 427 737, 1984.
313. O. M. Alvarez, *US Patent* 4 813 942, 1989.
314. T. Reichstein and A. Gruessner, *Helv. Chim. Acta*, 1935, **18**, 608.
315. H. S. Isbell, *Journal of Research of the National Bureau of Standards*, Vol. 33, Washington 1944, Research Paper RP 1594, 45.
316. J. Popova and M. Krachanova, *Maslo -Sapuncna Promis Lenost, Sofia*, 1977, **13/3**, 401.
317. K. D. Kulbe, A. Heinzler and G. Knopki, In *Annals of the New York Academy of Sciences*, Vol. 6, New York, 1987, p. 543.
318. K. D. Kulbe and G. Knopki, *German Patent* 3 502 141A1, 1985.
319. K. D. Kulbe, H. Czarnetzki, J. Giray, H. Schmidt, G. Miemietz, S. Novalin and R. Mattes, *4th Int. Workshop on Carbohydrates as Organic Raw Materials*, 1997, Vienna, Austria, March 20/21, p. 21.

CHAPTER 9

Hyaluronan: a Simple Molecule with Complex Character

KOEN P. VERCRUYSSE

Tennessee State University, Chemistry Department, 3500 John A. Merritt Blvd, Nashville, TN 37209, USA

9.1 Introduction

Hyaluronan (HA), also known as hyaluronic acid or hyaluronate, is a linear polysaccharide consisting of β-(1→4)-linked D-glucuronic acid β-(1→3)-*N*-acetyl-D-glucosamine disaccharide units (Figure 9.1). HA is ubiquitously present in the extracellular matrix (ECM) of the tissues in all higher animal species and is produced by some bacteria as an extracellular capsule.[1,2] HA was discovered decades ago and initially the study of this polysaccharide was focused on its "passive" hydrodynamic functions in the ECM. However, subsequent studies indicated that the presence of HA in the ECM has profound "active" cell-biological functions.[3] The discovery and study of multiple HA-binding proteins (HABPs) highlighted the importance of HA or HA–protein interactions in, *e.g.*, morphogenesis, cancer, inflammation or wound healing.[4] In recent years, the discovery of HA synthase (HAS) genes and the availability of $HAS^{-/-}$ knockout mice has further underscored the importance of HA in diverse cell-biological processes.[5] HA is considered to be non-toxic and non-immunogenic. In addition, multiple chemical strategies have been devised to modify its structure and alter its physico-chemical properties.[6] Hence, HA and its derivatives have found or are currently being actively investigated for potential applications in the fields of medicine, biotechnology, pharmaceutical

RSC Polymer Chemistry Series No. 1
Renewable Resources for Functional Polymers and Biomaterials
Edited by Peter A. Williams

Published by the Royal Society of Chemistry, www.rsc.org

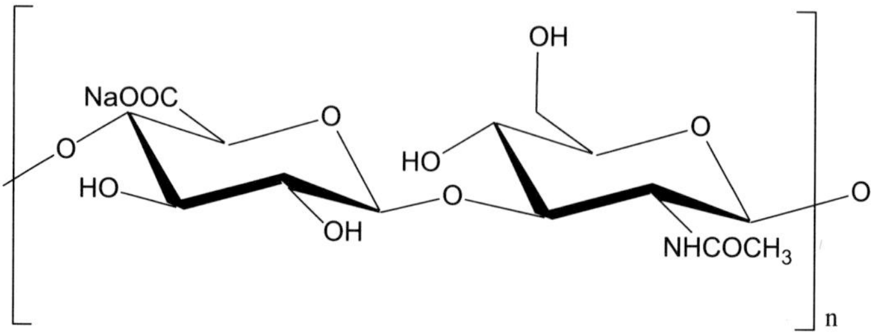

Figure 9.1 Chemical structure of the repeating unit of hyaluronan.

technology and cosmetics. This review aims to present an overview of the properties, functions and applications of this versatile polysaccharide and to emphasize that underneath its simple chemical structure HA hides a surprising and challenging complexity.

9.2 Physicochemical Properties

HA belongs to the family of glycosaminoglycan (GAG) polysaccharides which includes the various types of chondroitin sulfate (CS), heparan sulfate and heparin.[7] Unlike the other GAG polysaccharides, HA is unsulfated and not covalently linked to any protein.[8] Depending on its source, HA occurs as a linear polysaccharide with a molecular mass (MM) of several million Da (Table 9.1) and a broad molecular mass distribution (MMD),[9] although an accurate estimation of the average MM of HA is complicated by the fact that the HA molecules usually possess a tightly bound water mantle, skewing the measurements of its MM to higher estimates. No naturally occurring HA molecules have been isolated that contain chemical variations, *e.g.*, acetylation, phosphorylation or other, on the disaccharide unit shown in Figure 9.1. However, Longas *et al.* reported that, with increasing age, female breast skin contains *N*-deacetylated HA due to an increased activity of a specific HA *N*-deacetylase enzyme.[10,11]

Due to its high affinity for water, HA readily forms highly viscous solutions and the rheological properties of such solutions have been studied extensively.[9] The unique visco-elastic properties thus exhibited by HA have been exploited for various medical and cosmetic applications (see section 9.7). Given the polyelectrolyte nature of HA, numerous factors such as MM, pH, temperature, nature of counter ions, solvent, *etc.*, affect the molecular conformation adopted by HA in solution.[9] Sheehan *et al.*, *e.g.*, observed that HA aggregates were more likely to be formed in the presence of Na^+ or Ca^{2+} counter ions than in the presence of K^+ counter ions.[12]

Understanding the molecular conformation of HA is important as its cell-biological functions and biomedical applications depend strongly on this physico-chemical property. Most studies agree that, in solution, individual HA molecules exist as stiffened, random coils with a 2- to 4-fold helical secondary

Table 9.1 Molecular mass estimates of HA from various different tissues.

Source	*Estimated molecular mass (Da)*
Eye	
bovine	2–4 × 10^6
rabbit	5 × 10^6
Synovial fluid (human)	1.6–10.9 × 10^6
Skin (rat)	10^6 [a]

[a]Some degradation of the HA sample occurred prior to molecular mass estimation and it is assumed that the molecular mass of HA in the sample was several million Daltons.

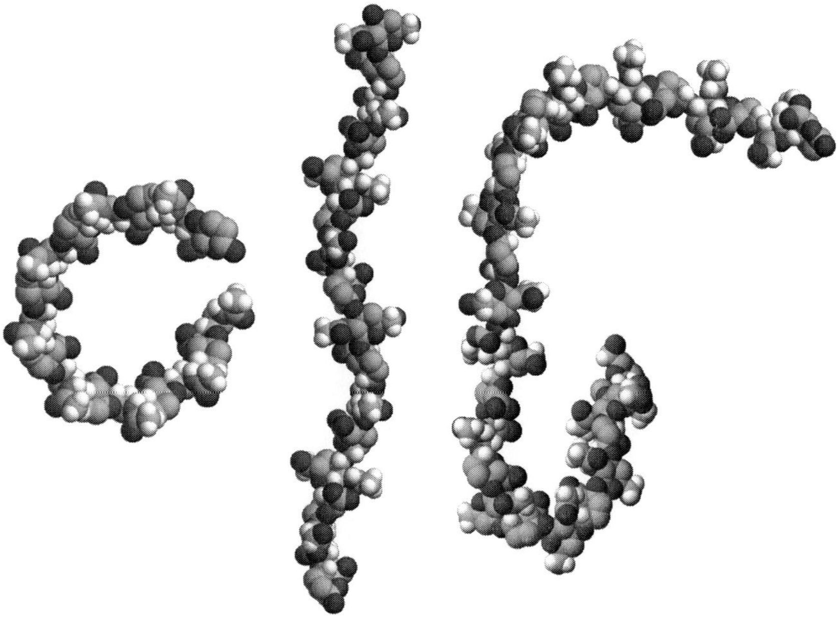

Figure 9.2 Molecular dynamics models of different conformations of HA. Reproduced from ref. 13, © 2001 with permission from Elsevier.

structure.[9,13,14] Based upon ^{13}C-NMR studies, Scott and Heatley proposed the existence of tertiary structures with β-sheet-like conformations in aqueous HA solutions.[14] Using capillary viscosimetry, low-angle laser light scattering and circular dichroism spectroscopy, Turner *et al.* investigated the possibilities of HA molecules forming intra- or intermolecular associations.[15] They provided evidence that, depending on its MM, HA molecules are present as monomers, dimers or higher order 3D-structures due to self-associations within a single molecule or between different molecules. Comparing computer modeling results with experimental evidence, Day and Sheehan argued[13] that HA molecules can adopt extended dynamic coil structures containing sharp kinks and folds (see Figure 9.2), resembling the proposed self-association structures by Turner *et al.* However, they envisioned HA molecules that have little tendency to form a 3D network through intermolecular associations.[13] A recent summary of the difficulty and ongoing discussion involving the determination of the molecular structure of HA was presented by Hargittai and Hargittai.[16]

9.3 Biosynthesis and Source

Traditionally, HA was extracted from animal sources, *e.g.*, rooster comb or bovine vitreous body, but the extraction and purification process regularly resulted in a reduction of the MM of HA. In addition, residual proteins, DNA or other GAG polysaccharides contaminations were usually present. Bacterial fermentation using extracellular HA-producing microorganisms, *e.g.*, *Streptococcus zooepidemicus,* has the advantage that the HA production can be controlled through manipulation of the culture conditions and the resulting HA product can easily be isolated, free from contaminating biomolecules.

The HA fermentation process by *S. zooepidemicus* has been extensively studied and modeled.[17–19] HA production can occur under anaerobic or aerobic conditions, the latter producing HA molecules with a higher MM. Culture pH, temperature and initial glucose concentration all affect the average MM and MMD of the final HA product (see Table 9.2).[18] Citing the benefit of using so-called Generally Recognized As Safe (GRAS) strains of bacteria, Izawa *et al.* investigated the fermentation conditions for the production of HA by a strain, YIT 2084, of *Streptococcus thermophilus*.[20]

HA is synthesized by a single enzyme, HA synthase or HAS, encoded by a single gene.[5,21] HAS enzymes or genes have been identified in group A and C

Table 9.2 Effects of growth conditions on the production of hyaluronan by *Streptococcus zooepidemicus* (from ref. 18).

Cell culture parameter	*Effect on cell growth and HA production*
Time	Maximum HA production is achieved within 8–10 hours of inoculation.
pH	HA production is optimal between pH = 6.3 and 8.0 with constant molecular mass average and polydispersity. At pH = 5.5 cell growth and HA production is poor.
Temperature	Rate of HA production increases with increasing temperature (32–40 °C). Amount of HA production decreases with increasing temperature. MM average of HA molecules decreases with increasing temperature.
Agitation rate	No effect on quantity and quality of HA produced.
Aeration	Compared to anaerobic growth conditions, aeration (0.2 or 1 vol/vol per min) increases HA production and its MM average without any effect on cell growth.
Initial glucose concentration	Increased HA production and MM average with increasing initial glucose concentration (20, 40 or 60 g mL^{-1}).

Streptococcus bacteria, in *Pasteurella multocida*, in mammals (*e.g.*, humans, mice), in frogs and in *Paramecium bursaria* chlorella virus.[5,21,22] In mammals, HAS contains seven putative transmembrane domains, two close to the *N*-terminal end and five close to the *C* terminal end. The cytoplasmic loop of the enzyme contains the putative glycosyltransferase catalytic sites and the UDP–carbohydrate binding motifs.[5] Unlike the biosynthesis of other GAG polysaccharides, biosynthesis of HA happens at the plasma membrane and not in the Golgi apparatus. In addition, biosynthesis of HA requires only one protein, *i.e.*, HAS is capable of binding both UDP-precursors, UDP–glucuronic acid and UDP–*N*-acetyl-glucosamine, and possesses dual glycosyltransferase capabilities.[5] So far three isoforms of HAS have been identified in mammals.[23] Despite their structural similarities, the three different types of HAS enzymes possess different enzymatic and functional properties (see Figure 9.3 and Table 9.3).[23]

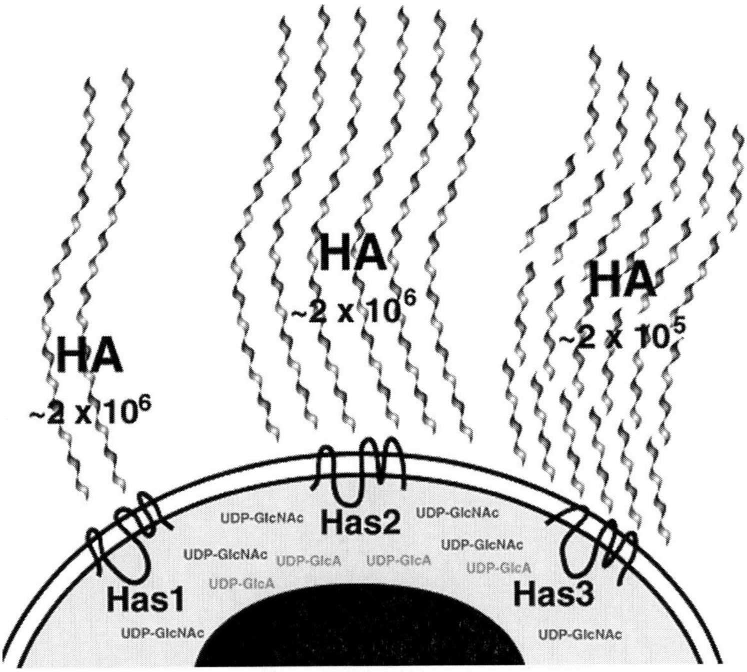

Figure 9.3 Model of the differential (amount and average chain length) of HA produced by the three different types of mammalian HAS. HAS1 produces small amounts of HMM HA, HAS 2 produces much higher amounts of HMM HA, while HAS 3 produces the highest amounts of HA but generally of much lower average MM. Reproduced, with permission, from "Eukaryotic Hyaluronan Synthases" by Andrew P. Spicer and John A. McDonald at http://www.glycoforum.gr.jp/science/hyaluronan/HA07/HA07E.html.

Table 9.3 Michaelis–Menten (K_m) constants for UDP-glucuronic acid (UDP-GlcUA) and UDP-*N*-acetyl-glucosamine (UDP-GlcNAc) of three different types of membrane-bound hyaluronan synthases (HAS). K_m values were estimated by varying the concentration of one UDP-precursor and keeping the concentration of the second UDP-precursor constant.

HAS	*UDP-GlcUA mM*	K_m *(UDP-GlcNAc) μM ± S.E.*	*UDP-GlcNAc mM*	K_m *(UDP-GlcUA) μM ± S.E.*
HAS1	0.05	799.0 ± 65.6	0.1	73.2 ± 0.7
	0.2	500.9 ± 32.3	0.5	53.3 ± 5.9
	1	1,011.4 ± 37.6	2.0	72.9 ± 4.3
HAS2	0.05	107.6 ± 27.2	0.1	32.9 ± 3.3
	0.2	330.6 ± 37.7	0.5	29.2 ± 4.3
	1	348.2 ± 15.1	2.0	30.0 ± 3.4
HAS3	0.05	82.1 ± 19.1	0.1	34.2 ± 4.4
	0.2	223.3 ± 11.3	0.5	35.1 ± 5.3
	1	247.2 ± 12.8	2.0	34.6 ± 2.0

Considerable differences among the three HAS enzymes exist in the MMD of the HA molecules they synthesize: their molecular stability, their affinity towards the UDP-precursors of glucuronic acid and *N*-acetyl-glucosamine, and their rates of HA synthesis. The existence of different isoforms of HAS implies that each might have different biological properties. Studies with $HAS2^{-/-}$ knockout mice indicate that HA synthesized by this gene is a critical ECM component needed for proper embryogenesis, as embryos of such mice died in mid-gestation with severe abnormalities, especially to the cardiac and vascular system.[24] In contrast, $HAS1^{-/-}$ or $HAS3^{-/-}$ knockout mice have been reported to be viable and fertile.[5,24] In addition to their differences in intrinsic enzymatic activity or genetic expression, the three HAS isoforms also differ in the regulation of their expression by growth factors or cytokines.[25] As for their genetic expression, the regulation of these genes by, *e.g.*, pro-inflammatory cytokines is cell-specific.[25]

A number of studies showed that pro-inflammatory cytokines like TNF-α or IL-1β enhance the production of HA through the activation of NF-κB.[26,27] Interestingly, HA degradation products have been shown to induce the NF-κB intracellular pathway (see section 9.5.5). These findings suggest the existence of an NF-κB-mediated tissue repair regulatory loop. Tissue injury leading to the degradation of ECM components like HA could stimulate the NF-κB intracellular pathway and promote *de novo* synthesis of HA. As mentioned earlier, the biosynthesis of HA requires the availability of UDP-precursors like UDP–glucuronic acid (UDP-GlcUA). UDP-GlcUA is derived from UDP–glucose through the action of the UDP–glucose dehydrogenase (UGDH) enzyme and the availability and regulation of UGDH is another factor regulating the biosynthesis of HA.[28]

9.4 Degradation and Turnover

In mammals, HA is degraded through non-enzymatic and enzymatic processes and it is cleared from tissues through catabolic mechanisms rather than excretion.[29] McCourt described three levels of HA clearance: (a) local degradation in tissues, (b) receptor-mediated uptake into lymph nodes followed by degradation and (c) receptor-mediated uptake by liver, kidneys or spleen followed by degradation.[29]

9.4.1 Non-enzymatic Degradation

Reactive oxygen species (ROS), similar to those released, *e.g.*, by activated leukocytes during acute or chronic inflammation events, are capable of depolymerizing HA.[30,31] However, Monzon *et al.* provided evidence that ROS may increase the expression of HA-degrading enzymes (see section 9.4.2) and as such enhance the degradation of HA.[32] Because of its polyelectrolyte nature, HA is capable of chelating redox-active metal ions like Fe^{2+}, Fe^{3+} or Cu^{2+}, which can catalyze the localized production of ROS, leading to site-specific degradation within HA molecules.[33] Other metal ions can protect HA from oxidative damage either through radical-scavenging or through the displacement of Fe^{2+}or Fe^{3+} ions from HA molecules.[31] Ascorbic acid (Vitamin C) has long been recognized to be capable of depolymerizing HA.[34] A free radical mechanism, possibly catalyzed in the presence of trace amounts of redox-active ions, has been proposed to describe this interaction.[35]

9.4.2 Enzymatic Degradation

Hyaluronidases (HAses), capable of depolymerizing HA, are present in various animal species and are produced by selected bacteria.[36,37] Unlike other types of HAses, bacterial HAses degrade HA through an elimination reaction, yielding unsaturated di- or oligosaccharide fragments (see Figure 9.4), and are termed HA lyases.[7] In humans, six HAse-like sequences have been identified.[38] HAse enzymes are used by spermatozoa during the process of fertilization to penetrate the HA-rich ECM surrounding oocytes (see section 9.5.1).[38] In other tissues, HAse is found in the cell's lysosomes, while a glycosylphosphia tidylinositol (GPI)-anchored isoform of the enzyme, although without exhibiting any HAse activity, is present on the outer cell membrane.[38] The properties and cell-biological importance of the HAse family of enzymes have been reviewed elsewhere[36,39] and will not be discussed in this review.

9.4.3 Receptor-mediated Clearance

The clearance of circulating plasma HA is achieved by liver endothelial cells, expressing a receptor termed HARE, or HA receptor for endocytosis; this is sometimes also referred to as Stabilin-2.[40–42] This receptor is non-specific as it also binds other GAG molecules like CS and heparin.[43] The clearance of HA

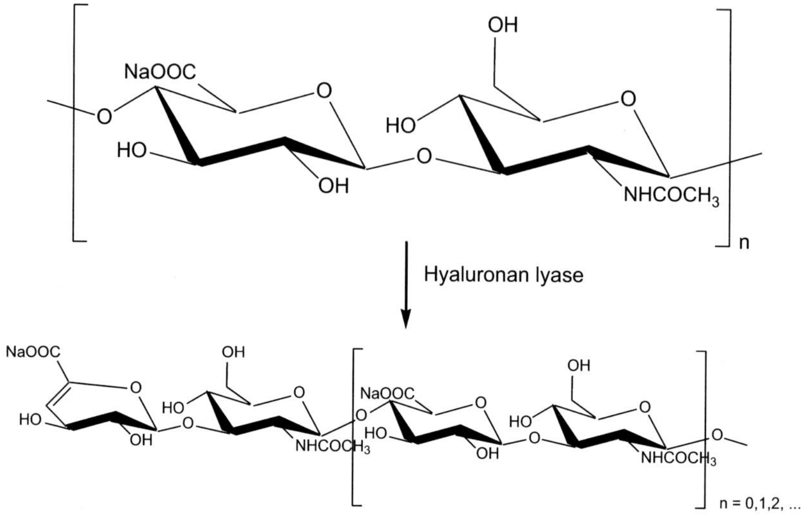

Figure 9.4 Chemical structure of HA di- and oligosaccharides generated from HA polysaccharide by the bacterial lyase-type of hyaluronidases.

by the lymphatic system has been attributed to the presence of HA-binding receptors on the cell surface of lymphatic endothelial cells, termed LYVE-1.[44] LYVE-1 is related to CD44, the common HA receptor found on many cells (see section 9.5.3). In addition to lymphatic endothelial cells, LYVE-1 is also expressed by some activated tissue macrophages and in the sinusoidal endothelium of the liver and spleen.[44,45] The clearance of HA from other tissues is achieved through receptor-mediated uptake *via* the ubiquitous CD44 receptor (see section 9.5.3).[46] The intracellular fate of HA following uptake is not entirely clear. Parts of the molecules undergo enzymatic degradation in the lysosomes, but the presence of a variety of intracellular HA-binding proteins (see section 9.5.4) provides tantalizing clues that HA might have specific intracellular functions.[46] In addition, HA uptake is tightly regulated by the modulation of the HA-binding capacity of CD44 or through the differential expression of different isoforms of this receptor (see section 9.5.3).[46]

9.5 Hyaluronan-binding Proteins (HABPs)

In addition to receptors involved in the clearance of HA (see section 9.4.3), multiple other HABPs, collectively termed hyaladherins, have been identified. They are broadly classified in two groups: Link module hyaladherins and non-Link module hyaladherins.[47] The HA-binding domain termed the Link module was first identified in the link protein, a protein found in cartilage tissues, and subsequently identified in the structure of other ECM proteoglycans. Non-Link module hyaladherins share little homology among themselves or with Link module hyaladherins.[47]

9.5.1 Extracellular Matrix Hyaluronan-binding Proteins

Many ECM proteoglycans possess multiple functional motifs or domains, often including so-called proteoglycan tandem repeats, also known as HA binding motifs or Link modules.[48] The structure of this Link module has been studied in detail and consists of two α-helices and two triple-stranded anti-parallel β-sheets, creating a sandwich conformation.[47] One such ECM Link-module containing HABP is TSG-6, the protein product of tumor necrosis factor-stimulated gene 6.[49] This protein appears to be involved in the process of inflammation and ovulation, particularly in the organization of the ECM surrounding cumulus oocyte complexes (COC).[49] Studies suggest that TSG-6 influences the formation and remodeling of this ECM through interactions with HA (via its Link module) and through the formation of covalent complexes with inter-alpha-inhibitor (IαI or ITI), a serine protease inhibitor.[49] IαI is a serum glycoprotein consisting of two so-called heavy chain proteins and bikunin, covalently linked together by a single CS molecule.[50] IαI interacts with HA in a non-covalent matter, but in the ECM of COC HA is covalently bound to IαI (see Figure 9.5). This covalent attachment of HA to IαI stabilizes the expanded ECM of the COC and provides rigid visco-elastic properties to the cumulus mass to protect it from mechanical damage during the extrusion process at ovulation. Lack of appropriate amounts of high molecular mass (HMM)

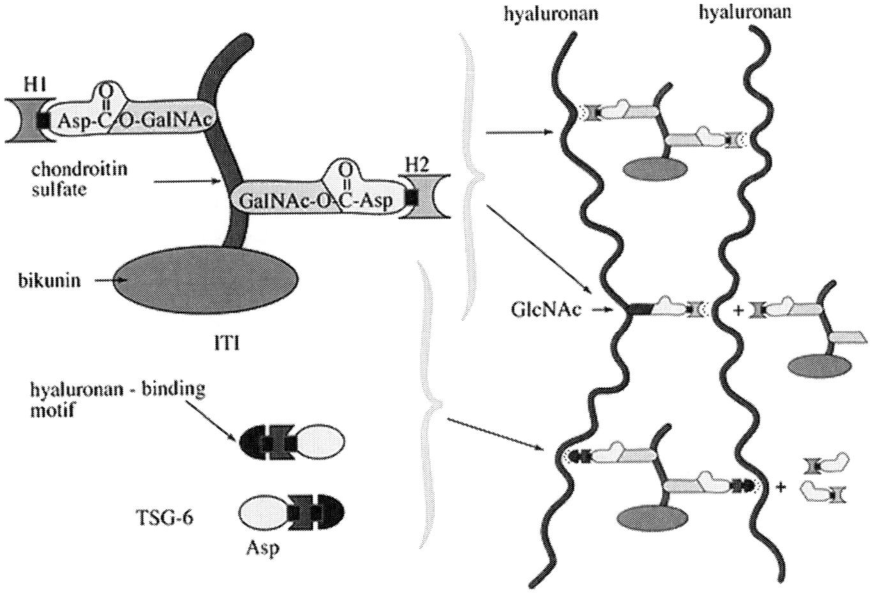

Figure 9.5 Model of the organization of the ECM in COC through covalent linkages between TSG-6, IαI (or ITI) and HA. Reproduced, with permission, from "Role of Hyaluronan during Ovulation and Fertilization" by Antonietta Salustri and Csaba Fulop at http://www.glycoforum.gr.jp/science/hyaluronan/HA03/HA03E.html.

HA in the ECM of COC can lead to disruptions in the cumulus environment surrounding the oocyte and to significant disturbances at the 2- or 4-cell stage as well as in the number of fragmented embryos developing.[51]

9.5.2 Plasma Hyaluronan-binding Proteins

An HA-binding protein has been isolated from human plasma by affinity chromatography using an HA-conjugated Sepharose™ column.[52] Subsequent studies showed this protein to be a serine protease.[53] However, the importance of the HA-binding capacity in these protease activities is not clear. Hence, HA is capable of binding a serine protease and members of the IαI family of protease inhibitors (see section 9.5.1). It is tantalizing to envision that HMM HA could regulate protease activity by physically bringing together the above-mentioned serine protease and protease inhibitors. However, no information appears to be available to indicate that such a mechanism of HA-mediated protease activity regulation exists.

9.5.3 Cell Surface Hyaluronan-binding Proteins

Multiple cell surface HABPs have been discovered, but CD44 and RHAMM have been studied most extensively.[47] CD44 is a type I transmembrane protein ubiquitously found on the cell surface of a wide variety of cells. The structure of CD44, its HA-binding capacity and its cell-biological functions have been studied extensively. Consequently, multiple review manuscripts on CD44 have been published in recent years.[54,55] Genomic analysis of CD44 indicates the presence of at least 20 exons and multiple CD44 isoforms have been identified.[54,55] The most abundant, standard or hematopoietic, isoform of CD44 (CD44s or CD44H) contains 12 exons and consists of an extracellular *N*-terminal domain (containing the HA binding site), a transmembrane domain and a cytoplasmic domain.[54,55] The non-standard isoforms of CD44 contain alternative exons inserted at a single site within the membrane-proximal region of the extracellular part of the protein (see Figure 9.6).[54] In addition, CD44 is extensively modified through posttranslational *N*- and *O*-glycosylation and modification with GAG molecules like CS.[56]

Adding to this complexity, the transmembrane protein can be proteolytically cleaved into two parts: the soluble, extracellular domain (termed sCD44) and the intracellular cytoplasmic domain.[54] Numerous studies, working with a broad variety of cell types, have indicated that CD44 is closely associated with or covalently linked (through disulfide bridges) to protein tyrosine kinases (PTKs) like the Src family of kinases, p185^{HER2} (also known as c-erbB-2 or neu) or focal adhesion kinase (FAK).[57–62] The stimulation of CD44 and the subsequent activation of the associated PTKs can lead to (1) activation of mitogen-activated protein kinase (MAPK) *via* the Ras-signaling pathway,[59,63] (2) activation of the PI3K-Akt pathway,[63] (3) activation of the Rac-1 or Ras signaling pathways mediated by the association of CD44 with accessory

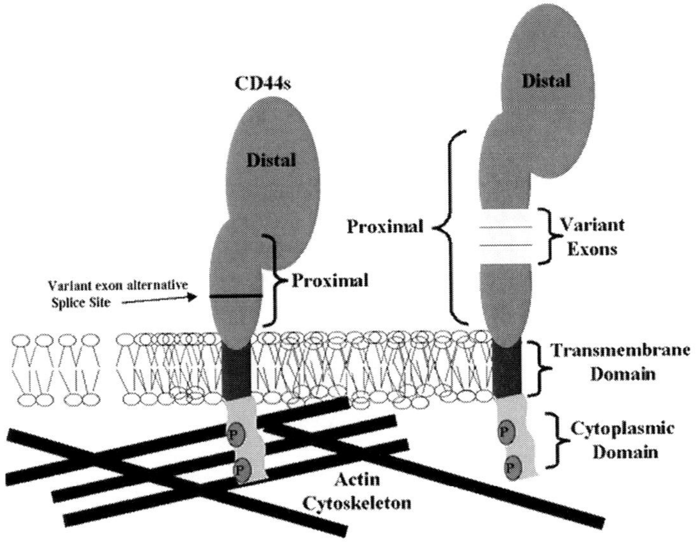

Figure 9.6 Model of the HA receptor CD44. The model illustrates the four principal domains of the protein: (1) distal extracellular domain, (2) membrane proximal extracellular domain, (3) transmembrane domain and (4) intracellular or cytoplasmic domain interacting with the cytoskeleton. Also shown is an isoform of CD44 carrying variant exons in the membrane proximal extracellular domain. Reproduced, with permission, from "The Hyaluronan Receptor, CD44" by Warren Knudson and Cheryl B. Knudson at http://www.glycoforum.gr.jp/science/hyaluronan/HA10/HA10E.html.

proteins like Tiam-1 (ref. 64 and 65) or Vav-2 (ref. 66) and (4) phosphorylation of cytoskeletal proteins.[67]

Particular attention has been paid to the interactions between CD44 and cytoskeletal proteins like ankyrin[68–70] or the moesin–ezrin–radixin (ERM) family of proteins.[71,72] CD44–cytoskeleton interactions seem to be mediated through Rho-like proteins.[73,74] In metastatic breast cancer cells, immunoblotting experiments (see Figure 9.7) indicated the existence of such CD44–Rho interactions. Such interactions lead to the activation of Rho–kinase (ROK) and the phosphorylation of CD44 and other intracellular proteins.[74] The phosphorylation of CD44 then promotes the binding of CD44 to ankyrin. In lipid rafts of the plasma membrane of mouse mammary epithelial cells, CD44 has been shown to interact with the actin cytoskeleton, an interaction mediated by annexin II proteins.[75]

All these studies seem to suggest a critical function of CD44 or CD44–HA interactions in the regulation of a cell's motility, particularly in tumor cells.[74]

Other studies have revealed links between HA, CD44 and matrix metalloproteinases (MMPs). CD44 has been shown to be closely associated with MMP-9 at the cell membrane.[76,77] Zhang *et al.* showed that in lung carcinoma cells, HA stimulates the secretion of MMP-2, a process that involved tyrosine

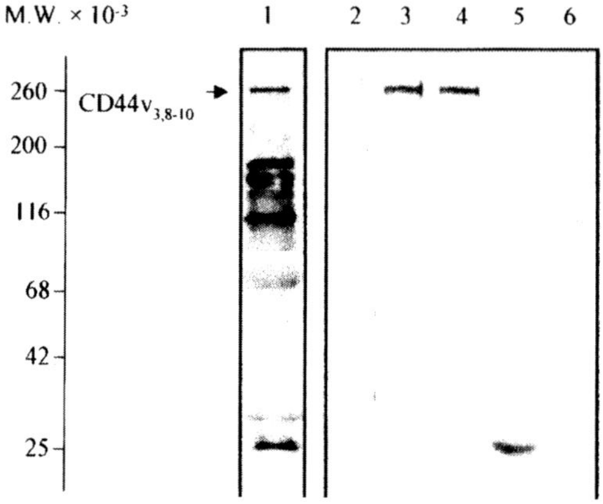

Figure 9.7 Analysis of $CD44_{v3,8\text{-}10}$ and the $CD44_{v3,8\text{-}10}$–RhoA complex in mouse breast tumor cells using immunoblotting techniques. Lane 1 indicates the biotinylated total solubilized plasma membrane-associated membranes. Lane 3 identifies the $CD44_{v3,8\text{-}10}$ isoform through immunoblotting with an anti-$CD44_{v3}$-specific antibody. Lane 4 identifies the $CD44_{v3,8\text{-}10}$ isoform in the $CD44_{v3,8\text{-}10}$-RhoA complex through anti-RhoA-mediated immunoprecipitation followed by immunoblotting with anti-$CD44_{v3}$-specific antibody. Lane 5 identifies RhoA in the $CD44_{v3,8\text{-}10}$-RhoA complex through anti-$CD44_{v3}$-mediated immunoprecipitation followed by immunoblotting with an anti-RhoA-specific antibody. Lanes 2 and 6 are control experiments to verify the specificity of the immunoprecipitation/immunoblotting techniques used.[74] Reprinted ref. 74, © 1999 Wiley-Liss, Inc with permission of Wiley-Liss, Inc a subsidiary of John Wiley & Sons, Inc.

phosphorylation of FAK and the MAPK signaling pathway.[78] On the other hand, Kawano *et al.* showed that the oncogene product Ras promotes a MMP-induced shedding of CD44, solubilizing the cell surface receptor (see section 9.4.3) and suppressing the HA/CD44-mediated cell motility.[79] They showed that this process required the activation of Rac, Cdc24 and PI3K. Independent information on the importance of HA in cell motility came from studies with *ras*-transformed fibroblasts. HA was shown to affect the cell's motility through an interaction with a cell surface protein termed RHAMM (Receptor for Hyaluronic Acid Mediated Motility).[80] Subsequent studies showed that this HA–RHAMM initiated locomotion involved the activation of PTK-like Src-family PTKs followed by the phosphorylation of FAK and/or the activation of protein kinase C (PKC).[81–83]

Similarly, in human pulmonary artery and human lung microvessel endothelial cells, expression of RHAMM was shown to mediate HA binding, cell adhesion, cell proliferation and appeared to involve the activation and phosphorylation of key intracellular components.[84] Stabilin-1 and stabilin-2 have been described as a novel family of fasciclin-like HA receptors.[85] Politz *et al.* speculated that stabilin-1 might interact with HA oligosaccharides and induce anti-inflammatory pathways in alternatively activated macrophages, in contrast to the HA oligosaccharide-induced pro-inflammatory response observed from conventionally activated macrophages (see section 9.5.5).[85] Kumarswamy *et al.* reported on the detection of a novel HABP in grade II human breast cancer tissue.[86]

9.5.4 Intracellular Hyaluronan-binding Proteins

Apart from its ubiquitous presence in the ECM, HA has also been detected intracellularly[87,88] and a variety of intracellular HABPs or IHABPs have been identified.[89–93] At present, little information exists concerning the intracellular function or relevance of HA and the IHABPs. However, Hascall *et al.* recently speculated that they might play a role in inflammatory processes.[94]

As mentioned earlier, in addition to being a cell-surface receptor, RHAMM has been identified as a cytoplasmic protein and as such has been termed intracellular hyaluronic acid binding protein or IHABP.[95] Subsequent studies have revealed that RHAMM/IHABP interacts with microtubules and actin filaments and is involved in early embryonic development.[96]

9.5.5 Hyaluronan Oligosaccharides and Hyaluronan-binding Proteins

Low molecular mass (LMM) fragments of HA have been shown to possess unique cell-biological properties, often not found within the HMM parent molecule.[97] Such LMM fragments can be generated from HA through reaction with locally-generated ROS (see section 9.4.1) or through enzymatic digestion (see section 9.4.2).[97] In this regard HMM HA can be considered as a reserve source of unique, cell-biologically active molecules since a single HMM HA molecule could generate hundreds to thousands of LMM HA oligosaccharides. LMM HA oligosaccharides have been observed to be angiogenic[98] and to induce the expression of several pro-inflammatory genes in macrophages.[99,100] These latter effects are comparable to lipopolysaccharide (LPS)-induced gene expression in macrophages. The stimulation of inflammatory gene expression was found to be mediated by CD44 and to involve the activation of NF-κB signaling pathway.[99,100] Apparently, the degradation of HA into LMM fragments and subsequent activation of pro-inflammatory genes serves as an early signal of ECM degradation and a need for an appropriate response.[97] However, it has been argued that the chemically-distinct oligosaccharide fragments derived from GAG molecules by a bacterial, lyase-type of enzymes (see section 9.4.2 and Figure 9.4), might possess anti-inflammatory properties as opposed to the pro-inflammatory oligosaccharides generated with a mammalian, hydrolase-type of enzymes.[101]

In addition to their effect on macrophages, LMM HAs have been shown to induce maturation of and to induce cytokine secretion from dendritic cells, an effect that appears to be mediated by Toll-like Receptor 4 (TLR-4) and independent of CD44 or RHAMM.[102,103] Interestingly, TLR-4 is the cell surface receptor for LPS, again illustrating the "LPS-like" properties of LMM HA oligosaccharides. Fieber *et al.* reported that LMM HA fragments can up-regulate MMP expression, an induction that apparently does not involve CD44, RHAMM or TLR-4.[104] Working with K562 cells, expressing very low levels of CD44, Xu *et al.* showed that HA tetrasaccharides can induce heat shock protein 72 (Hsp72) expression under hyperthermic conditions.[105] No such stimulation was observed with HMM HA, HA disaccharides, hexa- or higher order saccharides or with other GAG oligosaccharides. Interestingly, these stimulatory effects were observed with both saturated and unsaturated tetrasaccharides. The fact that K562 cells express no or very low levels of cell surface CD44 implied that some other receptor, yet unidentified, is responsible for this unique cell-biological activity of HA tetrasaccharides.

9.6 Cell-biological Functions

9.6.1 Cell Behavior and Morphogenesis

The presence of HA in the ECM profoundly affects the physico-chemical nature of the cell's surroundings. The contribution of HA to the hydration of the cell's environment has long been recognized as an important factor in providing a suitable environment for cell migration and mitosis.[106,107] Through specific interactions with other components of the ECM, HA contributes to the structural integrity of the ECM and affects the diffusion of other HMM biochemical components, *e.g.*, plasma proteins, into tissues.[1]

Several studies have shown that HA has a crucial function in morphogenetic events.[108] A combination of its unique biophysical properties (*e.g.*, contributing to increased tissue hydration), its interactions with ECM proteins (*e.g.*, contributing to the enhanced structural integrity of the ECM or pericellular matrix) and its capability to interact with select cell surface receptors (*e.g.*, affecting the cell's development or locomotion) have been invoked to explain these potent cell-biological properties of HA.[108] The polysaccharide is an important factor in the expansion of the cumulus oophorus during ovulation[51] (see section 9.4.1), the early gestational development and the formation of the villous tree in the placenta.[109] In addition, HA is involved in the morphogenesis of the epidermis,[110] endothelial cell proliferation and migration,[111] cardiogenesis,[112] renal organogenesis[113] and limb morphogenesis.[114] More extensive reviews regarding the effect of HA, its biosynthesis and biodegradation on embryogenesis have been published recently.[115–117]

9.6.2 Hyaluronan and Cancer

Several histological studies have shown that many human cancers, *e.g.*, breast cancer,[118] squamous cell carcinoma,[119] prostate cancer,[120,121] cutaneous myxoma,[122] thyroid carcinoma,[123] bladder cancer,[124] epithelial ovarian cancer,[125] minor salivary gland cancer,[126] gastric cancer,[127] colon cancer,[128] or non-small-cell lung cancer[129] possess an HA-enriched ECM. The existence of these high

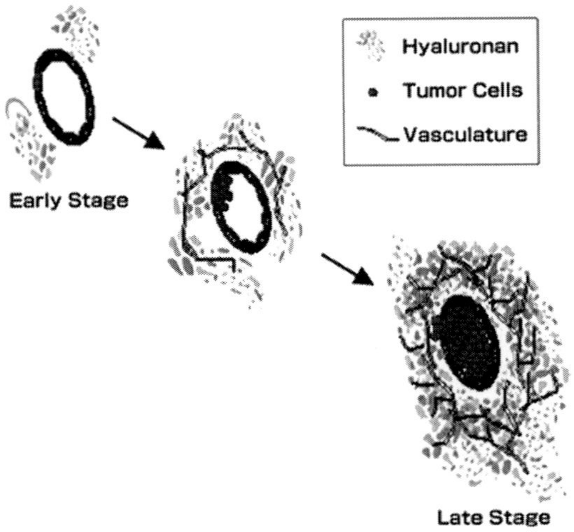

Figure 9.8 Schematic diagram of the function of HA in the progression of primary prostate tumors. Cancer progression is characterized by elevated levels of HA in the surrounding stroma and increased vascularization of the tissue due to the generation of small HA fragments with angiogenic potential. Reproduced, with permission, from "Hyaluronan in Prostate Cancer Progression" by James B. McCarthy and Melanie A. Simpson at http://www.glycoforum.gr.jp/science/hyaluronan/HA26/HA26E.html.

HA concentrations in the cancerous stroma is usually a negative marker for malignancy, metastasis, recurrence or patient survival (see Figure 9.8). HA seems to promote tumor growth and metastasis by enhancing the locomotion of cancer cells (see section 9.5.3).[130] In addition, HMM HA harbors angiogenic potential as the degradation products of HA were shown to be angiogenic.[98] The interactions between HA and cancer cells are often enhanced due to an enhanced expression of CD44 or the expression of unique CD44 isoforms on the cell surface (see section 9.5.3) of the cancer cells.[131] Y-box binding protein-1 (YB-1) is an oncogenic transcription factor frequently expressed in breast cancer cells. Binding of YB-1 to its promoter enhances the expression of CD44 among other proteins.[132] In prostate cancer cells, hepatocyte growth factor has been shown to enhance HA signaling *via* an isoform of CD44, CD44v9.[133] Such observations have led to the argument that CD44 is a potential therapeutic target for the treatment of tumors.[134]

9.6.3 Hyaluronan and Inflammation

In addition to the various studies that attribute potential pro-inflammatory properties to HA oligosaccharides (see section 9.5.5), other reports describe the contributions of HMM HA or its principal receptor, CD44, to inflammatory pathologies in general.

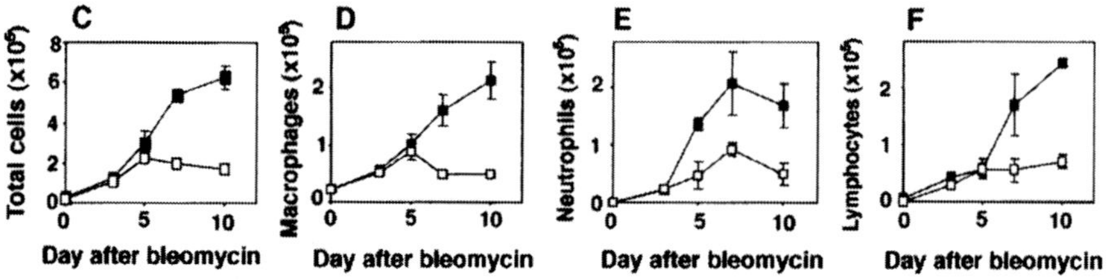

Figure 9.9 Cellular profiles obtained from bronchoalveolar lavage (panel C: total cells; panel D: macrophages; panel E: neutrophils; panel F: lymphocytes) following bleomycin treatment in wild type (□) and CD44-deficient (■) mice.[139] Reprinted with permission from ref. 139, © 2002 AAAS.

Studies have shown that the extravasation process of leukocytes into an infected or inflamed tissue often depends on HA–CD44 interactions.[135] Other *in vitro* and *in vivo* studies suggested that HA and CD44 are involved in rheumatoid arthritis.[136] In mice, monoclonal antibodies against CD44 were shown to increase the levels of serum interferon-γ (IFN-γ) and subsequently improve the severity of collagen-induced arthritis.[137] However, CD44 knockout mice were not protected from developing arthritis, although the severity of the arthritic events was reduced compared to wild-type mice.[138] These studies suggest that CD44 might not be an essential factor for the development of arthritic pathologies. Similarly, CD44 knockout mice were not protected from lung inflammation following exposure to bleomycin.[139] However, CD44 knockout mice succumbed to unremitted inflammatory conditions (impaired clearance of apoptotic neutrophils, continuous accumulation of HA fragments, *etc.*) not observed in wild-type mice (see Figure 9.9).[139]

These studies suggested that CD44 could represent a crucial component needed for the resolution of inflammatory conditions; an important signaling component of the switch between "killing" or "healing".[140] HA was also shown to be a crucial factor in the inflammatory phase of vascular diseases like atherosclerosis.[141,142] As with arthritis, HA, through interactions with its cell surface receptors CD44 or RHAMM, was shown to modulate the extravasation of leukocytes from the blood into the vascular wall.[84,143] In addition, CD44-null mice showed a marked reduction in development of atherosclerosis compared to wild-type mice.[144] Interestingly, LMM HA fragments, through a CD44-dependent mechanism, were shown to induce the proliferation of smooth muscle cells, another characteristic of atherosclerotic lesions. HMM HA molecules on the other hand inhibited such proliferation.[144] The presence of pro-inflammatory cytokines like IL-1β or TNFα and β,[145] the presence or absence of sialic acid in the carbohydrate moiety of the CD44 glycoprotein,[146] the degree of polymerization of the HA molecule,[147] *etc.*, all add their contribution to the complexity of the involvement of CD44–HA interactions in the inflammatory process. The role of HA–CD44 interactions in inflammation or inflammatory diseases has been discussed at length more recently.[148]

9.6.4 Hyaluronan and Wound Healing

The potential beneficial functions of HA in wound healing have long been observed and recognized.[1] Following tissue injury, HA concentrations increase and LMM fragments are generated.[149] Such LMM fragments contribute to scar formation and are pro-inflammatory (see section 9.5.5), while the HMM polysaccharide promotes wound healing by promoting tissue integrity.[150] However, Gao *et al.* discussed the promotion of wound healing by LMM HA fragments through its capacity to enhance angiogenesis and formation of lymph vessels.[151] Fetal cutaneous wounds heal through a regeneration process that results in little to no scarring.[152,153] This property has been attributed to the presence of higher HA concentrations in fetal skin compared to adult skin.

9.7 Applications

Because of its unique physico-chemical properties and because it is generally considered to be biocompatible, non-toxic, non-immunogenic and biodegradable, HA has found and is currently being actively investigated for multiple applications in the fields of pharmaceutical technology, medicine, biotechnology and cosmetics. HA and HA-based products have received US Food and Drug Administration (FDA) approval, *e.g.*, in the treatment of osteoarthritis, for wound healing purposes or for the reduction of wrinkles. In addition, HA can be chemically modified (see Figure 9.10) to tailor its physico-chemical properties, *e.g.*, to render it more hydrophobic or less biodegradable.[6] The applications of HA and HA-based materials have been the subject of several extensive reviews in recent years,[6,154,155] therefore, this review will focus on more recent reports.

9.7.1 Tissue Engineering

HA and its chemical derivatives or composite materials containing the polysaccharide have been manufactured into gels, sponges, films and other structures to serve as a matrix for cell adhesion and cell growth.[156–158] Such HA-based materials have been seeded with cells and transplanted to restore organ or tissue function, for wound healing purposes or to reduce scar formation in skin wounds (see Figure 9.11).[159,160] Autologous human keratinocytes were cultured onto membranes made from HA derivatives and applied as a graft to diabetic foot ulcers to enhance wound healing.[161] Transplants of myoblast cells in a HA hydrogel matrix have been evaluated in a rat model and demonstrated to be capable of forming multinucleated myotubes, opening the possibility of restoring muscle function, *e.g.*, in muscular dystrophy diseases.[162] HA material seeded with keratinocytes was applied to skin wounds in rats and found to greatly reduce scar formation.[159]

The biocompatibility of existing surgical devices, *e.g.*, their metallic surfaces, can be enhanced following chemical modification with HA.[163,164] The coating

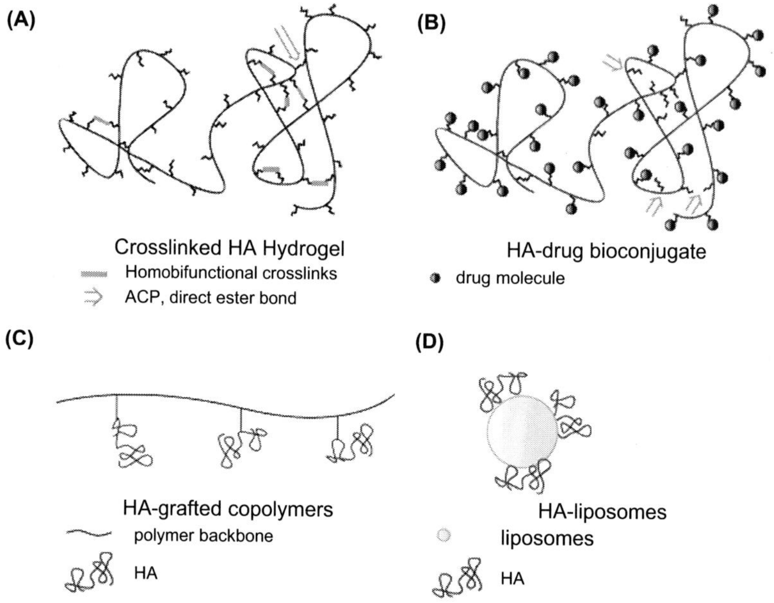

Figure 9.10 Schematic overview of different approaches in the chemical modification of HA, including (A) HA hydrogels (ACP = autocross-linked polymer formed through internal ester bonds between hydroxyl and carboxyl groups of HA), (B) HA–drug conjugates, (C) HA-grafted copolymer and (D) HA–liposome composite, for tissue engineering or drug-delivery purposes. Reproduced, with permission, from "Biomaterials from Chemically-Modified Hyaluronan" by Glenn D. Prestwich at http://www.glycoforum.gr.jp/science/hyaluronan/HA18/HA18E.html.

of endovascular stents with HA has been found to enhance thromboresistance and, because of its anti-inflammatory and wound healing properties (see sections 9.6.3 and 9.6.4), coating endovascular stents with HA is believed to reduce the risk of neointimal hyperplasia which is often observed following stent implantation.[165–167] The ever-expanding developments of new HA-based formulations and applications are the subject of numerous reviews.[117,168–171]

9.7.2 Drug Delivery

HA, chemically modified HA and HA-containing preparations have been extensively studied as vehicles for drug delivery purposes.[6,172–174] Table 9.4 presents an overview of some such applications developed and tested *in vitro* or *in vivo* in recent years. The presence of HA in drug delivery devices enhances their bioadhesion properties, resulting in enhanced bioavailabilities of the drugs.[175–177] In addition, the HA component of the drug delivery device can be used as a targeting agent towards, *e.g.*, tumor cells overexpressing HA receptors (see section 9.6.2), yielding more selective anti-tumor activity.[178–183]

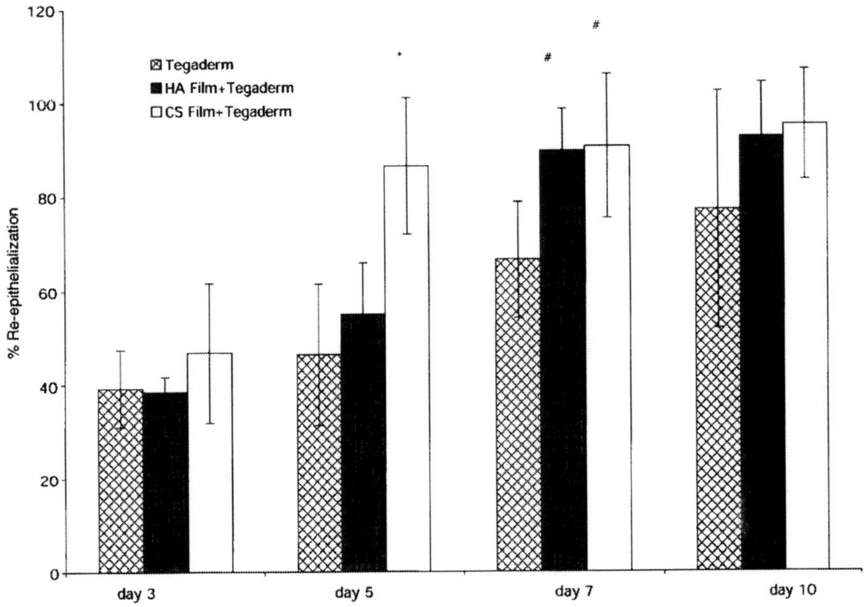

Figure 9.11 Percentage of wound re-epithelialization for *in vivo* wound healing experiments using a mouse model. Results are shown for two experimental groups, hyaluronan (HA) film plus Tegaderm™ and chondroitin sulfate (CS) film plus Tegaderm™ and the Tegaderm™-alone control group (mean ± s.d., $n = 6$). (*) significant at $p < 0.001$ *vs.* Tegaderm™ only and (#) significant at $p < 0.05$ *vs.* Tegaderm™ only.[160] Reprinted from ref. 160, © 2002, with permission from Elsevier.

HA-based microspheres have been studied for gene delivery purposes. Yun *et al.* conjugated such spheres with antibodies against E- and P-selectin and demonstrated that selective binding to cells expressing these receptors could be achieved, opening up the possibility of targeted gene delivery.[184] Oh *et al.* discussed the use of chemically modified HA as a carrier for protein, peptide or nucleotide therapeutics.[185] Topical formulations based on HA and containing diclofenac have been approved to treat actinic keratoses in the US and other topical applications for HA are being developed.[186]

9.7.3 Disease Marker

As mentioned earlier (see section 9.6.2) many cancers have been observed to possess an HA-enriched stroma or to over-express HABP on their cell surface. Hence, HA, CD44 or RHAMM have been explored as potential prognostic indicators of, *e.g.*, epithelial ovarian cancer,[187] head and neck tumors,[188] prostate cancer,[189] endometrial carcinomas[190] or breast cancer.[118] The evaluation of HA serum levels is currently investigated as a potential, non-invasive marker for progressive liver disease, especially liver cirrhosis.[191–195]

Table 9.4 Overview of some HA-based preparations used for drug-delivery purposes developed and tested *in vitro* or *in vivo* in recent years.

HA-based formulation	*Drug or biological agent*	*Reference*
Liposomes containing HA oligosaccharide-modified lipids	Doxorubicin	178
Liposomes surface-modified with HA	Mitomycin C	179
Microspheres made from chemically-modified HA	Influenza vaccine	177
Microparticles made from HA and chitosan	Gentamicin	175
Spray dried powder containing HA	Insulin	
Implantable pellets made from HA and chitosan	Insulin	
Hydrogel made from HA-modified poloxamer polymers	Ciprofloxacin	
Hydrogel made from HA–heparin conjugate	Basic fibroblast growth factor-2	
Hydrogel made from HA–drug conjugate crosslinked with poly(ethylene glycol)	Mitomycin C	
Drug-polymer conjugate made from *N*-(2-hydroxypropyl)methacrylamide and HA	Doxorubicin	180
Scaffold made from chemically-modified HA	Bone morphogenetic protein-2	
Pessaries made from solid triglycerides containing HA	Clotrimazole	176

9.7.4 Cryopreservation

HA has a very important function in the ECM of oocytes (see section 9.5.1). It serves a structural role, providing mechanical strength and support to the oocyte ECM, and it appears to be essential during the early stages of embryonic development. Therefore, HA has found applications as a supplement to the culture medium for the maturation and fertilization of oocytes and subsequent *in vitro* embryonic development.[196–198]

In view of these properties and its importance in morphogenesis (see section 9.6.1), it is tantalizing to speculate that HA might find similarly critical applications in the maintenance and differentiation of embryonic stem cells.

9.7.5 Wound Healing and Adhesion Prevention

The cell-biological functions of HA in the wound healing process have been discussed earlier (see section 9.6.4). Based upon such observations, HA and HA-based materials have been evaluated for their potential benefits in wound healing[160,199,200] or the prevention of adhesion formation following surgery.[201,202] Adhesions are caused by the formation of fibrin at the site of injury or surgery and by the failure of the activation of the fibrinolytic system. If microorganisms are entrapped in the fibrin clots, abscess formation can result.[203] The application of HA at the wound site is believed to provide a mechanical barrier between different wound surfaces, to improve peritoneal healing by stimulating mesothelial cell growth, and to reduce the inflammatory condition that might exist.[203] Although HA is generally considered safe for such applications, some reports warning about potential complications following the use of HA-based preparations have emerged.[204]

9.7.6 Anti-inflammation and Viscosupplementation Therapy

Together with corticosteroids, HA is the most commonly used agent for intra-articular injection treatment of osteoarthritis, especially in patients for whom other therapeutic possibilities have failed.[205] Apart from its potential effect on the inflammatory process (see section 9.6.3), injection of HA solutions can restore the elasticity and viscosity of the synovial fluid and provide extra "cushioning" for the joint, a therapeutic approach termed viscosupplementation. Some reports describe long-term benefits from intra-articular injection therapy with HA and rank it as at least as beneficial as intra-articular corticosteroid injection.[206,207] Others claim that there is little strong and reproducible evidence that such injections alter the progression of the disease.[205] In a meta-analysis study of the use of HA in the treatment of knee osteoarthritis, Lo *et al.* commented that intra-articular administration of HA has only limited, possibly overestimated effects compared to a placebo injection.[208] Because HA is generally considered to be non-toxic, its perceived beneficial effects are still considered a valuable adjuvant in the treatment of chronic inflammatory conditions.[209–211] However, recent reports have raised an awareness of potential side-effects or complications resulting from intra-articular HA therapy.[212,213]

9.7.7 Viscosurgery

Because of its unique viscoelastic properties, HA has found applications as an adjuvant during cataract surgery.[214] Cataract surgery involves the removal of the lens and replacing it with a new one. Such procedures have a high risk of injury to the fragile intraocular tissues. By injecting the anterior chamber with a visocelastic material like HA, the operative space is maintained and the risk of tissue damage significantly reduced (see Figure 9.12). This strategy is termed viscosurgery and in principle could be applied to other types of surgical procedures other than cataract surgery.

9.7.8 Soft-tissue Filler

HA has found applications as a soft-tissue filling agent to provide wrinkle reduction, contour improvement or volume augmentation, *e.g.*, of the lips.[215–222] Different formulations are available depending on the depth in the skin to which they will be injected and the products are designed to last for several months. Although HA is generally considered safe for this type of application, cases of allergic reactions or other complications following HA injection have been reported.[223–226]

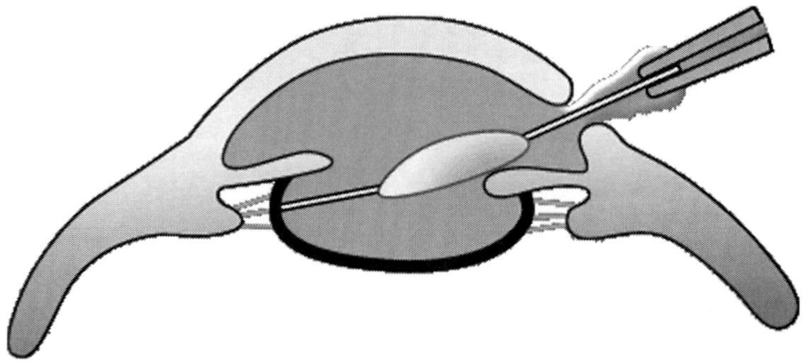

Figure 9.12 Application of HA in the surgery of intraocular lens insertion. During the procedure to remove and replace the lens, the depth of the anterior chamber is maintained by in the injection of a high viscosity solution of HA. Reproduced, with permission, from "Medical Applications of Hyaluronan" by Akira Asari and Satoshi Miyauchi at http://www.glycoforum.gr.jp/science/hyaluronan/HA12/HA12E.html.

9.8 Concluding Remarks

None of the topics discussed in this review are "closed" as the ongoing research on HA will lead to a continuous expansion of the knowledge and understanding within each section of this review.

Structure–activity relationships between HA and HABPs need to be established to further our understanding of their interactions and their relevance in normal or pathological cell physiology. This requires a more detailed knowledge of the HA-binding sites within these HABPs and an increased understanding of the conformations HA molecules can adopt under physiological conditions. Continuing research into cell surface or intracellular HABPs, more of which are likely to be discovered, or HAS will increase the appreciation of the importance of HA in cell physiology and increase its potential applications in medicine or biotechnology. In addition, the existing applications of HA or HA-based materials in the fields of pharmaceutical technology or bioengineering are bound to be expanded. So far, most approaches for the chemical derivatization of HA have involved chemical reactions on the readily available hydroxyl or carboxylate functionalities. The family of modified HA materials could be greatly expanded if a reliable enzymatic or chemical strategy could be developed to generate partially or fully deacetylated HA molecules without affecting the molecular mass of the molecules, providing readily available amino functionalities for further chemical derivatization. However, any development or evaluation of HA or HA-based materials for medical or cosmetic purposes should be aware of the reports of potential complications following HA administration that have emerged. In addition, the possibility of HA being biodegraded into cell-biologically active, *e.g.*, pro-inflammatory oligosaccharide fragments needs to be taken into consideration.

References

1. T. C. Laurent and J. R. Fraser, *Faseb J*, 1992, **6**, 2397.
2. P. L. DeAngelis, *Cellular and Molecular Life Sciences*, 1999, **56**, 670.
3. E. A. Balazs, in *Why hyaluronan has so many biological activities*, (Eds) G. Abatangelo and P. H. Weigel, Elsevier, Amsterdam, 2000, p 3.
4. P. Ghosh, *Clin Exp Rheumatol*, 1994, **12**, 75.
5. N. Itano and K. Kimata, *IUBMB Life*, 2002, **54**, 195.
6. K. P. Vercruysse and G. D. Prestwich, *Crit Rev Ther Drug Carrier Syst*, 1998, **15**, 513.
7. S. Ernst, R. Langer, C. L. Cooney and R. Sasisekharan, *Crit Rev Biochem Mol Biol*, 1995, **30**, 387.
8. R. L. Jackson, S. J. Busch and A. D. Cardin, *Physiol Rev*, 1991, **71**, 481.
9. J. Lapcik, Lubomir, L. Lapcik, S. De Smedt, J. Demeester and P. Chabrecek, *Chem Rev*, 1998, **98**, 2663.
10. M. O. Longas, C. S. Russell and X. Y. He, *Carbohydr Res*, 1987, **159**, 127.
11. M. O. Longas, J. D. Burden, J. Lesniak, R. M. Booth, J. A. McPencow and J. I. Park, *Biomacromolecules*, 2003, **4**, 189.
12. J. K. Sheehan, C. Arundel and C. F. Phelps, *Int J Biol Macromol*, 1983, **5**, 222.
13. A. J. Day and J. K. Sheehan, *Curr Opin Struct Biol*, 2001, **11**, 617.
14. J. E. Scott and F. Heatley, *Proceedings of the National Academy of Sciences of the United States of America*, 1999, **96**, 4850.
15. R. E. Turner, P. Y. Lin and M. K. Cowman, *Arch Biochem Biophys*, 1988, **265**, 484.
16. I. Hargittai and M. Hargittai, *Structural Chemistry*, 2008, **19**, 697.
17. D. C. Armstrong, M. J. Cooney and M. R. Johns, *Appl Microbiol Biotechnol*, 1997, **47**, 309.
18. D. C. Armstrong and M. R. Johns, *Applied and Environmental Microbiology*, 1997, **63**, 2759.
19. M. J. Cooney, L. Goh, P. L. Lee and M. R. Johns, *Biotechnol Prog*, 1999, **15**, 898.
20. N. Izawa, T. Hanamizu, T. Sone and K. Chiba, *Journal of bioscience and bioengineering*, **109**, 356.
21. P. H. Weigel, *IUBMB Life*, 2002, **54**, 201.
22. P. L. DeAngelis, W. Jing, M. V. Graves, D. E. Burbank and J. L. Van Etten, *Science*, 1997, **278**, 1800.
23. N. Itano, T. Sawai, O. Miyaishi and K. Kimata, *Cancer Res*, 1999, **59**, 2499.
24. T. D. Camenisch, A. P. Spicer, T. Brehm-Gibson, J. Biesterfeldt, M. L. Augustine, A. Calabro, Jr., S. Kubalak, S. E. Klewer and J. A. McDonald, *J Clin Invest*, 2000, **106**, 349.
25. A. D. Recklies, C. White, L. Melching and P. J. Roughley, *Biochem J*, 2001, **354**, 17.
26. S. Jones and A. O. Phillips, *Kidney International*, 2001, **59**, 1739.
27. T. Ohkawa, N. Ueki, T. Taguchi, Y. Shindo, M. Adachi, Y. Amuro, T. Hada and K. Higashino, *Biochim Biophys Acta*, 1999, **1448**, 416.

28. M. Viola, D. Vigetti, A. Genasetti, M. Rizzi, E. Karousou, P. Moretto, M. Clerici, B. Bartolini, F. Pallotti, G. De Luca and A. Passi, *Connect Tissue Res*, 2008, **49**, 111.
29. P. A. G. McCourt, *Matrix Biology*, 1999, **18**, 427.
30. B. M. Praest, H. Greiling and R. Kock, *Carbohydr Res*, 1997, **303**, 153.
31. G. T. Balogh, J. Illes, Z. Szekely, E. Forrai and A. Gere, *Arch Biochem Biophys*, 2003, **410**, 76.
32. M. E. Monzon, N. Fregien, N. Schmid, N. Santos-Falcon, M. Campos, S. M. Casalino-Matsuda and R. Malbran Forteza, *J Biol Chem*, 2010.
33. Z. Lurie, T. Offer, A. Russo, A. Samuni and D. Nitzan, *Free Radic Biol Med*, 2003, **35**, 169.
34. D. A. Swann, *Biochem J*, 1967, **102**, 42c.
35. R. M. Fink and E. Lengfelder, *Free Radic Res Commun*, 1987, **3**, 85.
36. G. Kreil, *Protein Sci*, 1995, **4**, 1666.
37. W. L. Hynes and S. L. Walton, *FEMS Microbiol Lett*, 2000, **183**, 201.
38. A. B. Csoka, G. I. Frost and R. Stern, *Matrix Biol*, 2001, **20**, 499.
39. G. I. Frost, T. Csoka and R. Stern, *Trends in Glycoscience and Glycotechnology*, 1996, **8**, 419.
40. B. Zhou, J. A. Weigel, L. Fauss and P. H. Weigel, *J Biol Chem*, 2000, **275**, 37733.
41. B. Zhou, J. A. Oka, A. Singh and P. H. Weigel, *J Biol Chem*, 1999, **274**, 33831.
42. E. N. Harris and P. H. Weigel, *Glycobiology*, 2008, **18**, 638.
43. E. N. Harris, J. A. Weigel and P. H. Weigel, *J Biol Chem*, 2008, **283**, 17341.
44. D. G. Jackson, R. Prevo, S. Clasper and S. Banerji, *Trends Immunol*, 2001, **22**, 317.
45. D. G. Jackson, *Trends Cardiovasc Med*, 2003, **13**, 1.
46. W. Knudson, G. Chow and C. B. Knudson, *Matrix Biol*, 2002, **21**, 15.
47. A. J. Day and G. D. Prestwich, *J Biol Chem*, 2002, **277**, 4585.
48. C. Kiani, L. Chen, Y. J. Wu, A. J. Yee and B. B. Yang, *Cell Res*, 2002, **12**, 19.
49. C. M. Milner and A. J. Day, *J Cell Sci*, 2003, **116**, 1863.
50. J. J. Enghild, I. B. Thogersen, F. Cheng, L. A. Fransson, P. Roepstorff and H. Rahbek-Nielsen, *Biochemistry*, 1999, **38**, 11804.
51. K. A. Hess, L. Chen and W. J. Larsen, *Biology of Reproduction*, 1999, **61**, 436.
52. N. H. Choi-Miura, T. Tobe, J. Sumiya, Y. Nakano, Y. Sano, T. Mazda and M. Tomita, *J Biochem (Tokyo)*, 1996, **119**, 1157.
53. N. H. Choi-Miura, K. Saito, K. Takahashi, M. Yoda and M. Tomita, *Biol Pharm Bull*, 2001, **24**, 221.
54. J. Cichy and E. Pure, *J Cell Biol*, 2003, **161**, 839.
55. C. M. Isacke and H. Yarwood, *Int J Biochem Cell Biol*, 2002, **34**, 718.
56. J. Bajorath, *Proteins*, 2000, **39**, 103.
57. T. E. I. Taher, L. Smit, A. W. Griffioen, E. J. M. Schilder-Tol, J. Borst and S. T. Pals, *J Biol Chem*, 1996, **271**, 2863.

58. S. Ilangumaran, A. Briol and D. C. Hoessli, *Blood*, 1998, **91**, 3901.
59. M. Serbulea, S. Kakumu, A. A. Thant, K. Miyazaki, K. Machida, T. Senga, S. Ohta, K. Yoshioka, N. Hotta and M. Hamaguchi, *Int J Oncol*, 1999, **14**, 733.
60. Z. Rozsnyay, *Immunology Letters*, 1999, **68**, 101.
61. L. Y. W. Bourguignon, H. Zhu, A. Chu, N. Iida, L. Zhang and M.-C. Hung, *Journal of Biological Chemistry*, 1997, **272**, 27913.
62. Y. Fujita, M. Kitagawa, S. Nakamura, K. Azuma, G. Ishii, M. Higashi, H. Kishi, T. Hiwasa, K. Koda, N. Nakajima and K. Harigaya, *FEBS Lett*, 2002, **528**, 101.
63. Y. Sohara, N. Ishiguro, K. Machida, H. Kurata, A. A. Thant, T. Senga, S. Matsuda, K. Kimata, H. Iwata and M. Hamaguchi, *Molecular Biology of the Cell*, 2001, **12**, 1859.
64. S. Oliferenko, I. Kaverina, J. V. Small and L. A. Huber, *J Cell Biol*, 2000, **148**, 1159.
65. L. Y. W. Bourguignon, H. B. Zhu, L. J. Shao and Y. W. Chen, *J Biol Chem*, 2000, **275**, 1829.
66. L. Y. Bourguignon, H. Zhu, B. Zhou, F. Diedrich, P. A. Singleton and M. C. Hung, *J Biol Chem*, 2001, **276**, 48679.
67. L. Y. Bourguignon, H. Zhu, L. Shao and Y. W. Chen, *J Biol Chem*, 2001, **276**, 7327.
68. L. Y. Bourguignon, V. B. Lokeshwar, J. He, X. Chen and G. J. Bourguignon, *Mol Cell Biol*, 1992, **12**, 4464.
69. L. Y. W. Bourguignon, V. B. Lokeshwar, X. Chen and W. G. L. Kerrick, *Journal of Immunology*, 1993, **151**, 6634.
70. L. Y. W. Bourguignon, *Current Topics in Membranes*, 1996, **43**, 293.
71. S. Tsukita, K. Oishi, N. Sato, J. Sagara, A. Kawai and S. Tsukita, *Journal of Cell biology*, 1994, **126**, 391.
72. M. Sainio, F. Zhao, L. Heiska, O. Turunen, M. den Bakker, E. Zwarthoff, M. Lutchman, G. A. Rouleau, J. Jaaskelainen, A. Vaheri and O. Carpen, *J Cell Sci*, 1997, **110**, 2249.
73. M. Hirao, N. Sato, T. Kondo, S. Yonemura, M. Monden, T. Sasaki, Y. Takai and S. Tsukita, *J Cell Biol*, 1996, **135**, 37.
74. L. Y. W. Bourguignon, H. B. Zhu, L. J. Shao, D. Zhu and Y. W. Chen, *Cell Motility and the Cytoskeleton*, 1999, **43**, 269.
75. S. Oliferenko, K. Paiha, T. Harder, V. Gerke, C. Schwarzler, H. Schwarz, H. Beug, U. Gunthert and L. A. Huber, *J Cell Biol*, 1999, **146**, 843.
76. L. Y. W. Bourguignon, Z. GunjaSmith, N. Iida, H. B. Zhu, L. J. T. Young, W. J. Muller and R. D. Cardiff, *J. Cell Physiol.*, 1998, **176**, 206.
77. Q. Yu and I. Stamenkovic, *Genes Dev.*, 1999, **13**, 35.
78. Y. Zhang, A. A. Thant, Y. Hiraiwa, Y. Naito, T. T. Sein, Y. Sohara, S. Matsuda and M. Hamaguchi, *Biochem Biophys Res Commun*, 2002, **290**, 1123.
79. Y. Kawano, I. Okamoto, D. Murakami, H. Itoh, M. Yoshida, S. Ueda and H. Saya, *J Biol Chem*, 2000, **275**, 29628.

80. E. A. Turley, L. Austen, K. Vandeligt and C. Clary, *J Cell Biol*, 1991, **112**, 1041.
81. C. L. Hall, C. Wang, L. A. Lange and E. A. Turley, *J Cell Biol*, 1994, **126**, 575.
82. C. L. Hall, L. A. Lange, D. A. Prober, S. Zhang and E. A. Turley, *Oncogene*, 1996, **13**, 2213.
83. C. L. Hall, L. A. Collis, A. J. Bo, L. Lange, A. McNicol, J. M. Gerrard and E. A. Turley, *Matrix Biol*, 2001, **20**, 183.
84. V. B. Lokeshwar and M. G. Selzer, *J Biol Chem*, 2000, **275**, 27641.
85. O. Politz, A. Gratchev, P. A. McCourt, K. Schledzewski, P. Guillot, S. Johansson, G. Svineng, P. Franke, C. Kannicht, J. Kzhyshkowska, P. Longati, F. W. Velten and S. Goerdt, *Biochem J*, 2002, **362**, 155.
86. S. B. Kumarswamy, H. N. Appaiah, R. K. Boregowda, A. Thomas, S. D. Banerjee and M. K. Kumar, *Adv Biol Res*, 2008, **2**, 6.
87. R. K. Margolis, C. P. Crockett, W. L. Kiang and R. U. Margolis, *Biochim Biophys Acta*, 1976, **451**, 465.
88. W. L. Kiang, C. P. Crockett, R. K. Margolis and R. U. Margolis, *Biochemistry*, 1978, **17**, 3841.
89. N. Grammatikakis, A. Grammatikakis, M. Yoneda, Q. Yu, S. D. Banerjee and B. P. Toole, *J Biol Chem*, 1995, **270**, 16198.
90. L. Huang, N. Grammatikakis, M. Yoneda, S. D. Banerjee and B. P. Toole, *J Biol Chem*, 2000, **275**, 29829.
91. T. B. Deb and K. Datta, *Journal of Biological Chemistry*, 1996, **271**, 2206.
92. S. Das, T. B. Deb, R. Kumar and K. Datta, *Gene*, 1997, **190**, 223.
93. B. K. Jha, D. M. Salunke and K. Datta, *J Biol Chem*, 2003, **278**, 27464.
94. V. C. Hascall, A. K. Majors, C. A. De La Motte, S. P. Evanko, A. Wang, J. A. Drazba, S. A. Strong and T. N. Wight, *Biochim Biophys Acta*, 2004, **1673**, 3.
95. V. Assmann, J. F. Marshall, C. Fieber, M. Hofmann and I. R. Hart, *J Cell Sci*, 1998, **111**, 1685.
96. M. Stojkovic, O. Krebs, S. Kolle, K. Prelle, V. Assmann, V. Zakhartchenko, F. Sinowatz and E. Wolf, *Biology of Reproduction*, 2003, **68**, 60.
97. A. Rossler and H. Hinghofer-Szalkay, *Horm Metab Res*, 2003, **35**, 67.
98. D. C. West, I. N. Hampson, F. Arnold and S. Kumar, *Science*, 1985, **228**, 1324.
99. C. M. McKee, M. B. Penno, M. Cowman, M. D. Burdick, R. M. Strieter, C. Bao and P. W. Noble, *J Clin Invest*, 1996, **98**, 2403.
100. C. M. McKee, C. J. Lowenstein, M. R. Horton, J. Wu, C. Bao, B. Y. Chin, A. M. Choi and P. W. Noble, *J Biol Chem*, 1997, **272**, 8013.
101. K. L. Jobe, S. O. Odman-Ghazi, M. M. Whalen and K. P. Vercruysse, *Immunology Letters*, 2003, **89**, 99.
102. C. C. Termeer, J. Hennies, U. Voith, T. Ahrens, J. M. Weiss, P. Prehm and J. C. Simon, *J Immunol*, 2000, **165**, 1863.
103. C. Termeer, F. Benedix, J. Sleeman, C. Fieber, U. Voith, T. Ahrens, K. Miyake, M. Freudenberg, C. Galanos and J. C. Simon, *J Exp Med*, 2002, **195**, 99.

104. C. Fieber, P. Baumann, R. Vallon, C. Termeer, J. C. Simon, M. Hofmann, P. Angel, P. Herrlich and J. P. Sleeman, *J Cell Sci*, 2004, **117**, 359.
105. H. Xu, T. Ito, A. Tawada, H. Maeda, H. Yamanokuchi, K. Isahara, K. Yoshida, Y. Uchiyama and A. Asari, *J Biol Chem*, 2002, **277**, 17308.
106. B. P. Toole, *J Intern Med*, 1997, **242**, 35.
107. S. P. Evanko, J. C. Angello and T. N. Wight, *Arterioscler Thromb Vasc Biol*, 1999, **19**, 1004.
108. B. P. Toole, *Semin Cell Dev Biol*, 2001, **12**, 79.
109. D. Marzioni, C. Crescimanno, D. Zaccheo, R. Coppari, C. B. Underhill and M. Castellucci, *European Journal of Histochemistry*, 2001, **45**, 131.
110. H. Gu, L. Huang, Y. P. Wong and A. Burd, *Experimental dermatology*, **19**, e336.
111. A. Genasetti, D. Vigetti, M. Viola, E. Karousou, P. Moretto, M. Rizzi, B. Bartolini, M. Clerici, F. Pallotti, G. De Luca and A. Passi, *Connect Tissue Res*, 2008, **49**, 120.
112. Y. Ito, S. Seno, H. Nakamura, A. Fukui and M. Asashima, *Dev Biol*, 2008, **319**, 34.
113. E. Rosines, H. J. Schmidt and S. K. Nigam, *Biomaterials*, 2007, **28**, 4806.
114. Y. Li, B. P. Toole, C. N. Dealy and R. A. Kosher, *Dev Biol*, 2007, **305**, 411.
115. N. Itano, *J Biochem*, 2008, **144**, 131.
116. P. Vabres, *Annales de dermatologie et de venereologie*, 2010, **137**(Suppl 1), S9.
117. N. Volpi, J. Schiller, R. Stern and L. Soltes, *Current medicinal chemistry*, 2009, **16**, 1718.
118. M. Wernicke, L. C. Pineiro, D. Caramutti, V. G. Dorn, M. M. Raffo, H. G. Guixa, M. Telenta and A. A. Morandi, *Mod Pathol*, 2003, **16**, 99.
119. S. Karvinen, V. M. Kosma, M. I. Tammi and R. Tammi, *Br J Dermatol*, 2003, **148**, 86.
120. S. Aaltomaa, P. Lipponen, R. Tammi, M. Tammi, J. Viitanen, J. P. Kankkunen and V. M. Kosma, *Urol Int*, 2002, **69**, 266.
121. P. Lipponen, S. Aaltomaa, R. Tammi, M. Tammi, U. Agren and V. M. Kosma, *Eur J Cancer*, 2001, **37**, 849.
122. E. Calikoglu, E. Augsburger, I. Masouye, P. Chavaz, J. H. Saurat and G. Kaya, *Dermatology*, 2002, **205**, 122.
123. J. Bohm, L. Niskanen, R. Tammi, M. Tammi, M. Eskelinen, R. Pirinen, S. Hollmen, E. Alhava and V. M. Kosma, *Journal of Pathology*, 2002, **196**, 180.
124. S. H. Hautmann, V. B. Lokeshwar, G. L. Schroeder, F. Civantos, R. C. Duncan, R. Gnann, M. G. Friedrich and M. S. Soloway, *J Urol*, 2001, **165**, 2068.
125. M. A. Anttila, R. H. Tammi, M. I. Tammi, K. J. Syrjanen, S. V. Saarikoski and V. M. Kosma, *Cancer Res*, 2000, **60**, 150.
126. R. Xing, J. A. Regezi, M. Stern, S. Shuster and R. Stern, *Oral Dis*, 1998, **4**, 241.
127. L. P. Setala, M. I. Tammi, R. H. Tammi, M. J. Eskelinen, P. K. Lipponen, U. M. Agren, J. Parkkinen, E. M. Alhava and V. M. Kosma, *British Journal of Cancer*, 1999, **79**, 1133.

128. K. Ropponen, M. Tammi, J. Parkkinen, M. Eskelinen, R. Tammi, P. Lipponen, U. Agren, E. Alhava and V. M. Kosma, *Cancer Res*, 1998, **58**, 342.
129. R. Pirinen, R. Tammi, M. Tammi, P. Hirvikoski, J. J. Parkkinen, R. Johansson, J. Bohm, S. Hollmen and V. M. Kosma, *International Journal of Cancer*, 2001, **95**, 12.
130. B. P. Toole, *Glycobiology*, 2002, **12**, 37R.
131. S. Jothy, *Clinical & Experimental Metastasis*, 2003, **20**, 195.
132. K. To, A. Fotovati, K. M. Reipas, J. H. Law, K. Hu, J. Wang, A. Astanehe, A. H. Davies, L. Lee, A. L. Stratford, A. Raouf, P. Johnson, I. M. Berquin, H. D. Royer, C. J. Eaves and S. E. Dunn, *Cancer Res*, 2010, **70**, 2840.
133. S. Ghatak, V. C. Hascall, R. R. Markwald and S. Misra, *J Biol Chem*, 2010, **285**, 19821.
134. V. Orian-Rousseau, *Eur J Cancer*, 2010, **46**, 1271.
135. H. C. DeGrendele, P. Estess and M. H. Siegelman, *Science*, 1997, **278**, 672.
136. D. Naor and S. Nedvetzki, *Arthritis Res*, 2003, **5**, 105.
137. M. Verdrengh, R. Holmdahl and A. Tarkowski, *Scand J Immunol*, 1995, **42**, 353.
138. R. Stoop, H. Kotani, J. D. McNeish, I. G. Otterness and K. Mikecz, *Arthritis Rheum*, 2001, **44**, 2922.
139. P. Teder, R. W. Vandivier, D. Jiang, J. Liang, L. Cohn, E. Pure, P. M. Henson and P. W. Noble, *Science*, 2002, **296**, 155.
140. C. Nathan, *Nature*, 2002, **420**, 846.
141. H. Levesque, N. Girard, C. Maingonnat, A. Delpech, C. Chauzy, J. Tayot, H. Courtois and B. Delpech, *Atherosclerosis*, 1994, **105**, 51.
142. S. P. Evanko, E. W. Raines, R. Ross, L. I. Gold and T. N. Wight, *Am J Pathol*, 1998, **152**, 533.
143. A. Nandi, P. Estess and M. H. Siegelman, *J Biol Chem*, 2000, **275**, 14939.
144. C. A. Cuff, D. Kothapalli, I. Azonobi, S. Chun, Y. Zhang, R. Belkin, C. Yeh, A. Secreto, R. K. Assoian, D. J. Rader and E. Pure, *J Clin Invest*, 2001, **108**, 1031.
145. D. Vigetti, A. Genasetti, E. Karousou, M. Viola, P. Moretto, M. Clerici, S. Deleonibus, G. De Luca, V. C. Hascall and A. Passi, *J Biol Chem*, 2010, **285**, 24639.
146. S. Katoh, S. Maeda, H. Fukuoka, T. Wada, S. Moriya, A. Mori, K. Yamaguchi, S. Senda and T. Miyagi, *Clin Exp Immunol*, 2010, **161**, 233.
147. G. M. Campo, A. Avenoso, S. Campo, A. D'Ascola, G. Nastasi and A. Calatroni, *Biochem Pharmacol*, 2010, **80**, 480.
148. P. Johnson and B. Ruffell, *Inflammation & allergy drug targets*, 2009, **8**, 208.
149. G. M. Turino and J. O. Cantor, *Am J Respir Crit Care Med*, 2003, **167**, 1169.
150. P. W. Noble, *Matrix Biol*, 2002, **21**, 25.
151. F. Gao, Y. Liu, Y. He, C. Yang, Y. Wang, X. Shi and G. Wei, *Matrix Biol*, 2010, **29**, 107.
152. J. A. Mack, S. R. Abramson, Y. Ben, J. C. Coffin, J. K. Rothrock, E. V. Maytin, V. C. Hascall, C. Largman and E. J. Stelnicki, *Faseb J*, 2003, **17**, 1352.

153. E. P. Buchanan, M. T. Longaker and H. P. Lorenz, *Advances in Clinical Chemistry*, 2009, **48**, 137.
154. G. Abatangelo and P. H. Weigel, in *Redefining Hyaluronan*, (Eds) G. Abatangelo and P. H. Weigel, Elsevier, Padua, Italy, 1999, p.
155. G. D. Prestwich and K. P. Vercruysse, *Pharm Sci Technol Today*, 1998, **1**, 42.
156. X. Z. Shu, Y. Liu, Y. Luo, M. C. Roberts and G. D. Prestwich, *Biomacromolecules*, 2002, **3**, 1304.
157. X. Z. Shu, Y. Liu, F. Palumbo and G. D. Prestwich, *Biomaterials*, 2003, **24**, 3825.
158. Y. Liu, X. Z. Shu, S. D. Gray and G. D. Prestwich, *J Biomed Mater Res*, 2004, **68A**, 142.
159. M. Hu, E. E. Sabelman, Y. Cao, J. Chang and V. R. Hentz, *J Biomed Mater Res*, 2003, **67B**, 586.
160. K. R. Kirker, Y. Luo, J. H. Nielson, J. Shelby and G. D. Prestwich, *Biomaterials*, 2002, **23**, 3661.
161. R. Lobmann, D. Pittasch, I. Muhlen and H. Lehnert, *J Diabetes Complications*, 2003, **17**, 199.
162. F. S. Kamelger, R. Marksteiner, E. Margreiter, G. Klima, G. Wechselberger, S. Hering and H. Piza, *Biomaterials*, 2004, **25**, 1649.
163. M. Mason, K. P. Vercruysse, K. R. Kirker, R. Frisch, D. M. Marecak, G. D. Prestwich and W. G. Pitt, *Biomaterials*, 2000, **21**, 31.
164. W. G. Pitt, R. N. Morris, M. L. Mason, M. W. Hall, Y. Luo and G. D. Prestwich, *J Biomed Mater Res*, 2004, **68A**, 95.
165. B. Heublein, E. G. Evagorou, R. Rohde, S. Ohse, R. R. Meliss, S. Barlach and A. Haverich, *Int J Artif Organs*, 2002, **25**, 1166.
166. B. Thierry, F. M. Winnik, Y. Merhi, J. Silver and M. Tabrizian, *Biomacromolecules*, 2003, **4**, 1564.
167. S. Verheye, C. P. Markou, M. Y. Salame, B. Wan, S. B. King, 3rd, K. A. Robinson, N. A. Chronos and S. R. Hanson, *Arterioscler Thromb Vasc Biol*, 2000, **20**, 1168.
168. D. D. Allison and K. J. Grande-Allen, *Tissue Engineering*, 2006, **12**, 2131.
169. A. Almond, *Cellular and Molecular Life Sciences*, 2007, **64**, 1591.
170. G. Kogan, L. Soltes, R. Stern and P. Gemeiner, *Biotechnology Letters*, 2007, **29**, 17.
171. E. Tognana, A. Borrione, C. De Luca and A. Pavesio, *Cells, Tissues, Organs*, 2007, **186**, 97.
172. J. Gaffney, S. Matou-Nasri, M. Grau-Olivares and M. Slevin, *Molecular bioSystems*, 2010, **6**, 437.
173. Y. H. Liao, S. A. Jones, B. Forbes, G. P. Martin and M. B. Brown, *Drug Deliv*, 2005, **12**, 327.
174. D. A. Ossipov, *Expert opinion on drug delivery*, **7**, 681.
175. S. T. Lim, B. Forbes, D. J. Berry, G. P. Martin and M. B. Brown, *Int J Pharm*, 2002, **231**, 73.
176. G. C. Ceschel, P. Maffei, S. L. Borgia, C. Ronchi and S. Rossi, *Drug Dev Ind Pharm*, 2001, **27**, 541.

177. M. Singh, M. Briones and D. T. O'Hagan, *J Control Release*, 2001, **70**, 267.
178. R. E. Eliaz and F. C. Szoka, Jr., *Cancer Res*, 2001, **61**, 2592.
179. D. Peer and R. Margalit, *International Journal of Cancer*, 2004, **108**, 780.
180. Y. Luo, N. J. Bernshaw, Z. R. Lu, J. Kopecek and G. D. Prestwich, *Pharm Res*, 2002, **19**, 396.
181. Y. Luo and G. D. Prestwich, *Bioconjugate Chemistry*, 1999, **10**, 755.
182. V. M. Platt and F. C. Szoka, Jr., *Molecular pharmaceutics*, 2008, **5**, 474.
183. A. K. Yadav, P. Mishra and G. P. Agrawal, *J Drug Target*, 2008, **16**, 91.
184. Y. H. Yun, D. J. Goetz, P. Yellen and W. Chen, *Biomaterials*, 2004, **25**, 147.
185. E. J. Oh, K. Park, K. S. Kim, J. Kim, J. A. Yang, J. H. Kong, M. Y. Lee, A. S. Hoffman and S. K. Hahn, *J Control Release*, 2010, **141**, 2.
186. M. B. Brown and S. A. Jones, *J Eur Acad Dermatol Venereol*, 2005, **19**, 308.
187. S. Sillanpaa, M. A. Anttila, K. Voutilainen, R. H. Tammi, M. I. Tammi, S. V. Saarikoski and V. M. Kosma, *Clin Cancer Res*, 2003, **9**, 5318.
188. E. J. Franzmann, G. L. Schroeder, W. J. Goodwin, D. T. Weed, P. Fisher and V. B. Lokeshwar, *International Journal of Cancer*, 2003, **106**, 438.
189. J. T. Posey, M. S. Soloway, S. Ekici, M. Sofer, F. Civantos, R. C. Duncan and V. B. Lokeshwar, *Cancer Res*, 2003, **63**, 2638.
190. D. T. Rein, K. Roehrig, T. Schondorf, A. Lazar, M. Fleisch, D. Niederacher, H. G. Bender and P. Dall, *J Cancer Res Clin Oncol*, 2003, **129**, 161.
191. P. Tangkijvanich, P. Kongtawelert, P. Pothacharoen, V. Mahachai, P. Suwangool and Y. Poovorawan, *Asian Pac J Allergy Immunol*, 2003, **21**, 115.
192. F. Stickel, G. Poeschl, D. Schuppan, C. Conradt, A. Strenge-Hesse, F. S. Fuchs, W. J. Hofmann and H. K. Seitz, *Eur J Gastroenterol Hepatol*, 2003, **15**, 945.
193. L. A. Kopke-Aguiar, J. R. Martins, C. C. Passerotti, C. F. Toledo, H. B. Nader and D. R. Borges, *Acta Trop*, 2002, **84**, 117.
194. A. Grzeszczuk, S. Chlabicz and A. Panasiuk, *Rocz Akad Med Bialymst*, 2002, **47**, 80.
195. H. A. Wyatt, A. Dhawan, P. Cheeseman, G. Mieli-Vergani and J. F. Price, *Arch Dis Child*, 2002, **86**, 190.
196. M. Stojkovic, S. Kolle, S. Peinl, P. Stojkovic, V. Zakhartchenko, J. G. Thompson, H. Wenigerkind, H. D. Reichenbach, F. Sinowatz and E. Wolf, *Reproduction*, 2002, **124**, 141.
197. M. Lane, J. M. Maybach, K. Hooper, J. F. Hasler and D. K. Gardner, *Molecular Reproduction and Development*, 2003, **64**, 70.
198. J. Block, L. Bonilla and P. J. Hansen, *Theriogenology*, 2009, **71**, 1063.
199. K. Ballard and A. J. Cantor, *Ostomy Wound Manage*, 2003, **49**, 37.
200. J. R. Vazquez, B. Short, A. H. Findlow, B. P. Nixon, A. J. Boulton and D. G. Armstrong, *Diabetes Res Clin Pract*, 2003, **59**, 123.
201. G. Acunzo, M. Guida, M. Pellicano, G. A. Tommaselli, A. Di Spiezio Sardo, G. Bifulco, D. Cirillo, A. Taylor and C. Nappi, *Human Reproduction*, 2003, **18**, 1918.

202. M. Pellicano, S. Bramante, D. Cirillo, S. Palomba, G. Bifulco, F. Zullo and C. Nappi, *Fertil Steril*, 2003, **80**, 441.
203. M. M. Reijnen, R. P. Bleichrodt and H. van Goor, *Br J Surg*, 2003, **90**, 533.
204. F. H. Remzi, M. Oncel, J. M. Church, A. J. Senagore, C. P. Delaney and V. W. Fazio, *Am Surg*, 2003, **69**, 356.
205. I. Uthman, J. P. Raynauld and B. Haraoui, *Postgrad Med J*, 2003, **79**, 449.
206. F. Tasciotaoglu and C. Oner, *Clin Rheumatol*, 2003, **22**, 112.
207. S. S. Leopold, B. B. Redd, W. J. Warme, P. A. Wehrle, P. D. Pettis and S. Shott, *J Bone Joint Surg Am*, 2003, **85-A**, 1197.
208. G. H. Lo, M. LaValley, T. McAlindon and D. T. Felson, *Jama*, 2003, **290**, 3115.
209. M. I. Hamburger, S. Lakhanpal, P. A. Mooar and D. Oster, *Semin Arthritis Rheum*, 2003, **32**, 296.
210. J. F. Hammesfahr, A. B. Knopf and T. Stitik, *Am J Orthop*, 2003, **32**, 277.
211. J. P. Raynauld, C. Buckland-Wright, R. Ward, D. Choquette, B. Haraoui, J. Martel-Pelletier, I. Uthman, V. Khy, J. L. Tremblay, C. Bertrand and J. P. Pelletier, *Arthritis Rheum*, 2003, **48**, 370.
212. M. Chen, B. Qiu and J. Kong, *Chin Med J (Engl)*, 2000, **113**, 189.
213. S. S. Leopold, W. J. Warme, P. D. Pettis and S. Shott, *J Bone Joint Surg Am*, 2002, **84-A**, 1619.
214. K. L. Goa and P. Benfield, *Drugs*, 1994, **47**, 536.
215. D. R. Jordan, *Can J Ophthalmol*, 2003, **38**, 285.
216. M. H. Gold, *Clinical interventions in aging*, 2007, **2**, 369.
217. R. J. Rohrich, A. Ghavami and M. A. Crosby, *Plastic and Reconstructive Surgery*, 2007, **120**, 41S.
218. I. Bogdan Allemann and L. Baumann, *Clinical interventions in aging*, 2008, **3**, 629.
219. R. S. Narins, S. H. Dayan, F. S. Brandt and E. K. Baldwin, *Dermatol Surg*, 2008, **34**(Suppl 1), S2.
220. J. S. Dover, M. G. Rubin and A. C. Bhatia, *Dermatol Surg*, 2009, **35**(Suppl 1), 322.
221. C. D. Humphrey, J. P. Arkins and S. H. Dayan, *Aesthetic surgery journal/ the American Society for Aesthetic Plastic surgery*, 2009, **29**, 477.
222. T. I. Kwak, M. Oh, J. J. Kim and D. G. Moon, *The journal of sexual medicine*, 2010.
223. M. J. Fernandez-Acenero, E. Zamora and J. Borbujo, *Dermatol Surg*, 2003, **29**, 1225.
224. J. R. Lupton and T. S. Alster, *Dermatol Surg*, 2000, **26**, 135.
225. S. Y. Huh, S. Cho, K. H. Kim, J. S. An, C. H. Won, S. E. Chang, M. W. Lee, J. H. Choi and K. C. Moon, *Annals of dermatology*, 2010, **22**, 81.
226. S. T. Arron and I. M. Neuhaus, *Journal of cosmetic dermatology*, 2007, **6**, 167.

CHAPTER 10

Chitin and Chitosan: Sources, Production and Medical Applications

THOMAS KEAN[a] AND MAYA THANOU[b]

[a] Benaroya Research Institute at Virginia Mason Hospital, Seattle, WA, 98101, USA; [b] Institute of Pharmaceutical Science, King's College London, Franklin-Wilkins Building, 150 Stamford Street, London, SE1 9NH, UK

10.1 Introduction

Chitin is the structural component found in the exoskeleton of many arthropods including insects and crustaceans; it is also found in the cell wall of fungi.[1–3] It is the second most abundant biopolymer after cellulose.[4] Chitin was originally termed fungine when discovered in mushrooms by Braconnot in 1811.[3,5] Later, in 1823, when found in the elytrum of the cockchafer beetle by Odier, it was named chitin.[3,6] This is derived from the Greek word *chitos,* meaning coat. The main commercial source of chitin is shell waste from the sea food industry.[1] Chitin is a structural polysaccharide that is almost always associated with protein.[7] Chitosan is made from chitin through a process of deacetylation; usually hot alkali is used to remove proteins followed by demineralization using acid at room temperature.[8] Although chitosan is reported in fungi in nature,[9] the predominant source is through the deacetylation of shell waste from the canning industry. The chemical structures of chitin and chitosan are represented in Figure 10.1.

RSC Polymer Chemistry Series No. 1
Renewable Resources for Functional Polymers and Biomaterials
Edited by Peter A. Williams

Published by the Royal Society of Chemistry, www.rsc.org

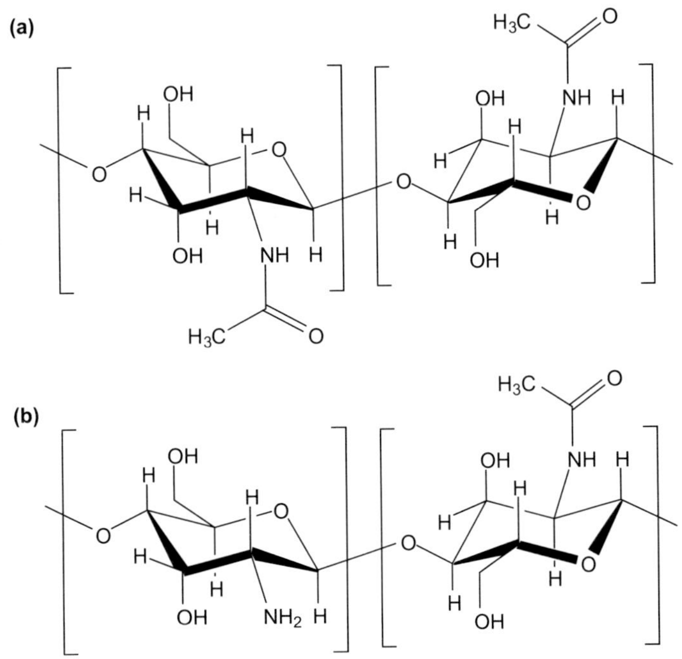

Figure 10.1 Chemical structure of chitin and chitosan. The structures of chitin (a) and chitosan (b) differ only in their degree of *N*-acylation; chitosan is generally more than 40% deacetylated whilst chitin is generally less than 40% deacetylated.

The terms chitin and chitosan may be confusing as they refer to a broad range of different but structurally similar polymers. Chitins (Figure 10.1a) are homopolymers of β-(1→4) linked *N*-acetyl-glucosamine ((1,4)-linked 2acetamido-2-deoxy-β-D-glucan) which have various molecular weights (M_w) (1.03×10^6 Da to 2.5×10^6 Da).[1] Chitosans (Figure 10.1b) are co-polymers of β-(1→4) linked glucosamine ((1,4)-linked 2-amino-2-deoxy-β-D-glucan) and *N*-acetylglucosamine, with varying degrees of deacetylation (40–99%) and molecular weights.[1,10] Chitosan is approved in Japan, Italy and Finland in weight-loss products,[11] and in a wound healing application in the USA.[12] It is widely accepted as a biodegradable, biocompatible[13,14] and 'safe' material[15,16] although applications for "generally regarded as safe" status in the USA have been withdrawn. Chitosan's biodegradability through lysozyme digestion in humans is dependent on the degree of deacetylation, with highly deacetylated chitosans being more resistant to degradation.[17] Interest, expressed by the volume of published research, in chitosan has increased exponentially over the last 20 years (Figure 10.2).

By far the most widely researched and hottest topics involving chitosan are in the field of medicine. The medical applications of this versatile biopolymer are diverse. These include gene delivery,[18–20] drug delivery,[21–23] bone/wound/tissue repair,[24–26] artificial skin,[27,28] pharmaceutical excipients[11] and dietary supplements.[29,30] Chitosan has also been utilized in waste water treatment,[31,32] cosmetics,[33] textiles,[34,35] agriculture[36] and anti-microbial food packaging.[37]

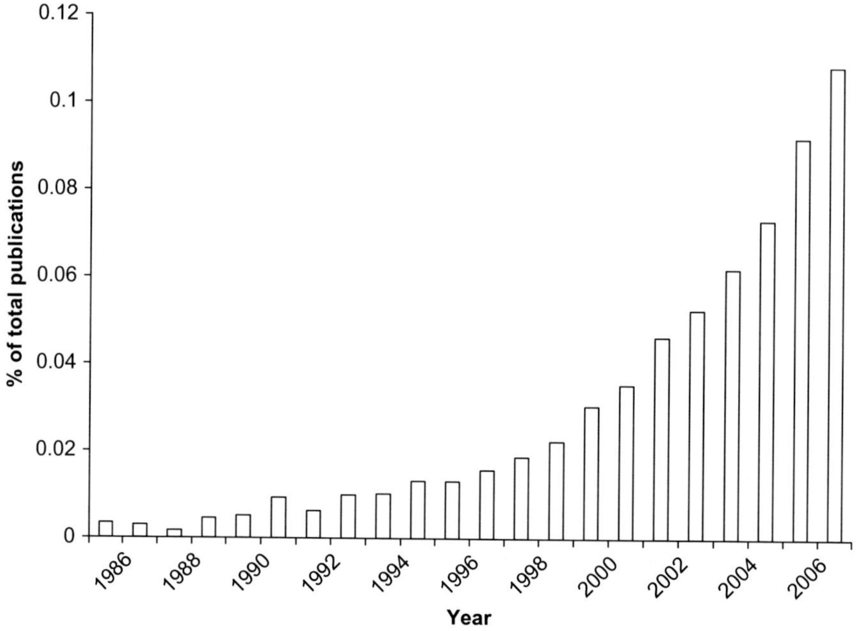

Figure 10.2 Chitosan publications over the past 20 years. Chart of chitosan publications returned when queried with "chitosan", expressed as a percentage of the total publications indexed. Total number of "chitosan" publications = 3070 with 619 in 2005 (www.ncbi.nlm.nih.gov/entrez).

10.2 Biomedical Applications of Chitin and Chitosan Materials

10.2.1 Chitosan-based Gene Delivery Systems

Gene delivery represents one of the most interesting applications of chitosan materials. Chitosan-based DNA carriers have achieved substantial gene transfer efficiency both *in vitro* and *in vivo*.[14,38–48] Chitosan forms polyelectrolyte complexes (polyplexes) with DNA due to the electrostatic interactions of the positive charge with the negative charge of phosphate groups in DNA. It is suggested that the transfection efficiency of chitosan polyplexes is related to the nitrogen : phosphorus (N : P) ratio, degree of chitosan deacetylation, polymer chain length, salt concentration (and type) in solutions used for complex formation, complex size/shape, cell type transfected, duration of transfection and time from transfection to harvesting or analysis of the protein product.

Chitosan has cell membrane-perturbing properties as it destabilizes the phospholipid bilayer.[49] Polyplexes of plasmid DNA (pDNA) with chitosan were investigated by Erbacher *et al.* who found that pDNA was condensed to 50–100 nm sized particles having donut or rod-like structures.[20] Chitosans with a higher degree of deacetylation interact more effectively with DNA, with 6–9 positive charges per chain necessary for interaction.[50] However, efficient interaction does not translate to efficient transfection as pDNA must later be released from the complex for transcription.

The N : P ratio is based on the number of nitrogen residues in the polymer and the number of phosphate residues in the DNA; it is approximately the same as a monomer molar ratio. However, it should not be assumed that every phosphate or nitrogen is charged or available for interaction. Investigation of N : P ratios has found 3 : 1 (N : P) to be the most efficient for 40 kDa chitosan (>85% deacetylated) transfection of human pancreatic cancer cells.[48] Similar results were found in human embryonic kidney (HEK293) cells with 190 kDa chitosan where 3 : 1 (N : P) was the most efficient ratio tested (85% deacetylated).[43] This was also true in human cervical cancer (HeLa) cells with 70 kDa chitosan.[20] Polymer chain length has some influence on transfection efficiency, as with other polymers: 40 kDa was more efficient compared to 1 kDa and 84 kDa chitosans.[48] However, a similar gene expression was found across a large M_w range (31 kDa and 170 kDa) chitosan with 99% deacetylation.[43]

A pioneering application of chitosan was reported in 1999 where Roy *et al.* used chitosan to form polyplexes with a plasmid containing the dominant peanut allergen gene (pCMVArah2). This was used in oral vaccination to provide immune protection in a murine model.[44] Since this study, mucosal immunization using chitosan–DNA polyplexes has been well accepted.[38,42,51–53]

For local lung immunization, intra-nasal delivery of polyplexes is the preferred route of administration. This has achieved vaccination of BALB/c mice with a plasmid encoding the CTL (cytotoxic T lymphocyte) epitope of respiratory syncytial virus; after immunization, a substantial reduction in viral load was observed.[53]

Fully deacetylated chitosan oligomers (1.2–10 kDa; $n = 10$–50) with low polydispersity (1.01–1.09) were investigated for their transfection efficiency.[39] Chitosan oligomers of <14 monomers formed unstable polyplexes.[39,41] Chitosans with 36–50 monomers were the most efficient in transfecting HEK293 cells.[39] However, *in vivo*, chitosans of 15–21 monomers (2.4–3.4 kDa) were more efficient.[39] Oligomeric chitosan's (5 kDa) interaction with pDNA was investigated at pH 5.5 and pH 12.0 by circular dichroism and atomic force microscopy.[40] It was found that at pH 5.5 strong interactions occurred, but at pH 12.0 only weak interactions were present. Spherical particles were formed at charge ratios greater than 2 : 1 (N : P).[40] Chitosan oligomers have recently been described as non-mutagenic and non-genotoxic with a maximum tolerated dose >10 g kg^{-1} in mice.[54] When administered at 3 g kg^{-1} no adverse effects and no significant hematological, clinical or histopathological changes were observed.[54]

Chitosan's main weakness is its limited solubility at physiological pH; this has been overcome through chemical derivatization. For instance, chitosan's quaternization (also known as trimethylation) of the amino group has yielded water-soluble derivatives across a range of molecular weights.[10,55,56] The availability of the amino group for modification[57,58] is one of chitosan's strengths; derivatives applied in gene delivery are listed in Table 10.1. From the table it can be seen that chemical modifications have been performed to

Table 10.1 Modifications of chitosan for gene delivery applications.

Modification	*R group*	*Improvement conferred*	*Reference*
Quaternization/ trimethylation	$(CH_3)_3$	Solubility at a wide pH range, controlled cationic character	10, 41, 134
Urocanic acid	—	pH dependent endosomal membrane rupture	61
Deoxycholic acid	—	Hydrophobic modification to achieve self aggregates	13, 62
5β-cholanic acid	—	Hydrophobic modification to achieve self aggregates	63
Alkylation (C8–16 optimum)	Series from C_4H_9 to $C_{16}H_{33}$	Hydrophobic interaction with DNA	64
Poly(ethylene glycol) (PEG)	$H(OCH_2CH_2)_nOH$ 2000, 5000, 5100, 3800 Da	Prevents aggregation of polyplexes after freeze drying, also used as a spacer	67, 123
PEG	$H(OCH_2CH_2)_nOH$ M_w 5000	Shielding of positive charge and improvement of serum polyplex stability	135
Galactosylation	Lactobionic acid attached through 6-*O*-carboxyl group on trimethyl chitosan as well as through *N*-modification	Asialoglycoprotein receptor targeting	18, 19, 65
Transferrin	Attachment through PEG spacer	Transferrin receptor targeting	67 (60)
KNOB (C-terminal globular domain of adenoviral capsid)	Attachment through PEG spacer	Coxsackie Virus and adenovirus receptor targeting	67 (60)

improve physicochemical properties, to attach a targeting ligand or as a stealth layer (poly(ethylene glycol) modification).

The quaternization of chitosan oligomers (Figure 10.3) produces a non-viral vector with controllable cationic character which is able to transfect mammalian cell lines.[10,41] Oligomeric chitosan derivatives are named trimethylated oligomers (TMOs) and polymeric derivatives are trimethylated chitosan (TMC). The degree of trimethylation is the percentage of glucosamine monomers which are trimethylated and is assessed by ^1H and ^{13}C NMR.[59] Chitosan oligomers were chosen because of their potential for faster cell clearance after transfection. The oligomer derivatives were found to be more efficient than unmodified oligomeric chitosan, and transfection was not affected by the presence of serum.[41] The optimum oligomer transfection efficiency was achieved at 44% degree of trimethylation.[10] Polyplexes formed with trimethyl oligomers were 45–160 nm spherical particles (Figure 10.4).

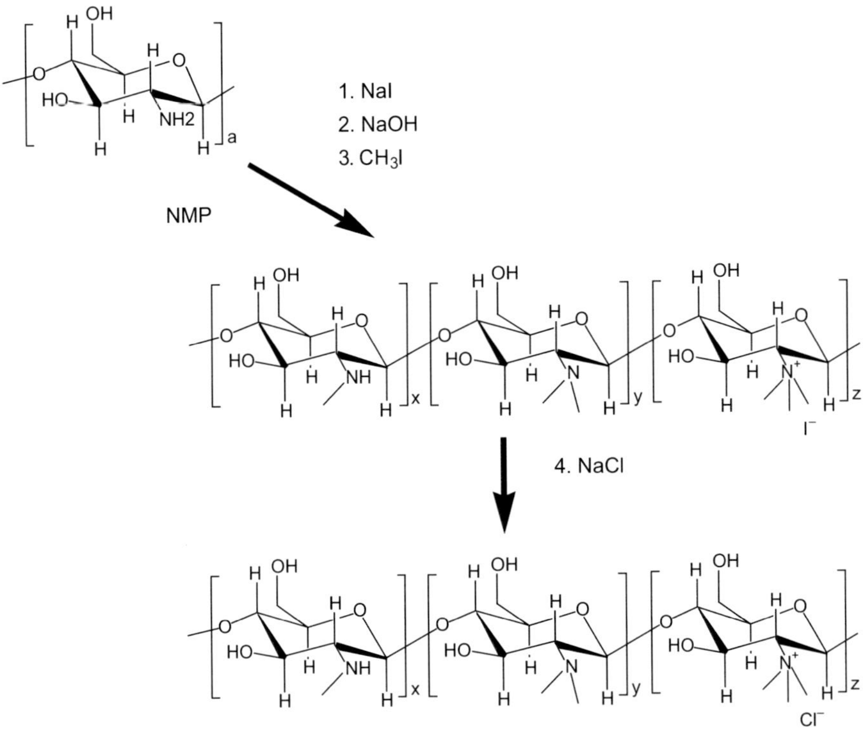

Figure 10.3 Trimethylation of chitosan. Reaction scheme of chitosan trimethylation followed by counter ion replacement. The reaction was performed at 60 °C and stirred under reflux conditions (adapted from Kean *et al.*[10]).

TMOs interact with DNA at low ratios and plasmid retardation in agarose gel is seen (Figure 10.5). Derivatives were able to transfect human breast cancer cells (MCF7) with higher efficiency than polyethylenimine (PEI) at the same w/w ratio although this was not an optimum ratio for either TMOs or PEI[10] (Figure 10.6). Increasing trimethylation altered the effect of the chitosan on cell viability but not significantly in oligomer (3–6 kDa) derivatives.[10]

The modification of chitosan with urocanic acid (Table 10.1, Figure 10.7) produces a derivative that contains an imidazole ring; this has a pK_a of 6.9 which is slightly higher than that of chitosan (p$K_a \sim 6.5$).[60] The addition of the imidazole ring provides a proton sponge effect enhancing endosomal escape. Cell viability was not detrimentally affected by incubation with urocanic chitosan (up to 30 μg ml^{-1}).[61] The transfection efficiency of human kidney cells (293T) was increased 10–100 fold over the native chitosan at similar N : P ratios.[61]

The hydrophobic modification of chitosan with alkyl/deoxycholic acid or cholanic acid produced self-aggregating chitosans; the size of the complexes was strongly molecular weight-dependent. The optimum transfection efficiency with deoxycholic acid derivatives was found with a 40 kDa derivative although transfection was found to be highly-serum dependent.[62] A ten-fold increase in

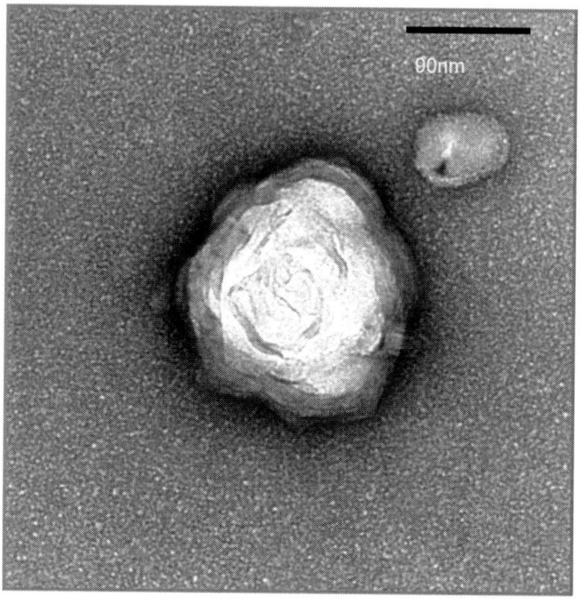

Figure 10.4 Scanning electron micrograph of TMO–plasmid polyplex. Negatively stained pictures of TMO : pGL3 luc (10 : 1 w/w) polyplexes having approximate sizes of 45–160 nm. Scale bar represents 10 nm.

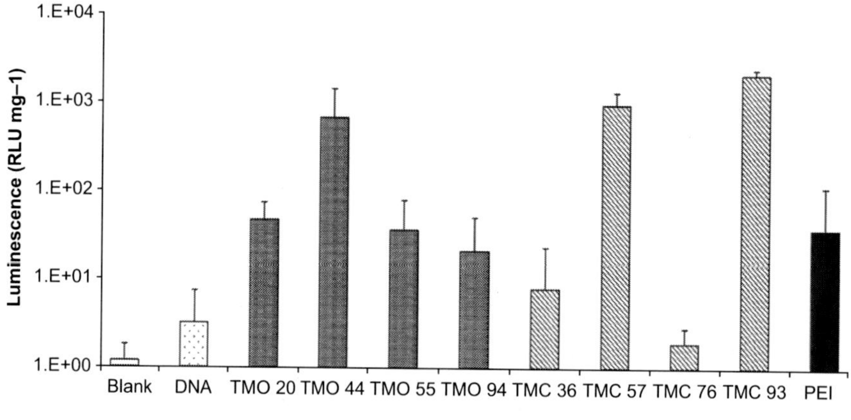

Figure 10.5 Transfection of MCF-7 cells with TMO. Transfection efficiencies of chitosan derivatives and PEI in MCF-7 cells. The ratio of the cationic vector : DNA is 10 : 1 (w/w). Average transfection efficiencies are expressed as relative light units (RLU) per mg protein. Data represent mean ($n = 4$) ± standard deviation. From Kean *et al.*[10]

the transfection efficiency of African green monkey kidney fibroblast (COS-1) cells was observed in cholanic acid-modified glycol chitosan.[63] Alkylation (with carbon chains 8–16 carbons long) of chitosan doubled its transfection efficiency in mouse skeletal muscle (C_2C_{12}) cells.[64]

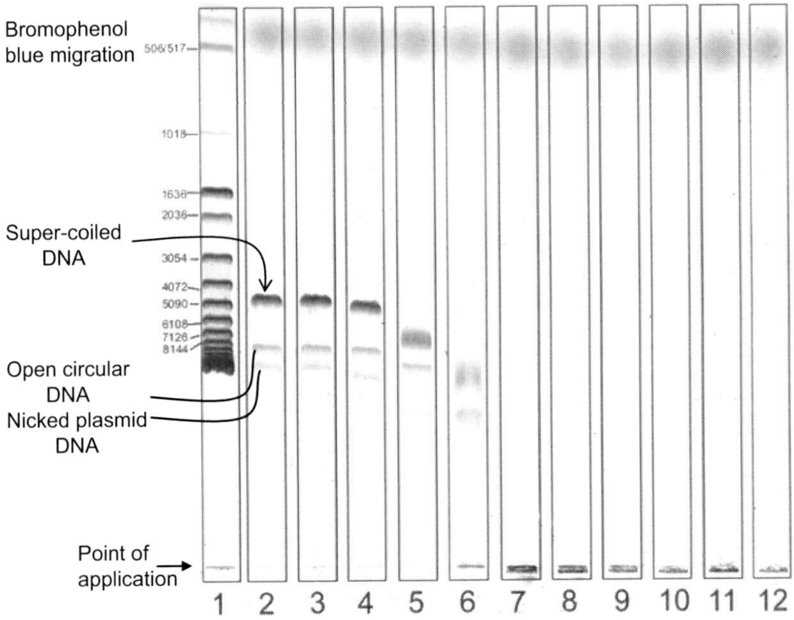

Figure 10.6 Polyplex analysis by agarose gel electrophoresis. The data show polyplexes produced with increasing ratio of TMO51 : pGL3 luc (w/w). 1 = 1 kb ladder (Invitrogen). 2 = pGL3 luc. 3 = 0.01 : 1 TMO51 : pGL3 luc. 4 = 0.1 : 1 TMO51 : pGL3 luc. 5 = 0.5 : 1 TMO51 : pGL3 luc. 6 = 1 : 1 TMO51 : pGL3 luc. 7 = 2 : 1 TMO51 : pGL3 luc. 8 = 3 : 1 TMO51 : pGL3 luc. 9 = 4 : 1 TMO51 : pGL3 luc. 10 = 14 : 1 TMO51 : pGL3 luc. 11 = 20 : 1 TMO51 : pGL3 luc. 12 = 30 : 1 TMO51 : pGL3 luc. Plasmid retardation can be seen at 0.5 : 1 TMO : DNA and complete retardation occurs at ratios higher than 2 : 1 TMO : DNA.

Within gene therapy chitosan has been targeted to the asialoglycoprotein receptor,[18,19,65,66] the Coxsackie Adenovirus Receptor (CAR) and the transferrin receptor.[67]

Trimethylated chitosan polymer derivatives were further modified with chloroacetic acid and the effect of galactose conjugates[18] and antennary galactose conjugates[19] on transfection of human hepatoma (HepG2) cells studied. Galactose-targeted trimethyl chitosan produced a small (although statistically insignificant) increase in transfection over trimethyl chitosan with a larger increase being seen in tetra-antennary galactose trimethyl chitosan-transfected cells.[18,19] The competitive inhibition of transfection with 50 mM lactose showed a statistically significant decrease indicating that specific targeting had been achieved.[19] Transferrin conjugation increased transfection by up to 4-fold whereas KNOB-protein (to target the CAR) conjugation increased transfection up to 130-fold.[67]

Chitosan and its derivatives are still less efficient than viral vectors but they provide a much more controllable delivery system. They have a better

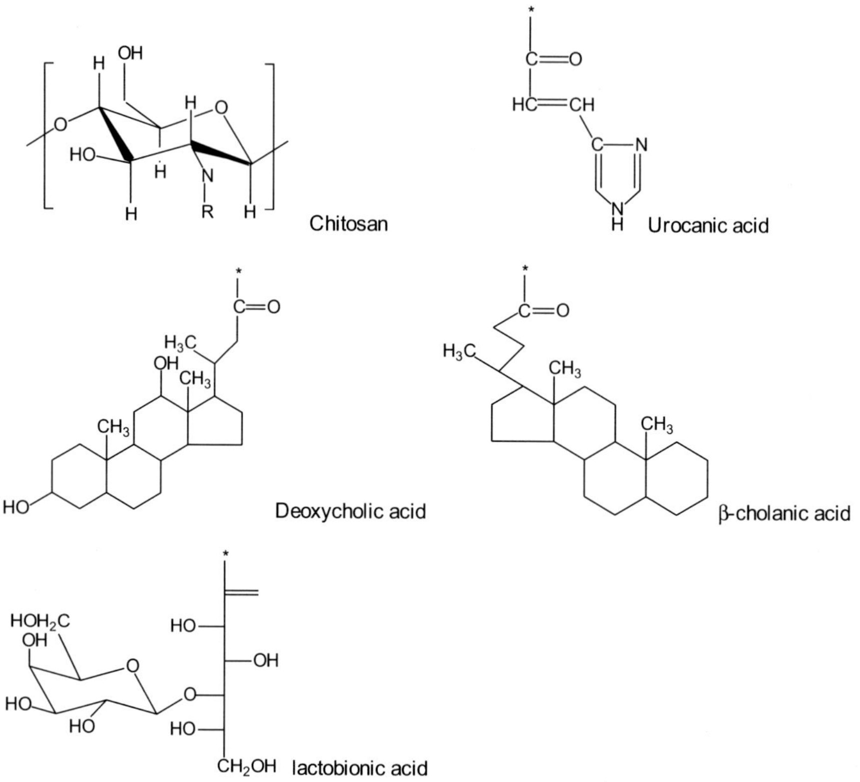

Figure 10.7 Chemical structures of chitosan and main substituents for derivatives used in gene delivery.

biocompatibility than viral vectors and have less potential to cause harm. It is expected that further modifications to delivery systems formulated from chitosan will increase transfection efficiency and ultimately result in a robust clinical application.

10.2.2 Chitosan-based Materials for Wound Repair

Chitosan is a structural biomaterial of natural origin; as such it is perhaps unsurprising that it has found broad application in tissue and bone repair. Chitin was identified as a suitable material for degradable sutures in 1986.[68] Since then chitosan has been considered as a functional biopolymer and investigated in veterinary applications.[17] It has been utilized to achieve tissue regeneration in periodontal one-wall intrabony defects in beagle dogs using non-woven chitosan membranes.[69] This follows on from research into open wounds in beagle dogs, in which chitosan was applied after wounding.[70] Chitosan was found to promote wound healing which was attributed to the enhancement of inflammatory cell function such as polymorphonuclear leukocytes (PMN), macrophages and fibroblast cells.[71] Chitosan improved the cellular granulation and the thickness of tissue covering the wounds.[70,71]

The US Food and Drug Administration (FDA) approval of chitosan as a wound dressing[72] was made through an expedient process after this dressing had shown efficacy in the treatment of liver hemorrhage in swine.[73] HemCon® bandages are made from chitosan fibers in the form of a patch that is used as a wound dressing. Their mechanism of action is due to their adhesive-like properties when in contact with blood, which promotes wound sealing and controls bleeding. The manufacturers state that this mechanism is due to chitosan's positive charge as it attracts red blood cells. The red blood cells create a cover over the wound as they are drawn into the bandage. In the treatment of severe injury, chitosan (in the form of HemCon®) may not be the most appropriate battlefield dressing for arterial bleeds as it failed in several tests. However, it may be more effective in surface injuries.[12] An advantage of the chitosan bandage over others was that it had no side effects, whilst others may cause tissue damage. Other advantages such as its bacteriostatic properties were not discussed. In the battlefield, in 97% (62 cases) of the treatments which used HemCon®, bleeding was completely stopped or it greatly improved bleeding control; the wounds ranged in severity.[74]

The mechanism of chitosan's haemostatic potential is not fully understood. When chitosan (80% deacetylated)-coated aminopropyltriethoxy-silane-coated silicon was exposed to serum or plasma, a weak, transient activation of the complement system and no activation of the intrinsic pathway were observed. Upon further acetylation of the polysaccharide coating it turned into a potent activator of the alternative complement pathway.[75] It is hypothesized that the haemostatic potential of chitosan is independent of the classical coagulation mechanism and that coagulation is produced through interaction of the chitosan with erythrocytes.[76]

Chitosan has also been employed as a composite material with collagen[28] and tropocollagen.[77] These combinations serve to increase the biocompatibility of the collagen scaffold through decreasing collagen degradation. In the case of collagen, the chitosan–collagen was coated with glutaraldehyde (up to 0.25%) to form cross-links and used as a skin substitute.[28] This cross-linked scaffold provides a stable, biocompatible surface that enhances cell growth. In the case of the tropocollagen–chitosan combination, good biocompatibility was achieved with the mixture having lower thrombogenicity. The tropocollagen–chitosan fibers were formed by spinning the mixture through a viscose-type spinneret into an aqueous solution of ammonia (5%) and ammonium sulfate (40–43%). These fibers were then N-modified with several acyl chains which resulted in changes in fiber strength (tenacity), density (titer) and elongation.[77]

There are several reports on the antibacterial properties of chitosan[34,78–81] which appear to be related to glucosamine's positive charge interacting with the negative cell wall[80], but there are probably several mechanisms through which chitosan inhibits bacteria including sequestering lipoteichoic acid.[82] Chitosan-coated cotton was an effective inhibitor of the bacterial growth of

Pseudomonas aeruginosa and methicillin-sensitive *Staphylococcus aureus* under both wet and dry conditions. Inhibition of methicillin-resistant *S. aureus* was only effective under wet conditions. The addition of organic matter to the system showed little effect on the antimicrobial activity of the chitosan-coated cotton.[34]

10.2.3 Chitosan-based Materials for Artificial Skin

Chitosans have found applications in most areas of tissue engineering. In the treatment of burns, epidermal growth factor (EGF) was embedded in a chitosan gel and used to treat burns on the backs of Sprague–Dawley rats.[83] This produced increased cell proliferation compared with an EGF solution alone and the epithelization of the tissue was increased.[83]

An increased hydrophilicity of chitosan films was achieved through treatment with argon-plasma. This treatment increased human skin-derived fibroblast cell attachment and proliferation on the chitosan films. Chitosan has a similar structure to glycosaminoglycans which can promote scarless wound healing, and it is suggested that the chitosan films could be used for this purpose.[27]

Collagen–chitosan scaffolds were prepared through cross-linking with glutaraldehyde. A dermal human fibroblast culture was supported *in vitro* and cell infiltration and proliferation were promoted. The scaffold retained the cytocompatibility of collagen with increased stability due to the cross-linkage with chitosan. *In vivo* data showed excellent integration of the scaffold but after 28 days, biodegradation of a 0.25% glutaraldehyde cross-linked scaffold was incomplete.[28]

10.2.4 Chitosan-based Materials for Bone and Cartilage Repair

Bone and cartilage have also been treated with chitosans with some success. For instance, when chitosan solution (0.1%) was injected into the rat knee articular cavity it provoked intra-articular fibrous formation and increased chondrocyte proliferation.[84] These factors suggest that chitosan promotes wound healing and growth of cartilage. An *in situ* gelling chitosan–glycerol phosphate was developed and employed in rabbit cartilage defects and was found to support chondrocyte formation of collagen. Chondrocytes were both delivered embedded in the chitosan gel and allowed to grow upon it. The gel was stable in defects for 1 day *in vivo* and up to 1 week in rabbit osteochondral defects.[85] Whilst neither of these studies looked at durable regeneration, it can be envisioned that chitosan, perhaps in combination with growth factors, could achieve an effective solution in a difficult, avascular area.

A poly(lactide co-glycolide) (PLGA) scaffold, a material which is widely used in bone repair, was coated with chitosan (880 kDa, 90% deacetylated), *N*-succinyl chitosan and collagen.[86] The attachment and differentiation of osteoblasts from rat calvaria stroma was studied; chitosan decreased cell attachment but increased the differentiation of osteoblasts.[86] It is possible that a composite material would produce a synergistic effect, increasing attachment and promoting cell differentiation.

10.2.5 Chitosan's Application as a Functional Material in Mucosal Drug Delivery

Chitosan's characteristics have seen it employed as a drug delivery vehicle both as a controlled release matrix and as a functional biomaterial. Chitin is known to have at least 3 structural forms (α, β, γ; see Figure 10.8) which may influence the properties of the chitosan which is derived from it. These forms relate to the packing of the polymer chain: α-chitin in an antiparallel arrangement which is the most stable and common, β-chitin has a parallel arrangement and γ-chitin has a mixed arrangement *i.e.* two parallel chains for each antiparallel chain or *vice versa* (Figure 10.8).[2,87,88] The physicochemical properties of chitinous materials allow several technologies to be utilized. These include its use as a tablet excipient/diluent, film, gel, fiber, matrix and microsphere/nanoparticle. These dosage forms are then utilized in several ways: oral and topical delivery, injectables and in implantable devices.

Different chitosan salt forms (aspartic, glutamic, citric, hydrochloric and lactic) were tested for their effect on the release of the widely used anti-inflammatory agent, sodium diclofenac. In this study, glutamic and aspartic chitosan salts gave the best release profiles for colon delivery (*i.e.* low release at acidic pH and high at alkaline pH).[89] Wet or dry granular formulation of

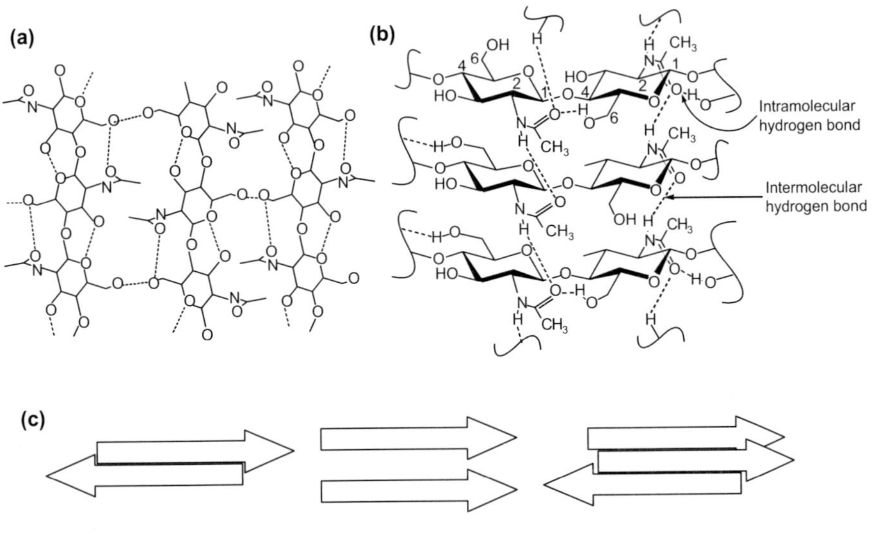

Figure 10.8 Structural forms of chitin: (a) *xy* projection of α-chitin; (b) *zy* projection of α-chitin; (c) general diagram of three chitin forms. Compiled and adapted from 2 and 88.

chitosan with theophylline found that the lower the initial viscosity of the polymer/drug blend in the wet state, the faster the release of theophylline.[90] The formulation of spray-dried chitosan acetate produced amorphous spherical particle agglomerates which were smaller than 75 μm in diameter.[91] The cumulative release of salicylic acid and theophylline from tablets prepared using spray-dried chitosan acetate (1–3%) was studied in distilled water, phosphate buffer (pH 6.8) and 0.1 M HCl; release was complete in 6, 16 and 24 h respectively.[91]

A recent phase II clinical trial using chitosan–holmium-166 for percutaneous injection in the treatment of poor surgical candidates with hepatocellular carcinoma reported safe and efficacious results. This study used chitosan as a delivery vehicle which, when injected into the tumor, produced a gel due to the neutral pH of the environment. This retains the holmium-166 at the site of injection. Tumor necrosis was observed in 77.5% of patients and the 1, 2 and 3 year survival rates were 87.2, 71.8 and 65.3% respectively.[92]

10.2.5.1 Chitosan for Colon-specific Delivery

An interesting function of chitosan is its susceptibility to degradation in the colon due to bacterial enzymes. Degradation by bacterial enzymes is related to the M_w and the degree of deacetylation, with lower M_w and degree of deacetylation producing the quickest degradation. This degradation was found to be similar to that achieved by almond emulsin β-glucosidase.[93] Chitosan is not digested in the upper gastro-intestinal tract by human digestive enzymes.[22] Colon-specific delivery using pectin–chitosan–hydroxypropyl tablets was achieved by Macleod *et al.*[94] The transport of radiolabelled tablets was studied in four patients and successful control of delivery to the colon was achieved.[94] The tablets were, however, quite complex in their structure. Tablets coated with pectin–chitosan in a 1 : 10 ratio were able to protect indomethacin and paracetamol tablets more efficiently than pectin alone *in vitro*. It was noted that the tablets must first be exposed to acidic conditions, modeling the upper gastro-intestinal tract, for enzyme-stimulated release. Chitosan coated with Eudragit® formed a chitosan core with pH-dependent release. In addition, it was found that the chitosan ionically cross-linked with Eudragit® which further controlled the release profile in the small intestine.[22] Chitosan capsule delivery to large intestine in rats was achieved in 3.5 h and a greater concentration of 5-aminosalicylic acid than that found with carboxymethyl cellulose was attained; therefore, a better therapeutic effect in the treatment of colitis was reported.[95]

Chitosan has also been reported to increase the permeability of Caco-2 cell monolayers through opening tight junctions between the cells.[96] This could be beneficial in the delivery of large molecules through the intestine. The effect of the degree of deacetylation and M_w on intestinal absorption at pH 5.5 was studied by Schipper *et al.*[97] A high degree of deacetylation increased paracellular transport most effectively at both high (170 kDa) and low (31 kDa) M_w.[97] However, a clear dose-dependent toxicity was observed which was not present with lower degrees of deacetylation.[97]

10.2.5.2 Gastric-specific Drug Delivery

With a view to increasing the gastric residence time of a formulation, floating microcapsules were made using chitosan and sodium dioctyl sulfosuccinate.[98] In simulated gastric fluid the microcapsules exhibited near zero-order release kinetics over several hours.[98] Similar microcapsules containing melatonin were employed to reduce the apoptotic effect of aflatoxin B1 on the liver; results showed a significant reduction in liver apoptosis in rats co-administered microcapsules and aflatoxin B1.[99]

Microspheres were also produced by Hejazi and Amiji, in this case to increase the efficiency of tetracycline against *Helicobacter pylori*. As in the above case, ionically crosslinked microspheres were prepared. A substantially greater loading of preformed microspheres (69%) was achieved than by mixing the drug prior to crosslinking (8%).[100] However, no increase in tetracycline concentration over that of an aqueous solution was found in a gerbil model. This has since been overcome through covalent crosslinking using glyoxal: microspheres were produced between chitosan and Tween 80, followed by crosslinking with glyoxal. These microspheres have a greater loading capacity (85%), and an increased residence time in the stomach. This served to increase the tetracycline concentration within the stomach and almost doubled the area under the curve (0.5–10 h).[101]

10.2.5.3 Ocular Drug Delivery

Ocular delivery using chitosan and trimethylchitosan is considered promising because of the ability of these substances to open tight junctions and their mucoadhesiveness.[102] The mucoadhesiveness of chitosan is greater in neutral or alkaline solutions.[103] Thus, the neutral pH of tear fluid along with the presence of lysozyme, which would degrade chitosan, could provide a controlled release mechanism from a chitosan matrix. Timolol maleate was combined with chitosan into a niosomal formulation. This formulation was found to be superior to both the commercial product Timolet® and a Carbopol formulation *in vivo*.[104] A chitosan microsphere ($\leq$25 μm) suspension gave prolonged concentrations of acyclovir in *in vivo* tests on rabbits.[105] The precorneal residence time of chitosan formulation containing tobramycin was significantly increased (at least 3-fold) compared to a commercial formulation and was well tolerated.[106]

10.2.5.4 Nasal Drug Delivery

Chitosan has been shown to increase the nasal absorption of several classes of compounds: proteins,[107] peptides,[11] low M_w drugs[108,109] and vaccines.[110,111]

In vivo studies have shown immune responses after intra-nasal delivery of chitosan containing the diphtheria antigen.[52] In a separate study both humoral and systemic immunity to hepatitis-B were achieved in a murine model using chitosan-coated PLGA nanoparticles.[112] Clinically, an effective immune response was seen in humans using a nasally administered inactivated trivalent influenza vaccine.[113] Although lower serum levels were achieved than that achieved with an intramuscularly delivered vaccine, optimization could realize an easily administered alternative form of vaccination.

10.2.5.5 Localization in the Oral Cavity

Local delivery of drugs in the oral cavity is improved through the mucoadhesive properties of chitosan.[114] Not only that, chitosan also has anti-fungal properties making its use synergistic.[115] Chitosan has excellent adherent qualities and was found to adhere to an *in vitro* culture model and periodontal pocket more efficiently than poly(ethylene oxide) and xanthan gum, but not in an oral mucosal adhesion test.[116] Gel and film devices containing nystatin were prepared for the oral treatment of mucositis. In hamsters, gel treatment resulted in the complete recovery of all animals and a significant absence of weight loss.[117] In healthy human volunteers, treatment with chitosan gels containing nystatin resulted in greater than minimum inhibitory concentrations of nystatin at the application site, for a longer duration, than the contralateral site in the oral cavity.[117]

Non-covalently crosslinked palmitoyl glycol chitosan was produced through freeze-drying the polymer, creating hydrophobic crosslinking interactions between the palmitoyl chains. The hydrophobicity of the palmitoyl glycol chitosan was controlled by the percentage of palmitoylation; as the hydrophobicity decreased, the erosion of the gel and rate of FITC-dextran release increased.[118] Denbufylline was loaded into the most hydrophobic of the gels (20.3 mol% palmitoylated glucosamine) and the buccal absorption was analyzed in rabbits. Denbufylline was detected after 30 min and sustained release was achieved for at least 5 h following dosing.[119] Overall, chitosan and its derivatives suggest themselves as promising agents for delivery in the oral cavity.

10.2.6 Chitosan Conjugates in Cancer Therapy

Chitosan has many advantages in the design of drug delivery vehicles for cancer. In addition to the biocompatibility already mentioned, it is possible to create conjugated drugs (both covalently and non-covalently linked) which have defined release characteristics. Nanoparticles of a chitosan-containing dextran–doxorubicin conjugate had an average size of 100 ± 10 nm and were localized within the tumor due to the enhanced permeability and retention (EPR) effect. They also had reduced side-effects associated with doxorubicin and improved therapeutic efficacy, as shown by a decrease in tumor size and longer survival rates.[120]

Palmitoyl glycol chitosan forms vesicles when sonicated with cholesterol. These vesicles were covalently labeled with transferrin for the targeting of receptors over-expressed in some tumors. An increase in the uptake of FITC-dextran-labeled transferrin–chitosan vesicles by human epidermoid carcinoma (A431) cells was seen.[121] Following this, transferrin-targeted chitosan vesicles were loaded with doxorubicin. Although an increased activity against a doxorubicin resistant cell line was observed, this did not translate into an *in vivo* increase in therapeutic activity which was lower than that of the free drug.[122]

Transferrin targeting was also developed by Aktas *et al.* In these experiments chitosan was modified using heterobifunctional PEG and then an antibody against transferrin receptor (OX26) was attached through biotin–streptavidin interaction. Uptake was seen within the mouse brain, using FITC-labeled nanoparticles, suggesting that this may be a potential route for an application of brain active peptides. These antibody targeted pegylated chitosan nanoparticles (636 nm) were able to entrap an anti-caspase peptide (Z-DEVD-FMK).[123]

In vivo translation of transferrin receptor-targeted delivery requires some or all of these attributes: avoidance of non-target (systemic, often recticular endothelial system) uptake, high tumor : normal tissue ratio of transferrin receptor expression and intracellular release characteristic optimization.

As mentioned, chitosan has attracted interest in biomedical applications. Several novel therapeutics have been developed and a number of novel drug delivery systems proposed. As fewer small molecules achieve improved efficacy over current drugs and fail to achieve therapy of new or existing diseases, pharmaceutical research is increasingly looking towards macromolecules. The majority of these are either susceptible to degradation in body fluids or characterized by poor permeability across biological membranes. In most cases both challenges exist and the delivery material in question must both protect the macromolecule and facilitate its passage across membranes. Chitosan and its modifications have been investigated in most of the modern drug delivery systems (Table 10.2). It has been found that these materials can facilitate the

Table 10.2 Chitosan's use in therapeutic systems.

Delivery system	*Characteristics*	*Applications*
Polyelectrolyte complex	Complex plasmid DNA[18,44]	Gene delivery Antisense delivery DNA vaccination
Microencapsulated cells	Chitosan/alginate encapsulation of cells[136,137]	Cell therapy Nerve growth
Solution, gel, microparticle	Excipient, binder, mucoadhesive, permeation enhancer[11,23,117]	Peptide, protein, small molecule delivery and vaccination Ocular, oral, topical and nasal delivery.
Nanoparticle	Encapsulation, protection[138,139]	Peptide and protein delivery and vaccination Ocular, oral and nasal delivery
Gel	Hydrogel, controlled release, protective film/coating[37,83,117,140]	Tissue engineering, implants, wound healing applications
Micelles	Hydrophobically modified chitosans[63,64,119]	Delivery of lipophilic compounds

delivery of therapeutic peptides, proteins, and nucleic acids. Chitosan's pH-dependent solubility, biocompatibility and mucoadhesiveness have led to the above-mentioned examples. It is expected that if chitosan is given full approval by the FDA as 'generally regarded as safe' such dosage forms will appear on the market.[3] Some of the *in vivo* applications reported are summarized in Table 10.3.

10.3 Modified Chitosans: Trimethylated Chitosan Applications in Drug Delivery

One method of increasing the aqueous solubility of chitosan is through quaternization or trimethylation of the amine residues of glucosamine monomers. Trimethylated chitosan has been found to enhance paracellular transport across Caco-2 cells.[56] This was dependent on the degree of trimethylation (DTM), with 12.8% DTM failing to increase transport but 61.2% DTM significantly increasing it at neutral pH.[124] Using confocal laser scanning microscopy, it was visualized that TMC opens the paracellular pathway without damaging the cell membrane.[125] The toxicity of TMC towards African green monkey kidney fibroblast (COS-7) cells and human breast cancer (MCF-7) cells was assessed and toxicity was seen to increase with increasing DTM and was also M_w-dependent as oligomeric (3–6 kDa) derivatives were not significantly toxic (Table 10.4).[10]

Highly quaternized TMC polymers were studied for their ability to increase the paracellular permeation of peptide drugs across intestinal monolayers *in vitro*. Caco-2 cell monolayers were used as an intestinal epithelium model to investigate the ability of two TMC polymers with degrees of substitution 40 and 60% (designated TMC40 and TMC60 respectively) to increase the paracellular transport of the peptide drug buserelin at neutral pH value. Both TMCs managed to significantly increase the transport of buserelin compared to the control; however, the effect of TMC60 was more prominent. TMC60 increased the transported amount of the peptide by up to 6% of the apically applied buserelin.[126] The effect of TMC60 concentration on the transport of hydrophilic compounds was investigated using the somatostatin analog octreotide at neutral pH values (7.4). In the presence of TMC60 octreotide permeation showed a TMC concentration dependent increase, indicating a specific interaction of the cationic polymer with components of the tight junctions; this was not saturated within the concentration range used (0.25–1.5% w/v).[126]

The effect of TMC60 polymers was studied *in vivo* in rats, using peptide drugs that find application in clinical use; these were administered subcutaneously and/or nasally. Buserelin (pI = 6.8) and octreotide (pI = 8.2) peptide analogs were used for this purpose. Buserelin formulations with or without TMC60 (pH = 7.2) were compared with chitosan dispersions at neutral pH values after intraduodenal administration in rats. A remarkable increase in buserelin serum concentrations was observed after co-administration of the peptide with TMC60, whereas buserelin alone was poorly absorbed. In the presence of TMC60 buserelin was rapidly absorbed from the intestine having T_{max} at

Table 10.3 *In vivo* applications of chitosan and its derivatives.[a]

Chitin/chitosan type	*Form*	*Animal model*	*Application*	*Effect*	*Result*	*Ref.*
Chitosan 82 deacetylated 80 kDa from crab shells	Cotton type (chitopack C)	Small and large wild and zoo animals	Topicalinjection into pleural cavity	Wound/ infection healing	Re-epithelization, decrease in treatment frequency	17
Chitin 9% deacetylated >100 kDa from squid pen	Sponge-like chitin (Chitipack S)	Small animals	Topical	Wound healing	minimal scar formation	17
	Composite chitin sheet (Chitipack P)	Small and large animals	Topical implantation	Wound healing	Good wound contraction, re-epithelization,	17
Chitin 370 kDa from squid pen	Non-woven fiber and chitin complex (Chitipack P)	Sheep	Implantation	Tendon healing	Increased immature and intermediate fibroplasia	17
Methylpyrrolidone chitosan	Freeze dried soft sponge	Rabbit	topical	Bone healing	induced bone repair	17
Phosphorylated chitosan 63% deacetylated 17 kDa	Phosphorylated chitosan with calcium phosphate cements	Rabbit	Implantation	Bone healing	osteoinduction	141
N,N-dicarboxymethyl chitosan	Soft spongy and hydrophilic material	Sheep	Topical	Bone healing	Accelerated regeneration of bone tissue	17
Chitosan 72% deacetylated from cuttlefish	Powder	Dog	Topical	Bone healing	Enhancement of bone healing	17
Chitosan 70 kDa from prawn shells	Solution	Rat	Injection into knee	Tissue regeneration (chondrogenic)	Intra-articular fibrous tissue formation, articular chondrocyte proliferation, reduced decrease in the epiphyseal cartilage thickness	84
Chitin <10% deacetylated 300 kDa Chitosan >80% deacetylated 80 kDa	Suspension	Mouse	Intraperitoneal injection	Analgesic	A dose-dependent decrease of the abnormal behaviors due to pain	17

Table 10.3 (*Continued*)

Chitin/chitosan type	*Form*	*Animal model*	*Application*	*Effect*	*Result*	*Ref.*
Chitin < 30% deacetylated	Granule	Dog	Topical	Wound healing	Re-epithelization in chitin and chitosan groups	17
Chitosan 82% deacetylated 30–80 kDa	Cotton type (Chitopack C)	Dog	Topical	Wound healing	Accelerated infiltration of PMN cells, increase in number of white blood cells	17
Chitosan 82% deacetylated 80 kDa from crab shells	Cotton type (Chitopack C)	Dog	Implantation	Immuno-potentiator	Enhanced activity of PMN cells, increase in the number of white blood cells	17
Phosphated chitin	Solution	Mouse	Intravenous infusion	Anti-inflammatory	Potentially useful in chitosan-induced acute respiratory distress syndrome	17
Chitosan	Chitosan-H gel containing EGF	Rat	Topical	Burn replacement of skin/ wound healing	Increased cell proliferation and epithelization	83
Chitosan 100% deacetylated 540 kDa	Non-woven fibers	Beagle Dogs	Topical	Tissue/ bone regeneration, wound closure	Suprabony and intrabony cementum	69
Photocrosslinkable chitosan (azide-chitosan-lactose) 800 to 1000 kDa	Azide-chitosan-lactose gel and Azide-chitosan-lactose gel containing FGF	Diabetic mice	Topical	Contraction of wound, promotion of healing and controlled release of FGF	Accelerated wound healing, closure and contraction with an increased filling rate	26, 142
Chitosan crosslinked with glutaraldehyde	Gels	Rat	Topical – Brachytherapy	Biodegradable implant for brachytherapy (radiotherapy)	brachytherapy	140

[a]Modified from table in Senel and McClure.[17]

Table 10.4 Effect of chitosan trimethylation on MCF7 and COS7 cell viability.[a]

			$IC_{50}/\mu g\ ml^{-1}$	
Product	*DTM/%*	*Exposure time/h*	*MCF -7*	*COS-7*
TMO	20	6	>10 000	>10 000
		24	>10 000	>10 000
	44	6	>10 000	>10 000
		24	>10 000	>10 000
	55	6	>10 000	>10 000
		24	5959 ± 943	661 ± 205
	94	6	1402 ± 210	2207 ± 381
		24	417 ± 210	430 ± 116
TMC	36	6	823 ± 324	>10 000
		24	285 ± 100	>10 000
	57	6	393 ± 259	676 ± 329
		24	265 ± 53	161 ± 50
	76	6	55 ± 10	40 ± 87
		24	59 ± 30	30 ± 8
	93	6	293 ± 68	79 ± 16
		24	118 ± 28	36 ± 3
PEI (25 kDa)	—	6	<20	30 ± 0
	—	24	<20	<20

[a]DTM: degree of trimethylation. Curves were fitted according to the Hill equation and the IC_{50} value was calculated; each concentration in the IC_{50} curve had $n = 6$ and concentrations in the range 20–10^4 μg ml^{-1} were tested for all cationic vectors. (Adapted from Kean *et al.*[10])

40 min, whereas chitosan dispersion (at pH 7.2) showed a slight increase in buserelin absorption compared to the control, but did not increase the buserelin concentrations to that achieved with TMC60. The absolute bioavailability of buserelin after co-administration with TMC60 was 13.0%.[126] Similar to the buserelin studies, octreotide absorption after intrajejunal administration was substantially increased resulting in a peptide bioavailability of 16%.[127] In pigs, octreotide oral bioavailability increased to 25%, indicating that substantial intestinal absorption of peptide analogs can take place when the therapeutic is co-administered with the trimethylated material. Bronchial absorption of octreotide in rats was increased 3.9-fold by TMC (60% DTM).[128]

The absorption or permeation of ibuprofen across a Caco-2 cell monolayer was studied using a mixture of chitosan and TMC (17% DTM) in citric acid crosslinked microparticles.[129] These ionically crosslinked microparticles were prepared at up to a 50 : 50 chitosan : TMC ratio which was able to encapsulate 36.5% ibuprofen.[129] No increase in permeation was observed, but the low DTM may account for this.

A library of PEGylated TMCs (~40% DTM) was prepared with different molecular weights: 5, 25, 50, 100 and 400 kDa.[130] The oligomeric TMC (5 kDa) was not significantly toxic, confirming previous work.[10,130] The 50 kDa, 100 kDa and 400 kDa TMCs were PEGylated with varying percentages and PEGylation with a 5 kDa PEG was found to be superior over a 550 Da PEG modification when compared at similar graft ratios. PEGylation of TMCs increased the cell viability, as did complexion with insulin.

The effect of DTM and M_w on buccal penetration (in an *ex vivo* porcine model) was studied.[131] TMCs of high (1460 kDa) and intermediate (580 kDa) M_w were used with three DTM: high M_w with 4%, 35% and 90% DTM; and an intermediate M_w with 3%, 46% and 78% DTM.[131,132] The viscosity of the solutions decreased with increasing DTM, an effect that was attributed to their decreased ability to form hydrogen bonds but which could have been due to degradation of the polymer chain giving lower M_w derivatives.[10,131] The intermediate derivatives had higher mucoadhesion and transport of fluorescein isothiocyanate dextran (4.4 kDa) increased with increasing DTM.

These findings verify the hypothesis that TMC, being a soluble polymer, efficiently increases the absorption of hydrophilic compounds at neutral pH values similar to the pH values found physiologically. In mucoadhesion studies using pig intestinal mucosae it was found that TMC at pH 7 showed similar adhesion strength as chitosan pH 6 (unpublished data). Therefore, it is unlikely that this effect of permeation enhancement can be attributed to differences in mucoadhesion. From *in vivo* studies in rats and pigs it is clear that the chitosan and the quaternized modification are efficient absorption enhancers of such peptide analog drugs.

10.4 Concluding Remarks

In this chapter we presented the most important pharmaceutical applications, as they appeared in the literature, in the two last decades. Improvements in purification and characterisation of chitosans and identification of appropriate sources mean that it is now possible to have well characterized material, leading to the ability to rationally choose the "right" chitosan for the right application.

However, in our opinion the immunological and toxicological profile is not yet fully clarified, although the recent report in mice is a step in the right direction.[54] Given the number of different chitosans, it is certain that significant disparities will occur. The effect of different types of chitosans on cytokine release, tissue compatibility and chronic toxicity could give information for structure–activity relationships, safety aspects and aid the further approval of such materials from regulatory authorities.

We have placed emphasis on the already-approved chitosan applications in wound healing and we presented the injectable forms of this biomaterial related to drug delivery. In this aspect, chitosan's potential for cell encapsulation for the production of therapeutic bioactives (implants) shows promise for future applications. Chitosan can form membranes around cells that allow the passage of molecules. Its biocompatibility and controllable biodegradability are additional advantages that could promote such applications.

We mainly focused on chitosan and some of the derivatives which found pharmaceutical application, mainly in drug delivery. It should be noted that there are many derivatives of chitosan that have not been mentioned, such as the dendrimerized chitosan sialic acid hybrid,[133] which may present suitable nanostructures for drug and DNA delivery.

Current chitosan research is very exciting with encouraging results being seen in clinical trials and still much potential involving this abundant resource. Chitosan certainly is proving to be the versatile biomaterial that was expected.

References

1. M. N. V. R. Kumar, *Reactive & Functional Polymers*, 2000, **46**, 1–27.
2. R. A. A. Muzzarelli, *Anal. Biochem.*, 1998, **260**, 255–257.
3. M. Thanou and H. E. Junginger. In *Polysaccharides, Structural Diversity and Functional Versatility*, (Ed.) S. Dumitriu, Marcel Dekker, 2nd edn, 2004, pp. 661–678.
4. C. Jeuniaux and M. F. Voss-Foucart, *Biochem. Syst. Ecol.*, 1991, **19**, 347–356.
5. H. Braconnot, *Ann. Chim. (Paris)*, 1811, **79**, 265–304.
6. A. Odier, *Mémoires de la Société D'Histoire Naturelle de Paris*, 1823, **1**, 29–42.
7. G. Falini, S. Weiner and L. Addadi, *Calcif. Tissue Int.*, 2003, **72**, 548–554.
8. USA Pat., 2040879, 1936.
9. C. T. Su, K. T. Teck, M. W. Sek and E. Khor, *Carbohydr. Polym.*, 1996, **30**, 239–242.
10. T. Kean, S. Roth and M. Thanou, *J. Control. Release*, 2005, **103**, 643–653.
11. L. Illum, *Pharm. Res.*, 1998, **15**, 1326–1331.
12. H. B. Alam, D. Burris, J. A. DaCorta and P. Rhee, *Military Medicine*, 2005, **170**, 63–69.
13. K. Y. Lee, I. C. Kwon, Y. H. Kim, W. H. Jo and S. Y. Jeong, *J. Control. Release*, 1998, **51**, 213–220.
14. K. Corsi, F. Chellat, L. Yahia and J. C. Fernandes, *Biomaterials*, 2003, **24**, 1255–1264.
15. H. Onishi and Y. Machida, *Biomaterials*, 1999, **20**, 175–182.
16. T. Honma, L. Zhao, N. Asakawa and Y. Inoue, *Macromol. Biosci.*, 2006, **6**, 241–249.
17. S. Senel and S. J. McClure, *Adv. Drug Del. Rev.*, 2004, **56**, 1467–1480.
18. J. Murata, Y. Ohya and T. Ouchi, *Carbohydr. Polym.*, 1996, **29**, 69–74.
19. J. Murata, Y. Ohya and T. Ouchi, *Carbohydr. Polym.*, 1997, **32**, 105–109.
20. P. Erbacher, S. M. Zou, T. Bettinger, A. M. Steffan and J. S. Remy, *Pharm. Res.*, 1998, **15**, 1332–1339.
21. M. Prabaharan and J. F. Mano, *Drug Deliv*, 2005, **12**, 41–57.
22. M. K. Chourasia and S. K. Jain, *Drug Delivery*, 2004, **11**, 129–148.
23. I. P. Kaur and R. Smitha, *Drug Development and Industrial Pharmacy*, 2002, **28**, 353–369.
24. S. E. Noorjahan and T. P. Sastry, *J Biomed Mater Res B Appl Biomater*, 2005, **75**, 343–350.

25. S. R. Frenkel, G. Bradica, J. H. Brekke, S. M. Goldman, K. Ieska, P. Issack, M. R. Bong, H. Tian, J. Gokhale, R. D. Coutts and R. T. Kronengold, *Osteoarthritis Cartilage*, 2005, **13**, 798–807.
26. K. Obara, M. Ishihara, M. Fujita, Y. Kanatani, H. Hattori, T. Matsui, B. Takase, Y. Ozeki, S. Nakamura, T. Ishizuka, S. Tominaga, S. Hiroi, T. Kawai and T. Maehara, *Wound Repair Regen*, 2005, **13**, 390–397.
27. X. Zhu, K. S. Chian, M. B. Chan-Park and S. T. Lee, *J. Biomed. Mater. Res.*, 2005, **73**, 264–274.
28. L. Ma, C. Gao, Z. Mao, J. Zhou, J. Shen, X. Hu and C. Han, *Biomaterials*, 2003, **24**, 4833–4841.
29. C. Ni Mhurchu, C. A. Dunshea-Mooij, D. Bennett and A. Rodgers, *Cochrane Database Syst Rev*, 2005, CD003892.
30. R. N. Schiller, E. Barrager, A. G. Schauss and E. J. Nichols, *Journal of the American Nutraceutical Association*, 2001, **4**, 42–49.
31. S. Babel and T. A. Kurniawan, *Chemosphere*, 2004, **54**, 951–967.
32. C. Jeon and W. H. Holl, *Water Res.*, 2003, **37**, 4770–4780.
33. S. Hirano, *Polym. Int.*, 1999, **48**, 732–734.
34. K. Takai, T. Ohtsuka, Y. Senda, M. Nakao, K. Yamamoto, J. Matsuoka and Y. Hirai, *Microbiol. Immunol.*, 2002, **46**, 75–81.
35. E. Pascual and M. R. Julia, *J. Biotechnol.*, 2001, **89**, 289–296.
36. S. O. Duke, S. R. Baerson, F. E. Dayan, A. M. Rimando, B. E. Scheffler, M. R. Tellez, D. E. Wedge, K. K. Schrader, D. H. Akey, F. H. Arthur, A. J. De Lucca, D. M. Gibson, H. F. Harrison, Jr., J. K. Peterson, D. R. Gealy, T. Tworkoski, C. L. Wilson and J. B. Morris, *Pest Manag Sci*, 2003, **59**, 708–717.
37. A. Cagri, Z. Ustunol and E. T. Ryser, *J. Food Prot.*, 2004, **67**, 833–848.
38. J. L. Chew, C. B. Wolfowicz, H. Q. Mao, K. W. Leong and K. Y. Chua, *Vaccine*, 2003, **21**, 2720–2729.
39. M. Koping-Hoggard, K. M. Varum, M. Issa, S. Danielsen, B. E. Christensen, B. T. Stokke and P. Artursson, *Gene Ther.*, 2004, **11**, 1441–1452.
40. W. Liu, S. Sun, Z. Cao, X. Zhang, K. Yao, W. W. Lu and K. D. Luk, *Biomaterials*, 2005, **26**, 2705–2711.
41. M. Thanou, B. I. Florea, M. Geldof, H. E. Junginger and G. Borchard, *Biomaterials*, 2002, **23**, 153–159.
42. W. Xu, Y. Shen, Z. Jiang, Y. Wang, Y. Chu and S. Xiong, *Vaccine*, 2004, **22**, 3603–3612.
43. M. Koping-Hoggard, I. Tubulekas, H. Guan, K. Edwards, M. Nilsson, K. M. Varum and P. Artursson, *Gene Ther.*, 2001, **8**, 1108–1121.
44. K. Roy, H. Q. Mao, S. K. Huang and K. W. Leong, *Nat. Med.*, 1999, **5**, 387–391.
45. M. Koping-Hoggard, Y. S. Mel'nikova, K. M. Varum, B. Lindman and P. Artursson, *J Gene Med*, 2003, **5**, 130–141.
46. T. Kiang, J. Wen, H. Lim and K. W. Leong, *Proc. Int. Symp. Control. Release Bioact. Mater.*, Seoul, Korea, 2002.
47. R. Hejazi and M. Amiji. In *Polymeric Biomaterials*, (Ed.) S. Dimitriu, Marcel Dekker, 2nd edn, 2002, pp. 213–236.

48. T. Ishii, Y. Okahata and T. Sato, *Biochim. Biophys. Acta-Biomembr.*, 2001, **1514**, 51–64.
49. V. Chan, H.-Q. Mao and K. W. Leong, *Langmuir*, 2001, **17**, 3749–3756.
50. S. P. Strand, S. Danielsen, B. E. Christensen and K. M. Varum, *Biomacromolecules*, 2005, **6**, 3357–3366.
51. H. O. Alpar, S. Somavarapu, K. N. Atuah and V. W. Bramwell, *Adv Drug Deliv Rev*, 2005, **57**, 411–430.
52. I. M. van der Lubben, J. C. Verhoef, G. Borchard and H. E. Junginger, *Adv. Drug Del. Rev.*, 2001, **52**, 139–144.
53. M. Iqbal, W. Lin, I. Jabbal-Gill, S. S. Davis, M. W. Steward and L. Illum, *Vaccine*, 2003, **21**, 1478–1485.
54. C. Qin, J. Gao, L. Wang, L. Zeng and Y. Liu, *Food Chem. Toxicol.*, 2006.
55. A. Domard, M. Rinaudo and C. Terrassin, *International Journal of Biological Macromolecules*, 1986, **8**, 105–107.
56. M. Thanou, A. F. Kotze, T. Scharringhausen, H. L. Luessen, A. G. de Boer, J. C. Verhoef and H. E. Junginger, *J. Control. Release*, 2000, **64**, 15–25.
57. H. Sashiwa and Y. Shigemasa, *Carbohydr. Polym.*, 1999, **39**, 127–138.
58. H. Sashiwa, N. Kawasaki, A. Nakayama, E. Muraki, H. Yajima, N. Yamamori, Y. Ichinose, J. Sunamoto and S. Aiba, *Carbohydr. Res.*, 2003, **338**, 557–561.
59. A. B. Sieval, M. Thanou, A. F. Kotze, J. C. Verhoef, J. Brussee and H. E. Junginger, *Carbohydr. Polym.*, 1998, **36**, 157–165.
60. N. Fang, V. Chan, H. Q. Mao and K. W. Leong, *Biomacromolecules*, 2001, **2**, 1161–1168.
61. T. H. Kim, J. E. Ihm, Y. J. Choi, J. W. Nah and C. S. Cho, *J. Control. Release*, 2003, **93**, 389–402.
62. Y. H. Kim, S. H. Gihm, C. R. Park, K. Y. Lee, T. W. Kim, I. C. Kwon, H. Chung and S. Y. Jeong, *Bioconj. Chem.*, 2001, **12**, 932–938.
63. H. S. Yoo, J. E. Lee, H. Chung, I. C. Kwon and S. Y. Jeong, *J. Control. Release*, 2005, **103**, 235–243.
64. W. G. Liu, X. Zhang, S. J. Sun, G. J. Sun, K. De Yao, D. C. Liang, G. Guo and J. Y. Zhang, *Bioconj. Chem.*, 2003, **14**, 782–789.
65. S. Gao, J. Chen, X. Xu, Z. Ding, Y. H. Yang, Z. Hua and J. Zhang, *Int. J. Pharm.*, 2003, **255**, 57–68.
66. I. K. Park, H. L. Jiang, S. E. Cook, M. H. Cho, S. I. Kim, H. J. Jeong, T. Akaike and C. S. Cho, *Archives of Pharmaceutical Research*, 2004, **27**, 1284–1289.
67. H. Q. Mao, K. Roy, V. L. Troung-Le, K. A. Janes, K. Y. Lin, Y. Wang, J. T. August and K. W. Leong, *J. Control. Release*, 2001, **70**, 399–421.
68. M. Nakajima, K. Atsumi, K. Kifune, K. Miura and H. Kanamaru, *Jpn J Surg*, 1986, **16**, 418–424.
69. Y. J. Yeo, D. W. Jeon, C. S. Kim, S. H. Choi, K. S. Cho, Y. K. Lee and C. K. Kim, *J Biomed Mater Res B Appl Biomater*, 2005, **72**, 86–93.
70. H. Ueno, H. Yamada, I. Tanaka, N. Kaba, M. Matsuura, M. Okumura, T. Kadosawa and T. Fujinaga, *Biomaterials*, 1999, **20**, 1407–1414.

71. H. Ueno, T. Mori and T. Fujinaga, *Adv. Drug Del. Rev.*, 2001, **52**, 105–115.
72. Hemcon, http://www.hemcon.com/Products/HemConBandageOverview/HemConBandage15x15.aspx, accessed in 2010.
73. A. E. Pusateri, S. J. McCarthy, K. W. Gregory, R. A. Harris, L. Cardenas, A. T. McManus and C. W. Goodwin, Jr., *J Trauma*, 2003, **54**, 177–182.
74. I. Wedmore, J. G. McManus, A. E. Pusateri and J. B. Holcomb, *J Trauma*, 2006, **60**, 655–658.
75. J. Benesch and P. Tengvall, *Biomaterials*, 2002, **23**, 2561–2568.
76. S. B. Rao and C. P. Sharma, *J. Biomed. Mater. Res.*, 1997, **34**, 21–28.
77. S. Hirano, M. Zhang, M. Nakagawa and T. Miyata, *Biomaterials*, 2000, **21**, 997–1003.
78. A. Di Martino, M. Sittinger and M. V. Risbud, *Biomaterials*, 2005, **26**, 5983–5990.
79. M. Jumaa, F. H. Furkert and B. W. Muller, *Eur J Pharm Biopharm*, 2002, **53**, 115–123.
80. F. S. Kittur, A. B. Vishu Kumar, M. C. Varadaraj and R. N. Tharanathan, *Carbohydr. Res.*, 2005, **340**, 1239–1245.
81. Y.-J. Jeon, P.-J. Park and S.-K. Kim, *Carbohydr. Polym.*, 2001, **44**, 71–76.
82. D. Raafat, K. von Bargen, A. Haas and H. G. Sahl, *Appl. Environ. Microbiol.*, 2008, **74**, 3764–3773.
83. C. Alemdaroglu, Z. Degim, N. Celebi, F. Zor, S. Ozturk and D. Erdogan, *Burns*, 2006, **32**, 319–327.
84. J. X. Lu, F. Prudhommeaux, A. Meunier, L. Sedel and G. Guillemin, *Biomaterials*, 1999, **20**, 1937–1944.
85. C. D. Hoemann, J. Sun, A. Legare, M. D. McKee and M. D. Buschmann, *Osteoarthritis Cartilage*, 2005, **13**, 318–329.
86. Y. C. Wu, S. Y. Shaw, H. R. Lin, T. M. Lee and C. Y. Yang, *Biomaterials*, 2006, **27**, 896–904.
87. M. Jaworska, K. Sakurai, P. Gaudon and E. Guibal, *Polym. Int.*, 2003, **52**, 198–205.
88. R. Minkc and J. Blackwell, *J. Mol. Biol.*, 1978, **120**, 167–181.
89. I. Orienti, T. Cerchiara, B. Luppi, F. Bigucci, G. Zuccari and V. Zecchi, *Int. J. Pharm.*, 2002, **238**, 51–59.
90. F.-L. Mi, N.-L. Her, C.-Y. Kuan, T.-B. Wong and S.-S. Shyu, *J. Appl. Polym. Sci.*, 1997, **66**, 2495–2505.
91. J. Nunthanid, M. Laungtana-Anan, P. Sriamornsak, S. Limmatvapirat, S. Puttipipatkhachorn, L. Y. Lim and E. Khor, *J. Control. Release*, 2004, **99**, 15–26.
92. J. K. Kim, K. H. Han, J. T. Lee, Y. H. Paik, S. H. Ahn, J. D. Lee, K. S. Lee, C. Y. Chon and Y. M. Moon, *Clin Cancer Res*, 2006, **12**, 543–548.
93. H. Zhang and S. H. Neau, *Biomaterials*, 2002, **23**, 2761–2766.
94. G. S. Macleod, J. T. Fell, J. H. Collett, H. L. Sharma and A.-M. Smith, *Int. J. Pharm.*, 1999, **187**, 251–257.

95. H. Tozaki, T. Odoriba, N. Okada, T. Fujita, A. Terabe, T. Suzuki, S. Okabe, S. Muranishi and A. Yamamoto, *J. Control. Release*, 2002, **82**, 51–61.
96. P. Artursson, T. Lindmark, S. S. Davis and L. Illum, *Pharm. Res.*, 1994, **11**, 1358–1361.
97. N. G. Schipper, S. Olsson, J. A. Hoogstraate, A. G. deBoer, K. M. Varum and P. Artursson, *Pharm. Res.*, 1997, **14**, 923–929.
98. I. El-Gibaly, *Int. J. Pharm.*, 2002, **249**, 7–21.
99. I. El-Gibaly, A. M. Meki and S. K. Abdel-Ghaffar, *Int. J. Pharm.*, 2003, **260**, 5–22.
100. R. Hejazi and M. Amiji, *Int. J. Pharm.*, 2002, **235**, 87–94.
101. R. Hejazi and M. Amiji, *Pharm Dev Technol*, 2003, **8**, 253–262.
102. M. J. Alonso and A. Sanchez, *J. Pharm. Pharmacol.*, 2003, **55**, 1451–1463.
103. M. N. Ravi Kumar, U. Bakowsky and C. M. Lehr, *Biomaterials*, 2004, **25**, 1771–1777.
104. D. Aggarwal and I. P. Kaur, *Int. J. Pharm.*, 2005, **290**, 155–159.
105. I. Genta, B. Conti, P. Perugini, F. Pavanetto, A. Spadaro and G. Puglisi, *J. Pharm. Pharmacol.*, 1997, **49**, 737–742.
106. O. Felt, P. Buri and R. Gurny, *Drug Development and Industrial Pharmacy*, 1998, **24**, 979–993.
107. L. Illum, N. F. Farraj and S. S. Davis, *Pharm. Res.*, 1994, **11**, 1186–1189.
108. L. Illum, P. Watts, A. N. Fisher, M. Hinchcliffe, H. Norbury, I. Jabbal-Gill, R. Nankervis and S. S. Davis, *J. Pharmacol. Exp. Ther.*, 2002, **301**, 391–400.
109. K. I. Roon, P. A. Soons, M. P. Uitendaal, F. de Beukelaar and M. D. Ferrari, *Br. J. Clin. Pharmacol.*, 1999, **47**, 285–290.
110. A. Bacon, J. Makin, P. J. Sizer, I. Jabbal-Gill, M. Hinchcliffe, L. Illum, S. Chatfield and M. Roberts, *Infect. Immun.*, 2000, **68**, 5764–5770.
111. I. Jabbal-Gill, A. N. Fisher, R. Rappuoli, S. S. Davis and L. Illum, *Vaccine*, 1998, **16**, 2039–2046.
112. K. S. Jaganathan and S. P. Vyas, *Vaccine*, 2006.
113. R. C. Read, S. C. Naylor, C. W. Potter, J. Bond, I. Jabbal-Gill, A. Fisher, L. Illum and R. Jennings, *Vaccine*, 2005, **23**, 4367–4374.
114. I. G. Needleman and F. C. Smales, *Biomaterials*, 1995, **16**, 617–624.
115. J. Knapczyk, A. B. Macura and B. Pawlik, *Int. J. Pharm.*, 1992, **80**, 33–38.
116. I. G. Needleman, G. P. Martin and F. C. Smales, *J Clin Periodontol*, 1998, **25**, 74–82.
117. P. Aksungur, A. Sungur, S. Unal, A. B. Iskit, C. A. Squier and S. Senel, *J. Control. Release*, 2004, **98**, 269–279.
118. L. Martin, C. G. Wilson, F. Koosha, L. Tetley, A. I. Gray, S. Senel and I. F. Uchegbu, *J. Control. Release*, 2002, **80**, 87–100.
119. L. Martin, C. G. Wilson, F. Koosha and I. F. Uchegbu, *Eur J Pharm Biopharm*, 2003, **55**, 35–45.
120. S. Mitra, U. Gaur, P. C. Ghosh and A. N. Maitra, *J. Control. Release*, 2001, **74**, 317–323.

121. C. Dufes, A. G. Schatzlein, L. Tetley, A. I. Gray, D. G. Watson, J. C. Olivier, W. Couet and I. F. Uchegbu, *Pharm. Res.*, 2000, **17**, 1250–1258.
122. C. Dufes, J. M. Muller, W. Couet, J. C. Olivier, I. F. Uchegbu and A. G. Schatzlein, *Pharm. Res.*, 2004, **21**, 101–107.
123. Y. Aktas, M. Yemisci, K. Andrieux, R. N. Gursoy, M. J. Alonso, E. Fernandez-Megia, R. Novoa-Carballal, E. Quinoa, R. Riguera, M. F. Sargon, H. H. Celik, A. S. Demir, A. A. Hincal, T. Dalkara, Y. Capan and P. Couvreur, *Bioconjug Chem*, 2005, **16**, 1503–1511.
124. A. F. Kotze, M. M. Thanou, H. L. Luebetaen, A. G. de Boer, J. C. Verhoef and H. E. Junginger, *J. Pharm. Sci.*, 1999, **88**, 253–257.
125. M. M. Thanou, J. C. Verhoef, S. G. Romeijn, J. F. Nagelkerke, F. W. H. M. Merkus and H. E. Junginger, *Int. J. Pharm.*, 1999, **185**, 73–82.
126. M. Thanou, B. I. Florea, M. W. Langemeyer, J. C. Verhoef and H. E. Junginger, *Pharm. Res.*, 2000, **17**, 27–31.
127. M. Thanou, J. C. Verhoef, P. Marbach and H. E. Junginger, *J. Pharm. Sci.*, 2000, **89**, 951–957.
128. B. I. Florea, M. Thanou, H. E. Junginger and G. Borchard, *J. Control. Release*, 2006, **110**, 353–361.
129. Z. Lu, J. H. Steenekamp and J. H. Hamman, *Drug Dev. Ind. Pharm.*, 2005, **31**, 311–317.
130. S. Mao, X. Shuai, F. Unger, M. Wittmar, X. Xie and T. Kissel, *Biomaterials*, 2005, **26**, 6343–6356.
131. G. Sandri, S. Rossi, M. C. Bonferoni, F. Ferrari, Y. Zambito, G. D. Colo and C. Caramella, *Int. J. Pharm.*, 2005, **297**, 146–155.
132. G. Di Colo, Y. Zambito, S. Burgalassi, I. Nardini and M. F. Saettone, *Int. J. Pharm.*, 2004, **273**, 37–44.
133. H. Sashiwa, H. Yajima and S. Aiba, *Biomacromolecules*, 2003, **4**, 1244–1249.
134. C. A. Jansma, M. Thanou, H. E. Junginger and G. Borchard, *STP Pharma Sci.*, 2003, **13**, 63–67.
135. X. Jiang, H. Dai, K. W. Leong, S. H. Goh, H. Q. Mao and Y. Y. Yang, *J Gene Med*, 2006.
136. L. Baruch and M. Machluf, Biopolymers, 2006, **82**, 570–579.
137. D. Maysinger, O. Berezovskaya and S. Fedoroff, *Exp. Neurol.*, 1996, **141**, 47–56.
138. M. Huang, Z. Ma, E. Khor and L. Y. Lim, *Pharm. Res.*, 2002, **19**, 1488–1494.
139. M. Huang, E. Khor and L. Y. Lim, *Pharm. Res.*, 2004, **21**, 344–353.
140. A. K. Azab, B. Orkin, V. Doviner, A. Nissan, M. Klein, M. Srebnik and A. Rubinstein, J. Control. Release, 2006, **111**, 281–289.
141. X. Wang, J. Ma, Y. Wang and B. He, *Biomaterials*, 2002, **23**, 4167–4176.
142. K. Ono, Y. Saito, H. Yura, K. Ishikawa, A. Kurita, T. Akaike and M. Ishihara, *J. Biomed. Mater. Res.*, 2000, **49**, 289–295.

CHAPTER 11
β-Glucans

STEVE W. CUI, QI WANG AND MEI ZHANG

Guelph Food Research Centre, 93 Stone Road West, Guelph, Ontario, N1G 5C9, Canada

11.1 Introduction

β-Glucan is a broad term covering any polysaccharide that is composed of β-D-glucopyranosyl (β-D-Glc*p*) unit as the primary building block. β-Glucans occur widely in plants, fungi and yeasts as structural components of cell walls or as reserve polysaccharides. Several β-glucans are significant due to their beneficial physiological effects and functional properties, such as cereal β-glucan, curdlan, β-glucans from edible mushrooms and yeast cell walls, *etc.* Cellulose, of course, also belongs to this group. However, due to its unique position in the field of biopolymers, it is covered in another chapter in the current volume.

Much research interest has been focused on (1→3)(1→4)-linked β-D-glucans and (1→3)-β-D-glucans due to their significant physiological benefits and bioactivities toward human health. Animal studies and clinical trials have led to the health claim of oats and barley based products specifically attributed to the levels of (1→3)(1→4)-linked β-D-glucans.[1–3] (1→3)-β-D-Glucans from certain mushrooms and bacteria fermentations exhibit biological activities, such as immunomodulation, antitumoral and anti-inflammatory activities. These bioactive β-glucans have great potential in the fields of biomedicine, functional foods and nutraceuticals.[4] Such biological activities are dependent on specific features of the (1 → 3)-β-D-glucan molecules, especially their conformation,

RSC Polymer Chemistry Series No. 1
Renewable Resources for Functional Polymers and Biomaterials
Edited by Peter A. Williams

Published by the Royal Society of Chemistry, www.rsc.org

molecular weight, and the presence or absence of branches, which usually consist of β-glucans linked to the main chain through β-(1→6) linkages.[5]

The current chapter reviews the most recent developments in β-glucans, covering structural characteristics, functional properties and bioactivities, and potential applications in the biomedical, nutraceutical and functional food industries.

11.2 Cereal β-Glucans

11.2.1 Sources and Structural Features

Cereal β-glucan is a polysaccharide which occurs in the subaleurone and endosperm cell walls of the seeds of cereals, including oats, barley, rye and wheat. A typical distribution of the β-glucans in grain seeds is illustrated in Figure 11.1. β-Glucan in oats is more concentrated in the subaleurone layers than the endosperms; in contrast, a more even distribution of β-glucan is observed for barley and rye grains. Although there is a lack of information regarding where the β-glucan resides in the seed of wheat, it is likely concentrated in the walls of the subaleurone cells as evidenced by the enrichment of β-glucans in the bran fractions using debranning or conditional milling processes.[6] The level of β-glucans in cereals varied from as low as 0.5–1% in wheat to as high as 3–11% in barley, as shown in Table 11.1.

The structure of cereal β-glucans is typical of a linear homopolysaccharide as it only contains a single type of sugar unit, *i.e.* β-D-glucopyranose (β-D-Glc*p*). Over 90% of the β-D-Glc*p* residues in cereal β-glucans are arranged as blocks of two or three consecutive (1→4)-linked units separated by a single

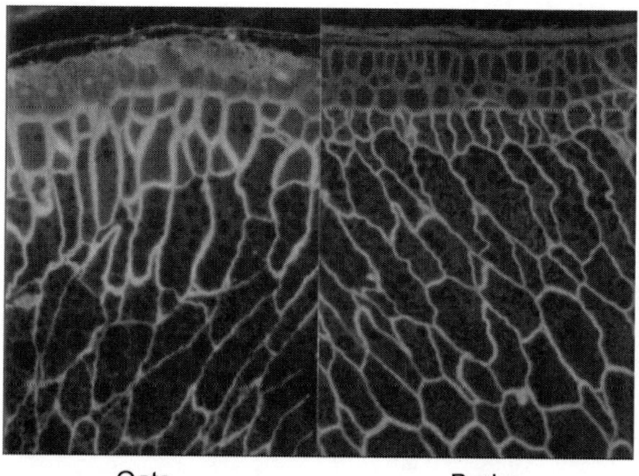

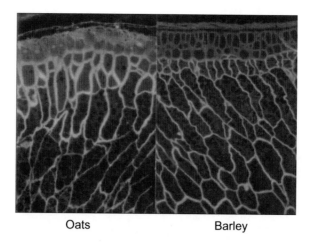

Figure 11.1 Fluorescence micrographs of sections of oat (left) and barley (right), showing the fluorescence (white) of calcofluor-stained β-glucan in the endosperm cell walls.[3]

Table 11.1 Amount (w/w) of β-glucans in four common cereals.

	β-glucan content
Barley	3–11%
Oat	3.2–6.8%
Rye	1–2%
Wheat	0.5–1%

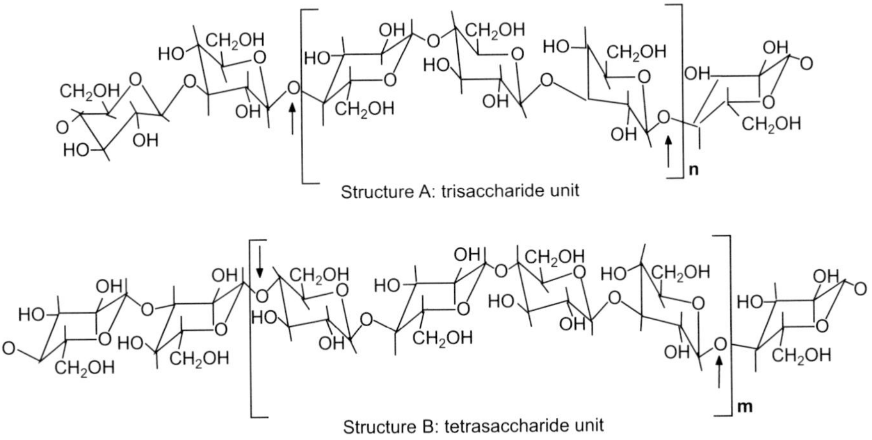

Figure 11.2 Chemical structure of cereal β-glucans.

(1→3)-linkage, which forms the two building blocks of cereal β-glucans: a cellotriosyl unit and a cellotetraosyl unit, as shown in Figure 11.2. The remaining 10% of the polymer chain is mainly composed of longer cellulosic sequences ranging from 5 up to 14 β-D-Glc*p* residues. The fingerprint of structural features of cereal β-glucans is represented by the oligosaccharide profiles released by a specific enzymatic hydrolysis (lichenase: EC 3.2.1.73., *endo*-1,3(4)-β-D-glucanase). Of the oligomers released, the trisaccharide and tetrasaccharide units are particularly important and their ratio constitutes the fingerprint of a particular grain.[7] For example, the tri/tetrasaccharide ratios for β-glucan from wheat, barley and oats are 4.5, 3.0 and 2.3, respectively (Table 11.2). A β-glucan, lichenan, extracted from *Cetraria islandica* (Iceland moss) almost exclusively contains the cellotriosyl unit: the ratio of the tri/tetrasaccharides is ~32, much higher than that in cereal β-glucans.

11.2.2 Functional Properties

11.2.2.1 Molecular Weight and Conformational Properties

Cereal β-glucans in aqueous solution adopt a disordered random coil conformation. Recent studies seem to suggest that there are no significant

Table 11.2 Oligosaccharide compositions of β-glucan from different sources.[a]

	Peak Area Percent%		*Total%*		*Molar ratio*
β-D-Glucan Source	*Tri*	*Tetra*	*Tri + tetra*	*Penta-nona*	*Tri/tetra*
Wheat	72.3	21.0	93.3	6.7	4.2–4.5
Barley	63.7	28.5	92.2	7.8	3.0
Oat	58.3	33.5	91.9	8.1	2.3
Rye	73.7	21.4	95.1	4.9	3.4
Lichenan	86.3	2.7	89.1	10.9	31.9

[a]Ref. 7 and 19.

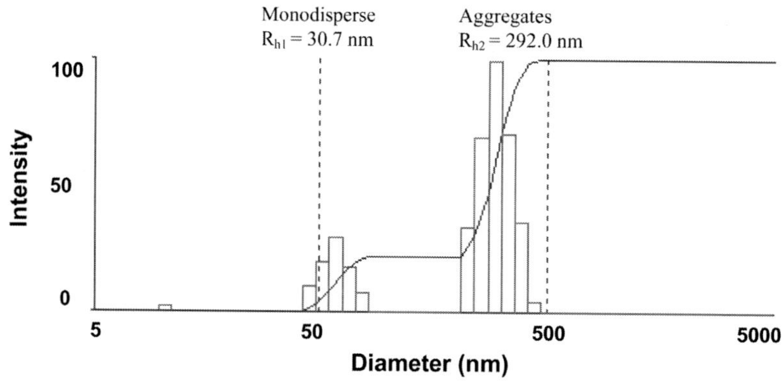

Figure 11.3 Particle size distribution of wheat β-glucan demonstrating the presence of aggregates in aqueous solutions.[10]

differences in molecular conformation between cereal β-glucans when measured in extreme dilute solution in a good solvent. The Mark–Houwink–Sakurada exponents obtained for oat, barley and wheat fall in the range of 0.70 ± 0.05, which implies an expanded, semi-flexible chain conformation. With such a conformation, the estimated persistence length, which is a measure of the chain stiffness, should not be much bigger than ~4 nm.[8] However, the measured values of persistence length were sometimes 10–20 times higher than 4 nm.[9] This is the result of molecular aggregation among β-glucan molecules through hydrogen bonding, which imparts great stiffness (and thus high solution viscosity) to the polymer. Recently, the existence of molecular aggregates in water solution was clearly demonstrated by dynamic light scattering measurements (Figure 11.3).[10] These aggregates could be effectively eliminated in 0.5 M sodium hydroxide, but not with other treatments, including the use of 6 M urea as a solvent and a repeated filtration process or ultrasonic treatment. The successes of obtaining an aggregate-free solution allowed the measurement of true molecular weight and conformational properties of single molecules; it confirmed an extended random coil conformation for all cereal β-glucans and the persistence length of wheat β-glucan was determined to be in the range of ~5 nm.[10]

The molecular weight of β-glucans in the cell wall matrix has so far not been measured. The apparent molecular weights obtained for isolated β-glucans were scattered in the range of 10^4–10^6 g mol^{-1}, depending on sources, methods

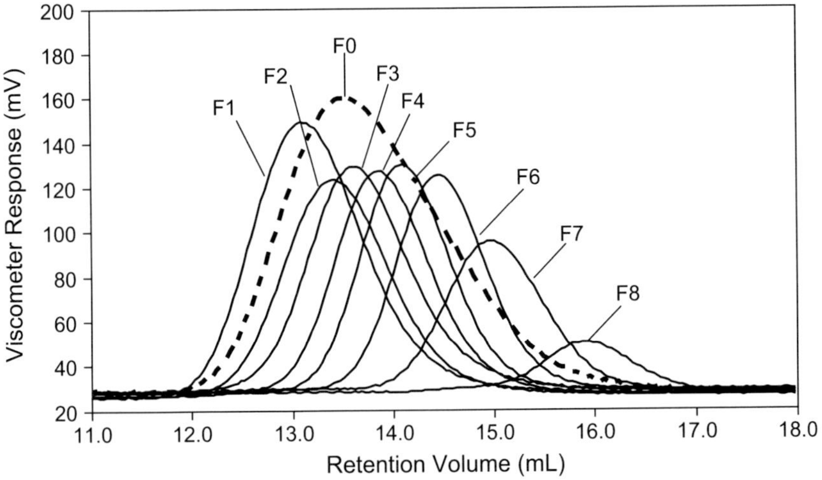

Figure 11.4 Molecular weight distribution of 6 fractions and unfractionated β-glucans isolated from barley.[11]

of isolation, degree of aggregation and determination techniques. Generally, the molecular weights of carefully isolated β-glucans follow the trend of oat >barley>rye>wheat, coinciding with their ease of extractability. One of the popular methods for determining molecular weight and its distribution is size exclusion chromatography (SEC). In the absence of a molecular weight detector, molecular weight standards are required to calibrate the columns. Isolated β-glucans are highly polydisperse in molecular weight, giving a broad molecular weight distribution.[11] Wang and co-workers developed a method which allowed separations of β-glucans into many fractions with low polydispersity, which are particularly suitable for the use as molecular weight standards (Figure 11.4).[11]

11.2.2.2 Viscosity and Gelation Properties

Semi-dilute and concentrated solutions of cereal β-glucans are appreciably different from one another as a consequence of varied chemical structures and molecular weight. For example, high molecular weight oat β-glucan solutions exhibit typical viscoelastic flow behavior, and do not form gels within a reasonable time period. Freshly prepared barley and wheat β-glucan solutions are also viscoelastic fluids as demonstrated in Figure 11.5. It is interesting to observe that a thermo-reversible gel could be formed if the β-glucan solutions are allowed to stand for a period of time under lower temperature.[12,13] Partially hydrolyzed oat β-glucans can also form a gel, but the gel development time is much longer[14,15]. Figure 11.6. shows the mechanical spectrum of a gel formed by 5% wheat β-glucan fraction (Mw = 340KD). It is interesting to note that the gel strength of cereal β-glucans increased with the decrease of molecular weight (unpublished data).

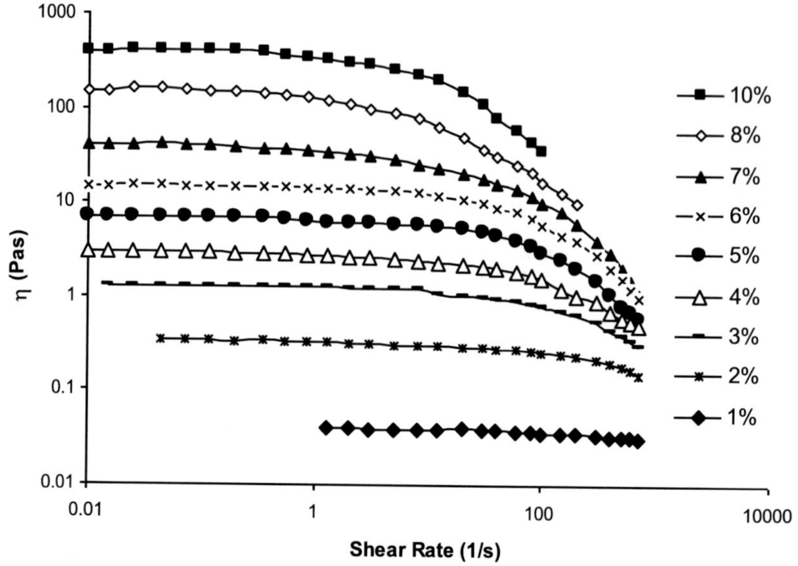

Figure 11.5 Flow behaviour of freshly prepared cereal β-glucans (barley β-glucan, $M_w = 210$ kDa) at different concentrations. η, viscosity; $\dot{\gamma}$, shear rate.[3]

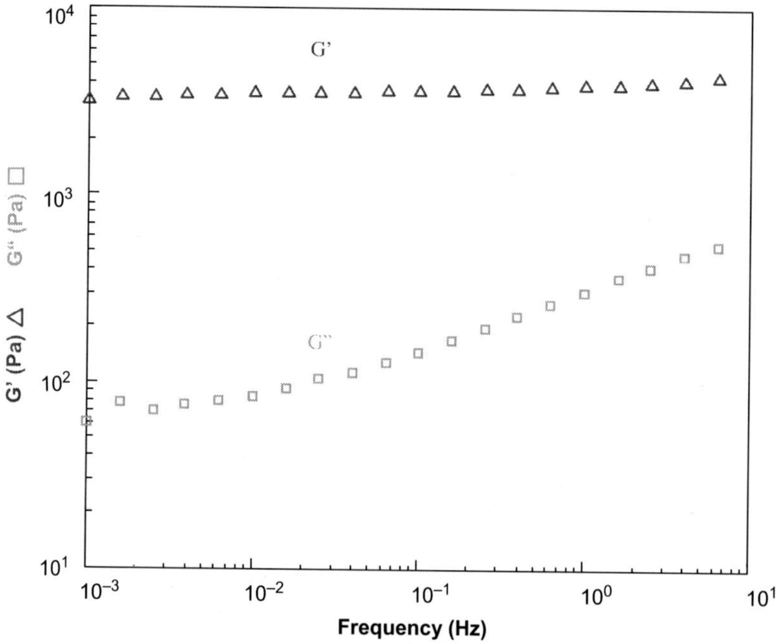

Figure 11.6 Frequency sweep of 5.0% wheat β-glucans fraction β (WF1, $M_w = 3.4 \times 10^4$) at 5 °C.[10]

Accumulating evidence suggests that the arrangement of cellotriosyl and cellotetraosyl units and their ratio in the polymer chain are important factors controlling the solution properties of these polysaccharides,[7,12,14,16] although earlier studies ascribed this mainly to the long runs of cellulosic segments.[17] Recent evidence from our laboratory suggested that sections of consecutive cellotriose units were mainly responsible for forming stable junction zones that lead to aggregation and gelation.[18] According to this mechanism, the gelation ability of cereal β-glucans is in the order of wheat > barley (rye) > oat, which has been repeatedly demonstrated experimentally.[7,14,19] Molecular weight is another important factor that exerts a significant influence on the solution properties of β-glucans. The gelation rate generally increases with a decrease in molecular weight, as shown in Figure 11.7 (it must be above the critical gelation chain length); this phenomenon is in disagreement with regular gelling polysaccharides in which longer polymer chains favor the formation of gels. The hypothesis for this phenomenon is that small β-glucan molecules have a higher mobility (less restriction of diffusion) than their large counterparts, and thus are more readily to interact with each other to form a stable junction zone; hence, the formation of aggregates. The further aggregation of the aggregates forms the three-dimensional gel networks under a concentrated solution regime.

11.2.3 Health Benefits and Applications

11.2.3.1 Health Benefits

Oat and barley β-D-glucans have demonstrated health benefits, including lowering cholesterol levels and attenuating postprandial glycemic response. The Food and Drug Administration (FDA) of the USA allowed a health claim for both oat and barley β-D-glucans for lowering the risk of coronary heart disease.[1,2] These claims were based on substantial scientific evidence from both animal models and human clinical trials. For example, Kalra and Jood[20] showed that barley β-glucan lowered the levels of total cholesterol, low-density lipoprotein (LDL)-cholesterol and triglycerides in rats, while Kahlon *et al.*[21] demonstrated that barley as well as oat β-glucan significantly lowered the cholesterol levels of hamsters. Bourdon *et al.*[22] reported that barley β-glucan lowered the cholesterol concentration and attenuated the insulin response in humans. Cavallero *et al.*[23] studied the effect of barley β-glucan on human glycemic response and found a linear decrease in glycemic index with the increase of β-glucan content. The presence of 5 g of oat β-glucan in extruded breakfast cereals caused a 50% decrease in glycemic response.[24] In a drinking model, both oat gum and guar gum significantly decreased the postprandial glucose rise and reduced the total and LDL-cholesterol levels in human subjects.[25,26] Wood later demonstrated that the observed physiological effect of oat β-glucan was positively correlated to the viscosity of the drink which is in turn determined by the molecular weight and concentration of the β–glucan used.[3] Cereal β-glucans also demonstrated other health benefits. For example, oat fibre prolongs satiety after meals and alleviates constipation.[27] It has been shown that the solubility of

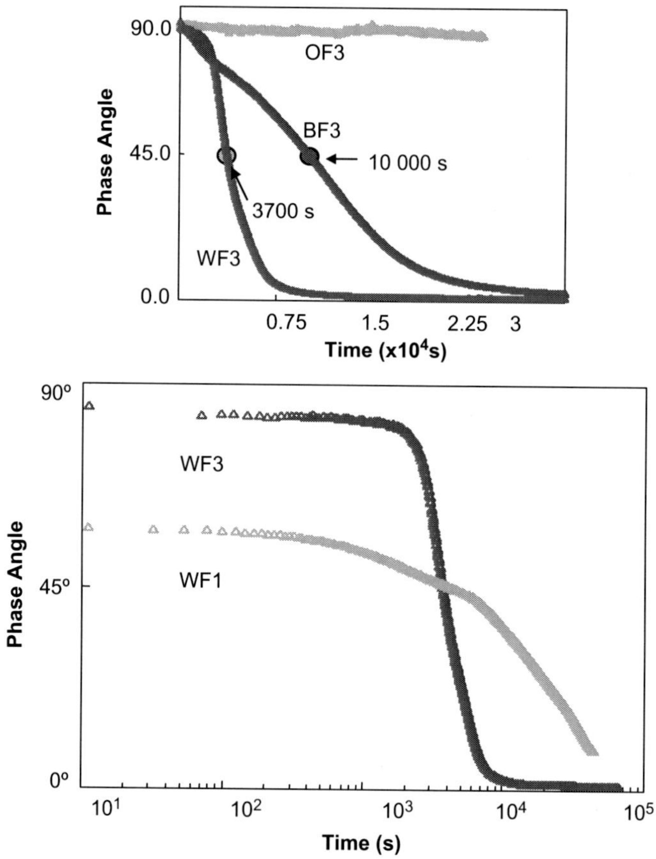

Figure 11.7 Effect of structure (top) and molecular weight (bottom) on the gelation rate of cereal β-glucans. A phase angle of less than 45° indicates the formation of a gel. OF-3, Oat β-glucan fraction 3; BF-3, Barley β–glucan fraction 3; WF1 and WF3, Wheat β-glucan fractions 1 and 3.[18]

the β-glucans is critical for exerting the reported positive physiological effect. By the same principle, the structural features, molecular weight, or any characteristics that could affect the viscosity producing properties of β-glucans, would have an impact on their physiological effect. However, there is lack of information on whether β-glucan gels have any significant influence on physiological effects.

11.2.3.2 Applications

Oat and barley flours and minimum processed bran products are readily available as food ingredients in North America, Europe and other parts of the world. Advanced processing and fermented oat products have emerged during the last decade, including Oatrim from oats, Glucagel™ from barley, and more recently *Viscofiber*® from both oats and barley. Oatrim is a product prepared by treating oat bran or oat flour with a thermo-stable α-amylase at high temperatures. The application of Oatrim in food includes bakery products, frozen desserts, processed meats, sauces and beverages. Oatrim is also used as a fat replacer.[28] Glucagel™ is based on barley β-D-glucans prepared by extraction and partial hydrolysis. The low molecular weight β-D-glucans (15 000 to 150 000 Daltons) form gels at 2% polymer concentration and above. The major functionality of Glucagel™ is its gelling and fat mimetic properties, which is used as fat substitutes in bakeries, dairy products, dressings, and edible films.[29] β-D-Glucan isolated from oats can also be used in the personal care industry, as a moisturizer in lotions and hand creams. *Viscofiber*® is a product recently developed by a Canadian company. It is a highly viscous product produced by a new technology using a combination of water–alcohol treatment of barley and oat flours with or without enzymes. The product is basically the intact cell walls of the endosperms. The current primary market for this product is as dietary supplements for the functional food and nutraceutical industry.[30] In the Scandinavian countries, oat-based dairy substitute products can be found on the market, such as beverages (Oatly), fermented oat products (*e.g.* oat yoghurt) and oat ice creams. With more evidence of the health benefits of cereal β-glucans and the development of novel processing technologies, it is predictable that the world-wide applications of β-glucans in the food, nutraceutical and cosmetic industries will surge in the near future.

11.3 Mushroom Glucans

11.3.1 Sources and Structural Features

Mushroom glucans exist as a structural component of fungal cell walls. A fungal cell wall is composed of two major glucan-type polysaccharides: one is a rigid fibrillar of chitin (or cellulose), the other one is a matrix of several polymers including β-glucan, α-glucan and glycoproteins.[31] Table 11.3 lists the source, type and bioactivities of some distinctive fungal glucans. Bioactive glucans can be isolated from mycelium, the fruiting body and sclerotium, which represent the three forms of a macrofungi in a life-cycle. All the species listed in Table 11.3 have been extensively studied in the past thirty years. Among them, schizophyllan from *Schizophyllum commune*, lentinan from *Lentinus edodes* and grifolan from *Grifola frondosa* have been commercialized as immunopotentiators, and have been used in clinical treatment of patients undergoing anticancer therapy. Their primary structure has been characterized as having a β-(1→3)-D-glucan as the backbone and β-(1→6)-glucose units as side chains. Other than β-(1→6)-branched β-(1→3)-D-glucans, the glucan-type polysaccharides isolated from mushroom species include glucose units with different types of glycosidic linkages, such as β-(1→4), β-(1→6), or α-(1→3) linkages. In most of the cases, the main chain consists of β-(1→3), β-(1→4), or mixed β-(1→3),

Table 11.3 Source, type and main activities of mushroom glucans.

Mushroom source	*Type*	*Main bioactivity*
Pleurotus tuberregium	β-D-glucan	Hepato-protective; anti-breast cancer
Auricularia auricula	Glucan	Hyperglycemia, immunomodulating, antitumor, anti-flammary, anti-radiative
Schizophyllum commune	(1→6)-branched β-(1→3)-Glucan (schizophyllan)	Antitumor
Lentinus edodes	(1→6)-branched β-(1→3)-Glucan (lentinan)	Immunomodulating, antitumor, antiviral
Sclerotinia sclerotiorum	(1→6)-branched β-(1→3)-Glucan scleroglucan (SSG)	Antitumor
Grifola frondosa	(1→6)-branched β-(1→3)-Glucan (grifolan)	Immunomodulating, antitumor, antiviral, hepatoprotective
Inonotus obliquus	Glucan	Antitumor, immunomodulating
Agaricus blazei	Glucan, glucan–protein, Glucomannan–protein complex	Antitumor
Flammulina velutipes	Glucan–protein complex	Antitumor, antiflammatory, antiviral, immunomodulating
Ganoderma applanatum	Glucan	Antitumor
Polypours umbellatus	Glucan	Antitumor, immunomodulating
Clitopilus caespitosus	Glucan	Antitumor
Omphalia lapidescens	Glucan	Antiflammory, immunomodulating
Phellinus linteus	Glucan	Antitumor
Dictyophora indusiata	Glucan	Antitumor, hyperlipidemia
Peziza vericulosa	Glucan	Immunomodulating, antitumor
Tricholoma mongolium	Glucan	Antitumor
Cordyceps sp	Glucan	Antitumor, immunomodulating, antitumor, heperglycemia

β-(1→4) with β-(1→6) side chains. In some mushroom species, glucans are found to bind with proteins or peptides in the form of glucan–protein/peptide complexes, which frequently show potent bioactivities.[32–34]

11.3.2 Helical Conformation

The helical conformation is an important structural feature found in antitumor mushroom glucans. A number of mushroom polysaccharides which have been used for treatment of certain cancers adopt a triple helical conformation in solution. They are schizophyllan, scleroglucan, lentinan, curdlan and cinerean.[25–42] A linear water-soluble (1→3)-β-D-glucan isolated from *Auricularia auricula* exists as a single helical chain in solution and shows very strong antitumor activity.[43] Since the helical conformational has been considered an important factor and plays a significant role in the biological recognition within cells, a basic understanding of the conformation and conformation transition is essential for medicinal applications. It has been shown that schizophyllan, which has the primary structure of (1→3)-β-D-glucan with one (1→6) branch for every three β-(1→3)-glucopyranoside linear linkages, adopts a triple helical conformation in water with the molecular parameters of 2170 nm^{-1} for molecular weight per contour length (M_L), 150 nm for persistence length (q) and 0.30 nm for distance per turn of helix (h).[36] It was found that the triple helices of schizophyllan could be changed to random coil by adding DMSO and this denaturation process is somewhat irreversible.[44] Such dissociation of triple helices and their reorganization has inspired researchers to study the interaction between a triple helical polysaccharide and DNA/RNA.[45–48] Sakurai *et al.* found that two single chains of schizophyllan could include one poly(C) chain, a polypeptide, to form a new triple helical complex. The complexation proceeds in a highly stoichiometric manner so that two schizophyllan repeating units bind to three poly(C) units.[49] Results from X-ray crystallographic studies indicated that the newly formed triple helical complex was quite similar in molecular parameters to that of the original polysaccharide, but with higher rigidity.[50] This interesting finding helps scientists to connect the biological function of triple helical polysaccharides to their interaction with DNA and RNA.[51]

11.3.3 Bioactivities

Extensive research on mushroom glucans has contributed to the wide application of these active biopolymers in pharmaceuticals and nutraceuticals. The reported bioactivities of mushroom glucans include antitumor, antiviral, antimicrobial activities and antioxidant properties. Other physiological effects, such as liver protection, antifibrotic effects, anti-inflammatory, antidiabetic and hypoglycemic activity, and hypocholesterolemic effects have also been reported. Among these reported bioactivities, the most promising ones are their immunomodulation and antitumor activities.

The involvement and importance of mushroom glucans in cancer treatment was first recognized more than 100 years ago when it was found that certain glucans could induce complete remission in patients with cancer.[52] Ever since the antitumor activity of macrofungal glucan was first reported by Chihara and his coworkers in the 1960s,[53] researchers have isolated structural diversified glucans with strong antitumor activity. Unlike traditional antitumor drugs, these substances produce an antitumor effect by activating various immune responses in the host and cause no harm to the body.[54]

Mushroom glucans have shown widely inhibitory effects towards many kinds of tumors including Sarcoma 180 solid cancers, Ehrlich solid cancer,

Sarcoma 37, Yoshida sarcoma and Lewis lung carcinoma.[54] The proposed mechanisms by which mushroom polysaccharides exert antitumor effect include: (1) the prevention of the oncogenesis (**cancer-preventing activity**); (2) enhancement of immunity against the bearing tumors (**immuno-enhancing activity**); and (3) direct antitumor activity to induce the apoptosis of tumor cells (**direct tumor inhibition activity**).

11.3.4 Structure–Bioactivity Relationship

Structural features such as (1→3)-β linkages in the main chain of the glucan and additional (1→6)-β branch points, have been indicated as important factors in antitumor action. β-Glucans containing mainly (1→6) linkages exhibit less activity, possibly due to their inherent flexibility of having too many possible conformations. However, antitumor polysaccharides may have other chemical structures, such as hetero-β-glucans,[55] β-glucan–protein,[56] α-glucan–protein.[55] A glucan receptor has been found on human macrophages, which has demonstrated high specificity for glucose in specific linkages.[57]

The triple-helical conformation of (1→3)-β-glucans is regarded as an important structural feature for the immuno-stimulating activity. A variety of biological and immuno-pharmacological activities of (1→3)-β-glucans are related to their triple helical conformation. For example, when lentinan was denatured with DMSO, urea, or sodium hydroxide, its tertiary structure was lost while its primary structure was maintained; its tumor inhibitory effect was lowered with progressive denaturation.[58] The same correlation between antitumor activity and triple helical structure was obtained for schizophyllan.[59] However, how exactly the triple helical conformation of (1→3)-β-glucan affects its antitumor action still remains unclear. Many of the biological and immuno-pharmacological activities, such as macrophage nitrogen oxide synthesis and limulus factor G activation, are dependent on the triple helical conformation; while other activities, such as synthesis of interferon-γ and colony stimulating factor, are independent on the triple helical conformation.[60] It has also been observed that the (1→3)-β-glucan backbone structure is of more importance than the tertiary structure of the molecule, and that helps to explain why (1→3)-α-mannan, having a similar backbone conformation to (1→3)-β-glucan, has shown comparable antitumor action to the latter. Mizuno *et al.* showed that high molecular weight (1→3)-β-glucans appeared to be more effective than those of low molecular weight in the molecular weight range of 500 to 2000 kDa.[61]

11.4 Curdlan: Microbial Produced β-Glucans

11.4.1 Preparation of Curdlan

Curdlan is a microbial fermented extracellular polysaccharide composed of β-(1→3) glucopyranosyl repeating units. It was first reported by Harada *et al.* in 1966 that a bacterial strain, *Alcaligenes faecalis* var. *myxogenes*, produced a water-insoluble and a thermo-gelling homoglucan.[62] Currently, it is commercially produced from a mutant strain of *Alcaligenes faecalis* var. *myxogenes* in pure culture fermentation, then extracted and purified. Curdlan has been extensively studied in Japan for its unique gel forming properties and is used as a key ingredient in various types of processed food, such as noodles, tofu and processed meat products.[63–70]

11.4.2 Chemical Properties and Molecular Characteristics

As shown in Figure 11.8, curdlan is an unbranched linear (1→3)-β-D-glucan with moderate molecular weights (degree of polymerization, DP ~ 450 to 12 000). Curdlan and other (1→3)-β-D-glucans can specifically bind with a number of dyes, including triphenylmethane dye, aniline blue,[71] and a benzophenone fluorochrome in the dye mixture.[72] It can also bind with Calcofluor and Congo Red to induce fluorescence; however, the interactions are not specific for (1→3)-β-glucans.[71,73]

Curdlan is insoluble in water, alcohols and most organic solvents, but dissolves in dilute bases (0.25 M NaOH), dimethylsulfoxide (DMSO) and formic acid. It is also soluble in some strong solvents such as *N*-methylmorpholino-*N*-oxide and lithium chloride in dimethylacetamide.[74]

The molecular structure of curdlan has been investigated either in solution, as gels or crystalline solids using X-ray diffraction, high-resolution solid-state ^{13}C NMR spectroscopy and atomic force microscopy (AFM). It is generally believed that curdlan forms junction zones consisting of parallel in-phase triple right-handed six-fold helices in the solid state, and forms an uncharged rigid rod-like conformation in solution.[75] Different morphological forms can be obtained when curdlan is precipitated from NaOH or DMSO solutions, ranging from endless microfibrils to spindle-shaped fibrils of various lengths, depending on the method of preparation.[76] For example, three forms of crystalline allomorphs have been regenerated from curdlan: one anhydrous form, and two hydrated forms. The anhydrous form was obtained by vacuum heat annealing to produce crystallites consisting of parallel molecules of (1→3)-β-D-glucan intertwined into triple helices.[77] The three chains in the triple helices are linked through strong intermolecular hydrogen bonds. The first hydrate form was obtained when annealing was performed under hydrothermal conditions. The hydration process simply expands the hexagonal unit cell, and permits the water molecules to enter into the inter-triplex space.[75] In this case, each

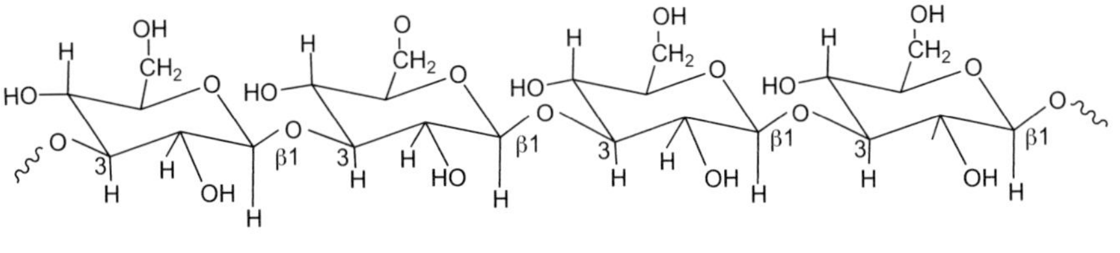

Figure 11.8 Structure of curdlan: linear (1→3)-β-D-glucan.

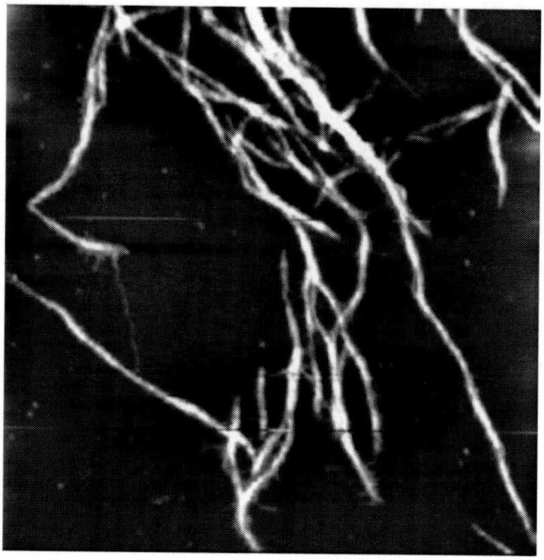

Figure 11.9 Atomic force micrograph of heat-treated curdlan deposited onto mica by AFM.[82]

crystalline hydrate contains two molecules of water per glucosyl residue; the presence of the water molecules causes the loss of symmetry of the crystallite. The second hydrated form was obtained by dialyzing alkaline solutions of curdlan against water or neutralizing alkaline solutions of the spray-dried polymer. Two conflicting propositions for curdlan structure were proposed for this allomorph. In the first case, it was believed that curdlan molecules occurred as highly hydrated single helices;[78,79] in the second case, curdlan was believed to form loosely intertwined triple helices.[80,81] Further experiments are needed to confirm the crystalline structures of the second hydrated form. In a recent study, Ikeda and Shishido[82] showed AFM images of curdlan solubilized in 0.01 M NaOH, which revealed that the majority of the molecules were in the form of heterogeneous supramolecular assemblies (as microfibrils) (Figure 11.9). The lengths of the microfibrils were in the order of micrometres and their cross-sectional heights were ~2–3 nm. In the same system, single molecular chains were also observed, which were partially dissociated from the microfibrils.[82]

11.4.3 Gelation Properties and Mechanisms

Curdlan can form gels with different characteristics.[74] Curdlan is insoluble in water, but its aqueous dispersion forms two types of gels upon heating and subsequent cooling: a low-set gel and a high-set gel. The low-set gel is thermo-reversible, and is typically obtained by heating the aqueous dispersion to a temperature between 55 and 60 °C, then cooling to below 40 °C. The high-set gel is thermo-irreversible, and is prepared by heating the aqueous dispersion to

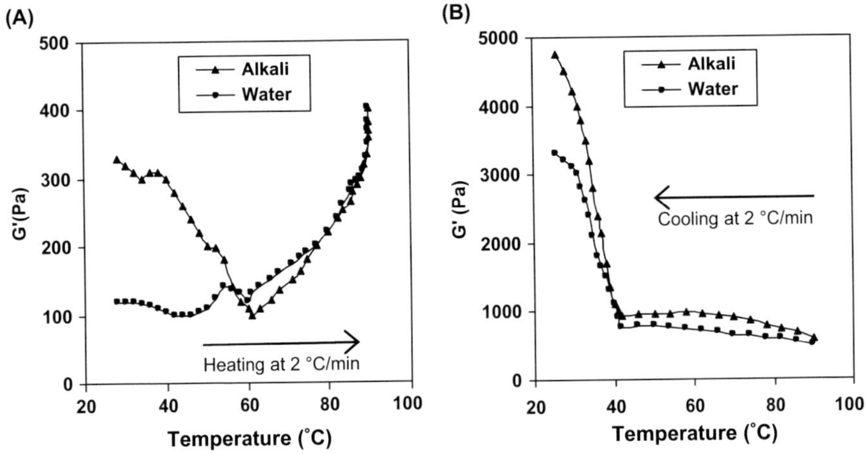

Figure 11.10 Heating (A) and cooling (B) curves of curdlan dispersions (•, water; ×, alkali) in water at 25 °C at 3%. G' is the storage modulus.[84]

above 80 °C.[83] Figure 11.10 demonstrates the heating and cooling curves of 3% curdlan dispersions in water and in alkaline solutions at 25 °C.

Many factors, including concentration, heating temperature, heating time, heating rate, and dispersing method, affect the mechanical properties of curdlan gels.[84,85] For example, the thermo-irreversibility cannot be 100% achieved if the heating temperature is below 100 °C.[63] An absolute thermo-irreversibility was reported when heating temperature was above 120 °C[86] or above 145 °C.[87] Swelling or hydration is also an important step for curdlan gels in the initial gelation process. The behavior of curdlan at this stage determines the subsequent gelling characteristics.[84] Salts can influence the gelling characteristics of curdlan in an aqueous system by affecting the mobility of water molecules through hydration.[88]

Curdlan gels are formed by the aggregation of rod-like triple helices through non-covalent bonding, such as hydrogen bonds, to form extended junction zones. At high temperatures, the triple-helical strands may unwind to give single chains, which will anneal to form triple helices when the temperature is lowered. In low-set gels, curdlan molecules are partially unwound and form a sort of helix *via* hydrogen bonding upon cooling. This was evidenced by the presence of some single helical chains[78,89] and supported by ^{13}C NMR data.[90] In high-set gels, high temperatures favor the unwinding of the triple helices into single chains, which may get involved in more than one complex interconnecting the triple helices.[89] In alkaline solution, the curdlan triple helix also unwinds and, on neutralization or dialysis against water, a low-set gel can be formed without heating. Such neutralised gels can be converted to irreversible high-set gels by heating the gel to above 80 °C.[63,91] The transition temperature of curdlan from thermo-reversible to irreversible gels is concentration- and molecular weight-dependent. The increase in concentration not only lowers the transition temperature from thermo-reversible to irreversible gels, but also decreases the amount of thermoreversible component in the final gel when heated at 90 °C.[63,92] The effect of molecular weight on transition temperature has been demonstrated by Nishinari's group: the lower the molecular weight of curdlan, the more significant an increase of the storage modulus G' can be observed.[93,94] This observation coincides with an earlier report by Harada *et al.* that lower molecular weight material annealed more successfully compared to its higher molecular weight counterparts.[95] It is worth noting that cereal β-glucans also exhibit similar gelation properties: a lower molecular weight favors the gelation process as described in section 11.2.2.2 of this chapter. Based on data from rheology and DSC, and the spin–spin relaxation time T_2 in proton NMR spectroscopy, Nishinari and Zhang concluded that the annealing of triplex structures is the main time-dependent phenomenon during the heating and cooling process.[93] By holding a curdlan gel close to its transition temperature, some of the "imperfect" triplexes can be melted out; at the same time, more "perfect" triplexes are formed by the annealing process. This annealing process will essentially lead to the formation of thermo-irreversible gels by converting the "imperfect" triplex into "perfect" triplexes. This conversion process was studied by X-ray diffraction analysis which showed that the single stranded helices in curdlan gels could be converted into triple stranded helices by heating at higher temperature, then cooling.[78,96]

11.4.4 Bioactivity and Physiological Effect

Like other (1→3)-β-D-glucans, curdlan and its derivatives have medical and pharmaceutical potential. These polysaccharides are known as biological response modifiers which can enhance or restore normal immune defences.[97] Broad immunomodulating and pharmacological responses have also been reported, including anti-inflammatory and anti-infective activities, antitumor activities,[98,99] wound repair, protection against radiation and anticoagulant activity.[5,100] Although the mechanism of its antitumor action is not completely clear, because curdlan cannot exert any direct action on tumor cells, it is believed that its antitumor action should rise from the interaction with the immunocompetent cells of the host. Curdlan derivatives have also been conjugated with other compounds in order to increase their immunological enhancement activity.[101] A curdlan sulfate demonstrated strong anti-HIV-1 activity *in vitro* due to its interference with the membrane fusion process during HIV-1 infection.[102]

The type of bioactivity and the effectiveness of curdlan and other (1→3)-β-D-glucans depends on their structural features, molecular weight and conformation, as described in section 11.3. However, studies on structure–activity relationships of curdlan revealed some controversial results. For example, several studies suggested that the intactness of the triple helical structure was important for receptor binding.[103–106] In another study, the partially opened triple helix was believed to be the biologically active form which induced inflammatory responses in rats.[107] Earlier studies showed that the single helix in curdlan was more potent than the triple helix as an anti-tumor agent.[108–112] It is apparent that further studies are still required in order to fully understand the mechanisms by which curdlan and its derivatives exert their bioactivities.

In order to improve the bioactivity of curdlan, attempts have been made to improve its solubility, such as partial hydrolysis and introduction of charged groups to the backbone chain. Partial hydrolysis can improve the solubility of curdlan, however, if the degree of hydrolysis is too severe (the $DP < 50$), curdlan loses its antitumor activity.[113] The introduction of charged groups could improve curdlan solubility and its bioactivity. For example, carboxymethyl ether and sulfate and phosphate esters of curdlans exhibit increased water solubility and enhanced biological activity.[45,114,115] A water-soluble aminated curdlan derivative demonstrated tumorigenic properties.[116] Curdlan sulfates of varying chain lengths and degrees and position of sulfation also show anticoagulant (antithrombotic) activity by interfering with the coagulation-dependent cascade at several points.[117] A curdlan sulfate has demonstrated anti-HIV activity[102] and inhibitory effects on the development of malarial parasites *in vitro*.[118]

Curdlan also has the normal property of dietary fibre. When feeding rats with curdlan, the caecal mass increased significantly, as did the proportion of short chain fatty acids and lactic acid in the caecal content.[119] The significant increase in the mass of the caecum was accompanied by a decrease in faecal mass. The transit time of the gastrointestinal tract was extended by curdlan supplementation.[120] Significant decrease was observed in the total hepatic cholesterol and low values were measured in the proportion with secondary bile acids. All those parameters revealed that curdlan is easily degraded and fermented by intestinal bacteria in the caecum and lowers cholesterol concentration in the liver.[119] Based on these immunomodulating responses, curdlan is proposed for use in cosmetic formulations[121] and as a protective agent for farmed fish.[122]

11.4.5 Applications of Curdlan

Curdlan has been used in Japan to improve the texture and water-holding capacity of processed foods and/or to create new types of foods based on its ability to form two types of gel as described earlier in the chapter.[64,65] The safety of curdlan has been assessed in animal studies and *in vitro* tests and it is approved for food use in Korea, Taiwan and Japan as an inert dietary fibre. It was approved by the US FDA in December 1996 as a formulation aid, processing aid, stabilizer and thickener or texturizer for use in food.[121]

The physical nature of curdlan is somewhat like starch: it can adsorb moisture reversibly from the atmosphere and caking could occur at a moisture content of 13–14% at 40 °C and a relative humidity of 62%. In most food applications, curdlan is used in the high-set, thermo-irreversible gel form and the gel is stable during retorting, deep-fat frying and cycles of freeze–thawing. The high-set gel has a more firm and resilient texture than the low-set gel.[65] Another advantage of the high-set gel is that it is tasteless, odorless and

Table 11.4 Function of curdlan in different food products.[a]

Gelling agent	Desert, jelly, jelly-like food, pudding, dry mixes
Bulking agent	Dietetic foods, diabetic foods, low-calorie foods
Film, fibre-former	Edible film, edible fibre casings
Improvement of visco-elasticity	Noodle, hamburger, sausage
Binding agent	Hamburger, starch jelly
Water holding agent	Noodle, sausage, ham, starch jelly
Prevention of deterioration from freeze-thawing processes	Frozen egg products
Masking of malodours or aromas	Boiled rice
Retention of shape	Starch jelly, dry desert mixes
Thickeners and stabilizers	Salad dressing, low-calorie foods, frozen foods
Coating agent	Flavor

[a]Ref. 123.

colorless. Due to the above characteristics, curdlan has been widely used in frozen and retorted foods.[66] Curdlan gels have also been used to develop new food products, such as freezable tofu and noodles. Since there are no digestive enzymes for curdlan in the upper alimentary tract, curdlan can be used as a fat-substitute. Curdlan gels can be used as a fat mimetic system by itself or in conjunction with other hydrocolloids. The texture of a non-fat sausage using a curdlan-based fat mimetic system was found to be similar to that of the control containing 20% fat as evaluated by rheological measurements.[68,69] Pulverized curdlan gels have been applied in emulsion-type, low-calorie foods, such as mayonnaise, whipped cream and dressing-type seasonings. These products were described as having desirable appearance, taste and texture.[122] Table 11.4 lists some of the most common applications of curdlan in the food industry.[123]

In a US patent, broad applications of curdlan have been claimed, which include using curdlan gels as electrophoresis gel media, immunoprecipitation gel media, as carriers for biological and pharmaceutical substances, targeted release coatings and coatings for biological materials, preparation of toothpaste and disposable contact lenses.[124] Based on the immunomodulating function, curdlan was proposed for use in cosmetic formulations[125] and as a protective agent for farmed fish.[126]

11.5 Concluding Remarks

Increased interests in scientific research of bioactive carbohydrates are attributed to the potentials of this kind of materials for maintaining, improving and enhancing human health. Dietary fibre, non-starch polysaccharides, oligosaccharides, glycoproteins and other carbohydrates have been subjected to closer examination for this purpose. Naturally occurring β-glucans described in the current chapter are only one group of the bioactive carbohydrates from nature or produced by biotechnology. The bioactivities, such as immunomodulation, antitumoral and anti-inflammatory, anti-oxidant activities observed for some of the β-glucans also have been observed for other types of carbohydrates. The challenge ahead of us is how we can fully understand the action mechanisms of those polysaccharides. Basic researches in the following area are just a start to approaching the answers from a chemist's point of view: fine structure of the interested poly- (oligo)saccharides; conformational properties such as shape, size, ability to form single, double or triple helical structures. In the same time, extensive studies are required for understanding the mechanisms of interactions of those bioactive polysaccharides with host cells, using both *in vitro* and *in vivo* models. With a full understanding of the structures and bio-functionalities, and their interrelationship, one could find/develop new polysaccharides with desired functionalities.

References

1. FDA Talk Paper, FDA allows whole oat foods to make health claim on reducing the risk of heart disease, T97-5, 1997.
2. FDA News, FDA Finalizes health claim associating consumption of barley products with reduction of risk of coronary heart disease, P06-70, 2006.
3. P. J. Wood, Relationships between solution properties of cereal β-glucans and physiological effects-a review, *Trends Food Sci. Technol.*, 2004, **13**, 313–320.
4. V. Ooi and F. Liu, A review of pharmacological activities of mushroom polysaccharides. *Int. J. Med. Mushrooms*, 1999, **1**, 195–201.
5. J. A. Bohn and J. M. BeMiller, (1→3)-β-D-Glucans as biological response modifiers: a review of structure-functional activity relationships, *Carbohydr. Polym.*, 1995, **28**, 3–14.
6. J. E. Dexter and P. J. Wood, Recent applications of debranning of wheat before milling, *Trends Food Sci. Technol.*, 1996, **7**, 35–41.
7. W. Cui, P. J. Wood, B. Blackwell and J. Nikiforuk, Physicochemical properties and structural characterization by 2 dimensional NMR spectroscopy of wheat β-D-glucan-comparison with other cereal β-D-glucans, *Carbohydr. Polym.*, 2000, **41**, 249–258.
8. C. Gomez, A. Navarro, C. Garnier, A. Horta and J. V. Carbonell, Physical and structural-properties of barley (1→3),(1→4)-beta-D-glucan III-formation of aggregates analyzed through its viscoelastic and flow behavior, *Carbohydr. Polym.*, 1997, **34**(3), 141–148.
9. A. Grimm, E. Kruger and W. Burchard, Solution properties of β-D-(1→3)(1→4)—glucan isolated from beer, *Carbohydr. Polym.*, 1995, **27**, 205–214.
10. W. Li, S. Cui and Q. Wang, Solution and conformational properties of wheat β-D-glucan studied by light scattering and viscometry, *Biomacromolecules*, 2006, **7**, 446–452.

11. Q. Wang, P. J. Wood, X. Haun and S. Cui, Preparation and characterization of molecular weight standards of low polydispersity from oat and barley (1→3)(1→4)-β-D-glucan, *Food Hydrocolloids*, 2003, **17**, 845–853.
12. N. Bohm and W. M. Kulicke, Rheological studies of barley (1-3)(1-4)-β-glucan in concentrated solution: investigation of the viscoelastic flow behaviour in the sol-state, *Carbohydr. Res.*, 1999, **315**, 293–301.
13. S. Cui, Polysaccharide gums from agricultural products: processing, structures and applications, Boca Raton: CRC Press. 2001, 103–166.
14. A. Lazaridou and C. G. Biliaderis, Cryogelation od cereal b-glucans: structure and molecular size effects, *Food Hydrocolloids*, 2004, **18**, 933–947.
15. S. M. Tosh, P. J. Wood, Q. Wang and J. Weisz, Structural characteristics and rheological properties of partially hydrolyzed oat b-glucan: the effects of molecular weigand hydrolysis method, *Carbohydr. Polym.*, 2004, **55**, 425–436.
16. M. Izawa, Y. Kano and S. Koshino, Relationship between structure and solubility of (1-3)(1-4)-β-D-glucan from barley, *J. Am. Soc. Brew. Chem.*, 1993, **51**, 123–127.
17. J. L. Doublier and P. J. Wood, Rheological properties of aqueous-solutions of (1→3)(1→4)-beta-D-glucan from oats (Avena-Sativa L), *Cereal Chem.*, 1995, **72**, 335–340.
18. W. Cui and Q. Wang, Structure-Function Relationships of Cereal β-Glucans, Oral presentation at the 8th International Hydrocolloids Conference, held in Trondheim, Norway, 2006.
19. S. M. Tosh, Y. Brummer, P. J. Wood, Q. Wang and J. Weisz, Evaluation of structure in the formation of gels by structurally diverse (1→3)(1→4)-β-D-glucans from four cereal and one lichen species, *Carbohydr. Polym.*, 2004, **57**, 249–259.
20. S. Kalra and S. Jood, Effect of dietary barley β-glucan on cholesterol and lipoprotein fractions in rat, *J. Cereal Sci.*, 2000, **31**, 141–145.
21. T. S. Kahlon, F. I. Chow, B. E. Knuckles and M. M. Chiu, Cholesterol-lowering effects in hamsters of beta-glucan-enriched barley fraction, dehulled whole barley, rice bran, and oat bran and their combinations, *Cereal Chem.*, 1993, **70**, 435–440.
22. P. Bourdon, W. Yokoyama, P. Davis, C. Hudson, R. Backus and D. Richter, Postprandial lipid, glucose, insulin, and cholecystokinin responses in men fed barley pasta enriched with β-glucan, *Am. J. Clin. Nutr.*, 1999, **69**, 55–63.
23. A. Cavallero, S. Empilli, F. Brighenti and M. Stanca, High (1→3)-(1→4)β-glucan barley fractions in bread making and their effects on human glycemic response, *J. Cereal Sci.*, 2002, **36**, 59–66.
24. L. Tappy, E. Gugolz and P. Wursch, Effects of breakfast cereals containing various amounts of beta-glucan fibers on plasma-glucose and insulin responses in NIDDM subjects, *Diabetes Care*, 1996, **19**, 831–834.
25. P. J. Wood, J. T. Braaten, F. W. Scott, D. Riedel and L. M. Poste, Comparisons of viscous properties of oat and guar gum and the effects of

these and oat bran on glycemic Index, *J. Agric. Food Chem.*, 1990, **38**, 753–757.
26. J. T. Braaten, P. J. Wood, F. W. Scott, M. S. Wolynetz, M. K. Lowe and P. Bradleywhite, Oat beta-glucan reduces blood cholesterol concentration in hypercholesterolemic subjects, *Eur. J. Clin. Nutr.*, 1994, **48**, 465–474.
27. Y. Mälkki and E. Virtanen, Gastrointestinal effects of oat bran and oat gum: a review, *Lebensm-Wiss. Technol.*, 2001, **34**, 337–347.
28. S. Lee, S. Kim and G. E. Inglett, Effect of shortening replacement with oatrim on the physical and rheological properties of cakes, *Cereal Chem.*, 2005, **82**, 120–124.
29. K. R. Morgan, C. J. Roberts, S. J. B. Tendler, M. C. Davies and P. M. Williams, A ^{13}C CP/MAS NMR spectroscopy and AFM study of the structure of GlucagelTM, a gelling β-glucan from barley, *Carbohydr. Res.*, 1999, **315**, 169–179.
30. T. Vasanthan, Grain fiber compositions and methods of use. *US Pat.* 20050208145, 2005.
31. J. Ruiz-Herrera, Fungal glucans. In *Fungal Cell Wall*, (Ed.) J. Ruiz-Herrera, CRC Press, 1992, 59–67.
32. Z. Cun, T. Mizuno, H. Ito, K. Shimura, T. Sumiya and M. Kawade, Antitumor activity and immunological property of polysaccharides from the mycelium of liquid-cultured Grifola Frondosa, *J. Jpn. Soc. Food Sci. Technol.*, 1994, **41**, 724–733.
33. S. C. Jong, J. M. Birmingham and S. H. Pai, Immunomodulatory substances of fungal origin, *J. Immunol. Immunopharmacol.*, 1991, **11**, 115–122.
34. T. Mizuno and C. Zhuang, Grifola frondosa: pharmacological effects, *Food Res. Int.*, 1995, **11**, 135–149.
35. T. Yanaki, T. Norisuye and H. Fujita, Triple helix of *schizophyllum commune* polysaccharide in dilute solution. 3. Hydrodynamic properties in water, *Macromolecules*, 1980, **13**, 1462–1466.
36. Y. Kashiwagi, H. Fujita and T. Norisuye, Triple Helix of Schizophyllum-commune polysaccharide in dilute-solution. 4. Light-scattering and viscosity in dilute aqueous sodium-hydroxide, *Macromolecules*, 1981, **14**, 1220–1225.
37. T. Sato, T. Norisuye and H. Fujita, Trip helix of *schizophyllum commune* polysaccharide in dilute solution. 5. Light scattering and refractometry in mixtures of water and dimethyl sulfoxide, *Macromolecules*, 1983, **16**, 185–189.
38. H. Saito, Y. Yoshioka, M. Yakoi and J. Yamada, Distinct gelation mechanism between linear and branched (1→3)-β-D-glucans as revealed by high-resolution solid-state carbon-13 NMR, *Biopolymers*, 1990, **29**, 1689–1698.
39. M. Gawronski, G. Aguirre, H. Conrad, T. Springer and K. P. Stahmann, Molecular structure and precipitates of a rod-like polysaccharide in aqucous solution by SAXS experiments, *Macromolecules*, 1996, **29**, 1516–1520.

40. M. Gawronski, N. Donkai, T. Fukuda, T. Miyamoto, H. Conrad and T. Springer, Triple helix of the polysaccharide cinerean in aqueous solution, *Macromolecules*, 1997, **30**, 6994–6996.
41. M. Gawronski, H. Conrad, T. Springer and K. P. Stahmann, Conformational changes of the polysaccharide cinerean in different solvents from scattering methods, *Macromolecules*, 1996, **29**, 7820–7825.
42. T. Itou, A. Teramoto, T. Matsuo and H. Suga, Ordered structure in aqueous polysaccharide. 5. Cooperative order-disorder transition in aqueous schizophyllan, *Macromolecules*, 1986, **19**, 1234–1240.
43. L. Zhang and L. Q. Yang, Properties of auricularia-auricula-judae beta-D-glucan in dilute-solution, *Biopolymers*, 1995, **36**, 695–700.
44. M. Heinrich and B. A. Wolf, On the kinetics of phase separation-trapping of molecules in non-equilibrium phases, *Macromolecules*, 1990, **23**, 590–596.
45. K. Koumoto, M. Mizu, K. Sakurai, T. Kunitake and S. Shinkai, Polysaccharide/polynucleotide complexes. Part 6. Complementary-strand-induced release of single stranded DNA bound in the *schizophyllan complex*, *Chemistry and Biodiversity*, 2004, **1**, 520–529.
46. M. Mizu, K. Koumoto, T. Kimura, K. Sakurai and S. Shinkai, Protection of polynucleotides against nuclease-mediated hydrolysis by complexation with schizophyllan, *Biomaterials*, 2004, **25**, 3109–3116.
47. D. H. Yang, A. H. Bae, K. Koumoto, S. W. Lee, K. Sakuraia and S. Shinkai, *In situ* monitoring of polysaccharide-polynucleotide interaction using a schizophyllan-immobilized QCM device, *Sens. Actuators, B*, 2005, **105**, 490–494.
48. T. Anada, H. Matsunaga, R. Karinaga, K. Koumoto, M. Mizu and K. Nakano, Proposal of new modification technique for linear double-stranded DNAs using the polysaccharide schizopyllan, *Bioorg. Med. Chem. Lett.*, 2004, **14**, 5655–5659.
49. K. Sakurai, M. Mizu and S. Shinkai, Polysaccharide-polynucleotide complexes. 2. Complementary polynucleotide mimic behavior of the natural polysaccharide schizophyllan in the macromolecular complex with single-stranded RNA and DNA, *Biomacromolecules*, 2001, **2**, 641–650.
50. M. Mizu, K. Koumoto, T. Kimura, K. Sakurai and S. Shinkai, Polysaccharide-polynucleotide complexes, *Polym. J.*, 2003, **35**, 714–720.
51. M. Numata, T. Matsunoto, M. Umeda, K. Koumoto, K. Sakurai and S. Shinkai, Polysaccharide-polynucleotide complexes (15): thermal stability of schizophyllan (SPG)/poly(C) triple strands is controllable by a-amino acid modification, *Bioorg. Chem.*, 2003, **31**, 163–171.
52. H. C. Nauts, W. E. Swift and B. L. Coley, The treatment of malignant tumors by bacterial toxins as developed by the late William B. Coley, reviewed in the light of modern research, *Cancer Res.*, 1946, **6**, 205–214.
53. G. Chihara, The antitumor polysaccharide lentinan: an overview. In *Manipulation of Host Defense Mechanism,* (Eds) T. Aoki, *et al.*, Nature, 1969, **222**, 687–694.

54. S. P. Wasser and A. L. Weis, Medicinal properties of substances occurring in Higher Basidiomycetes Mushroom: current perspective, *Int. J. Med. Mushrooms*, 1999, **1**, 31–51.
55. T. Mizuno, H. Saito, T. Nishitoba and H. Kawagashi, Antitumor-active substances from mushrooms, *Food Res. Int.*, 1995, **11**, 23–61.
56. H. Kawagishi, T. Kanao, R. Inagaki, T. Mizuno, K. Shimura and H. Ito, Formulation of a potent antitumor (1-6)-beta-D-glucan-protein complex from Agaricus blazei fruiting bodies and antitumor activity of the resulting products, *Carbohydr. Polym.*, 1990, **12**, 393–404.
57. Y. J. Lombard, A new method for studying the binding and ingestion of zymosan particles by macrophages, *J. Immunol. Methods*, 1994, **174**, 155–163.
58. Y. Y. Maeda, S. T. Watanabe, C. Chihara and M. Rokutanda, Denaturation and renaturation of a beta-1,6-1,3-glucan, lentinan, associated with expression of T-cell-mediated responses, *Cancer Res.*, 1988, **48**, 671–675.
59. T. Yanaki, W. Ito and K. Tabata, Correlation between antitumor activity of schizophyllan and its triple helix, *Agr. Biol. Chem.*, 1986, **509**, 2415–2416.
60. T. Yadomae, N. Ohno, Sparassis crispa Fr. Extract, *Jpn. Pat.*, 2000-217543 2000, 8 August.
61. T. Mizuno, P. Yeohlui, T. Kinoshita, C. Zhuang, H. Ito and Y. Mayuzymi, Antitumor activity and chemical modification of polysaccharides from Niohshimeiji mushroom, Tricholoma giganteum, *Biosci., Biotechnol., Biochem.*, 1996, **60**, 30–33.
62. T. Harada, M. Masada, K. Fulimori and I. Maeda, Production of a firm, resilient gel-forming polysaccharide by a mutant of Alcaligenes faecalisvar. myxogenes 10C3, *Agr. Biol. Chem.*, 1966, **30**, 196–198.
63. T. Funami, M. Funami, T. Tawada and Y. Nakao, Decreasing oil uptake of doughnuts during deep-fat frying using curdlan, *J. Sci.*, 1999, **64**, 883–888.
64. T. Harada, M. Terasaki and A. Harada, Curdlan. In *Industrial gums*, (Eds) R. L. Whistler and J. N. BeMiller, 3rd edn, San Diego, CA: Academic Press, 1993, 427–445.
65. M. Miwa, Y. Nakao and K. Nara, Food applications of curdlan. In *Food Hydrocolloids, Structures, Properties, and Functions*, (Eds) K. Nishinary and E. Doi, Plenum Press, New York, 1994, 119–124.
66. Y. Nakao, A. Konno, T. Taguchi, T. Tawada, H. Kasai and J. Toda, Curdlan: Properties and application to foods, *J. Food Sci.*, 1991, **776**, 769–772.
67. T. Funami and Y. Nakao, Effects of curdlan on the rheological properties and gelling process of meat gels under a model system using minced pork, *Nipon Shokuhin Kogaku Kaishi*, 1996, **43**, 21–28.
68. T. Funami, H. Yada and Y. Nakao, Thermal and rheological properties of curdlan gel in minced pork gel, *Food Hydrocolloids*, 1998, **12**, 55–64.
69. T. Funami, H. Yada and Y. Nakao, Curdlan properties for application in fat mimetics for meat products, *J. Food Sci.*, 1998b, **63**, 283–287.

70. T. Funami, F. Yotsuzuka, H. Yada and Y. Nakao, Thermo-irreversible characteristics of curdlan gels in a model reduced fat pork sausage, *J. Food Sci.*, 1998, **63**, 575–579.
71. I. Nakanishi, K. Kimura, S. Kusui and E. Yamazaki, Complex formation of gel-forming bacterial (1-3)-β-D-glucans (curdlan-type polysaccharides) with dyes in aqueous solution, *Carbohydr. Res.*, 1974, **32**, 47–52.
72. N. A. Evans, P. A. Hoyne and B. A. Stone, Characteristics and specificity of the interaction of a fluorochrome from aniline blue (Sirofluor) with polysaccharides, *Carbohydr. Polym.*, 1984, **4**, 215–230.
73. P. J. Wood and R. G. Fulcher, Interaction of some dyes with cereal β-glucans, *Cereal Chem.*, 1978, **55**, 952–966.
74. F. Yotsuzuka, Curdlan. In *Handbook of dietary fiber*, (Eds) S. S. Cho and M. L. Dreher, Marcel Dekker, Inc., New York, 2001, 737–757.
75. C. T. Chuah, Y. Deslandes, R. H. Marchessault and A. Sarko, Packing analysis of carbohydrates and polysaccharides. 14. triple-helical crystalline-structure of curdlan and paramylon hydrates, *Macromolecules*, 1983, **16**, 1375–1382.
76. A. Koreeda, T. Harada, K. Ogawa, S. Sato and N. Kasai, Study of the ultrastrucuture of gel-forming $(1\rightarrow3)$β-D-glucan (curdlan-type polysaccharide) by electron microscopy, *Carbohydr. Res.*, 1974, **33**, 396–399.
77. Y. Deslandes, R. H. Marchessault and A. Sarko, Packing analysis of carbohydrates and polysaccharides. 13. triple-helical structure of (1-3)-beta-D-glucan, *Macromolecules*, 1980, **13**, 1466–1471.
78. K. Okuyama, A. Otsubo, Y. Fukuzawa, M. Ozawa, T. Harada and N. Kasai, Single-helical structure of native curdlan and its aggregation state, *J. Carbohydr. Chem.*, 1991, **10**, 645–656.
79. H. Saito, M. Yokoi and Y. Yoshida, Effect of hydration on conformational change or stabilization of (1-3)-β-D-glucans of various chain lengths in solid state as studied by high-resolution solid-state ^{13}C NMR spectroscopy, *Macromolecules*, 1989, **22**, 3892–3898.
80. W. S. Fulton and E. Atkins, The gelling mechanism and relationship to molecular structure of microbial polysaccharide curdlan. In *Fibre Diffraction Methods.*, eds. A. D. French, K. H. Gardner, American Chemical Society, Washington DC, 1980, 385–410.
81. A. J. Stipanovic, P. J. Giammatto and S. R. Vasconcellos, Characterization and applications of viscoelastic solutions and gels of water-soluble microbical polysaccharideds, *Polym. Mater.: Sci. Eng.*, 1987, **57**, 260–264.
82. S. Ikeda and Y. Shishido, Atomic force microscopy Studies on heat-induced gelation of curdlan, *J. Agr. Food Chem.*, 2005, **53**, 786–791.
83. A. Konno and T. Harada, Thermal properties of curdlan in aqueous suspension and curdlan gel, *Food Hydrocolloids*, 1991, **5**, 427–434.
84. T. Funami, M. Funami, H. Yada and Y. Nakao, A rheological study on the effects of heating rate and dispersing method on the gelling characteristics of curdlan aqueous dispersions, *Food Hydrocolloids*, 2000, **14**, 509–518.

85. I. Maeda, H. Saito, M. Masada, A. Misaki and T. Harada, Properties of gels formed by heat treatment of curdlan, a bacterial β-1, 3 glucan, *Agr. Biol. Chem.*, 1967, **31**, 1184–1188.
86. M. Hirashima, T. Takaya and K. Nishinari, DSC and rheological studies on aqueous dispersions of curdlan, *Thermochim. Acta*, 1997, **306**, 109–114.
87. A. Konno, K. Okuyama, A. Koreeda, A. Harada and Y. Kanazawa, Molecular association and dissociation in formation of curdlan gels. In *Food Hydrocolloids, Structures, Properties and Function*, (Eds) K. Nishinari and E. Doi, Plenum Press, New York, 1994, 113–118.
88. T. Funami and K. Nishinari, Gelling characteristics of curdlan aqueous dispersions in the presence of salts, *Food Hydrocolloids*, 2007, **21**, 59–65.
89. N. Kasai and T. Harada, Ultrastructure of curdlan, *Am. Chem. Soc. Symp.*, 1980, **141**, 363–383.
90. H. Saito, E. Miyata and Y. Sasaki, A 13C nuclear magnetic resonance study of gel-forming (1→3)-β-D-glucans: Molecular-weight dependence of helical conformation and of the presence of junction zones for association of primiary molecules, *Macromolecules*, 1978, **11**, 1244.
91. T. Tada, T. Matsumoto and T. Masyda, Influence of alkaline concentration on molecular association structure and viscoelastic properties of curdlan aqueous systems, *Biopolymers*, 1997, **42**, 479–487.
92. T. Funami, M. Funami, H. Yada and Y. Nakao, Rheological and thermal studies on gelling characteristics of curdlan, *Food Hydrocolloids*, 1999a, **13**, 317–324.
93. K. Nishinari and H. Zhang, Recent advances in the understanding of heat set gelling polysaccharides, *Trends Food Sci. Technol.*, 2004, **15**, 305–312.
94. H. Zhang, K. Nishinari, M. Williams, T. Forster and I. Norton, A molecular description of the gelation mechanism of curdlan, *Int. J. Biol. Macromol.*, 2002, **30**, 7–16.
95. T. Harada, K. Okuyama, A. Konno, A. Koreeda and A. Harada, Effect of heating on formation of curdlan gels, *Carbohydr. Polym.*, 1994, **24**, 101–106.
96. T. Takeda, N. Yasuoka, N. Kasai and T. Harada, X-ray structure studies of (1→3)-b-D-glucan (curdlan), *Polym. J.*, 1978, **10**, 365–368.
97. R. Medzhitov and C. A. Janeway, Decoding the patterns of self and nonself by the innate immune system, *Science*, 2002, **12**, 298–300.
98. Y. Sasaki, M. Ishiye, H. Goto and T. Kamikubo, Purification and subunit structure of RNA polymerase II from the pea, *Biochim. Biophys. Acta, Nucleic Acids Protein Synth.*, 1979, **564**, 437–447.
99. M. Takahashi, T. Hirato, S. Nakai and Y. Hirai, Investigation of structure-function relationship of human macrophage colony-stimulating factor(M-CS): C-terminally truncated recombinant human M-CSFS are biologically active, *Int. J. Immunopharmacol.*, 1988, **10**, 56–60.
100. B. A. Stone and A. E. Clarke, *Chemistry and Biology of (1-3)-β-glucans*, La Trobe University Press, Bundoora, Australia, 1992, 808.

101. Y. Ohya, T. Nishimoto, J. Murata and T. Ouchi, Immunological enhancement activity of muramyl dipeptide analogue/CM-curdlan conjugate, *Carbohydr. Polym.*, 1994, **23**, 47–54.
102. P. Jagodzinski, R. Wiaderkiewicz, G. Kurzawski, M. Kloczewiak, H. Nakashima and E. Hyjek, Mechanism of the inhibitory effect of curdlan sulfate on HIV-1 infection *in vitro*, *Virology*, 1994, **202**, 735–745.
103. A. Mueller, J. Raptis, P. J. Rice, J. H. Kalbfleisch, R. D. Stout and H. E. Ensley, The influence of glucan polymer structure and solution conformation on binding to (1→3)-β-D-glucan receptor in a human monocyte-like cell line, *Glycobiology*, 2000, **10**, 339–346.
104. K. Kataoka, T. Muta, S. Yamazaki and K. Takeshige, Activation of macrophages by linear (1→3)-β-D-glucans, *J. Biol. Chem.*, 2002, **277**, 36825–36851.
105. B. H. Falch, T. Espevik, L. Ryan and B. T. Stokke, The cytokine stimulating activity of (1→3)-β-D-glucan is dependent on the triple helix conformation, *Carbohydr. Res.*, 2000, **329**, 587–596.
106. T. Kojima, K. Tabata, W. Itoh and T. Yanaki, Molecular weight dependence of the antitumor activity of schizophyllan, *Agr. Biol. Chem.*, 1986, **50**, 231–232.
107. S. Young, V. A. Robinson, M. Barger, D. G. Frazer and V. Castranova, Partially opened triple helix is the biologically active conformation of 1,3-β-glucans that induces pulmonary inflammation in rats, *J. Toxicol. Environ. Health*, 2003, **66**, 551–563.
108. J. Aketagawa, S. Tanaka, H. Tamura, Y. Shibata and H. Saito, Actiivation of limulus coagulation factor G by several (1→3)-beta-D-glucans: comparison of the potency of glucans with identical degree of polymerization but different conformations, *J. Biochem.*, 1993, **113**, 683–686.
109. N. Nagi, N. Ohno, Y. Adachi, J. Adetagawa, H. Tamura and Y. Shibata, Application of limulus test (G pathway) for the detection of different conformers of (1→3)-beta-D-glucans, *Biol. Pharm. Bull.*, 1993, **16**, 822–828.
110. N. Ohno, T. Hashimoto, Y. Adachi and T. Yadomae, Corrigendum to: "conformation dependency of nitric oxide synthesis of murine peritoneal macrophages by β-glucans *in vitro*", *Immunol. Lett.*, 1996, **53**, 157–163.
111. H. Saito, Y. Yoshioka, N. Uehara, J. Aketagawa, S. Tanaka and Y. Shibata, Relationship between conformation and biological response for (1,3)-β-glucan in the activation of coagulation factor G from limulus amebocyte lysate and host-mediated antitumor activity. Demonstration of single-helix conformation as a stimulant, *Carbohydr. Res.*, 1991, **217**, 181–190.
112. T. Suzuki, N. Ohno, K. Saito and T. Yadomae, Activation of the complement system by (1→3)-beta-D-glucans having different degrees of branching and different ultrastructures, *J. Pharmacobio-dynam.*, 1992, **15**, 277–285.
113. T. Sasaki, N. Abiko, Y. Sugino and K. Nitta, Dependence on chain length of antitumour activity of (1,3)-β-D-glucan fromo alcaligenes faecalis var.

myxogenes IFO13140 and its acid-degraded products., *Cancer Res.*, 1978, 379–383.

114. S. Honda, T. Asano, A. Kakinuma and H. Sugino, Activation of the Alternative Pathway of Complement by an Antitumor (1-)3)-Beta-D-Glucan from Alcaligenes-Faecalis Var Myxogenes Ifo-13140, and Its Lower Molecular-Weight and Carboxymethylated Derivatives, *Immunopharmacology*, 1986, **11**, 29–37.
115. T. Toida, A. Chaidedgumjorn and T. J. Linhardt, Structure and bioactivity of sulphated polysaccharides, *Trends Glycosci. Glycotechnol.*, 2003, **15**, 29–46.
116. R. Seljelid, A water-soluble aminated beta-1-3-D-glucan derivative causes regression of solid tumors in mice, *Biosci. Rep.*, 1986, **6**, 845–851.
117. S. Alban and G. Franz, Partial synthetic glucan sulphates as potential new antithromboties: A review, *Biomacromolecules*, 2001, **2**, 354–361.
118. S. G. Evans, D. Morrison, Y. Kaneko and I. Havlik, The effect of curdlan sulphate on development in vitro of plasmodium falciparum, *Trans. R. Soc. Trop. Med. Hyg.*, 1998, **92**, 87–89.
119. J. Shimizu, M. Wada, T. Takita and S. Innami, Curdlan and gellan gum bacterial gel-forming polysaccharides exhibit different effects on lipid metabolism, cecal fermentation and fecal bile acid excretion in rats, *J. Nutr. Sci. Vitaminol.*, 1999, **45**, 251–262.
120. M. Tetsuguchi, S. Nomura, M. Katayama and Y. Sugawaka-tayama, Effects of curdlan and gellan gum on the surface structure of intestinal mucosa in rats, *J. Nutr. Sci. Vitaminol.*, 1997, **43**, 515–527.
121. W. B. Davis, Unique bacterial polysaccharide polymer gel in cosmetics. Pharmaceuticals and Foods, *US Pat.*, 1992, **5**, 772–773.
122. I. Y. Lee, E. J. Vandamme and A. Steinbuchel, *Boipolymers 5, Polysccharides I: Polysaccharides from prokaryotes*, Wiley, 2002, 135–158.
123. FDA, Food additives permitted for direct addition to food for human consumption: curdlan, 21 CFR Part 172, US Food and Drug Administration, 1996, **61**, 65941–65942.
124. Y. Okura, T. Tawada and Y. Nakao, Emulsion-type, low-calorie foods with desirable mouthfeel, appearance and taste properties, in which part or all of the fat is replaced by a polvmerized curdlan gel. (Takeda Chemical Indusffies Ltd, Japan), *US Pat.*, 5 360 624, 1994.
125. E. J. F. Spicer, E. I. Goldenthal and T. Ikeda, A toxicological assessment of curdlan, *Food Chem. Toxicol*, 1999, **37**, 455–479.
126. D. W. Renn, L. E. Dumont, W. C. Snow and F. P. Curtis, Beta.-1,3-glucan polysaccharides compositions, and their preparation and uses, *United States Pat.,* 5 688 775, 1997.

CHAPTER 12

Microbial Polyesters: Biosynthesis, Properties, Biodegradation and Applications

CHANG-SIK HA*[a] AND WON-KI LEE[b]

[a] Department of Polymer Science and Engineering, Pusan National University, Pusan, Korea; [b] Division of Chemical Engineering, Pukyong National University, Pusan, Korea

12.1 Introduction

Microbial polyesters produced by microorganisms as carbon and energy reserves have attracted much interest due to their biodegradability and biocompatibility.[1] Extensive research has been devoted to the biosynthesis and characterization of microbial polyesters, and regular reviews on the subject are also available.[2–9] Poly[(*R*)-3-hydroxybutyrate] (P3HB) and a copolymer of (*R*)-3-hydroxybutyrate and 3-hydroxyvalerate, [P(3HB–3HV)] are the two main microbial polyesters of current interest for their biodegradability, biocompatibility and useful mechanical properties. Besides P3HB and P(3HB–3HV), to date approximately 150 different hydroxyalkanoate homopolyesters and copolyesters have been identified, all of which are described by the general name polyhydroxyalkanoates (PHAs).[3,7,8] Among these, typical examples of the poly-3-hydroxyalkanoate (P3HA) family are listed in Figure 12.1.

Despite extensive reports on the biosyntheses, physical properties and biodegradation of microbial polyesters, the inherent brittleness and high production costs of these polymers limit their more widespread practical applications.

RSC Polymer Chemistry Series No. 1
Renewable Resources for Functional Polymers and Biomaterials
Edited by Peter A. Williams

Published by the Royal Society of Chemistry, www.rsc.org

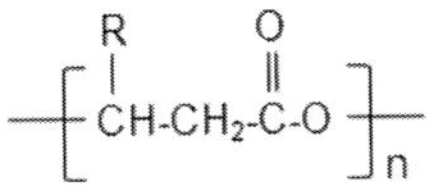

; P(3HA)

Figure 12.1 General chemical structure and names of typical members of poly-3-hydroxyalkanoate (P3HA) family.

For this reason two approaches are pursued to improve the physical properties of P3HB. One approach is the microbial synthesis of copolymers containing hydroxyalkanoate monomer units other than *(R)*-3HB (3HB). For example, P(3HB–3HV) has been produced from alkanoic acids by *Ralstonia eutropha* (formerly known as *Alcaligenes eutrophus)*.[1] In this approach, the physical and thermal properties of microbial copolyesters can be regulated by varying their molecular structure and copolymer compositions, for instance, P(3HB–3HV) with 28 mol% 3HV has a tensile strength of 30 MPa and an elongation to break of 700%, while a random copolyester of 3HB and 4-hydroxybutyrate [P(3HB–4HB)] with 44 mol% 4HB has a tensile strength of 10 MPa and an elongation of 500%. Therefore, the copolyesters are in general much more ductile and elastic than P3HB.[3]

The second approach is the blending of P3HB with other polymers. The blend partners of P3HB according to published reports include non-biodegradable polymers such as poly(vinyl acetate) (PVAc) and poly(methyl methacrylate) (PMMA), as well as biodegradable polymers such as poly(ethylene oxide) (PEO), poly(vinyl alcohol) (PVA), poly-ε-caprolactone (PCL) and polysaccharides. The thermal and physical properties and the rate of biodegradability of these blends are significantly affected by the nature of the blend partner with P3HB.

In this chapter, we discuss a general overview of the biosynthesis, properties and enzymic degradation of various microbial polyesters. We also highlight experimental results on the dependence of polymer properties on comonomer composition, surface plasma treatment and blending with different biodegradable and nonbiodegradable polymers. Various aspects of biodegradability, mechanical properties, surface modification and biomedical applications of microbial polyesters and other biodegradable polymers are covered in the preceding volume of this series.[10]

12.2 Biosyntheses of Microbial Homo- and Copolyesters

12.2.1 Syntheses of Microbial Homopolyesters

Research and development on biodegradable microbial polyesters have expanded rapidly in recent years, especially in the area of biomedical applications.[3,11–14] The simplest and most common member of the microbial PHA family is P3HB, first discovered by Lemoigne in 1925.[15] PHAs serve as a reserve material in the form of nanoparticles in single cell microorganisms (Figure 12.2), playing a role essentially analogous to that of starch in

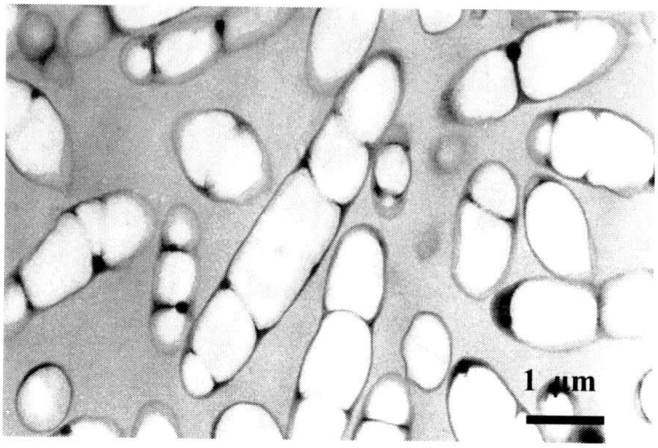

Figure 12.2 Transmission electron micrograph of thin section of a single cell recombinant, *Ralstonia eutropha,* containing 81% P3HB on dry cell weight basis.

plants. Under optimized conditions, up to 90% of the cell dry weight can be accumulated as P3HB. Inside the cell, the polymer remains in a non-crystalline amorphous state. When the polymer is extracted, rapid crystallization occurs with high levels of crystallinity.

These polymers are truly biodegradable, are enzymically degraded by a wide range of bacteria, fungi and algae, and their rates of degradation depend on their structure and the environment. Thus, in principle the application of such polymers is wide ranging and extends to many fields where biodegradability is important as a means of disposal.

PHA biosynthesis pathways are catalyzed by PHA synthases. More than 60 PHA synthase genes (*phaC*) from eubacteria have been cloned and sequenced, and many more PHA synthase sequences have been revealed.[8] The natural substrates of key PHA synthases are coenzyme A thioesters (CoA thioesters) of (*R*)-hydroxyalkanoic acid; the carbon chain length of the hydroxy acid varies widely with a large variety of substituents, and the hydroxy group may be at position 3, 4, 5, 6 or more of the acyl moiety.[3,4,7,8]

In PHA-accumulating cells, PHA synthases are bound to the surface of the PHA granules together with other proteins among which "phasins" and specific regulator proteins are probably the most important. Since the biosynthesis of PHAs is independent of any template, and since the processivity of the enzyme is high, polydisperse products with relatively high molecular weights of up to several million are formed.[8]

According to their subunit composition, and their substrate specificity, three different well-studied classes of PHA synthases are distinguished.[16,17] They are *Ralstonia eutropha*, *Pseudomonas aeruginosa* and *Allochromatium vinosum*, representing classes 1, 2 and 3, respectively; these are the subject of extensive research.[8]

R. eutropha and many other microorganisms synthesize P3HB from acetyl-CoA *via* *R*(–)-3-hydroxybutyryl-CoA, employing a 3-step pathway with β-ketothiolase, an NADPH-dependent acetoacetyl-CoA reductase, and PHA

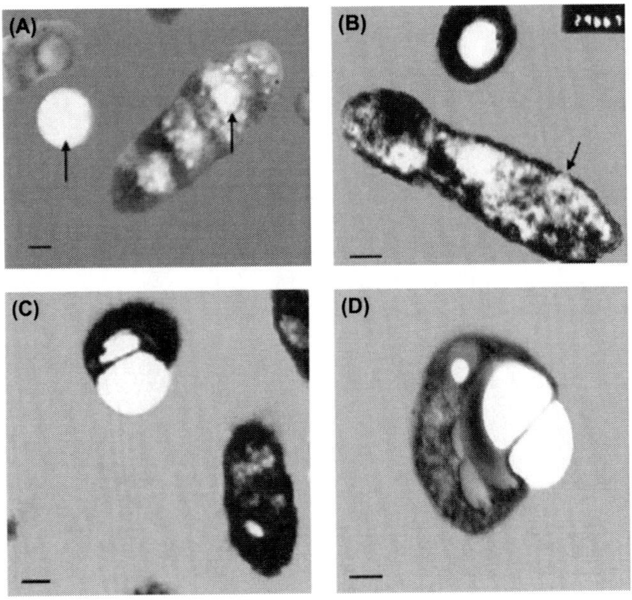

Figure 12.3 Transmission electron micrographs of *Rhodobacter sphaeroides* cells with PHB granules. (A) Cells after 3 days of the growth. Arrows indicate the PHB granules inside and outside of the cells. (B) Cells after 5 days of the growth. Arrow indicates lysis of cell wall. (C) and (D) Cells after 5 and 10 days of the growth showing PHB being released from the cells. Scale bars = 0.3 mm. (http://www.academicjournals.org/AJB)

synthase.[4a] One example is the photosynthetic bacterium *Rhodobacter sphaeroides* (O.U.001). It is used for photobiohydrogen production and can also accumulate P3HB as a by-product when cultivated anaerobically with a minimal medium containing L-malic acid, sodium glutamate and some vitamins under illumination. Figure 12.3 shows transmission electron micrographs of *R. sphaeroides* cells at different stages of producing and releasing PHB granules.[4b]

Figure 12.4 illustrates three pathways for the biosynthesis of P3HB from acetyl-CoA. In *Chromatium vinosum*, an NADH-dependent acetoacetyl-CoA reductase produces *R*(–)-3-hydroxybutyryl-CoA. In *Rhodospirillum rubrum*, the formation of *R*(–)-3-hydroxybutyryl-CoA is achieved *via* *L*(+)-3-hydroxybutyryl-CoA, employing two stereospecific enoyl-CoA hydrates.

Acetyl-CoA is a central intermediate in the metabolism of all organisms, and is therefore formed not only from carbohydrates and fatty acids, but also from many other carbon sources. Thus, carbon sources that are degraded *via* acetoacetyl-CoA or *R*(–)-3-hyrdoxybutyryl-CoA tend to shortcut the pathway to PHA. The synthesis of PHAs other than P3HB occurs only from precursor substrates in most microorganisms. These are structurally related to the constituents that are to be incorporated into PHAs, and fed as the sole carbon source or as a co-substrate,[4,8] as described in more detail in ref. 3–8.

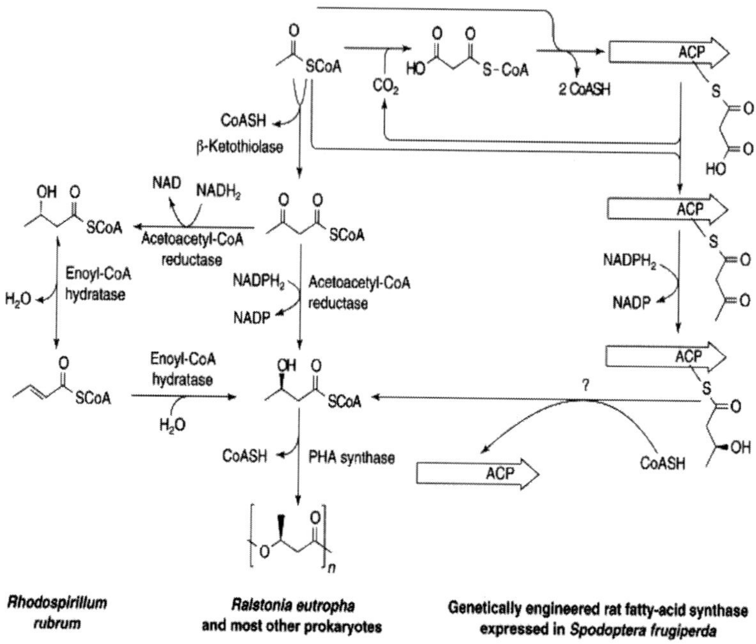

Figure 12.4 Three pathways for the biosynthesis of poly(3-hydroxybutyric acid) (P3HB) from acetyl-CoA (reproduced from ref. 4 with permission). Center: the 3-step PHB biosynthesis pathway with β-ketothiolase, NADPH-dependent acetoacetyl-CoA reductase and PHA synthase occurs in *Ralstonia eutropha* and in most other prokaryotes. Left: the pathway occurring in *Rhodospirillum rubrum*, including the formation of *L*(+)-3-hydroxybutyryl-CoA by an NADH-dependent acetoacetyl-CoA reductase and conversion of the *L*(+)-stereoisomer to an *R*(–)-stereoisomer by enoyl-CoA hydralases. Right: the pathway with a genetically engineered rat fatty acid synthase and the *R. eutropha* PHA synthase expressed in transgenic *Spodoptera frugiperda* insect cells. ACP, acyl carrier protein; PHA, polyhydroxyalkanoic acid. Open arrow represents the genetically modified rat fatty acid synthase with the intermediates of fatty acid *de novo* synthesis bound to ACP.

12.2.2 Syntheses of Microbial Copolyesters

The homopolymer P3HB is a brittle material with limited applications, but the incorporation of a second monomer into P3HB can significantly enhance its useful properties. Therefore, there is considerable interest in producing biosynthetic copolyesters for industrial applications,[18–22] and these efforts signify the beginning of the second developmental stage of PHA research.

Thus, copolyesters containing 3HA units with different chain lengths (C3 to C12) have been produced by 17 bacterial strains from various carbon substrates, such as alkanes, alcohols, and alkanoic acids.[1,3–5] The major copolyesters are P(3HB–3HV)s, which are marketed under the Biopol trademark. The copolymers in P(3HB–3HV) group are semicrystalline with melting temperatures (T_m) ranging from 120 to 170 °C, depending on the copolymer composition. Flexibility and ductility improves with increasing 3HV content in the copolymer. The copolymers are thermoplastic and truly biodegradable in the range from weeks to over a year, depending on the environment and polymer structure. Table 12.1 shows the structures for typical PHA copolymers.

The biosynthesis of P(3HB–3HV) requires, beside 3HB-CoA, also 3-hydroxyvaleryl-CoA (3HV-CoA). The latter is also required for the synthesis of P3HV homopolymer and other copolyesters containing 3HV. 3HV-CoA is produced from the condensation of acetyl-CoA and propionyl-CoA to 3-ketovaleryl-CoA, and a subsequent reduction of the condensation product.[8]

These two reactions are catalyzed by β-ketothiolase of *R eutropha*; acetoacetyl-CoA reductase of *R. eutropha* PHA biosynthesis operon also catalyzes these reactions. However, the β-ketothiolase *PhaA* enoded by the PHA biosynthesis operon of *R. eutropha* does not catalyze this reaction.[14–23] Acetyl-CoA is an obligate central intermediate occurring in all organism and under any physiological condition, but this is not the case for propionyl-CoA, which is only synthesized under special physiological conditions, and from only few substrates. Thus, processes aiming at the biosynthesis of copolyesters like

Table 12.1 Microbial synthesis of PHA copolymers containing (*R*)-3HB as the main constituent (adapted from ref. 3 with permission).

Bacterial strain	*Carbon substrate*	*Random copolyester*
Ralstonia eutropha	Propionic acid	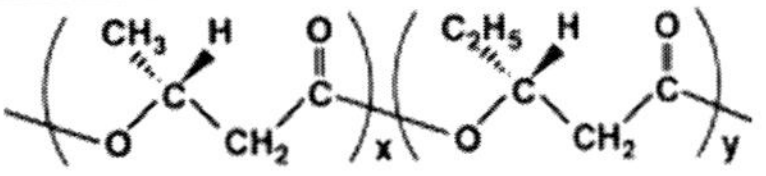 (*R*)-3HB (*R*)-3HV
Ralstonia eutropha	Pentanoic acid, 3-Hydroxypropionic acid	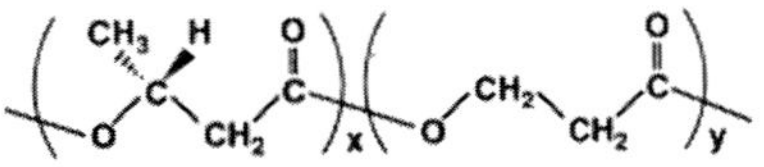 (*R*)-3HB 3HP
Alcaligenes latus *Aeromonas cavie*	1,5-Pentanediol Plant oils	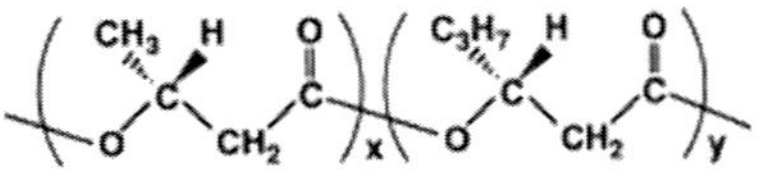 (*R*)-3HB (*R*)-3HHx
Pseudomonas sp *Ralstonia eutropha*	Sugar 4-Hydroxybutyric acid	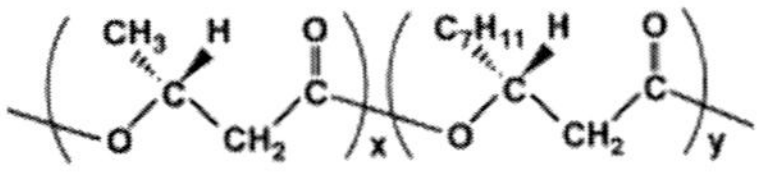 (*R*)-3HB (*R*)-3HD
Alcaligenes latus *Comamonas acidovorans*	γ-Butyrolactone 1,4-Butanediol 1,6-Hexanediol	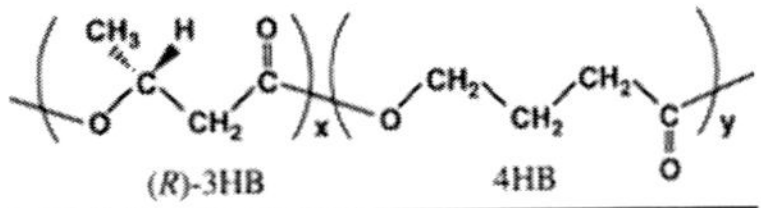 (*R*)-3HB 4HB

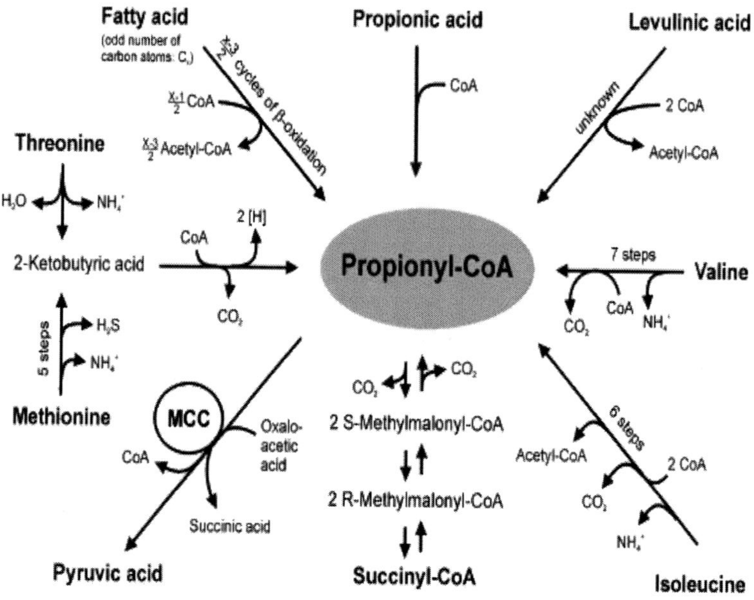

Figure 12.5 Sources of propionyl-CoA for biosynthesis of PHAs containing 3HV as constituent (reproduced from ref. 8 with permission).

poly(3HB–3HV) require the formation and presence of propionyl-CoA in the cells.[8] Figure 12.5 shows the sources of propionyl-CoA for biosyntheses of PHAs containing 3HV as a constituent.

On the other hand, several bacteria possessing PHA synthase of short carbon chain hydroxyalkanoic acids comprising 3–5 carbons can incorporate 4-hydroxybutyric acid (4HB) into PHAs, though this strongly depends on the use of precursor substrates used as a carbon source.[8] This is also true for all other PHAs that contain non-3HA constitutes, which are normally only synthesized if precursor substrates are used. No wild-type strain has so far been described which synthesizes 4HB-containing PHAs from unrelated carbon sources. P4HB is not only hydrolyzed by PHA depolymerases, but also by lipases and esterases, because no alkyl side chain is attached to the polyester backbone. Thus, because of their general biodegradation mechanism, these polyesters are considered suitable for biomedical and pharmaceutical applications.

Hydroxyalkanoic acids whose mean carbon chain length is 3–5 carbons, and 6 or more carbons, are designated by the editor as "short chain" and "medium chain" (Sc and Mc) hydroxyalkanoic acids. (The designations suggested in ref. 8 are complicated unnecessarily, see ref. 24.) In general, the biosynthesis of copolyesters of 3HASc and 3HAMc requires the presence of a PHA synthase with a substrate range combining those of both PHASc and PHAMc synthases.

In addition, the biosynthesis of poly(3HASc–3HAMc) also requires the provision of a PHAScMc synthase with CoA thioesters of 3HAs with carbon chains ranging from 3 to more than 6. This is because if only PHASc synthase and PHAMc synthase were present, they would result in the formation of a mixture of poly(3HASc) and poly(3HAMc), rather than the formation of poly(3HASc–3HMc). This is actually observed in the case of a recombinant strain of *P. oleovorans* expressing, in addition to its own class 2 PHA synthase, the class 1 PHA synthase of *R. eutropha*.[25]

The same would also be true for the synthesis of any other copolyester whose constituents are incorporated to the copolymer by PHA synthases that exclude any given substrate range, for example, one for xHA-CoA and another for yHA-CoA, but not one that incorporates both xHA and yHA. As long as two different PHA synthases are present, only two different types of PHA will be synthesized. Only if the cells possess "one" PHA synthase that contains both xHA-CoA and xHA-CoA, can a true copolyester be synthesized. For instance, copolyesters of 3HB with 3-hydroxyhexanoic acid (3HH), and 3HB with 3-hydroxyoctanoic acid (3HO), can be obtained from fatty acids by recombinant strains such as *Pseudomonas sp* 61-3, *R. eutropha* and others that meet these conditions.

The engineering aspects of PHA biosynthesis pathways have recently been reviewed in detail by Steinbüchel and Lütke-Eversloh.[8] The physiological and engineering aspects of PHA are also well summarized by Braunegg *et al.*[7] and by Sudesh *et al.*,[3] where recent strategies in the optimization of PHA production are discussed.

12.3 Properties and Biodegradation of Microbial Polyesters

12.3.1 Mechanical Properties of Microbial Polyesters

Crystalline P3HB is a compact right handed 2_1 helix, with a two-fold screw axis (*i.e.* two monomer units complete one turn of the helix), and a fiber repeat of 0.596 nm, which comes mainly from van der Waals interactions between the carbonyl oxygen and the methyl groups.[26,27] The stereoregularity of P3HB makes it a highly crystalline material. It is optically active, with the chiral carbon always in the R absolute configuration in biologically produced P3HB.

However, widespread use of P3HB is limited because of its stiffness and relatively high melting point ($T_m = 178\,°C$), at which significant thermal degradation occurs, thus limiting its use for possible applications that would require melt processing.

The elongation to break, Young's modulus and tensile strength of P3HB are reported as 7%, 3.5 GPa, and 40 MPa, respectively. These mechanical properties are close to those of isotactic polypropylene (PP), as shown in Table 12.2. The elongation to break (5%) for P3HB is, however, markedly lower than that of PP (400%). Therefore, P3HB is a stiffer and more brittle plastic material than PP.[3] These properties can be modified by changing the configuration of P3HB polymer chain, or introducing a comonomer in the polymer backbone (see above).

In addition to its inherent brittleness, the high production cost of P3HB also limits its more versatile practical applications. For this reason, two approaches

Table 12.2 Main properties of two examples of poly(3-hydroxybutyrate–3-hydroxyvalerate)s [P(3HB–3HV)s] in comparison with two common plastics (adapted from ref. 3 with permission).

Sample	*Melting temperature* (T_m)	*Glass transition temperature* (T_g)	*Young's modulus/ GPa*	*Tensile strength/ MPa*	*Elongation to break (%)*
P3HB	180	4	3.5	40	5
P(3HB–20 mol% 3HV)	145	−1	0.8	20	50
P(3HB–6 mol% 3HA)[a]	133	−8	0.2	17	680
Polypropylene	176	−10	1.7	38	400
Low density polyethylene	130	−30	0.2	10	620

[a]HA comprised 3HA, 3-hydroxydecanoate (3 mol%), 3-hydroxydodecanoate (3 mol%), 3-hydroxyoctanoate (<1 mol%), 3-hydroxy-cis-5-dodecenoate (<1 mol%).

have pursued to improve the physical properties of P3HB—microbial synthesis of copolymers containing other hydroxyalkonoate monomeric units, and blending with other polymers.[28–36]

Thus, copolyesters having up to 95 mol% of 3HV can be produced by microbial synthesis,[37] copolyesters with higher 3HV contents possess low Young moduli and are soft. For P(3HB–3HV) molecules of different compositions, crystallization rate generally decreases with increasing 3HV content. In copolyesters of intermediate compositions (40–55 mol% 3HV) cocrystallization takes place. At high 3HB or 3HV contents, the minor structural unit cocrystallizes in the major constituent lattice. T_m and the glass transition temperature (T_g) also strongly depend on the comonomer fraction.[38] Figure 12.6 shows the dependence of T_g on the comonomer composition.

T_g of P3HB is in the range of 0–4 °C, and that of P3HV in the range of −10 to −12 °C. The T_m of P3HB is 178 °C. For the copolymer P(3HB–3HV), T_m comes down to 71 °C for the copolymer with 40 mol% 3HV, it shows a minimum, and then increases to 107–112 °C for P3HV.[1,3] In the case of P(3HB–4HB), the crystallinity of the copolyester is reduced proportionally with increasing 4HB content. Copolymer properties can be adjusted ranging from rigid (with high crystallinity) to elastomeric by varying the composition. For instance, the mechanical properties of P(3HB–6 mol% 3HA), produced by genetically engineered *Pseudomonas sp* 61-3, is very similar to those of low density polyethylene, as shown in Table 12.2. Elongation to break of this copolyester reaches 680% while maintaining its tensile strength of 17 MPa.[3]

12.3.2 Molecular Weights of Microbial Polyesters

Detailed investigations into the molecular weight (M_w) of PHA are scarce, despite the fact that this is an important characteristic which could influence most applications of the polymer. This is to some extent because it is not possible to

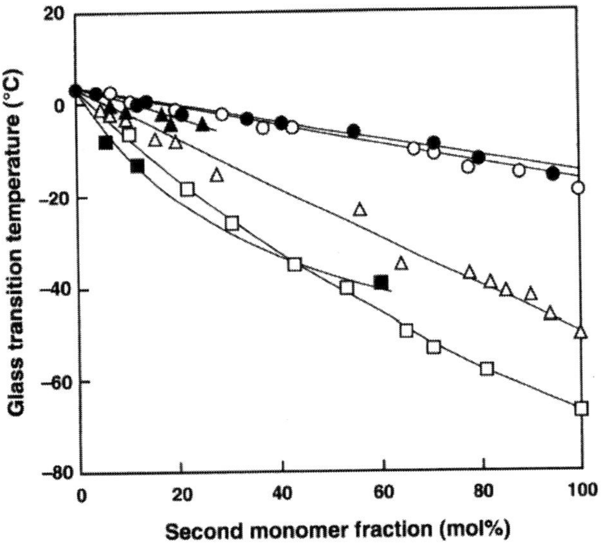

Figure 12.6 Glass transition temperatures (T_g) of random copolymers of (*R*)-3HB with other hydroxyalkanoates (HA) (reproduced from ref. 3 with permission). ●, P(3HB–3HV); ▲, P(3HB–3HX); ■, P(3HB–3HA); ○, P(3HB–3HP); △, P(3HB–4HB); □, P(3HB–6HH).

determine the true molecular weight of P3HB copolymers containing other monomers because of the necessity of purification from bacterial cells, and limited data on the intrinsic viscosity and Mark–Houwink–Sakurada parameters.[3]

The relative molecular weights of PHAs reported so far are based on polystyrene standards by gel permeation chromatography (GPC). A wide range of molecular weights is produced by PHAs from different microorganisms and also from different stages and conditions of cultivation. (R)-3HA monomers are polymerized into polymers with high molecular weights in the range of 200 000 to 3 000 000, depending on the microorganism and growth conditions.[39] The largest number average molecular weight (M_n) value reported for biologically synthesized P3HB is about 20 million, obtained from a recombinant strain of *Escherichia coli*.[40] The M_w of PHAs is also affected by the type and concentration of the carbon source.

12.3.3 Biodegradation of Microbial Polyesters

PHAs are degraded by microorganisms in soil, seawater and activated sludge.[1,3] During enzyme attacks on water-insoluble PHAs, the enzyme first adsorbs onto the polymer surface, followed by an enzymic reaction. It is reported that PHA depolymerases are more active in seawater than in soil and are most active at neutral pH ranges.[1]

The chain-folded solution-grown lamellar crystals (SGCs) of P3HB have been used as a model system for elucidating the mechanisms of both enzymic and hydrolytic degradation in the crystalline region,[26,27] but the enzymic degradation rate is not dependent on the spherulite size of P3HB. P3HB depolymerase is an excreted extracellular enzyme from *Alicaligenes faecalis* T1.[27]

The rate of enzymic degradation of P3HB films decreases with increasing crystallinity, and preferentially occurs first in the amorphous region and subsequently in the crystalline region. Thus, the lower the degree of crystallinity, the higher is the rate of enzymic degradation, but enzymic degradation rate is not dependent on the spherulite size of P3HB.

P3HB depolymerase initially hydrolyzes the amorphous interfibrillar chains on the surface, and subsequently erodes the chains in the crystalline fibrils (bundle of lamellae).[29] The erosion rates for PHA copolymer films are several times higher than for P3HB homopolymer films with the same degree of crystallinity. The rate of enzymic erosion of PHA films is influenced not only by the degree of crystallinity, but also by the structure of the lamellar crystals. In this sense, the rate of the formation of disordered PHA chains from the crystal edge by P3HB depolymerase is greater for PHA copolymers than that for the P3HB homopolymer.[41–43]

12.4 Biodegradability of Polymer Blends Containing Microbial Polyesters

The blending of PHAs not only affects polymer physical properties but also their biodegradability. For immiscible blends, the degradation depends on the intrinsic biodegradability of both components, phase distribution and substrate accessibility to the degrading enzyme. On the other hand, when the two polymers are miscible, the biodegradability is strongly affected by the mobility of the amorphous mixed phase which forms on blending. Both morphology and biodegradability are also influenced by the blending method, *e.g.* solvent casting or melt mixing.[44]

P3HB has been blended with several nonbiodegradable and biodegradable polymers.[34–36] The non-biodegradable polymer partners for blending with P3HB include PVAc, poly(vinyl acetate–vinyl alcohol), PMMA and poly(vinyl phenol). Biodegradable polymer partners for blending with P3HB include other PHAs, PEO, PCL, polylactide, other aliphatic polyesters, cellulose esters, poly(vinyl alcohol) (PVA), chitosan, chitin and polyglutamate. Fuller details of miscibility, properties, and biodegradability of these blends are reviewed elsewhere.[45]

As an example for improvement of properties of P3HB by blending, Table 12.3 lists the modulus, stress and elongation to break for P3HB homopolymer and a few P3HB–rubber blends.[46] The addition of rubber to P3HB improves the fracture toughness of the polymer, but a more substantial enhancement of toughness is obtained when a modified rubber, especially succinic acid-grafted ethylene–propylene rubber (EPR–g-SA), is used instead of EPR and ethylene–vinyl acetate copolymer (EVA). For EPR and EVA blends with P3HB, a decrease in particle size of the dispersed phase and increased adhesion to the

Table 12.3 Modulus (E), stress (σ_R) and elongation (ε_R), at rupture of P3HB homopolymer and P3HB–rubber blends (adapted from ref. 46).

Sample	$E \times 10^{-3}/kg\ cm^{-2}$	$\sigma_R/kg\ cm^{-2}$	ε_R (%)
P3HB	2.1	290	1.3
P3HB/EPR	1.5	170	2.0
P3HB/EVA	1.6	175	2.0
P3HB/EVAL	1.7	185	3.0
P3HB/EPR⁻g-DBM	1.6	175	4.0
P3HB/EPR⁻g-SA	1.6	180	6.5

matrix improve the ultimate tensile properties such as elongation to break and high speed fracture toughness of the material.

Generally microorganisms excrete extracellular P3HB depolymerases to degrade P3HB, and utilize the decomposed compounds as nutrients.[45] Although PHAs can possess two types of optical isomers, (*R*) and (*S*), because of the C3 methyl carbon in 3-hydroxyalkanoates, microbial synthesis produces the pure (*R*) form of optically active polyesters, *i.e.* a totally isotactic polymer. Recently, however, the chemical synthesis of P3HB containing monomeric units of both (*R*) and (*S*) 3HB has been reported, and also shown to be hydrolyzed by microbial P3HB depolymerase.[46]

In general, physical properties and enzymic degradation are influenced by polymer stereoregularity. For instance, blends of bacterial isotactic P3HB and synthetic atactic P3HB are miscible above 60% synthetic P3HB.[47] The phase behavior is highly sensitive to parameters such as blend preparation technique and molecular weight. In the case of bacterial and partially isotactic P3HB blends, the miscibility-enhancing isomorphism could be also invoked, which leads to a strongly interacting system.

When enzymic degradation of blends of P3HB with P[(R,S)-3HB] by P3HB depolymerase from *Pseudomonas pickettii* is measured in 0.1 M potassium phosphate buffer (pH 7.4), at 37 °C, enzymic hydrolysis takes place on the surface of P3HB films, while little hydrolysis occurs on the surface of atatic and syndiotactic P[(R,S)-3HB].[48] However, the enzymic erosion of the films is accelerated, and the highest rate of enzymic hydrolysis is observed at around 50 wt% P[(R,S)-3HB] content.[33]

Only limited work has yet been published on the miscibility, properties and biodegradability of blends containing P(3HB–3HV) or other copolyesters, since the structures and properties of these copolyesters have not been revealed in detail in comparison to those of P3HB, but one such blend is that of natural P(3HB–3HV) with synthetic atactic P3HB (a-P3HB), cast from chloroform.[49] The copolymer P(3HB–3HV) contains 10 mol% of 3HV units. Blends of bacterial P(3HB–3HV) with synthetic a-P3HB, in the explored range of 10–50% a-P3HB, are miscible in the melt and solidify with spherulitic morphology.

The influence of a-P3HB content on thermal and mechanical properties of the blends is significant. The degree of crystallinity decreases with increasing a-P3HB in the film, and elongation to break for a sample containing 50%

a-P3HB is 30-fold that of the pure copolymer. In degradation experiments with both alkali (pH 7.4, 70 °C) and enzyme (P3HB depolymerase A from *Pseudomonas lemoignei* in Tris HCl buffer pH 8, 37 °C), the rate of enzymic degradation of the blends was higher than that of the copolymer, and increased with a-P3HB content, whereas pure a-P3HB did not degrade under these conditions. The conclusion is that presence of a partially crystalline PHA (*i.e.* copolymer versus P3HB) in a blend is essential for enzymic degradation. Related work on blends containing P(3HB–3HV) is summarized in ref. 45.

12.5 Control of Enzymic Degradation of Microbial Polyesters

A recent trend in biodegradable polymer research is the desired life span of the polymer. Since the mechanical properties of microbial polyesters are catastrophically lost during the initial stage of degradation, *i.e.* 66% strength loss for 1.7 % weight loss.[50] Thus, the development of microbial polyesters with higher initial stability towards degradation is required for applications in disposable items such as packing materials and mulching films in agriculture.

The relationship between surface and bulk degradation and erosion of biodegradable polymers is discussed by Timmins and Liebmann-Vinson and others.[51,52] It is evident that the initial rate of degradation is dominated by the physical accessibility of the polymer structure to the degrading agent. Thus, the crystallinity of the polyester strongly influences its rate of hydrolysis due to the low permeability of crystalline regions with respect to the degradation medium. Hydrolysis is initially restricted to the amorphous phase and to the fringes of the crystallites. Since the amorphous chains tie the crystallites together, initial hydrolysis leads to catastrophic mechanical failure, even before any weight loss.[50,51] To design commercial degradable materials, therefore, it is necessary to control the rate of initial degradation.[10,51,53]

Since the degradation of PHAs proceeds *via* surface erosion and occurs first in amorphous regions, PHA-based blends can offer the opportunity to design biodegradable devices with the desired hydrolytic kinetics by controlling the surface organization of the device.[54–56] An alternative method is surface modification.[57]

12.5.1 Control of Enzymic Degradation of Microbial Polyesters by Blending

The enzymic degradation of P(3HB–3HV) can be controlled by blending it with small amounts of a non-degradable polymer, such as polystyrene (PS) or poly(methyl methacrylate) (PMMA) by solution casting. Since enzymic degradation of P(3HB–3HV) initially occurs by surface erosion, degradation can be observed by the surface structure of the blend film by, *e.g.* X-ray photoelectron spectroscopy (XPS). The surface of P(3HB–3HV)/PS blend films reveals an excess of PS, whereas the surface of P(3HB–3HV)/PMMA blend films is nearly covered by P(3HB–3HV). In case of P(3HB–3HV)/PS, PS exists within P(3HB–3HV) spherulites on the surface, and acts as a retardant towards enzymic attack on the surface of the blend film.

Figure 12.7 shows the atomic force microscope (AFM) images of P(3HB–3HV) and those of its blend films with 5 wt% PS before and after enzymic hydrolysis for 30 min.[56] All the cast films show clear volume-filled spherulite-like crystalline morphology throughout the film surface. The effect of adding small amounts of polystyrene on the surface (crystalline) morphology of the polyesters is not large (Figure 12.7B), since P(3HB–3HV) is completely immiscible with PS.[58] However, the enzymic attack leads to quite different morphologies on surfaces of the copolymer and its blend films.

The surface of P(3HB–3HV) film shows significant erosion due to enzymic attack (Figure 12.7C).[44] The remaining spherulites indicate that the preferential enzymic attack starts in disordered chains between the crystalline lamellas. However, no major difference is found on the surface morphologies of the blend films after the enzymic treatment for 30 min (Figure 12.7D). It may be attributed to the presence of enzymically inactive PS on the film surface. The retarding effect on the biodegradation of P(3HB–3HV)/PS (95 : 5 by wt%) blend film is more significant after 150 min of enzymic degradation. The weight loss in the surface layer cannot be directly measured, but it should be strongly correlated to surface erosion.

Figure 12.8 shows some typical weight loss profiles of cast films and thermally treated P(3HB–3HV) and its blends with PS and PMMA.[56] Weight loss

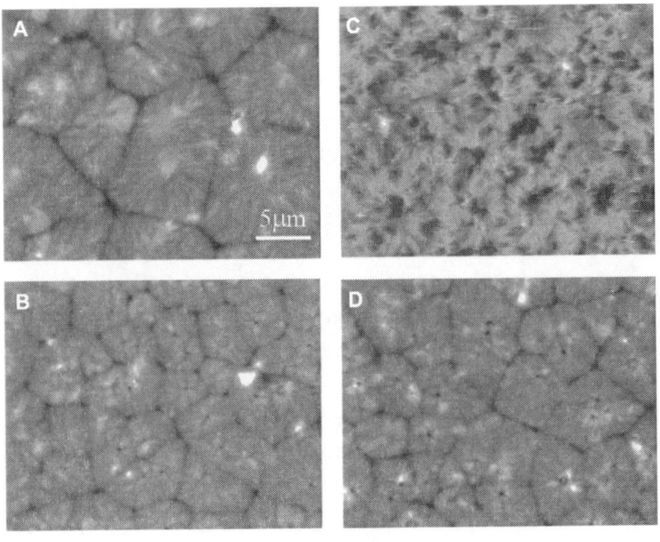

Figure 12.7 Atomic force microscope (AFM) images of cast films of P(3HB–3HV) and its blends (reproduced from ref. 56 with permission). A, P(3HB–3HV); B, blend of P(3HB–3HV) with PS; C and D, corresponding images after 30 min enzymic degradation.

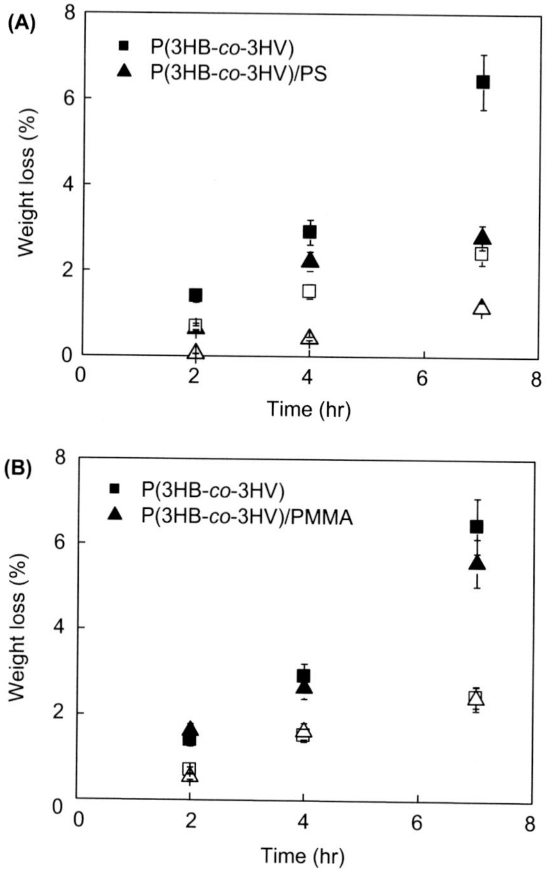

Figure 12.8 Weight loss profiles of (A) P(3HB–3HV) and its blend films with polystyrene (PS) and (B) poly(methyl methacrylate) (PMMA) (reproduced from ref. 56 with permission). Filled symbols, cast films; open symbols, thermally treated films.

in all the films increases proportionally with time, but the rate of enzymic erosion decreases markedly in P(3HB–3HV)/PS blends. The addition of PMMA to P(3HB–3HV) shows little difference in the erosion rate compared to P(3HB–3HV) itself. For thermally treated P(3HB–3HV) and its blends with PS and PMMA, the rate of enzymic erosion apparently decreases relative to that of the cast films due to their higher crystallinity as well as the size of the crystals in thermally treated samples. These results clearly imply that the surface-enriched hydrophobic PS retards the biodegradation of the less hydrophobic copolyester film at the initial stage of enzymic attack, whereas the effect is not so prominent in the blend of copolyester with PMMA due to their similar surface free energy.

12.5.2 Control of Enzymic Degradation of Microbial Polyesters by Surface Modification

Various modes of surface modification of biodegradable polymers without changing their bulk properties are discussed in detail by Thissen,[59] and plasma treatment in particular has been described by Chu.[60] In order to investigate the effect of surface modification for controlling the enzymic degradation of microbial polyesters we studied the degradation of PHB and P(3HB–3HV) films with an extracellular PHB depolymerase from *A. faecalis* T1, before and after CHF_3 or O_2 plasma treatment. Figure 12.9 shows the scanning electron micrographs of PHB films before (left) and after (right) enzymic degradation for 3 h.[57] After the degradation, the surfaces of the pure and O_2 plasma-treated P3HB films show similarly degraded morphologies. However, the surface morphology of CHF_3 plasma-treated P3HB film was little changed. Although the weight loss of PHB films before and after CHF_3 or O_2 plasma treatment increased proportionally with degradation time, the weight loss of CHF_3 plasma-treated P3HB after enzymic degradation was much slower than that of an untreated or O_2-treated one; the CHF_3 plasma-treated P3HB exhibited significant retardation of enzymic erosion because the surface components of

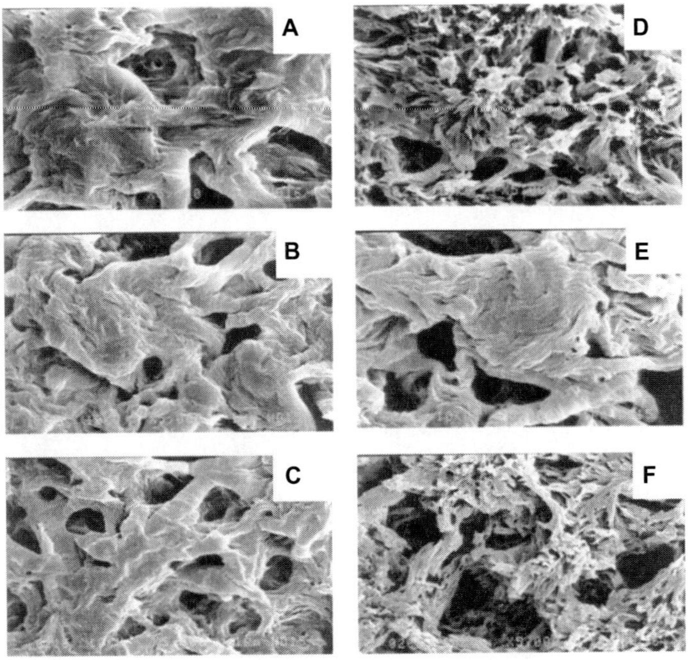

Figure 12.9 Scanning electron micrographs of PHB films before (left) and after (right) enzymic degradation for 3 h (reproduced from ref. 57 with permission). A and D, pure PHB; B and E, CHF_3 plasma treatment for 10 s; C and F, O_2 plasma treatment for 10 s.

CH_xF_y induced by CHF_3 plasma cause an increased surface hydrophobicity and the inactivity of the enzyme. The increased hydrophilicity of P3HB film induced by O_2 plasma, however, resulted in no significant change in the enzymatic erosion. Since the enzymic degradation of polyesters initially occurs by the surface erosion process and the surface layer modified by plasma is very thin, the oxidation layer would be removed at the very initial stage of degradation.

Thus, the biodegradation rate of microbial polyesters is strongly affected by surface wettability as controlled by plasma treatment; surface hydrophobicity reduces enzymic degradation, whereas surface hydrophilicity does not enhance degradation.[57]

The most common way to determine the degradation rate is a weight loss method using films. Recently, however, it should be noted here that a Langmuir technique has also been found to be a useful model to describe the *in situ* degradation behavior of polyester monolayers at the molecular scale.[61–63] The hydrolytic kinetics of polyester monolayers could be measured by the change of occupied areas of hydrolyzable monolayers in a subphase at a constant surface pressure, as shown in Figure 12.10, where A_0 and A represent the areas occupied by the film at time 0 and t, respectively. It is reasonable to assume that the area reduction with time reflects the dissolution of low molecular oligomers due to hydrolysis $(1 - A/A_0)$ assuming that the occupied area per repeating unit at constant surface pressure is a constant.[61,62]

12.6 Applications of Microbial Polyesters

Microbial polyesters can be applied in a variety of fields such as agricultural mulching films, bottles and containers, marine fishing nets and lines. They can also be applied as a matrix for the slow release of drugs, hormones, herbicides, insecticides, flavors and fragrances in medicine, pharmacy, agriculture and food industry, and as a scaffold in tissue engineering.[4,10,14,19,64,65] They can be applied as biocompatible surgical sutures, bone plates and tapes, surgical fabric, and hygienic diapers, as well as for piezoelectric materials.

In addition, the latex of PHAs may be applied to paper and cardboard to form a water-resistant layer as an alternative to non-biodegradable materials prepared from, for example, cardboard plus polyethylene or aluminium.[4,65] PHAs are also being considered as a resource for the synthesis of enantiomerically pure chemicals and as raw materials for the production of paints.[6] New copolyesters can also be used for packaging materials as substitute for polyethylene and polypropylene. The medical and pharmaceutical applications of P3HB are, however, restricted by its very slow biodegradation and high hydrolytic stability in sterile tissues, whereas other PHAs which are more readily hydrolyzed by enzymes such as lipases or esterases have not yet been examined in much detail.[4]

Among various types of PHAs, those containing 4HB may hold promise for potential use in pharmaceutical and biomedical fields.[3] 4HB, *i.e.* γ-hydroxybutyrate (GHB), was initially used as an intravenous anesthetic agent in Europe

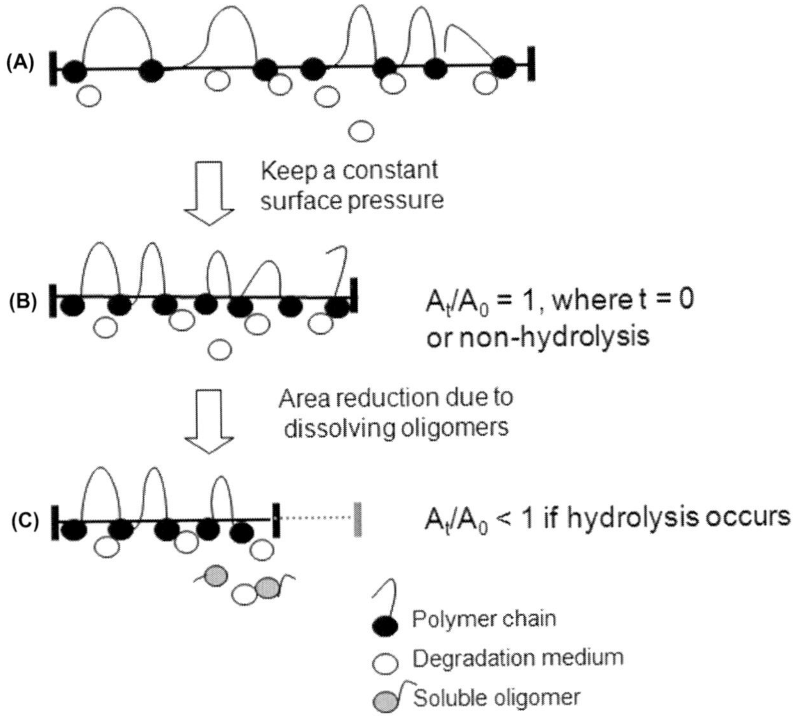

Figure 12.10 Schematic representation of monolayers on a subphase with degradation medium. A, as-spread; B compressed to a constant surface pressure; C, reduction of occupied area due to dissolving oligomers generated by degradation into a subphase (reproduced from ref. 61 with permission).

and Japan because it can cross the blood–brain barrier rapidly to induce a sleep-like state with cardiovascular stability.[3,66,67] In fact 4HB was found to be a normal metabolite in extracts of brain tissue of rat, pigeon and man;[68] 4HB was also used for the treatment of narcolepsy as it increased slow wave sleep (SWS) and rapid eye movement (REM) sleep by inducing the symptoms of narcolepsy and containing them at night.[69]

The Food and Drug Administration (FDA) in the United States has also approved the use of 4HB for investigational research in areas such as narcolepsy trials.[69,70] The use of 4HB has also been advocated in the treatment of alcohol addiction, including alcohol withdrawal syndrome.[71,72] P(3HB–4HB) may also have potential application in controlled release for therapeutic purposes, since the copolyester is biodegradable and biocompatible.[3]

However, the cost of production of PHAs is relatively high, and their mechanical properties deteriorate rapidly in certain applications at the very early stage of degradation. To overcome these drawbacks, research continues on the biosynthesis of PHAs. At the same time their biodegradation rate, and hence their life time, can also be influenced by plasma surface modification,[57] and by blending with small amounts of non-biodegradable polymers.[58]

12.7 Abbreviations

AFM	atomic force microscope
CoA	coenzyme A
EPR-g-SA	succinic acid grafted ethylene–propylene rubber
EVA	ethylene–vinyl acetate copolymer
FDA	Food and Drug Administration
GHB	γ-hydroxybutyrate
GPC	gel permeation chromatography
4HB	4-hydroxybutyrate
3HO	3-hydroxyoctanoic acid
3HV	3-hydroxyvalerate
3HV-CoA	3-hydroxyvaleryl-CoA
Mc	medium chain
M_n	number average molecular weight
M_w	molecular weight
PCL	poly-ε-caprolactone
PEO	poly(ethylene oxide)
PHA	polyhydroxyalkanoates
PMMA	poly(methyl methacrylate)
PP	polypropylene
P3HB	Poly[(R)-3-hydroxybutyrate]
PS	polystyrene
PVA	poly(vinyl alcohol)
PVAc	poly(vinyl acetate)
Sc	short chain
SGC	solution-grown lamellar crystals
T_g	glass transition temperature
T_m	melting temperature
XPS	x-ray photoelectron spectroscopy

Acknowledgements

The work was supported by the National Research Foundation of Korea (NRF) Grant funded by the Ministry of Education, Science and Technology, Korea (MEST) (Acceleration Research Program (No. 20110000385; Pioneer Research Center Program (No.2011-0001667/2011-0001668), and the Brain Korea 21 Project of the MEST. WKL is grateful for the award of grant No. 2011-0010106 from the Ministry of Education, Science and Technology.

References

1. Y. Doi, *Microbial Polyesters*, VCH Publishers, New York, 1990.
2. R. H. Marchessault, T. L. Bluhm, Y. Deslandes, G. K. Hamer, W. J. Orts, P. R. Sundararajan and M. G. Taylor, Poly(β-hydroxyalkanoates): Biorefinery polymers in search of applications, *Macromol. Symp.*, 1988, **19**, 235–254.
3. K. Sudesh, H. Abe and Y. Doi, Synthesis, Structure, and properties of polyhydroxyalkanoates: Biological polyesters, *Prog. Polym. Sci.*, 2000, **25**, 1503–1555.
4. (a) A. Steinbüchel and B. Füchtenbusch, Bacterial and other biological systems for polyester production, *Trends in Biotechnol (TIBTECH)*, 1998, **16**, 419–427; (b) D. Çetin, U. Gündüz, I. EroğLu, M. Yücel and L. Türker, Poly-β-hydroxybutyrate accumulation and releasing by hydrogen producing bacteria, Rhodobacter Sphaeroides, *Afric. J. Biotechnol.*, 2006, **5**, 2069–2072; http://www.academicjournals.org/AJB.
5. A. Steinbüchel, PHB and other polyhydroxyalkanoic acids, in *Biotechnology*, (Ed.) M. Roehr, 2nd edn, Wiley-VCH, Weinheim, 1996, pp. 403–464.
6. H. M. Müller and D. Seebach, Poly(hydroxyalkanoates) – A 5th class of physiologically important organic biopolymers, *Angew. Chem. Int. Ed. Eng.*, 1993, **32**(4), 477–502.
7. G. Braunegg, G. Lefebvre and K. F. Genser, Polyhydoxyalkanoates, biopolyesters from renewable resources: physiological and engineering aspects, *J. Biotechnol.*, 1998, **65**(1), 127–161.
8. A. Steinbüchel and T. Lütke-Eversloh, Metabolic engineering and pathway construction for biotechnological production of relevant polyhydroxyalkanoates in microorganisms, *Biochem. Eng. J.*, 2003, **16**, 81–96.
9. (a) Y. Tokiwa and C. U. Ugwu, Biotechnological production of (R)-3-hydroxybutyric acid Monomer, *J. Biotechnol.*, 2007, **132**(3), 264–272; (b) E. Zagar, A. Krzan, G. Adamus and M. Kowalczuk, Biopolyesters from microorganisms: Biochemical basis of microbial synthesis, properties and applications, *Biomacromolecules*, 2006, **7**(7), 2210–2216.
10. R. Arshady, (Ed.) *Biodegradable Polymers*, The PBM Series, Vol 2, Kentus Books, London, 2006.
11. M. Avella, E. Martuscelli and M. Raimo, Properties of blends and composites based on poly(3-hydroxybutyrate) and poly(3-hydroxybutyrate-hydroxyvalerate) copolymers, *J. Mater. Sci.*, 2000, **35**(3), 523–545.
12. R. Li, H. Zhang and Q. Qi, The production of polyhydroxyalkanoates in recombinant escherichia coli, *Bioresource Technol.*, 2007, **98**(12), 2313–2320.
13. P. R. Patnaik, Perspectives in the modeling and optimization of PHB production by pure and mixed cultures, *Crit. Rev. in Biotechnol.*, 2005, **25**(3), 153–171.
14. C. S. Ha and J. A. Gardella, Jr., Surface chemistry of biodegradable polymers for drug delivery Systems, *Chem. Rev.*, 2005, **105**(11), 4205–4232.

15. M. Lemoigne, Produit de déshydratation et de polymérisation de lacide β-oxybutyrate, *Bull. Soc. Chim. Biol.*, 1926, **8**, 770–786.
16. B. H. A. Rehm and A. Steinbüchel, Biochemical and genetic analysis of PHA synthases and other proteins required for PHA synthesis., *Int. J. Biol. Macromol.*, 1999, **25**, 3–19.
17. A. Steinbüchel, E. Hustede, M. Liebergesell, U. Pieper, A. Timm and H. Valentin, Molecular basis for biosynthesis and accumulation of polyhydroxyalkanoic acids in bacteria., *FEMS Microbiol. Rev.*, 1992, **103**, 217–230.
18. G. N. Bernard and J. K. Sanders, The poly-β-hydroxybutyrate granule in vivo: A new insight based on NMR spectroscopy of whole cells, *J. Biol. Chem.*, 1989, **264**(6), 3286–3291.
19. X. H. Zou and G. Q. Chen, Metabolic engineering for microbial production and applications of copolyesters consisting of 3-hydroxybutyrate and medium-chain-length 3- hydroxyalkanoates, *Macromol. Biosci.*, 2007, **7**(2), 174–182.
20. I. S. Lee, O. H. Kwon, W. Meng, I. K. Kang and Y. Ito, Nanofabrication of microbial polyester by electrospinning promotes cell attachment, *Macromol.Res.*, 2004, **12**(4), 374–378.
21. H. Sato, R. Murakami, J. Zhang, Y. Ozaki, K. Mori, I. Takahashi, H. Terauchi and I. Noda, X-ray diffraction and infrared spectroscopy studies on crystal and lamellar structure and CHO hydrogen bonding of biodegradable poly(hydroxyalkanoate), *Macromol. Res.*, 2006, **14**(4), 408–415.
22. E. Zagar, A. Krzan, G. Adamus and M. Kowalczuk, Sequence distribution in microbial poly(3-hydroxybutyrate–3-hydroxyvalerate) co-polyesters determined by NMR and MS, *Biomacromolecules*, 2006, **17**(7), 2210–2216.
23. S. Slater, K. L. Houmiel, M. Tran, T. A. Mitsky, N. B. Taylor, S. R. Padgette and K. J. Grays, Multiple β-ketothiolase mediate poly(β-hydroxyalkanoate) copolymer synthesis in Ralstonia eutropha, *J. Bacteriol.*, 1998, **180**, 1979–1987.
24. (a) A. Timm, A. Byrons and A. Steinbüchel, Formation of blends of various poly(3- hydroalkanoic acids) by a recombinant strain of Pscudomonas oleovorans, *Appl. Microbial. Biotechnol.*, 1990, **33**, 296–301; (b) H. Prusting, J. Kingma, G. Huisman, A. Steinbüchel and B. Witholt, Formation of polyester blends by a recombinant strain of Pseudomanas oleovorans; different poly(3-hydroxyalkanoates) are stored in separate granules, *J. Environ. Polym. Degrad.*, 1983, **1**, 45–53.
25. R. Arshady, *Science & Medical Style Guide: General Writing and Illustration Skills*, Kentus Books, London, 2006.
26. Y. Kuamgai, Y. Kanesawa and Y. Doi, Enzymic degradation of microbial poly(3-hydroxybutyrate) films, *Makromol. Chem.*, 1992, **193**, 53–57.
27. T. Iwata and Y. Doi, Crystal structure and biodegradation of aliphatic polyester crystals, *Macromol. Chem. Phys.*, 1999, **200**, 2429–2442.
28. P. A. Holmes, Biologically produced (R)-3-hydroxyalkanoate polymers and copolymers, in *Developments in Crystalline Polymers*, (Ed.) D. C. Bassett, Vol. 2, Elsevier, London, 1998, pp. 1–65.

29. N. Koyama and Y. Doi, Effects of solid-state structure on the enzymic degradability of bacterial poly(hydroxyalkanoic acids), *Macromolecules*, 1997, **30**, 826–832.
30. H. Abe, Y. Doi, H. Aoki and T. Akehata, Solid-state structures and enzymic degradabilities for melt-crystallized films of copolymers of (R)-3-hydroxybutyric acid with different hydroxyalkanoic acids, *Macromolecules*, 1998, **31**, 1791–1797.
31. N. Yoshie, M. Fujiwara, L. Kasuya, H. Abe, Y. Doi and Y. Inoue, Effect of monomer composition and composition distribution on enzymic degradation of poly(3-hydroxybutyrate–3-hydroxyvalerate), *Macromol. Chem. Phys.*, 1999, **200**, 977–982.
32. A. Cao, Y. Arai, N. Yoshie, K. Kasuya, Y. Doi and Y. Inoue, Solid structure and biodegradation of the compositionally fractionated poly(3-hydroxybutyric acid-co-3-hydroxypropionic acid)s, *Polymer*, 1999, **40**, 6821–6830.
33. D. L. Vander-Hart, W. J. Orts and R. H. Marchessault, 13C-NMR determination of the degree of cocrystallization in random copolymers of poly(β-hydroxybutyrate–β-hydroxyvalerate), *Macromolecules*, 1995, **28**, 6394–6400.
34. T. H. M. Abou-Aiad, Morphology and dielectric properties of polyhydroxybutyrate (PHB)/poly(methylmethacrylate) (PMMA) blends with some antimicrobial Applications, *Polym. Plast. Technol.and Eng.*, 2007, **46**(4), 435–439.
35. X. Songling, L. Rongcong, W. Linping, X. Kaitian and G. Q. Chen, Blending and characterizations of microbial poly(3-hydroxybutyrate) with dendrimers, *J. Appl. Polym. Sci.*, 2006, **102**(4), 3782–3790.
36. B. Yalcin, M. Cakmak, A. H. Arkin, B. Hazer and B. Erman, Control of optical anisotropy at large deformations in PMMA/chlorinated-PHB (PHB-Cl) blends: Mechano-optical behavior, *Polymer*, 2006, **47**(24), 8183–8193.
37. Y. Doi, A. Tamaki, M. Kunioka and K. Soga, Biosynthesis of terpolymer of 3- hydroxybutyrate, 3-hydroxyvalerate, and 5-hydroxyvalerate in Alcaligenes eutrophus from 5-chloropentanoic and pentanoic acid, *Macromol. Chem. Rapid Comm.*, 1987, **8**, 631–635.
38. Y. Doi, S. Kitamura and H. Abe, Microbial synthesis and characterization of poly(3-hydroxybutyrate–3-hydroxyhexanoate), *Macromolecules*, 1995, **28**, 4822–4828.
39. D. Byrom, Plastics from microbes: microbial synthesis of polymers and polymer precursors. In *Plastics from Microbes*, (Ed.) D. P. Mobley, Munich, Hanser, 1994, pp. 5–33.
40. S. Kusaka, H. Abe, S. Y. Lee and Y. Doi, Molecular mass of poly[(R)-3 hydroxybutyric acid] produced in a recombinant., *Escherichia coli. Appl. Microbiol Biotechnol.*, 1997, **47**, 140–143.
41. T. Iwata, Y. Doi, K. Kasuya and Y. Inoue, Visualization of enzymic degradation of poly[(R)-3-hydroxybutyrate] single crystals by an extracellular PHB depolymerase, *Macromolecules*, 1997, **30**, 833–839.
42. T. Iwata, Y. Doi, T. Tanaka, T. Akehata, M. Shiromo and S. Teramachi, Enzymic degradation and adsorption on poly [(R)-3-hydroxybutyrate]

single crystals with two types of extracellular PHB depolymerase., *Macromolecules*, 1997, **30**, 5290–5296.
43. P. J. Hocking, R. H. Marchessault, M. R. Timmins, R. W. Lenz and R. C. Fuller, Enzymic degradation of single crystals of bacterial and synthetic poly(β-hydroxybutyrate), *Macromolecules*, 1996, **29**, 2472–2478.
44. L. Finelli, M. Scandola and P. Sadocco, Biodegradation of blends of bacterial poly(3-hydroxybutyrate) with ethyl cellulose in activated sludge and in enzymic solution, *Macromol. Chem. Phys.*, 1998, **199**, 695–703.
45. C. S. Ha and W. J. Cho, Miscibility, properties and biodegradability of microbial polyester containing blends, *Prog. Polym. Sci.*, 2002, **27**, 759–809.
46. M. Abbate, E. Martuscelli, G. Ragosta and G. Sarinzi, Tensile properties and impact behavior of poly(D-hydroxybutyrate)/rubber blends, *J. Mater. Sci.*, 1991, **26**, 1119–1125.
47. R. Pearce, J. Jesudason, W. Orts, R. H. Marchessault and S. Bloembergen, Blends of bacterial and synthetic poly(β-hydroxybutyrate): effect of tacticity on melting behavior, *Polymer*, 1992, **33**, 4647–4649.
48. H. Abe, I. Matsubara and Y. Doi, Physical properties and enzymic degradability of polymer blends of bacterial poly [(R)-3-hydroxybutyrate] and poly[(R,S)-3-hydroxybutyrate]s stereoisomers, *Macromolecules*, 1995, **28**, 844–853.
49. M. Scandola, M. L. Focarete, G. Adamus, W. Silkorska, I. Baranowska, S. Swierczek, M. Gnatowski, M. Kowalczuk and Z. Jedlinski, Polymer blends of natural P(3-hydroxybutyrate–3-hydoxyvalerate) and a synthetic atactic poly(3-hydroxybutyrate) characterization and biodegradation studies, *Macromolecules*, 1997, **30**, 2568–2574.
50. R. J. Fredericks, A. J. Melveger and L. J. Dolegiewitz, Morphological and structural changes in a copolymer of glycolide and lactide occurring as a result of hydrolysis, *J. Polym. Sci. Polym. Phys. Ed.*, 1984, **22**, 57–66.
51. M. Timmins, A. Liebmann-Vinson, Biodegradable polymers: Degradation mechanisms, Part 1, (Ed.) R. Arshady,The PBM Series, Vol. 2, Kentus Books, London, 2006, pp. 285–328; Part 2, pp. 329–372.
52. (a) G. Scott and D. Gilead, *Degradable Polymers*, Chapman & Hall, London, UK,1995; (b) W. K Lee, T. Iwata, H. Abe and Y. Doi, Studies on the enzymic hydrolysis of solution-grown poly[(R)-3-hydroxybutyrate] crystals; defects in crystals, *Macromolecules*, 2000, **33**, 535–541.
53. W. K. Lee and J. A. Gardella, Jr, Hydrolytic kinetics of biodegradable polyester monolayers, *Langmuir*, 2000, **16**, 3401–3406.
54. W. K. Lee, W. J. Cho, C. S. Ha, A. Takahara and T. Kajiyama, Surface enrichment of the solution-cast poly(methyl methacrylate)/poly(vinyl acetate) blends, *Polymer*, 1995, **36**, 1229–1234.
55. W. K. Lee, Y. Doi and C. S. Ha, Retardation effect of enzymic degradation of microbial polyesters at surface by blending with polystyrene, *Macromol. Biosci.*, 2001, **1**, 114–118.

56. W. K. Lee, J. H. Ryou and C. S. Ha, Retardation of enzymic degradation of microbial polyesters using surface chemistry: effect of addition of non-degradable polymers, *Surf. Sci.*, 2003, **542**, 235–243.
57. J. H. Ryou, C. S. Ha, J. W. Kim and W. K. Lee, Control of enzymic degradation of microbial polyesters by plasma modification, *Macromol. Biosci.*, 2003, **3**, 44–50.
58. P. Dave, R. A. Gross, C. Brucato, S. Wong and S. P. McCarthy, Biodegradation of blends containing poly (3-hydroxybutyrate–valerate), *Polym. Mater. Sci. Eng.*, 1990, **62**, 231–235.
59. H. Thissen, Surface modification of biodegradable polymers, The PBM Series, Vol. 2, In *Biodegradable Polymers*, (Ed.) R. Arshady, Citus Books, London, 2006, pp. 175–210.
60. C. C. Chu, Degradation phenomena of two linear aliphatic polyester fibers used in medicine and surgery, *Polymer*, 1985, **26**, 591–594.
61. N. J. Jo, T. Iwata, K. T. Lim, S. H. Jung and W. K. Lee, Degradation behaviors of polyester monolayers at the air/water interface: alkaline and enzymatic degradations, *Polym. Degrad. Stab.*, 2007, **92**, 1199–1203.
62. E. Kim, J. K. Lee and W. K. Lee, Interfacial degradation of biodegradable polyester monolayers at the air/enzyme-containing water interface, *J. Nanosci. Nanotechnol.*, 2008, **8**, 4830–4833.
63. H. K. Moon, Y. S. Choi, J. K. Lee, C. S. Ha, W. K. Lee and J. A. Gardella, Miscibility and hydrolytic behavior of poly(trimethylene carbonate) and poly(l-lactide) and their blends in monolayers at the air/water interface, *Langmuir*, 2009, **25**, 4478–4483.
64. R. Arshady, Biodegradable microcapsules: Basic criteria and properties, in Microspheres, Microcapsules and Liposomes, The MML Series, Vol. 2, *Medical and Biotechnology Applications*, (Ed.) R. Arshady, Citus (now Kentus) Books, London, 1999, pp. 227–257.
65. C. A. Lauzier, C. J. Monasterios, I. Saracovani, R. H. Marchessault and B. A. Ramsay, Film formation and paper coating with poly(β-hydroxyalkanoate), a biodegradable latex, *Tappi J.*, 1993, **76**(5), 71–77.
66. H. Laborit, Sodium 4-hydroxybutyrate, *Int. J. Neuropharmacol.*, 1964, **43**, 433–452.
67. A. S. Hunter, W. J. Long and C. C. Ryrie, An evaluation of gamma hydroxybutyric acid in pediatric practice., *Br. J. Anaesth*, 1971, **43**, 620–627.
68. S. P. Bessman and W. N. Fishbein, Gamma-hydroxybutyrate, a normal brain metabolite., *Nature*, 1963, **200**, 1207–1208.
69. M. Mamelak, M. B. Scharf and M. Woods, Treatment of narcolepsy with hydroxybutyrate. A review of clinical and sleep laboratory findings, *Sleep*, 1986, **9**(1), 285–289.
70. CDC, Gamma hydroxybutyrate use, *MMVR Morb. Mortal Wkly Rep.*, 1997, **46**, 281–283.
71. L. Gallimberti, M. Ferri, S. D. Ferrara, F. Fadda and G. L. Gessa, Gamma-hydroxybutyric acid in the treatment of alcohol dependence: a double-blind study, *Alcohol. Clin. Exp. Res.*, 1992, **16**, 673–676.

72. G. Addolorato, G. Balducci, E. Capristo, M. L. Attilia, F. Taggi, G. Gasbarrini and M. Ceccanti, Gamma-hydroxybutyric acid (GHB) in the treatment of alcohol withdrawal syndrome: a randomized comparative study versus benzodiazepine, *Alcohol Clin. Exp. Res.*, 1999, **23**(10), 1596–1604.

CHAPTER 13

Glycoproteins and Adhesion Ligands: Properties and Biomedical Applications

B.K. MANN*[a] AND S.D. TURNER[b]

[a] University of Utah, Department of Bioengineering, 72 South Central Campus Drive, Rm. 2646, Salt Lake City, Utah 84112, USA; [b] University of Toronto, Faculty of Medicine, 1 King's College Circle, Toronto, Ontario M5S 1A8, Canada

13.1 Introduction

The extracellular matrix (ECM) of a tissue is a complex milieu of water, proteins, growth factors, cytokines, hormones, proteases, protease inhibitors, and minerals that are specific for each tissue. The proportion of the various ECM components, as well as the type of proteins present, provides an indication of the state of the tissue. That is, the make-up of the ECM often changes during development, aging, wound healing, and the course of a disease. The glycoproteins comprise one class of proteins found in the ECM. It is a heterogeneous class, with the proteins exhibiting a variety of sizes, structures, and functions. Some of the glycoproteins serve primarily structural roles, others are signaling molecules, and some serve both roles. In general, glycoproteins help bind cells to their ECM, link other proteins to one another, help regulate cellular fate processes, and contribute to the overall mechanical properties of a tissue.

In this chapter we will focus on some of the structural glycoproteins. These macromolecules have binding sites for cells to attach, thereby providing support for the cells. Once ligated by cellular transmembrane receptors, these

RSC Polymer Chemistry Series No. 1
Renewable Resources for Functional Polymers and Biomaterials
Edited by Peter A. Williams

Published by the Royal Society of Chemistry, www.rsc.org

binding sites also provide signaling to the cells that may affect proliferation, migration, differentiation, and even viability. Due to their multiple functionalities, these glycoproteins are finding uses in the field of biomaterials, alone and in combination with other proteins or with synthetic materials. Here we will discuss some specific glycoproteins and their use in biomaterial applications, as well as specific binding sites found within the glycoproteins and how they may be used, apart from the entire glycoprotein, with biomaterials.

13.2 Prototypical Structural Glycoproteins

As previously mentioned, the glycoproteins comprise a large, heterogeneous class. Table 13.1 summarizes many of the known structural glycoproteins and their properties. Fibronectin and laminin are the two most well characterized structural glycoproteins. Each of them is found in large proportions in a variety of tissues. While the proteins are different, they possess similarities in cell adhesion properties and protein binding affinities, properties that can be exploited for use in biomaterial and tissue engineering applications. In general, less is known about the other structural glycoproteins, and thus they are less often used in biomaterial applications. As more is known about them it is likely they will fill particular niches for specific applications. This section will detail the structure, function and properties of fibronectin and laminin. Later sections of this chapter will detail the application of these glycoproteins to the fields of biomaterials and tissue engineering.

13.2.1 Fibronectin: A Model Structural Glycoprotein

Fibronectin (FN) is a structural glycoprotein that is produced by most mesenchymal and epithelial cells and plays an important role in cellular migration and adhesion.[4,14] Targeted inactivation of the fibronectin gene in mice is lethal and points to the importance of fibronectin in embryogenesis.[4,6,15] Primarily, fibronectin serves as a bridge between cells and the ECM.[1,6] While this protein is found in both plasma and the extracellular matrix, the soluble and insoluble forms of the protein are slightly different.[4] The creation and deposition of fibronectin fibrils is a highly regulated, cell-mediated process which leads to ECM development.[15] The versatility of fibronectin in forming complexes with other molecules is evident in the presence of binding domains for cells, collagen, heparin, some proteoglycans, DNA, hyaluronic acid and fibrin.[4] Fibronectin has played an important part in elucidating the mechanisms of cell-adhesive interactions because it can be cleaved into fragments that retain their cell adhesivity. Therefore, these fragments can serve as a system for isolating the molecular mechanisms of a specific sequence of the full-length protein.[4] Furthermore, the relative abundance of fibronectin in plasma (300 μg mL^{-1}) serves as valuable source of the protein.[1]

Fibronectin is a product of a single gene but can exist in multiple forms because of alternative splicing. Twenty human variants of fibronectin have been

Table 13.1 Examples of structural glycoproteins and their locations, functions, interactions and major cell adhesion sites. Table adapted from ref. 3, 4 and 13.[a]

Protein	*Location*	*Function*	*Interactions*	*Cell adhesion peptides*	*Ref.*
Fibronectin	Plasma Cellular	Cell Attachment, Differentiation, Early Development, Wound Repair, Thrombosis, Cytoskeleton Organization, Opsonin	C(I) C(III) F H HA DNA	RGD IDAPS LDV REDV KNNQKSEPLIGRKKT ALNGR (FN-C/H II)	1–5
Laminin	Basement Membranes	Basement Membranes, Cell Development, Differentiation, Nerve Regeneration, Wound Healing	C(IV) E HS	α chain: RGD IKLLI SIYITRF IAFQRN LQVQLSIR IKVAV YFQRYLI β1 chain: LGTIPG RYVVLPRP PDSGR YIGSR γ1 chain: LRE (x2) RNIAEIIKDI	1–4, 6–8
Bone Sialoprotein BSP	Bone, Mineralized Connective Tissue	Bone Formation	Hap	RGD	1, 9, 10
Entactin Nidogen	Basement Membrane	Links L and C(IV) to Basement Membrane	L C(IV) FN	RGD	1, 3, 4
Fibrillins	Connective Tissue, Elastic Tissues	Component of Elastic Fibers		Fibrillin-1 & Fibrillin-2: RGD	1, 3, 4

Table 13.1 (*Continued*)

Protein	*Location*	*Function*	*Interactions*	*Cell adhesion peptides*	*Ref.*
Osteonectin BM40, SPARC	Bone, Basement Membrane	Mineralization Anti-adhesive	C(IV) Hap		1, 3, 4
Osteopontin OPN, 2ar, SPPI	Bone, Placenta, Kidney, CNS	Bone Development Bone Remodeling Osteogenesis		RGD	1, 10
Tenascins	Skin, Bone, Cartilage, Tendon, Adult Brain, Myotendinous Junction	Tissue Remodeling Anti-Adhesion Cell Rounding and Detachment	FN CS H	Tenascin C: RRGDM	1, 3, 4, 11
Thrombospondin-1	Platelet α-granuales	Cell Growth, Adhesion, Migration, Platelet Aggregation, Fibrin Deposition	C L Fibrinogen Plasminogen H FN	RGD VTCG (x2) Platelet Adhesion Sites	1, 3, 4
Vitronectin S-protein, Serum Spreading Factor	Plasma, Wound Formation, Connective Tissue	Cell Migration Tissue Remodeling Coagulation Inhibits Cytolysis Haemostasis Phagocytosis Immune Function	C H PAI-1 Thrombin-Antithrombin II Complex	RGDV	1, 3, 4
Von Willebrand Factor	Plasma Subendothelium	Carrier of Factor VIII Platelet Adhesion	Factor VIII C H	RGDS GP Ibα	1, 3, 12

[a]Abbreviations: C(type) = collagen(type), H = heparin, F = fibrin, HA = hyaluronic acid, L = laminin, FN = fibronectin, CS = chondroitin sulfate, Hap = hydroxyapatite, PAI-1 = plasminogen activator inhibitor, E = entactin, HS = heparan sulfate proteoglycan.

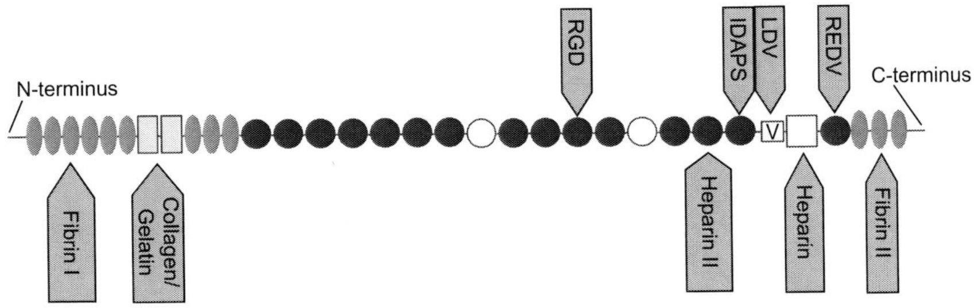

Figure 13.1 Schematic of fibronectin molecule indicating the location of FN-type repeats and ECM- and cell-binding domains. FN type I (gray ovals), type II (light gray rectangles) and type III (dark circles) repeats are indicated, as well as alternative splice regions (white regions) and a variable segment (white square with "V" inside). The general locations of the ECM- and cell-binding domains are shown by arrows. Adapted from ref. 1, 6, 15 and 17.

identified.[15] All fibronectins share the same basic functional domains; however, the variability created by alternative splicing in the cell adhesion domain leads to the cell-type specificity exhibited by the protein.[6] In general, the fibronectin protein is a dimer of two non-identical subunits, each with a molecular weight (M_W) of 250 kDa that are disulfide bonded at the carboxy-terminus.[1] Depending on the cell type, fibronectin can contain 4–10% carbohydrate, with both O-linked and N-linked residues found in FN.[4] The physiological role of the carbohydrates is still under investigation, but they appear to stabilize the molecule against hydrolysis.[15]

Figure 13.1 provides a schematic detailing the basic functional domains present in fibronectin. Roughly ninety percent of the fibronectin molecule is made up of a series of three types of FN repeats.[15] Alternative splicing at three locations yields the variants of fibronectin.[16,17] As shown in Figure 13.1, there are several cell binding domains located within the fibronectin sequence, along with collagen-binding, heparin-binding, and fibrin-binding sites. The heparin-binding domains interact with heparin sulfate proteoglycans, and may aid in cell adhesion.[13,15] The fibrin-binding sites may also be important for cell adhesion, as well as cell migration into fibrin clots.[15]

The presence of several different cell adhesion motifs, including RGD, LDV, REDV, and IDAPS, within fibronectin speaks to the broad cell adhesion properties of this glycoprotein.[6] Experiments with synthetic RGD and LDV motifs indicate that these short peptides can mimic some of the inherent adhesivity of the full-length protein; however, these sequences do not function alone, as competitive inhibition assays and avidity determinations show that these sequences are less active than the native protein and fully active fragments.[13] For instance, synthetic RGD has 100-fold less activity than the intact fibronectin.[13] However, another short peptide sequence, PHSRN found in the 9th type III repeat, has been shown to be a synergistic site to RGD, found in the 10th type III repeat, for cell binding.[18–20]

13.2.2 Laminin

Laminin is a structural glycoprotein that is ubiquitously distributed in basement membranes.[1,4] As with fibronectin, laminin is a multifunctional protein. It plays a key role in development, differentiation and migration. Laminin acts primarily as a bridging molecule between basal lamina and cells by anchoring cells to the basement membrane through type IV collagen.[6] Localized binding sites for collagen, heparin and entactin as well as distinct sites for cell attachment are present in laminin.[1,6] Laminin additionally contains several epidermal growth factor (EGF)-like domains that stimulate cell proliferation in a manner similar to EGF.[6]

Unlike fibronectin, variant forms of laminin are assembled from genes encoding isoforms of particular chains. The form present in the ECM is a function of the tissue type and developmental stage.[6] The prototypical laminin molecule is a disulfide-bonded heterotrimer containing three non-identical chains (designated α, β, and γ), which form a cruciform structure with globular domains at the end of each arm (see Figure 13.2).[4] The most well-characterized

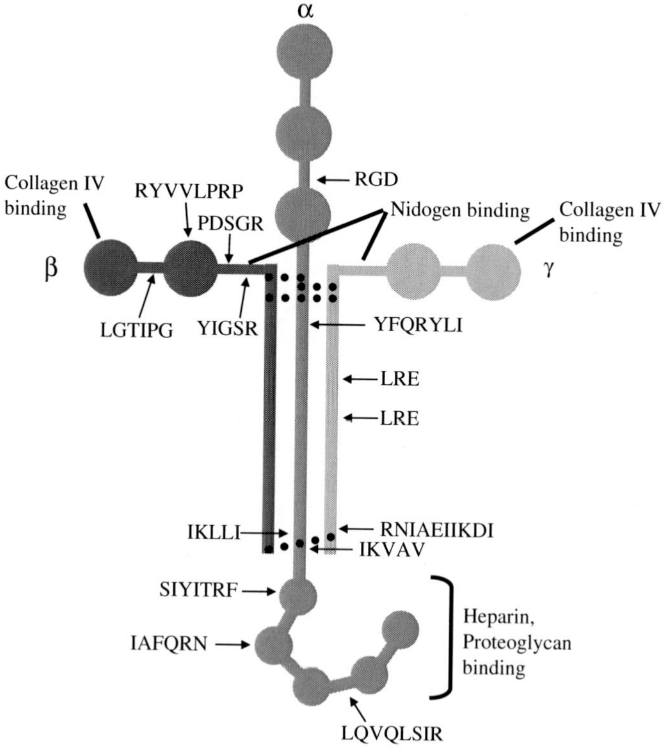

Figure 13.2 Schematic of laminin structure. Approximate locations of some cell adhesion peptides are indicated with arrows. Regions binding to other molecules are also indicated. Interchain disulfide bridges are indicated by dotted lines. Adapted from ref. 7, 28 and 29.

Table 13.2 Known isoforms of laminin. Table assembled from ref. 28–37.

	Chain composition		
Isoform	α	β	γ
Laminin-1	1	1	1
Laminin-2	2	1	1
Laminin-3	1	2	1
Laminin-4	2	2	1
Laminin-5	3	3	2
Laminin-6	3	1	1
Laminin-7	3	2	1
Laminin-8	4	1	1
Laminin-9	4	2	1
Laminin-10	5	1	1
Laminin-11	5	2	1
Laminin-12	2	1	3

of the laminins is laminin 1, made up of α1, β1, and γ1 chains.[7,21] Table 13.2 provides other known laminin isoforms and their chain compositions.

Laminin contains numerous adhesion sequences, some of which are listed in Table 13.1, including IKLLI, SIYITRF, IAFQRN, LQVQLSIR, RGDN and IKVAV in the α chain; YIGSR, LGTIPG, RYVVLPR, and PDSGR in the β1 chain; and RNIAEIIKDI, and two LRE sequences on the γ1 chain.[22,23] Approximate locations of these peptides on their respective chains are shown in Figure 13.2. While these are some of the more characterized sequences, other novel peptides and peptides with specific functions have also been determined.[7,21,24–27] As with peptides derived from fibronectin, laminin-derived peptides have been found to require higher molar concentrations to yield the same level of cell binding as the native protein.[13] It is known that IKVAV requires additional chain–chain polypeptide interactions for the adhesion of neuronal cells, and these additional interactions may explain the higher molar concentrations needed when the peptide is used alone.[13]

13.3 Glycoproteins for Biomaterial Applications

Glycoproteins have been utilized in various ways in biomaterial applications. In some cases they have been adsorbed or covalently attached to the surface of a material. Altering the surface with a glycoprotein provides a means of cell attachment and may provide an environment more suitable for cell and tissue growth. In other cases glycoproteins have been used alone or in combination with another protein or a synthetic polymer to create a scaffold for tissue engineering. Again, the glycoproteins provide a natural means of cell attachment and signaling for the cells.

Additionally, as glycoprotein-derived peptides have been found to mediate cell attachment by themselves, these peptides have been used as an alternative to the entire glycoprotein molecule. Harsher processing conditions may be used in the production of these peptides when compared to the conditions needed for the full-length protein and therefore the peptides may be suitable for a broader spectrum of applications. Moreover, the effect of the domain itself may be studied in isolation, for instance, specific data regarding the concentration of the peptide needed, the interactions that may be necessary for function and the types of cells targeted can be determined from the use of these peptides in isolation.

13.3.1 Surface Modification with Glycoproteins or Peptides

While cell attachment to some medical devices, such as biosensors, is undesirable as it may disrupt the function of the device, other devices rely on the adhesion of cells. For example, a complete endothelial cell monolayer is needed on the surface of synthetic vascular grafts in order to eliminate thrombosis. This monolayer can only be achieved if the endothelial cells can attach to and be maintained under physiological flow conditions on the graft. Another example would be in orthopedic applications, where an implant needs to be integrated with the surrounding bone to prevent movement of the implant and potential damage to both the implant and the bone. For proper osteointegration, cells from the surrounding bone need to be able to attach to the implant. Glycoproteins provide one method of altering a device surface in order to promote and enhance cell adhesion.

Glycoproteins, as with other types of proteins, will readily adsorb to many biomaterials, depending on the specific properties of the surface and the individual protein.[38] This provides a fast and easy method for altering the surface of a material. However, the conformation of the glycoproteins could be altered upon adsorption, which may disrupt the normal interaction of cells with the glycoprotein. Further, competitive adsorption by other proteins that may be present, whether during *in vitro* culture or once implanted, could cause the glycoproteins adsorbed to the surface to desorb. A more stable and predictable means of altering a device surface is to covalently attach the glycoprotein or an adhesion peptide from the glycoprotein. This means of attachment is sometimes termed bioconjugation. Bioconjugation can be achieved in a number of ways, depending on the glycoprotein being attached and the underlying material.

13.3.1.1 Fibronectin

One of the primary uses of covalent attachments to material surfaces thus far has been to encourage endothelialization of vascular devices. Small blood vessels with diameters less than about 6 mm, such as the coronary arteries, display poor patency rates with synthetic vascular grafts, primarily due to excessive thrombosis on the graft surface. One method that may help to counteract this effect is to develop a monolayer of endothelial cells on the surface of the vascular graft. The monolayer could be grown by either seeding or sodding the cells on the surface *in vitro*, or by encouraging migration and growth of cells from the surrounding tissue once the graft is placed *in vivo*. In either case, the cells require adhesion sites on the graft surface for both adhesion and migration. This has led researchers to investigate coatings and covalent attachment of ECM proteins or adhesion ligands to the surface of such vascular grafts. While coatings of ECM proteins, such as fibronectin and laminin, have encouraged initial attachment and growth of endothelial cells *in vitro*, the cells have a tendency to shear off the surface once exposed to pulsatile blood flow.[39] Covalent attachment is thus considered to be a more viable option for this application.

Fibronectin has been used to coat the surface of expanded polytetrafluoroethylene (ePTFE) vascular grafts and shown to improve endothelial cell adhesion.[40] The surfaces of poly(carbonate-urea)urethane vascular grafts have been altered by either coating with an ECM protein, including fibronectin, or by covalently attaching RGD.[41] While the simple protein coatings did not improve the retention of endothelial cells sodded on the surface, the covalently attached peptide did. RGD-containing peptides have also been covalently attached to gold-coated polyurethanes, mesylMPEO, and poly(vinyl amine) and found to promote the adhesion of endothelial cells.[42–44] REDV-containing peptides have also been used for cardiovascular applications due to their high specificity for the adhesion of endothelial cells.[45] Materials that have been modified by covalent attachment of REDV peptides to the surface include polyethylene oxide incorporated into polyethylene terephthalate and gold-coated polyurethane.[42,46] Another peptide that has been covalently attached to surfaces to promote adhesion of vascular endothelial cells is FN-C/H-V from the carboxy-terminal heparin-binding domain of fibronectin.[47] Attachment to both polystyrene and polyethylene terephthalate surfaces promoted cell adhesion and spreading. An alternative to using fibronectin has been to coat the surface of polyurethane vascular grafts with a fibronectin-like engineered protein polymer that contains multiple copies of RGD, just as the natural protein does.[39]

Orthopedic and dental implants can often fail due to poor integration of the implant with the surrounding tissue. These implants may be improved by enhancing cell adhesion to the surface of the implant and cell function to encourage tissue integration. Covalent attachment of an ECM protein or adhesion ligand on the surface of the implant could provide the necessary cell adhesion sites. Fibronectin has been both adsorbed and covalently attached to the surface of poly(vinylidenefluoride) (PVDF) for use in bone applications.[48] Adhesion of osteoblasts was enhanced with both adsorbed and immobilized fibronectin, but proliferation of the cells was only enhanced on the surfaces with the immobilized fibronectin.

As with the vascular applications, adhesion peptides have also been covalently attached to device surfaces for orthopedic applications. An RGD-containing peptide has been attached to polypyrrole-coated titanium and shown to enhance the attachment of osteoblasts.[49] Interpenetrating polymer networks of poly(acrylamide-co-ethylene glycol/acrylic acid) on titanium surfaces have also been functionalized with RGD peptides.[50] The modified surfaces were found to support osteoblast attachment and spreading, as well as significant mineralization. An intermediate solution to using either the whole fibronectin molecule or a short oligopeptide has been to use a fibronectin fragment, FNIII(7-10).[18] This fragment contains both the RGD site in the 10th type III repeat and the PHSRN synergy site in the 9th type III repeat of fibronectin. Osteoblast-like cells were able to adhere, spread, and assemble focal adhesions on surfaces functionalized with the fibronectin fragment.

13.3.1.2 Laminin

Peptides from laminin have been coupled to material surfaces in order to control neural cell adhesion and differentiation, and to try to direct and guide neurite extension. Micropatterns of IKVAV were attached to the surface of polylactide-polyethylene glycol (PL-PEG) in order to direct neurite extension.[51] YIGSR and IKVAV have both been coupled to fluorinated ethylene propylene films for the attachment of neuroblastoma and PC12 cells.[52] YIGSR and IKVAV have also been coupled to ePTFE fibers and examined for their ability to allow neurite extension from dorsal root ganglia (DRGs).[53] Short sequences (CYIGSR and CIKVAV) and extended sequences (CDPGYIGSR and CQAASIKVAV) were both used, and extended sequences allowed for the longest neurites. The extended sequences may place the peptides in a conformation that is more similar to their native conformation in laminin.

13.3.1.3 Glycoprotein-derived Peptides

Cell adhesion peptides from fibronectin and laminin have also been used to modify surfaces for other applications. RGD has been conjugated to PLL-g-PEG, which has then been adsorbed onto negatively charged surfaces.[54] This coating allowed the adhesion and spreading of human dermal fibroblasts, while effectively blocking non-specific adsorption of serum proteins to the surface. YIGSR and RGDS have been covalently attached to polydimethyl siloxane substrates in order to develop a synthetic keratoprosthesis.[55] Synergistic peptides for each adhesion peptide, PDSGR and PHSRN respectively, were also included and found to increase adhesion of human corneal epithelial cells, again indicating that synergistic peptides may be beneficial. RGD has also been attached to a glass surface to examine the influence of attached peptides on cell behavior.[56] While cell adhesion to the surface increased with increasing peptide concentration, both migration and proliferation decreased at the highest peptide concentration (see Figure 13.3).

13.3.2 Glycoprotein/Peptide Incorporation in Tissue Engineering Scaffolds

13.3.2.1 Fibronectin

Fibronectin has been used in conjunction with a variety of materials, both natural and synthetic, in order to encourage cell adhesion and proper functioning on tissue engineering scaffolds. Generally, the fibronectin is coated or

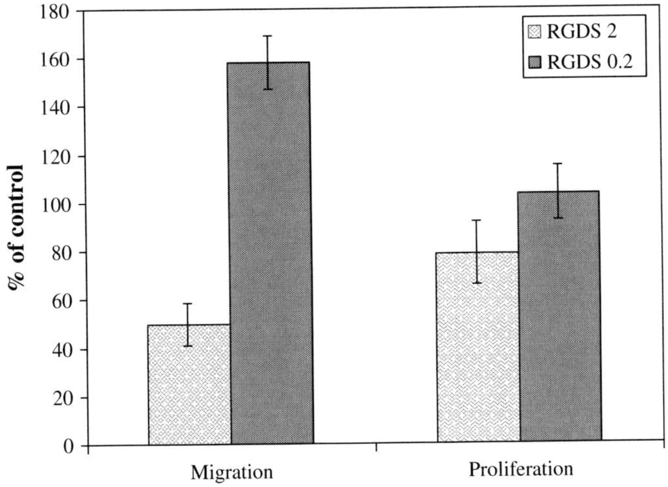

Figure 13.3 Migration and proliferation of smooth muscle cells on glass surfaces with covalently attached RGDS at 2.0 and 0.2 nmol cm^{-2} compared to control surfaces. Control surfaces for migration were surfaces with attached RGES, and for proliferation were surfaces with no peptide. Adapted from ref. 56.

adsorbed to the scaffold, although the glycoprotein has also been incorporated within scaffolds as well. Fibronectin coatings have been used with scaffolds for a variety of tissues, including cartilage, bone, nerve, adipose tissue, and blood vessels.[57–62] Such widespread use may be due to the fact that fibronectin is found in the ECM of so many tissues, as well as in blood plasma, and that there are multiple adhesion sites within a single fibronectin molecule. Fibronectin may be considered a model protein for incorporating in scaffolds due in part to the fact that it is readily available and routine assays methods are well characterized.[57]

Copolymers of trimethylene carbonate and glycolide have been coated with fibronectin and seeded with chondrocytes.[57] The cells were able to adhere to and infiltrate the porous scaffolds, remaining viable for up to four weeks in culture. Hyaluronan-based sponges have been coated with fibronectin and placed in osteochondral defects in rabbits in order to aid repair and integration of new cartilage with the surrounding tissue.[58] Hyaluronic acid-based polymers coated with fibronectin and seeded with mesenchymal progenitor cells have also been shown to encourage bone and cartilage formation when implanted subcutaneously in nude mice.[59] For peripheral nerve regeneration, collagen nerve guidance conduits (NGCs) have been coated with fibronectin.[60] Fibronectin coatings on ePTFE surgical mesh have increased the sodding efficiency of preadipocytes as compared to coatings with other ECM components and has led to differentiation of the cells to adipocytes.[61] Fibronectin has also been used as a coating on a mesh-stent to be deployed in a porcine coronary artery as a means of attaching engineered smooth muscle cells to the stent.[62]

In addition to using fibronectin as a coating on a scaffold, it has also been incorporated within scaffolds, for example, fibronectin has been combined with alginate in a hydrogel.[63,64] This hydrogel has been seeded with Schwann cells and coated on polyhydroxybutryate (PHB) fibers that were then grafted in a cervical spinal cord injury in adult rats.[64] The scaffold helped to support neuronal survival and regeneration. The alginate–fibronectin hydrogel seeded with Schwann cells has also been used for peripheral nerve regeneration by placing the hydrogel within a PHB conduit to bridge a gap in the sciatic nerve.[63] The added fibronectin enhanced viability of the Schwann cells both *in vitro* and *in vivo*, and led to an enhanced regeneration rate *in vivo*. Fibronectin has been covalently attached to poly(vinyl alcohol) (PVA) hydrogels in order to promote cell attachment and potential use of these hydrogels in tissue engineering applications.[65,66] Poly(D,L-lactide) grafted with polyacrylic acid has also been modified by covalently binding fibronectin as a means of functionalizing the graft polymer.[67]

While the structural glycoproteins are most often used in combination with other materials, fibronectin has been used alone to create scaffolds. Fibrous mats and large cables of fibronectin have both been investigated as potential scaffolds for tissue repair.[68–70] Heparin could be incorporated into the fibronectin mats, which would then bind and slowly release basic fibroblast growth factor.[68] Fibronectin cables up to 14 cm long and 1.5 cm in diameter were created, having pore sizes between 10 and 100 μm. Schwann cells were able to adhere and migrate on the cables, and both Schwann cells and fibroblasts were able to grow within the pores of the cables. These fibronectin cables may thus be useful for tissue engineering applications, such as peripheral nerve regeneration.[69]

13.3.2.2 Laminin

One of the primary uses thus far for incorporating laminin into a tissue engineering scaffold has been for peripheral nerve regeneration. Laminin is expressed during neural development and has been shown to stimulate neurite outgrowth.[24] One method of presenting laminin to neural cells has been to adsorb or coat a tissue engineering scaffold with the protein. For example, NGCs fabricated from collagen have been coated with laminin.[60] Collagen fibers coated with laminin and collagen sponges soaked with laminin have each been placed in the lumen of polyglycolic acid (PGA) conduits in order to enhance nerve regeneration.[71,72] Laminin has also been adsorbed into micro-patterned grooves on the surface of poly(D,L-lactic acid) and poly(lactide-co-glycolide) films.[73,74] The combination of having a chemical cue, the laminin, with a physical cue, the grooves, was found to have a synergistic effect in enhancing directional neurite outgrowth.

An alternative method to adsorption on the surface is to incorporate the laminin within the scaffold. Laminin, along with fibronectin and nerve growth factor (NGF), has been incorporated into gels made from styrene-derivatized gelatin that allowed neurite extension in an *in vitro* model.[75] Agarose hydrogels have also been modified to covalently incorporate laminin.[76,77] These laminin-modified agarose hydrogels, which also incorporated NGF, have significantly enhanced neurite extension in *in vitro* models compared to unmodified agarose gels.[76] Moreover, these hydrogels have demonstrated nerve regeneration potential after 2 months in a 10 mm sciatic nerve gap in rats that was comparable to autografts.[77] These studies not only demonstrate the importance of this structural glycoprotein in nerve regeneration, but also indicate that laminin may be suitable for inclusion in other types of tissue engineering scaffolds.

Laminin has also been adsorbed to surfaces for use with other tissues, including cardiac muscle, adipose tissue, and salivary glands. Cardiomyocytes have been seeded onto lanes of laminin that were microcontact-printed on bovine serum albumin-coated surfaces and were found to be constrained to those laminin lanes.[78] Additionally, the resulting lanes of cardiomyocytes were found to beat, and if some cells were able to bridge lanes, synchronously beating fields of cardiomyocytes were obtained. Laminin may also improve the function of preadipocytes, which have been shown to readily adhere to and migrate on laminin-1 coated surfaces.[79] Additionally, salivary epithelial cells have been seeded onto PLLA, PGA or PLGA disks coated with laminin or fibronectin with the goal of developing an artificial salivary gland.[80] In an interesting application of this glycoprotein, laminin has been incorporated into a synthetic extracellular matrix composed of a thiol-modified carboxymethylhyaluronic acid (CMHA-S) and a thiol-modified gelatin (Gtn-DTPH) that was crosslinked with polyethylene glycol-diacrylate (PEGda). MDCK cells encapsulated within the matrix began to form acini within 3 days of *in vitro* culture (see Figure 13.4a), and formed very large ring structures by day 10 (Figure 13.4b). Due to its prevalence in various tissues and its potential for aiding cell attachment and function, it is likely that laminin will continue to be investigated in conjunction with various polymers for tissue engineering applications.

13.3.2.3 Fibronectin-derived Peptides

The RGD peptide from fibronectin has been covalently attached to numerous types of materials to be used as tissue engineering scaffolds in order to facilitate the adhesion of a variety of cell types. For example, RGD has been covalently incorporated into alginate hydrogels for attaching fibroblasts,[81] preadipocytes,[82] osteoblasts,[83] and myoblasts.[84,85] Macroporous fragments of alginate hydrogels have also been conjugated with RGD and seeded with preadipocytes, which supported tissue and vascular ingrowth *in vivo*.[82] Fibrin gels with covalently attached RGD have been found to influence the angiogenic behavior of endothelial cells *in vitro*[86] and neurite extension from embryonic chick DRGs.[87] In the case of the DRGs, both a linear and a cyclic RGD sequence were investigated, and less of the cyclic peptide was required for maximum neurite extension compared to the linear peptide.[87] RGD conjugated to PVA hydrogels enabled the attachment and spreading of vascular smooth

(A)

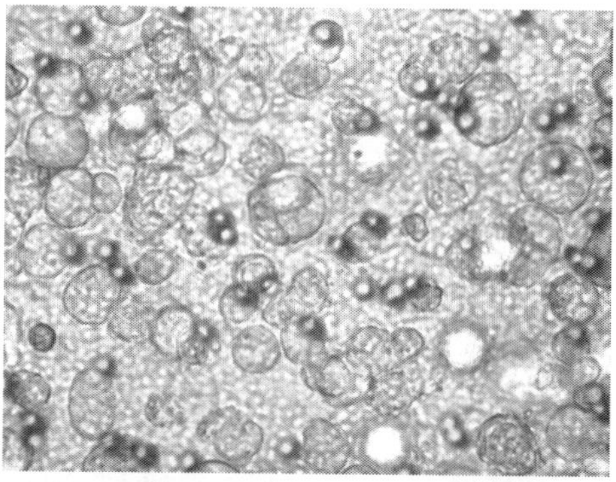

(B)

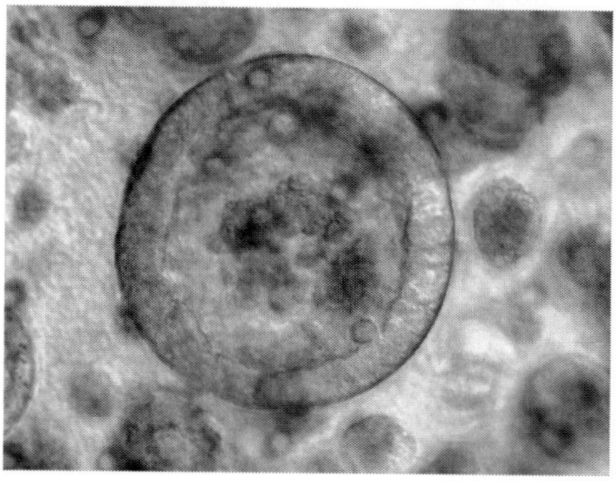

Figure 13.4 MDCK cells encapsulated within CMHA-S/Gtn-DTPH/PEGda hydrogels containing 100 μg ml^{-1} laminin. The cells begin forming acini by day 3 (A) and form large multi-cellular ring structures by day 10 (B). (Photos courtesy of Dr Terry Tandeski, Center for Therapeutic Biomaterials, University of Utah.)

muscle cells[88] and fibroblasts.[89] Polyethylene glycol hydrogels with covalently incorporated RGD support adhesion and growth of smooth muscle cells[56,90,91] and fibroblasts,[91] adhesion and cytoskeletal organization of osteoblasts,[92] and neurite extension from PC12 cells (Figure 13.5a).[93] It should be noted that the concentration of incorporated peptide has been found to affect several aspects of cell behavior, including attachment, spreading, migration, and proliferation (Figure 13.6),[56,90,91,93] with biphasic behavior often exhibited, indicating that too much peptide may interfere with desired cellular function. Hydrogels of *N*-(2-hydroxypropyl)methacrylamide chemically linked with RGD have been investigated as scaffolds for promoting axonal regeneration in brain and spinal cord injuries.[94,95] Another hydrogel that has covalently incorporated RGD is based on poly(propylene fumarate-co-ethylene glycol) and was shown to support the adhesion, proliferation, and migration of osteoblasts.[96]

RGD coated onto PLGA disks enhanced the early stages of osteocompatibility and osseous ingrowth,[97] while covalently attaching RGD to PLA scaffolds led to enhanced alkaline phosphatase activity and calcium levels of osteogenic precursor cells.[98] RGDSK has also been covalently attached to a

(A)

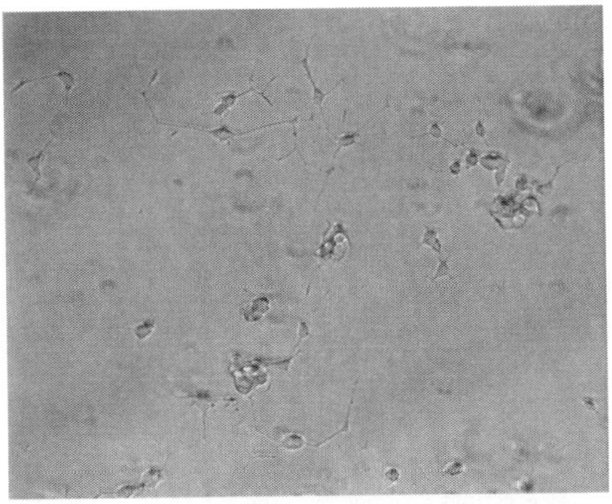

(B)

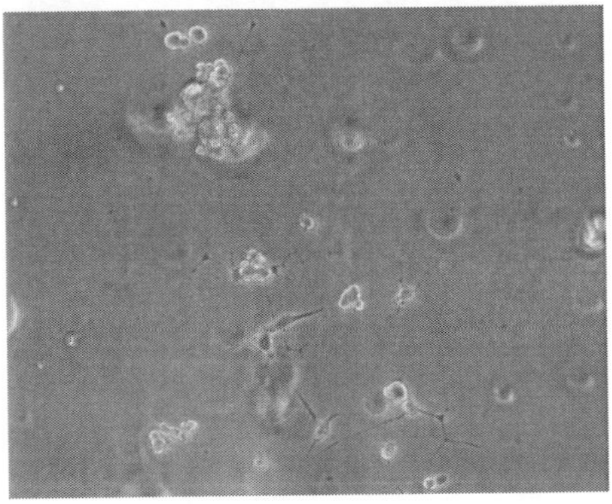

Figure 13.5 PC12 cells extending neurites 48 h after seeding on top of PEG-based hydrogels with covalently incorporated adhesion peptides. Media contained 50 ng ml^{-1} NGF. A: on gels with 1.3 μmol ml^{-1} RGDS; B: on gels with 0.7 μmol ml^{-1} CSRARKQAASIKVAVSADR.

DNA base analog used to create self-assembled rosette nanotubes.[99] These nanotubes were then used to enhance osteoblast interaction with hydrogels and the modification of poly(lactic acid-co-lysine) scaffolds with RGD enhanced attachment of bovine aortic endothelial cells.[100] RGD has also been used as part of amphiphilic block copolymers that may be used to modify the surfaces of other polymer scaffolds in order to render them bioactive.[101] Copolymers were created with the peptide attached through the C-terminus, GRGDSG-b-PEO-b-PLA, and with the peptide attached through the N-terminus, PLA-b-PEO-b-GRGDSG.[101] Additionally, a copolymer of RGD and sugar moieties has been developed for an insulinoma cell culture.[102] The copolymer, poly(*N*-p-vinylbenzyl-D-maltonamide-co-6-(p-vinylbenzamido)-hexanoic acid-g-GRGDS), led to enhanced proliferation of the cells and a greater insulin secretion.[102]

A more recent focus with FN and RGD has been placed on finding fragments or peptides with more specific affinity to particular integrins, such as $\alpha_5\beta_1$, $\alpha_9\beta_1$, or $\alpha_v\beta_3$, rather than general cell adhesion in order to stimulate a particular cell signaling cascade, thereby resulting in a particular desired cell behavior, such as differentiation or migration.[103–107] Studies have shown that

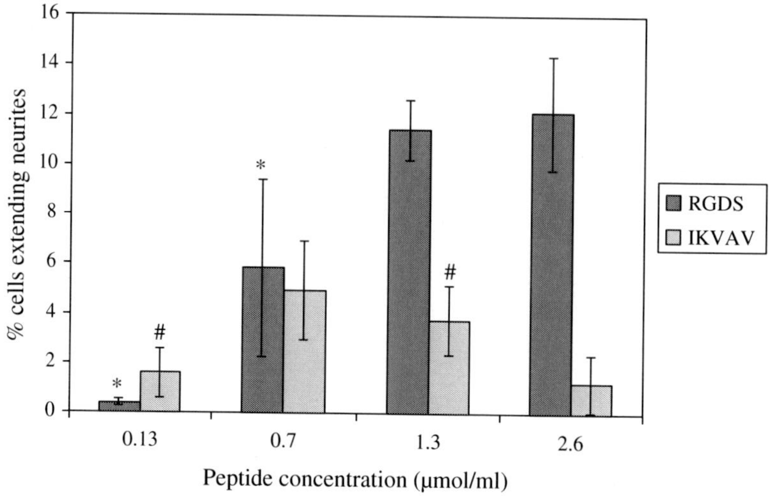

Figure 13.6 Neurite extension from PC12 cells on PEG hydrogels with covalently incorporated RGDS or CSRARKQAASIKVAVSADR. Adapted from ref. 93.

ligation of a particular adhesion receptor can lead to a change in cell response, such as calcification by valvular interstitial cells.[108] Further, simultaneously ligating different receptors, such as $\alpha_2\beta_1$ and $\alpha_5\beta_1$, through the use of multiple peptides can result in enhanced focal adhesion assembly, adhesion strength, FAK activation, and proliferation.[109]

13.3.2.4 Laminin-derived Peptides

Peptides derived from laminin have most often been used in conjunction with materials under investigation as scaffolds to enhance or promote nerve regeneration. Laminin peptides have been covalently incorporated into fibrin gels during coagulation through the use of Factor XIIIa. These constructs enhanced neurite extension *in vitro* and increased the number of regenerated axons in a severed dorsal root model *in vivo*.[110] The peptides used in the fibrin gels were RGD, YIGSR, and IKVAV and these peptides were used both individually and in combinations. These same peptides have also been incorporated individually and in combinations into PEG hydrogels[93] and agarose hydrogels[111] allowing neurite extension from PC12 cells and DRGs. Neurite extension from PC12 cells seeded onto PEG hydrogels with a covalently linked IKVAV-containing peptide is shown in Figure 13.5b, and the effect of peptide concentration on the neurite extension is shown in Figure 13.6. The different combinations of the peptides have been found to have different effects on neurite extension (see Table 13.3), likely owing to the fact that cells *in vivo* are normally exposed to multiple ligands and hinting at the complex nature of intracellular signaling following multiple receptor ligation.

Table 13.3 Effect of combinations of cell adhesion ligands on neurite extension. ↓: inhibitory, ↑: additive, ↔: noninteractive, ↑↑: synergistic, ND: not determined. Table adapted from 93 and 110.

Ligand combination	*Effect on PC12 cells on PEG gels*	*Effect on DRGs in fibrin gels*
RGD/IKVAV	↓	↓
RGD/YIGSR	↔	↑
IKVAV/YIGSR	↔	↔
RGD/IKVAV/YIGSR	↑	ND
RGD/IKVAV/YIGSR/RNIAEIIKDI	ND	↑↑

Chitosan membranes conjugated with two different laminin-derived peptides, AGTFALRGDNPQG (A99 peptide) and RKRLQVQLSIRT (AG73 peptide), have also been shown to promote neurite extension from PC12 cells.[22] While the A99 peptide contains the RGD sequence and interacts with an integrin receptor, the AG73 peptide interacts with a proteoglycan-type receptor. YIGSR has also been grafted to hydrogel copolymers of collagen and *N*-isopropylacrylamide and promoted *in vivo* regeneration of corneal epithelium, stroma, and nerves when implanted into pigs' corneas.[112]

While laminin-derived peptides have most often been used in neural applications, they have also been used with other cell types. For example, an IKVAV-containing peptide has been used as an alternative to the whole laminin protein for incorporating into disulfide crosslinked CMHA-S hydrogels. 3T3 fibroblasts seeded on top of these gels were found to attach and spread on both peptide-containing and laminin-containing hydrogels (see Figure 13.7), but could not attach to CMHA-S only gels. Additionally, YIGSR has been incorporated into a PEG and polyurethaneurea microporous scaffold and was found to promote endothelialization of the scaffold.[113,114] The presence of the YIGSR peptide in the scaffold also enhanced cell migration, proliferation, and matrix protein production (see Figure 13.8).

13.3.2.5 Protein Fragments

In certain applications, it may be more appropriate to use a protein fragment rather than the whole protein or even just a short peptide. In the case of laminin, the whole protein is very large (~500 kDa), yet short peptides from it often do not have the same level of bioactivity as the whole protein or fragments due to the lack of supporting sequences for the short peptides.[8] Examples of protein fragments that have been used for cell adhesion include a recombinant FN fragment that includes the FN(III)10 domain[115] and a recombinant LN fragment from the LG4 domain.[8] Several fibronectin functional domains have been incorporated into hyaluronic acid-based hydrogels.[116] The three functional domains used in the study contained: (1) the RGD binding and PHSRN synergy sites, (2) the heparin II binding site, and (3) the REDV

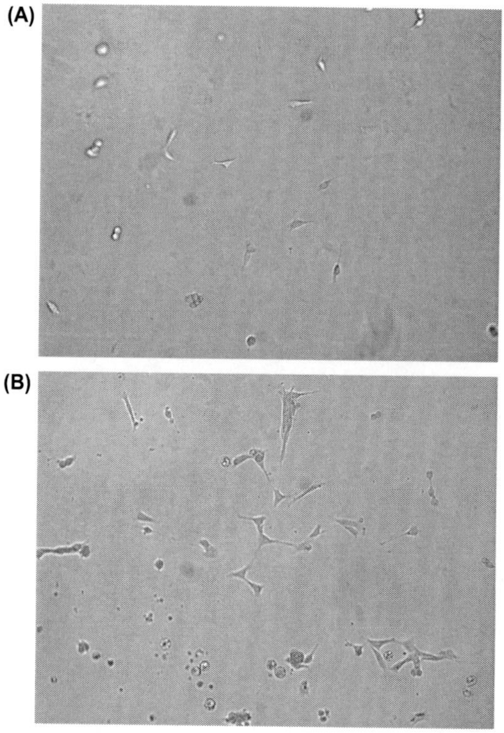

Figure 13.7 NIH 3T3 fibroblasts seeded on top of disulfide crosslinked CMHA-S hydrogels with covalently incorporated peptide or protein. A: on gels with 100 μg ml^{-1} CGIKVAVGY; B: on gels with 100 μg ml^{-1} laminin. (Photos courtesy of Dr Terry Tandeski, Center for Therapeutic Biomaterials, University of Utah.)

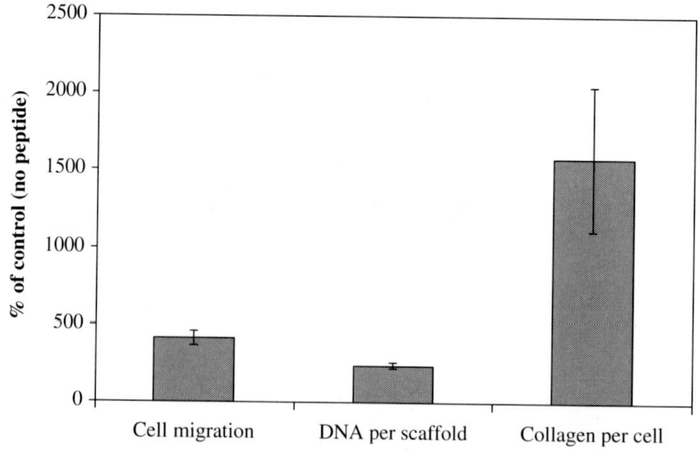

Figure 13.8 Migration, proliferation, and collagen production by endothelial cells on PEG/polyurethaneurea scaffolds modified with YIGSR compared to scaffolds without peptide. Adapted from ref. 114.

binding site. Incorporation of these domains into the hydrogels allowed fibroblast attachment and migration *in vitro*, as well as recruitment of stromal fibroblasts in porcine cutaneous wounds.

13.3.3 Other Applications

Both linear and cyclic RGD peptides have been investigated for use in targeting tumor cells, either for imaging or delivery of anti-cancer drugs. This is due to the specific expression of the integrin $\alpha_v\beta_3$ in tumors during growth and metastasis.[117] A significant focus in the past decade has been on radiolabeling RGD for imaging with SPECT or PET.[117] It has also been labeled with IRDye800 for fluorescent detection.[118] Further, RGD has been conjugated to various polymers for delivering such tumor treatments as paclitaxel, (90)Y (a radiotherapeutic agent), a gelanamycin derivative, and gold nanorods for use in near-infrared laser irradiation.[119–121]

Cell adhesion peptides have also been combined with polymers for other applications. An artificial protein has been developed by recombinant DNA methods.[122] The artificial protein contains repeating amino acid sequences, containing the RGD motif, 2 plasmin degradation sites, and a heparin-binding site. This artificial protein was then grafted with PEG-diacrylate and photo-polymerized to create protein-graft–PEG hydrogels. Although the RGD motif in this case was derived from fibrinogen, the RGD motif from other ECM proteins, such as fibronectin, could potentially also be used.

Other recombinant proteins containing RGD have also been created. For example, the peptide has been used in conjunction with either a starch binding module or a cellulose binding module to create bi-functional recombinant proteins.[123,124] These recombinant proteins were then adsorbed to dextrin-based hydrogels or cellulose fibers, respectively, to provide functionality to those materials. An artificial ECM protein has also been engineered using elastin-like structural repeats and fibronectin cell-binding domains,[125] as well as a synthetic spider silk that incorporated two RGD sequences.[126]

YIGSR-PEG-lipids have been created that could be included in liposomal formulations.[127] These liposomes could be useful as long-circulating, ligand-presenting materials for targeted drug delivery applications. RGD and YIGSR have each been conjugated to both ends of a linear PEG chain to produce bifunctional hybrid polymers: RGD-PEG-RGD and YIGSR-PEG-YIGSR.[23] These hybrid polymers promoted the aggregation of neural cells and may also be useful for the large-scale culture of anchorage-dependent cells.

13.3.4 Coupling Methods

For coupling proteins or peptides to a surface or another molecule, particular functional groups are generally targeted. These functional groups on the protein or peptide include amines, carboxyls, and thiols. Following are some specific coupling methods that have been used. One method that has been used to couple glycoproteins to another material is through the use of the crosslinking agent 1,19-carbonyldiimidazole (CDI). Laminin has been coupled to agarose hydrogels using CDI,[76] and fibronectin has been linked to PVA hydrogels using CDI after first linking an alkyl chain containing an acid group (11-bromoundecanoic acid) to the hydroxyl groups of the PVA.[65] Fibronectin has been attached to PVDF surfaces by first modifying the surface using plasma induced graft copolymerization of acrylic acid or CVD polymerization of 4-amino(2,2)paracyclophane, followed by immobilizing the fibronectin to the functionalized surface.[48] A review of surface modification of biodegradable polymers has previously been written.[128]

Adhesion peptides have been covalently incorporated into PEG hydrogels by linking the peptide through its N-terminus to a monoacrylated PEG spacer chain (see Figure 13.9).[56,90,91] During photopolymerization of the hydrogel, the acrylate end-group of the spacer chain links to acrylate groups of PEG-diacrylate, thereby attaching the peptide to the hydrogel. This coupling method has also been used to link growth factors to hydrogels,[129] and thus could likely also be used to attach a glycoprotein itself to such a material. An alternative method that has been used for coupling an adhesion peptide to a PEG spacer chain,[130] subsequently incorporated into a hyaluronic acid-based hydrogel, is shown in Figure 13.10. As shown, this method requires that the adhesion peptide have a cysteine in the sequence;[131] however, an acrylate will also react with primary amines, providing an alternative means for covalent attachment. Such a method has also been used to couple fragments of fibronectin to a hydrogel to promote cell adhesion.[116] RGD has also been acrylamide-terminated, by reacting acrylic acid with an amine group, thereby leaving out

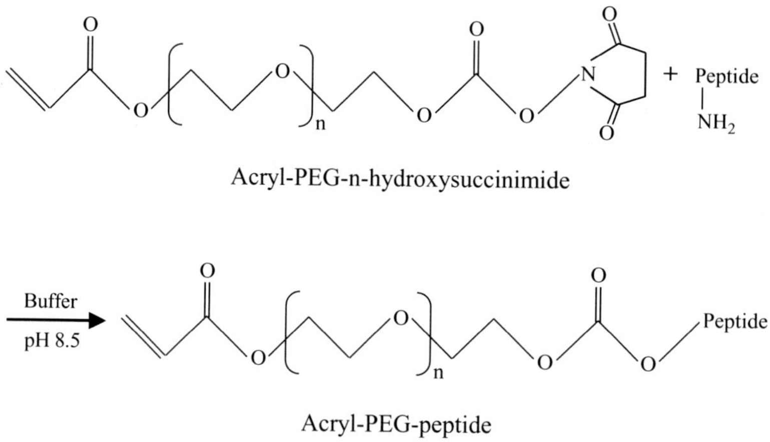

Figure 13.9 Coupling of a cell adhesion peptide *via* its N-terminus to a PEG chain. The reaction is usually allowed to proceed for 2 h at room temperature. The resultant acryl-PEG-peptide can be incorporated into biomaterials through the acrylate endgroup.

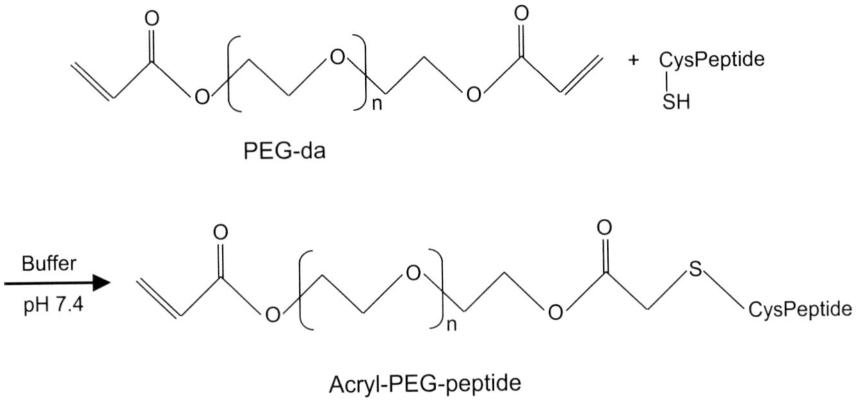

Figure 13.10 Coupling of a cysteine-containing cell adhesion peptide *via* the free thiol to a PEG chain. The reaction requires only 10 min at room temperature. The resultant acryl-PEG-peptide can be incorporated into biomaterials through the acrylate endgroup.

the PEG spacer chain described above.[132] Other methods for modifying surfaces with adhesion ligands[133] or attaching ligands to synthetic polymers[134] have been reviewed previously.

13.4 Concluding Remarks

Glycoproteins are an important part of the natural extracellular matrix of tissues, providing a means for cells to adhere to the matrix, linking matrix proteins to one another, and providing signals critical for cell behavior. The structural glycoproteins themselves generally exhibit multiple functionalities, reflected in the multiple domains of the molecules. These glycoproteins have found use in conjunction with biomaterials, used either alone or in combination with other natural or synthetic polymers. We have highlighted the use of fibronectin and laminin here, as they are the most widely used glycoproteins for biomaterial applications. They are also found in many tissues throughout the body, are readily obtained, and have been the most widely studied to date, making them useful even as model proteins. As more information is obtained about the other structural glycoproteins, both in terms of structure and function, it is likely they will also be used for specific applications or to elicit a particular cellular response.

While the simple adsorption of a glycoprotein to a device is often a fast and easy method of surface modification, there are some disadvantages to this approach. In particular, depending on the properties of the substrate to be modified, the glycoprotein is likely to alter its conformation upon adsorption, which may interfere with its ability to interact with cells. Additionally, such non-specific protein adsorption may be altered once placed in the body due to competitive adsorption/desorption by other proteins present. One method to combat these issues is to covalently attach, or conjugate, the glycoprotein to the surface through a particular site on the glycoprotein molecule. Both adsorption and conjugation of glycoproteins have been used to modify device surfaces, particularly for vascular and orthopedic or dental applications.

An alternative to using the entire glycoprotein molecule is to simply use a fragment or domain from the glycoprotein or even a short oligopeptide. These fragments or peptides typically contain one or more cell adhesion sites found in the protein, and are generally covalently linked to a polymer. RGD is a ubiquitous cell adhesion peptide found within many of the extracellular matrix proteins, including both fibronectin and laminin. Differences in the amino acids flanking either side of the tripeptide sequence lend some degree of specificity for particular integrin receptors. RGD is the adhesion sequence most often used with biomaterials, as it is the best characterized and most cells have at least one integrin receptor that will interact with it. RGD-containing peptides have been attached to many different types of biomaterials, including degradable and non-degradable, natural and synthetic, soft and hard materials. These RGD-modified materials have been used in a variety of applications, including cardiovascular, bone, nerve, and adipose tissue engineering.

Although RGD has been found useful for promoting adhesion and spreading of many different cell types, such non-specificity may be a disadvantage. For example, applications where the specific adhesion of one cell type over others is desired may require a peptide with much greater specificity. One such application would be for the surface of a vascular graft, where adhesion of endothelial cells is needed, but adhesion of other cells, such as platelets that would promote thrombosis, would be detrimental. Other peptides may therefore be more useful in some applications, and new adhesion sequences are continuously being discovered. Another consideration is that once an adhesion peptide is ligated to its receptor, intracellular signaling is initiated. Depending on the peptide sequence and the receptor, this signaling may not result in a particular desired cell behavior, such as the production of new extracellular matrix proteins, necessary for tissue engineering applications. Again, other peptide sequences, either alone or in combination, may be required in order to initiate a desired signaling pathway within the cell. Additionally, some cell adhesion sites within the glycoproteins are known to have synergistic sites. Using a fragment of the glycoprotein that includes both the adhesion site and the synergistic site may enhance the interaction of cells with a biomaterial.

As more information is obtained regarding the structure and function of the various glycoproteins, as well as how cells interact with portions of them, an even greater variety of biomaterials will be developed incorporating either the whole protein, fragments of the protein, or short peptide sequences. Studies regarding cell signaling pathways that are initiated following interaction of the cell with each of these, alone and in combinations, will help in the design of better biomaterials that are able to elicit very specific, desired cellular responses.

13.5 Abbreviations and Symbols

Peptides mentioned in this chapter conform to the standard one-letter abbreviation for the amino acids.

CDI	1,19-carbonyldiimidazole
CMHA-S	thiol-modified carboxymethylhyaluronic acid
CVD	chemical vapor deposition
DNA	deoxyribonucleic acid
DRG	dorsal root ganglia
ECM	extracellular matrix
EGF	epidermal growth factor
ePTFE	expanded polytetrafluoroethylene
FN	fibronectin
Gtn-DTPH	thiol-modified gelatin
MPEO	tri-block coupling-polymer composed of 4,4'-methylenediphenyldiisocyanate and poly(ethylene oxide)
NGC	nerve guidance conduit
NGF	nerve growth factor
PEG	polyethylene glycol
PEGda	polyethylene glycol-diacrylate
PGA	polyglycolic acid
PHB	polyhydroxybutryate
PLL	poly-L-lactide
PLGA	poly(lactic-co-glycolic acid)
PLLA	poly-L-lactic acid
PVA	polyvinyl alcohol
PVDF	poly(vinylidenefluoride)

References

1. S. Ayad, R. P. Boot-Handford, M. J. Humphries, K. E. Kadler, and C. A. Shuttleworth, *The Extracellular Matrix Facts Book*, 2nd edn, Academic Press, San Diego, CA 1998.
2. S. Drake, J. Varnum, K. H. Mayo, P. C. Letourneau, L. T. Furcht and J. B. McCarthy, *J. Biol. Chem.*, 1993, **268**, 15859.
3. B. O. Palsson and S. N. Bhatia, *Tissue Engineering*, Pearson Prentice Hall, NJ, 2004.
4. M. A. Haralson and J. R. Hassell, in *Extracellular Matrix: A Practical Approach*, (Eds) M. A. Haralson and J. R. Hassell, Oxford University Press, Oxford, 1995, p 1.
5. M. Okochi, S. Nomura, C. Kaga and H. Honda, *Biochem. Biophys. Res. Commun.*, 2008, **371**, 85.
6. J. M. Ross, in *Frontiers in Tissue Engineering*, Ed. C. W. Patrick Jr., A. G. Mikos, and L. V. McIntire, Elsevier Science Ltd, Oxford, 1998, p 15.

7. K. Tashiro, A. Monji, I. Yoshida, Y. Hayashi, K. Matsuda, N. Tashiro and Y. Mitsuyama, *Biochem. J.*, 1999, **340**, 119.
8. H. Yamashita, M. Tripathi, M. P. Harris, S. Liu, B. Weidow, R. Zent and V. Quaranta, *Biomaterials*, 2010, **31**, 5110.
9. B. Ganass, R. H. Kim and J. Sodek, *Crit. Rev. Oral Biol. Med.*, 1999, **10**, 79.
10. W. T. Butler, *J. Biol. Buccale*, 1991, **19**, 83.
11. E. H. Sage and P. Bornstein, *J. Biol. Chem.*, 1991, **266**, 14831.
12. Z. M. Ruggeri, *J. Clin. Invest.*, 1997, **99**, 559.
13. K. M. Yamada, *J. Biol. Chem.*, 1991, **266**, 12809.
14. R. O. Hynes, *Fibronectins*, Springer-Verlag, New York, 1990.
15. R. Pankov and K. M. Yamada, *J. Cell Science*, 2002, **115**, 3861.
16. P. Norton and R. O. Hynes, *Nucl. Acids Res.*, 1990, **18**, 4089.
17. I. Wierzbicka-Patynowski and J. E. Scharzbauer, *J. Cell Science*, 2003, **116**, 3269.
18. S. M. Cutler and A. J. Garcia, *Biomaterials*, 2003, **24**, 1759.
19. S. Aota, M. Nomizu and K. M. Yamada, *J. Biol. Chem.*, 1994, **269**, 24756.
20. S. D. Redick, D. L. Settles, G. Briscoe and H. P. Erickson, *J. Cell Biol.*, 2000, **149**, 521.
21. S. K. Powell, J. Rao, E. Roque, M. Nomizu, Y. Kuratomi, Y. Yamada and H. K. Kleinman, *J. Neurosci. Res.*, 2000, **61**, 302.
22. M. Mochizuki, Y. Kadoya, Y. Wakabayashi, K. Kato, I. Okazaki, M. Yamada, T. Sato, N. Sakairi, N. Nishi and M. Nomizu, *FASEB J.*, 2003, **17**, 875.
23. W. Dai, J. Belt and W. M. Saltzman, *Bio/technol.*, 1994, **12**, 797.
24. S. K. Powell and H. K. Kleinman, *Int. J. Biochem. Cell Biol.*, 1997, **29**, 401.
25. M. Nomizu, Y. Kuratomi, M. L. Ponce, S. Song, K. Miyoshi, A. Otaka, S. K. Powell, M. P. Hoffman, H. K. Kleinman and Y. Yamada, *Arch. Biochem. Biophys.*, 2000, **378**, 311.
26. M. Nomizu, Y. Kuratomi, K. M. Malinda, S. Song, K. Miyoshi, A. Otaka, S. K. Powell, M. P. Hoffman, H. K. Kleinman and Y. Yamada, *J. Biol. Chem.*, 1998, **273**, 32491.
27. M. L. Ponce, M. Nomizu, M. C. Delgado, Y. Kuratomi, M. P. Hoffman, S. Powell, Y. Yamada, H. K. Kleinman and K. M. Malinda, *Circ. Res.*, 1999, **84**, 688.
28. K. Beck, I. Hunter and J. Engel, *FASEB J.*, 1990, **4**, 148.
29. I. C. Teller and J.-F. Beaulieu, *Exp. Rev. Mol. Med.*, 2001, **3**, 1.
30. E. Engvall U. M. Wewer, *J. Cell. Biochem.*, 1996, **61**, 493.
31. E. Engvall, D. Earwicker, T. Haaparanta, E. Ruoslahti and J. R. Sanes, *Cell. Regul.*, 1990, **1**, 731.
32. M. Koch, P. A. Olson, A. Albus, W. Jin, D. D. Hunter, W. J. Brunken, R. E. Burgeson and M. F. Champliaud, *J. Cell. Biol.*, 1999, **145**, 605.
33. M. F. Champliaud, G. P. Lunstrum, P. Rousselle, T. Nishiyama, D. R. Keene and R. E. Burgeson, *J. Cell. Biol.*, 1996, **132**, 1189.
34. W. G. Carter, M. C. Ryan and P. J. Gahr, *Cell*, 1991, **65**, 599.

35. M. P. Marinkovich, G. P. Lunstrum, D. R. Keene and R. E. Burgeson, *J. Cell. Biol.*, 1992, **119**, 695.
36. J. H. Miner, B. L. Patton, S. I. Lentz, D. J. Gilbert, W. D. Snider, N. A. Jenkins, N. G. Copeland and J. R. Sanes, *J. Cell. Biol.*, 1997, **137**, 685.
37. P. Rousselle, G. P. Lunstrum, D. R. Keene and R. E. Burgeson, *J. Cell. Biol.*, 1991, **114**, 567.
38. T. A. Horbett, in *Biomaterials Science: An Introduction to Materials in Medicine*, (Eds) B. D. Ratner, A. S. Hoffman, F. J. Schoen, and J. E. Lemons, Academic Press, San Diego, 1996, p 133.
39. A. Tiwari, A. Kidane, H. Salacinski, G. Punshon, G. Hamilton and A. M. Seifalian, *Eur. J. Vasc. Endovasc. Surg.*, 2003, **25**, 325.
40. A. M. Schneider, G. Chandra, G. Lazarovici, I. Vlodavsky, G. Merin, G. Uretzky, J. B. Borman and H. Schwalb, *Thromb. Haemost.*, 1997, **78**, 1392.
41. B. Krijgsman, A. M. Seifalian, H. J. Salacinski, N. R. Tai, G. Punshon, B. J. Fuller and G. Hamilton, *Tissue Eng.*, 2002, **8**, 673.
42. R. McMillan, B. Meeks, F. Bensebaa, Y. Deslandes and H. Sheardown, *J. Biomed. Mater. Res.*, 2001, **54**, 272.
43. D. A. Wang, J. Ji, Y. H. Sun, J. C. Shen, L. X. Feng and J. H. Elisseeff, *Biomacromol.*, 2002, **3**, 1286.
44. G. Murugesan, M. A. Ruegsegger, F. Kligman, R. E. Marchant and K. Kottke-Marchant, *Cell Commun. Adhes.*, 2002, **9**, 59.
45. J. A. Hubbell, S. P. Massia, N. P. Desai and P. D. Drumheller, *Biotechnol.*, 1991, **9**, 568.
46. D. B. Holt, R. C. Eberhart and M. D. Prager, *ASAIO J.*, 1994, **40**, M858.
47. J. B. Huebsch, G. B. Fields, T. G. Triebes and D. L. Mooradian, *J. Biomed. Mater. Res.*, 1996, **31**, 555.
48. D. Klee, Z. Ademovic, A. Bosserhoff, H. Hoecker, G. Maziolis and H. J. Erli, *Biomaterials*, 2003, **24**, 3663.
49. E. De Giglio, L. Sabbatini, S. Colucci and G. Zambonin, *J. Biomater. Sci. Polym. Ed.*, 2000, **11**, 1073.
50. T. A. Barber, S. L. Golledge, D. G. Castner and K. E. Healy, *J. Biomed. Mater. Res.*, 2003, **64A**, 38.
51. N. Patel, R. Padera, G. H. Sanders, S. M. Cannizzaro, M. C. Davies, R. Langer, C. J. Roberts, S. J. Tendler, P. M. Williams and K. M. Shakesheff, *FASEB J.*, 1998, **12**, 1447.
52. J. P. Ranieri, R. Bellamkonda, E. J. Bekos, T. G. Vargo, J. A. Gardella Jr and P. Aebischer, *J. Biomed. Mater. Res.*, 1995, **29**, 779.
53. D. Shaw M.S. Shoichet, *J. Craniofac. Surg.*, 2003, **14**, 308.
54. S. VandeVondele, J. Voros and J. A. Hubbell, *Biotechnol. Bioeng.*, 2003, **82**, 784.
55. L. Aucoin, C. M. Griffith, G. Pleizier, Y. Deslandes and H. Sheardown, *J. Biomat. Sci. Polym. Ed.*, 2002, **13**, 447.
56. B. K. Mann and J. L. West, *J. Biomed. Mater. Res.*, 2002, **60**, 86.
57. R. S. Bhati, D. P. Mukherjee, K. J. McCarthy, S. H. Rogers, D. F. Smith and S. W. Shalaby, *J. Biomed. Mater. Res.*, 2001, **56**, 74.

58. L. A. Solchaga, J. Gao, J. E. Dennis, A. Awadallah, M. Lundberg, A. I. Caplan and V. M. Goldberg, *Tissue Eng.*, 2002, **8**, 333.
59. L. A. Solchaga, J. E. Dennis, V. M. Goldberg and A. I. Caplan, *J. Orthopaed. Res.*, 1999, **17**, 205.
60. X. Tong, K. Hirai, H. Shimada, Y. Mizutani, T. Izumi, N. Toda and P. Yu, *Brain Res.*, 1994, **663**, 155.
61. J. G. Kral and D. L. Crandall, *Plas. Recon. Surg.*, 1999, **104**, 1732.
62. C. J. Panetta, K. Miyauchi, D. Berry, R. D. Simari, D. R. Holmes, R. S. Schwartz and N. M. Caplice, *Hum. Gene Ther.*, 2002, **13**, 433.
63. A. Mosahebi, M. Wiberg and G. Terenghi, *Tissue Eng.*, 2003, **9**, 209.
64. L. N. Novikov, L. N. Novikova, A. Mosahebi, M. Wiberg, G. Terenghi and J. O. Kellerth, *Biomaterials*, 2002, **23**, 3369.
65. C. R. Nuttelman, D. J. Mortisen, S. M. Henry and K. S. Anseth, *J. Biomed. Mater. Res.*, 2001, **57**, 217.
66. M. B. Zajaczkowski, E. Cukierman, C. G. Galbraith and K. M. Yamada, *Tissue Eng.*, 2003, **9**, 525.
67. G. C. Steffens, L. Nothdurft, G. Buse, H. Thissen, H. Hocker and D. Klee, *Biomaterials*, 2002, **23**, 3523.
68. R. A. Brown, G. W. Blunn and O. S. Ejim, *Biomaterials*, 1994, **15**, 457.
69. Z. Ahmed, S. Underwood and R. A. Brown, *Tissue Eng.*, 2003, **9**, 219.
70. Z. Ahmed, S. Underwood and R. A. Brown, *Cell Motil. Cytoskel.*, 2000, **46**, 6.
71. K. Matsumoto, K. Ohnishi, T. Kiyotani, T. Sekine, H. Ueda, T. Nakamura, K. Endo and Y. Shimizu, *Brain Res.*, 2000, **868**, 315.
72. T. Toba, T. Nakamura, Y. Shimizu, K. Matsumoto, K. Ohnishi, S. Fukuda, M. Yoshitani, H. Ueda, Y. Hori and K. Endo, *J. Biomed. Mater. Res.*, 2001, **58**, 622.
73. C. Miller, S. Jeftinija and S. Mallapragada, *Tissue Eng.*, 2002, **8**, 367.
74. C. Miller, S. Jeftinija and S. Mallapragada, *Tissue Eng.*, 2001, **7**, 705.
75. E. Gamez, K. Ikezaki, M. Fukui and T. Matsuda, *Cell Transplant.*, 2003, **12**, 481.
76. X. Yu, G. P. Dillon and R. V. Bellamkonda, *Tissue Eng.*, 1999, **5**, 291.
77. X. Yu and R. V. Bellamkonda, *Tissue Eng.*, 2003, **9**, 421.
78. T. C. McDevitt, J. C. Angello, M. L. Whitney, H. Reinecke, S. D. Hauschka, C. E. Murry and P. S. Stayton, *J. Biomed. Mater. Res.*, 2002, **60**, 472.
79. C. W. Patrick, Jr. and X. Wu, *Ann. Biomed. Eng.*, 2003, **31**, 505.
80. D. J. Aframian, E. Cukierman, J. Nikolovski, D. J. Mooney, K. M. Yamada and B. J. Baum, *Tissue Eng.*, 2000, **6**, 209.
81. J. J. Marler, A. Guha, J. Rowley, R. Koka, D. Mooney, J. Upton and J. P. Vacanti, *Plast. Reconstr. Surg.*, 2000, **105**, 2049.
82. C. Halberstadt, C. Austin, J. Rowley, C. Culberson, A. Loebsack, S. Wyatt, S. Coleman, L. Blacksten, K. Burg, D. Mooney and W. Holder Jr, *Tissue Eng.*, 2002, **8**, 309.
83. E. Alsberg, K. W. Anderson, A. Albeiruti, R. T. Franceschi and D. J. Mooney, *J. Dent. Res.*, 2001, **80**, 2025.

84. J. A. Rowley, G. Madlambayan and D. J. Mooney, *Biomaterials*, 1999, **20**, 45.
85. J. A. Rowley and D. J. Mooney, *J. Biomed. Mater. Res.*, 2002, **60**, 217.
86. H. Hall, T. Baechi and J. A. Hubbell, *Microvasc. Res.*, 2001, **62**, 315.
87. J. C. Schense and J. A. Hubbell, *J. Biol. Chem.*, 2000, **275**, 6813.
88. M. J. Moghaddam and T. Matsuda, *Trans. Am. Soc. Artif. Intern. Organs*, 1991, **37**, M437.
89. R. H. Schmedlen, K. S. Masters and J. L. West, *Biomaterials*, 2002, **23**, 4325.
90. B. K. Mann, A. S. Gobin, A. T. Tsai, R. H. Schmedlen and J. L. West, *Biomaterials*, 2001, **22**, 3045.
91. A. S. Gobin and J. L. West, *FASEB J.*, 2002, **16**, 751.
92. J. A. Burdick and K. S. Anseth, *Biomaterials*, 2002, **23**, 4315.
93. J. W. Gunn, S. D. Turner and B. K. Mann, *J. Biomed. Mater. Res.*, 2005, **72A**, 91.
94. S. Woerly, E. Pinet, L. de Robertis, D. Van Diep and M. Bousmina, *Biomaterials*, 2001, **22**, 1095.
95. G. W. Plant, S. Woerly and A. R. Harvey, *Exper. Neurol.*, 1997, **143**, 287.
96. E. Behravesh, K. Zygourakis and A. G. Mikos, *J. Biomed. Mater. Res.*, 2003, **65A**, 260.
97. K. Eid, E. Chen, L. Griffith and J. Glowacki, *J. Biomed. Mater. Res.*, 2001, **57**, 224.
98. Y. Hu, S. R. Winn, I. Krajbich and J. O. Hollinger, *J. Biomed. Mater. Res.*, 2003, **64A**, 583.
99. L. Zhang, F. Rakotondradany, A. J. Myles, H. Fenniri and T. J. Webster, *Biomaterials*, 2009, **30**, 1309.
100. A. D. Cook, J. S. Hrkach, N. N. Gao, I. M. Johnson, U. B. Pajvani, S. M. Cannizzaro and R. Langer, *J. Biomed. Mater. Res.*, 1997, **35**, 513.
101. V. Proks, L. Machova, S. Popelka and F. Rypacek, *Adv. Exper. Med. Biol.*, 2003, **534**, 191.
102. K. Na, H. K. Choi, T. Akaike and K. H. Park, *Biosci. Biotechnol. Biochem.*, 2001, **65**, 1284.
103. M. M. Martino, M. Mochizuki, D. A. Rothenfluh, S. A. Rempel, J. A. Hubbell and T. H. Barker, *Biomaterials*, 2009, **30**, 1089.
104. J. A. Craig, E. L. Rexeisen, A. Mardilovich, K. Shroff and E. Kokkoli, *Langmuir*, 2008, **24**, 10282.
105. R. H. Kimura, A. M. Levin, F. V. Cochran and J. R. Cochran, *Proteins*, 2009, **77**, 359.
106. A. V. Shinde, C. Bystroff, C. Wang, M. G. Vogelezang, P. A. Vincent, R. O. Hynes and L. Van De Water, *J. Biol. Chem.*, 2008, **283**, 2858.
107. V. Rerat, G. Dive, A. A. Cordi, G. C. Tucker, R. Bareille, J. Amedee, L. Bordenave and J. Marchand-Brynaert, *J. Med. Chem.*, 2009, **52**, 7029.
108. X. Gu and K. S. Masters, *J. Biomed. Mater. Res. A*, 2010, **93**, 1620.
109. C. D. Reyes, T. A. Petrie and A. J. Garcia, *J. Cell Physiol.*, 2008, **217**, 450.
110. J. C. Schense, J. Bloch, P. Aebischer and J. A. Hubbell, *Nature Biotechnol.*, 2000, **18**, 415.

111. R. Bellamkonda, J. P. Ranieri and P. Aebischer, *J. Neurosci. Res.*, 1995, **41**, 501.
112. F. Li, D. Carlsson, C. Lohmann, E. Suuronen, S. Vascotto, K. Kobuch, H. Sheardown, R. Munger, M. Nakamura and M. Griffith, *Proc. Natl. Acad. Sci. USA*, 2003, **100**, 15346.
113. H.-W. Jun and J. L. West, *J. Biomed. Mater. Res. Part B: Appl. Biomater.*, 2005, **72B**, 131.
114. H.-W. Jun and J. L. West, *Tissue Eng.*, 2005, **11**, 1133.
115. H. W. Kim, W. Kang, E. Jeon and J. H. Jang, *Biotechnol. Lett.*, 2010, **32**, 29.
116. K. Ghosh, X.-D. Ren, X. Z. Shu, G. D. Prestwich and R. A. F. Clark, *Tissue Eng.*, 2006, **12**, 601.
117. S. Liu, *Bioconjug. Chem.*, 2009, **20**, 2199.
118. Z. Liu, S. Liu, G. Niu, F. Wang, S. Liu and X. Chen, *Mol. Imaging*, 2010, **9**, 21.
119. Z. Hu, F. Luo, Y. Pan, C. Hou, L. Ren, J. Chen, J. Wang and Y. Zhang, *J. Biomed. Mater. Res. A*, 2008, **85**, 797.
120. D. B. Pike and H. Ghandehari, *Adv. Drug Deliv. Rev.*, 2010, **62**, 167.
121. Z. Li, P. Huang, X. Zhang, J. Lin, S. Yang, B. Liu, F. Gao, P. Xi, Q. Ren and D. Cui, *Mol. Pharm.*, 2010, **7**, 94.
122. S. Halstenberg, A. Panitch, S. Rizzi, H. Hall and J. A. Hubbell, *Biomacromol.*, 2002, **3**, 710.
123. S. M. Moreira, F. K. Andrade, L. Domingues and M. Gama, *BMC Biotechnol.*, 2008, **8**, 78.
124. F. K. Andrade, S. M. Moreira, L. Domingues and F. M. Gama, *J. Biomed. Mater. Res. A*, 2010, **92**, 9.
125. J. C. Liu and D. A. Tirrell, *Biomacromol.*, 2008, **9**, 2984.
126. A. W. Morgan, K. E. Roskov, S. Lin-Gibson, D. L. Kaplan, M. L. Becker and C. G. Simon Jr, *Biomaterials*, 2008, **29**, 2556.
127. S. Zalipsky, N. Mullah, J. A. Harding, J. Gittelman, L. Guo and S. A. DeFrees, *Bioconjug. Chem.*, 1997, **8**, 111.
128. H. Thissen H, in *The PBM Series, Vol 2: Biodegradable Polymers*, (Ed.) R. Arshady, Citus Books, London, 2003, p 175.
129. B. K. Mann, R. H. Schmedlen and J. L. West, *Biomaterials*, 2001, **22**, 439.
130. X. Z. Shu, K. Ghosh, Y. Liu, F. S. Palumbo, Y. Luo, R. A. Clark and G. D. Prestwich, *J. Biomed. Mater. Res.*, 2004, **68A**, 365.
131. D. L. Elbert and J. A. Hubbell, *Biomacromol.*, 2001, **2**, 430.
132. X. He, J. Ma and E. Jabbari, *Langmuir*, 2008, **24**, 12508.
133. R. Arshady, in *The PBM Series, Vol 1: Introduction to Polymeric Biomaterials*, (Ed.) R. Arshady, Citus Books, London, 2003, p 1.
134. K. Shakesheff, S. Cannizzaro and R. Langer, *J. Biomater. Sci. Polym. Ed.*, 1998, **9**, 507.

CHAPTER 14

Nucleic Acid Polymers and Applications of Recombinant DNA Technology

IAN HOLT*[a] AND Y. CHAN N. PHAM[b]

[a] Wolfson Centre for Inherited Neuromuscular Disease, RJAH Orthopaedic Hospital, Oswestry, Shropshire, SY10 7AG, UK; [b] Countess of Chester Hospital, Chester, Cheshire, CH2 1UL, UK

14.1 Introduction

Deoxyribonucleic acid (DNA) biopolymers are the chemical stores of heritable genetic information in cells. This highly specific linear sequence of four types of monomer determines the amino acid sequence of proteins, which in turn determines the properties and functions of the proteins. The long chemical sequences of DNA, sometimes called the code of life, reflect the evolutionary pressures that a species has encountered. Some viruses use ribonucleic acid (RNA) as the information-carrying material, but most forms of life that have been studied use DNA. These naturally occurring nucleic acid polymers have the unique properties of acting as a template for their own replication and undergoing mutation and rearrangement to provide genetic variation upon which the evolutionary process depends.

The origin of molecular biology as a discipline can be traced to 1944 when it was shown that purified DNA from one strain of bacterium could be used to transfer virulent traits to a different strain,[1] thereby demonstrating that DNA is the material of inheritance. Nine years later the famous double helical structure

RSC Polymer Chemistry Series No. 1
Renewable Resources for Functional Polymers and Biomaterials
Edited by Peter A. Williams

Published by the Royal Society of Chemistry, www.rsc.org

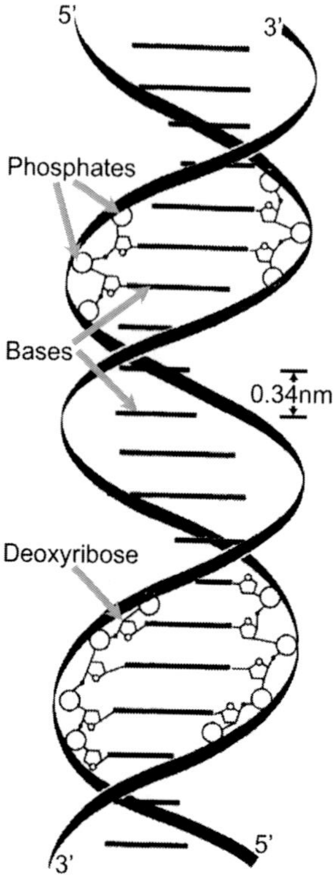

Figure 14.1 The double helical structure of DNA. Diagram adapted from: http://users.rcn.com/jkimball.ma.ultranet/BiologyPages/D/DoubleHelix.html with permission from Professor J.W. Kimball.

of DNA (Figure 14.1) was elucidated.[2] Discoveries in the following years revealed how sequences of DNA carry information for protein synthesis. Coding sequences of DNA (genes) were found to encode complementary transcripts of ribonucleic acid (RNA),[3] which in turn code for the amino acid sequences of proteins.[4,5]

Techniques of genetic engineering or recombinant DNA technology began to be developed in the mid-1970s which allowed the precise manipulation of DNA. The studies included finding methods for specifically cutting[6] and joining DNA sequences, detecting a specific sequence with a complementary probe,[7] sequencing sections of the genetic code[8,9] and site-directed mutagenesis to make specific changes to the code.[10] Another major landmark was the development of the polymerase chain reaction (PCR)[11–13] which was conceptualized in 1983 and is a method to amplify exponentially a specific sequence of DNA from a preparation containing low levels of a target material. PCR has undoubtedly hastened the development of recombinant DNA technology.

These techniques for manipulating DNA are the basic tools for recombinant DNA technology, whereby foreign gene sequences may be expressed and perpetuated in a host organism. The advancement of automated DNA sequencing in the mid to late 1980s led to the launch of the publicly funded Human Genome Project and the establishment of the Genome Database in 1990. The mapping of DNA which makes up the human chromosomes was divided between laboratories in the USA, UK, Germany, France, Japan and China and was the largest international collaboration in biology ever undertaken.

In 2000 the publicly funded International Human Genome Sequencing Consortium and the commercial Celera Genomics jointly announced the completion of the first drafts of the human genome sequence. The two versions of the draft sequence were published in 2001[14,15] with revisions subsequently occurring and a "finished grade" human reference genome being published in 2004.[16]

The GenBank database is an open access collection of nucleotide sequences, with protein translations from more than 100 000 distinct organisms. In 2010 the complete genomes of more than 3800 organisms had been sequenced, the vast majority of which were bacterial or viral.[17] Nucleotide sequences are added to the database by direct submission from individual laboratories and by bulk submissions from large scale sequencing centres, with the size of the database doubling approximately every 18 months. Technical improvements are leading to the replacement of automated Sanger sequencing by next generation sequencing technologies.[18] The identification of specific gene sequences, which in the past may have taken years to obtain, are now available by a quick computer search. The analysis of the nucleotide sequences and their protein products in health and disease is a massive ongoing task.

14.2 Structure, Location and Properties of DNA[19–21]

14.2.1 Structure of DNA

DNA consists of a backbone of repeating pentose sugar rings (2′-deoxyribose) and phosphate groups. The phosphate is a bridge between the 3′-OH of one sugar and the 5′-OH of the next. One of four possible bases (heterocyclic aromatic rings) is covalently attached to each sugar by a glycosidic link (Figure 14.2a). The four DNA bases (monomers) are as follows.

- Bicyclic purines: adenine (A) and guanine (G)
- Monocyclic pyrimidines: cytosine (C) and thymine (T).

Therefore, the DNA molecule consists of deoxyribonucleotides (monomeric units) joined by phosphodiester bonds to form a long chain polymer. At physiological pH the phosphate groups are negatively charged, so nucleic acids are highly charged polymers. Thus, the sugar–phosphate backbone is strongly hydrophilic, whereas the bases are strongly hydrophobic.

The primary structure of the RNA molecule is similar to that of DNA, except that the sugar is ribose (with a hydroxy group at the 2′ position) instead of deoxyribose and the pyrimidine base thymine (T) in DNA is replaced by uracil

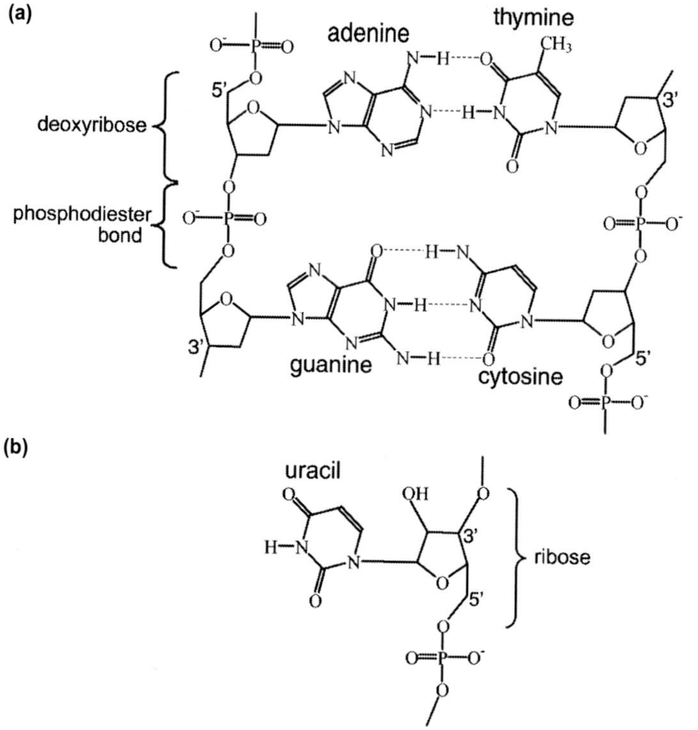

Figure 14.2 (a) Schematic structure of a small section of double stranded DNA. Backbones of deoxyribose sugars are joined by phosphodiester bonds with one of four possible bases joined at the 1′ position of each sugar by a glycosidic bond. Bases on one strand form hydrogen bonds (dashed lines) with complementary bases on the other strand. Two hydrogen bonds "pair" adenine with thymine and three hydrogen bonds "pair" guanine with cytosine. (b) Schematic structure of a nucleotide from a single stranded RNA molecule. Ribose sugars are joined by phosphodiester bonds. In this example the base uracil is joined to the 1′ position of the ribose. RNA contains the four bases adenine, uracil, guanine and cytosine.

(U, no methyl group at the 5-position) (Figure 14.2b). However, these apparently minor differences in the monomeric units between DNA and RNA have a profound effect on their higher structural orders. RNA usually occurs as a single stranded molecule, whereas DNA commonly occurs in a double stranded state.

The bases can bind to one another in a specific way. In DNA, two hydrogen bonds are formed between A and T bases and three hydrogen bonds between G and C. These geometrically matched bases pair in this way along a length of DNA and the single strands hybridize to a complementary strand to give the stable double stranded DNA (see Figures 14.1 and 14.2a).

The DNA strands are oriented in opposite directions and are said to be antiparallel. In other words, the coding or sense strand of DNA is represented with the sequence of bases written in a 5′ to 3′ direction whereas the complementary antisense strand will be in the 3′ to 5′ direction. With the usual 'B form' of DNA, the two separate chains of DNA are wound around each other forming a right-handed double helix with approximately 10 nucleotides per turn (see Figure 14.1). The negatively charged sugar–phosphate backbones are on the outside and the paired bases of the two strands stack one above the other, 0.34 nm apart, in the centre of the helix.[2] RNA is readily broken down by ribonuclease enzymes whereas DNA is a more stable store of genetic information. DNA of an organism may persist long after death. For example, DNA sequence information was used for identification of skeletal remains of Tsar Nicholas II and his family members who were killed in 1918.[22] Recently, a draft sequence of the neandertal genome has been reported.[23]

14.2.2 Location of DNA

In eukaryotic (with nucleus) autosomal (non-sex) cells, the vast majority of DNA is found in the nucleus as discrete bodies called chromosomes. The exception to this being the small amount of DNA found in organelles outside the nucleus (mitochondria and chloroplasts). In human cells, more than 99% of the total DNA is nuclear genomic DNA and the remainder mitochondrial. Each nuclear chromosome, of which there are 46 in human autosomal cells, is thought to be a single linear molecule consisting of up to 280 mega base pairs (280 000 000 base pairs).

There are 22 pairs of autosomal chromosomes with one of each pair inherited from the mother and one from the father and two sex chromosomes (XY in males and XX in females). The chromosomes are packaged by the formation of an organized complex of DNA and protein. Most of the proteins are positively charged histones which bind strongly to DNA to produce nucleosomes, which supercoil to form chromatin.

During the non-dividing phase of the cell cycle (interphase), individual chromosomes cannot be distinguished. Relatively dispersed chromatin is called euchromatin and regions of very densely packed fibers of transcriptionally inactive DNA are called heterochromatin. At the ends of chromosomes are repeat sequences called telomeres which form hairpin loops to protect the ends of the chromosomes. The double helical structure, the nuclear packaging and the presence of telomeres provide stability and reduce damage to the DNA. However, frequent mutation damage does occur to these massive nucleotide sequences which are constantly under surveillance and repair by a number of mechanisms.

14.2.3 DNA Transcription and Translation

A gene is a segment of DNA that contains exons (regions that code for amino acid sequences) and introns (which do not contain amino acid sequence information). There are more than 20 000 human protein-coding genes with exons occupying as little as 1.1% of the genome and introns somewhere in the region of 24%.[15,16] The remaining 75% of the genome is still poorly understood. Much of this DNA consists of repeat sequences of "transposons" which are sequences of DNA that can move to different positions in the genome and can cause changes in gene expression. It has been suggested that transposons may have important roles in driving cellular differentiation and evolutionary change.[24]

Transcription occurs in the nucleus and is the first step in the mechanism by which genes give rise to protein synthesis. Transcription factors are important in controlling which genes are expressed as proteins,[25] for example, cells in the pancreas of an organism are often required to produce a very different set of proteins than those produced in muscle cells in the same organism. Protein production is controlled at the level of DNA transcription by gene regulators, activators and repressors. These DNA binding transcription factors and the proteins with which they interact provide fine control of gene expression which is dependent, for example, on the location and function of the cell and the stage of development of the organism.[25]

For transcription to occur, the duplex DNA strands are transiently separated to provide a single stranded template and RNA polymerase II catalyses the transcription in a 5′ to 3′ direction and synthesizes the pre-messenger RNA (pre-mRNA) complementary to the antisense template strand. The ribose nucleotides containing the bases A, G, C and U align with the DNA bases T, C, G and A, respectively.

The eukaryotic pre-mRNA undergoes a number of processing events in the nucleus which includes splicing out the noncoding introns and joining the two flanking exons. Often, "alternative splicing" of pre-mRNA may occur to give multiple protein isoforms with different functions that are derived from the same gene. Indeed, at least 74% of human multiexon genes are alternatively spliced.[26] Cell specific regulation of alternative splicing by RNA-binding proteins contributes to a large, diverse and dynamic human proteome expressed from quite a small number of genes. Messenger RNA is transported to the cytoplasm for translation into its amino acid sequence. The mRNA contains the "transcript" of the genetic code of the DNA.

The genetic code of DNA from 5′ to 3′ consists of nucleotide triplets (codons) corresponding to the amino acid sequence of a protein from the N- to the C-terminus. In the genetic code there are 64 possible codons, 61 of which encode the 20 amino acids from which proteins are made. The code is said to be degenerate, with some amino acids coded by up to 6 different codons. "Initiation" of protein synthesis occurs with a methionine amino acid (RNA codon AUG), while three "stop" codons mark the termination of protein synthesis (UAG, UAA or UGA). Whilst most of the organisms that have been studied use the same general amino acid coding sequences, the genetic code is not universal. Exceptions to the code have been identified, for example in yeast[27] and in mitochondrial DNA, in which a small number of codons have been found to have functions which differ from the usual genetic code. Also, naturally occurring amino acids which are not the standard 20 amino acids have been identified.[28]

Translation, the process by which mRNA determines the amino acid sequence during polypeptide synthesis, occurs at subcellular structures called ribosomes. Ribosomes consist of ribosomal RNA and proteins which have structural and catalytic roles and control the interaction between mRNA and aminoacyl-transfer RNA (tRNA). tRNAs are adaptor molecules that deliver specific amino acids to the ribosome and decode the information in mRNA, a process in which the "anticodon" of tRNA binds to the appropriate codon on mRNA. However, a tRNA molecule may sometimes recognize more than one codon, because the third base in the codon is less stringent.

Amino acids are carried on the tRNA molecules and are aligned in the order determined by the genetic code. Amino acids are joined together by the formation of peptide bonds, while at the same time detached from the tRNA. Aminoacyl-tRNA enters the ribosome and deacylated tRNA leaves the ribosome. Elongation of the polypeptide chain occurs as the mRNA moves along the ribosome until a termination codon is encountered, at which point the ribosome dissociates from the RNA. The completed polypeptide chain is then released at this stage and goes through its own set of processes such as folding, cleavage and chemical modifications to form the corresponding protein.

14.2.4 DNA Replication

Before a cell can divide by the process of mitosis, the entire genome must be replicated. Replication follows a "semi-conservative" mechanism. The strands of the double helix separate and each acts as a template to direct the synthesis of complementary daughter strands. The DNA is synthesized in a 5′ to 3′ direction. The "leading strand" of the DNA fork, which goes from 5′ to 3′, is synthesized in one continuous stretch. The "lagging strand" is on the opposite side of the replication fork from the leading strand and has 3′ to 5′ orientation. Nucleotides are added to the lagging DNA strand as short fragments in the reverse direction.

Eukaryotic chromosomes have multiple origins of replication. Helicase enzymes separate the strands. The leading strand and the lagging strand fragments are "primed" by the synthesis of a short piece of RNA which provides a 3′ hydroxy end from which DNA polymerase enzymes extend new DNA strands. The RNA primers are removed, the gaps filled with DNA and the fragments joined by ligase enzyme to give two double stranded DNA molecules that are identical to each other and to their parent molecule. Thus, cell division by mitosis results in two (daughter) cells that are genetically identical.

14.2.5 DNA Recombination

Recombination is the process of "crossing over" of DNA fragments between "precisely corresponding sequences" in paired chromosomes, which occurs during meiosis in eukaryotic cells during the production of germ cells (sex cells) in males (spermatogenesis) and females (oogenesis). During meiosis the chromosomes condense and paired chromosomes associate closely side by side and adhere together at particular locations. At the stage of division, the paired chromosomes separate and breaks in the original strands occur where the chromosomes were joined. The broken ends of one chromatid fuse with the broken ends of the other. Following division, the result in humans is germ cells that have only one set of 23 chromosomes (haploid) and which contain information from both maternally and paternally inherited chromosomes. Therefore, subsequent generations inherit genes from all four grandparents.

Recombination along with other processes such as mutation leads to variations between individuals and the evolutionary change in characteristics of organisms in successive generations by the forces of natural selection.

14.3 Chemistry of Nucleic Acids

14.3.1 Isolation and Physicochemical Properties of DNA[21,29]

Chemical formulas of some of the bases, nucleosides and nucleotides are shown in Table 14.1. A number of methods are employed for the isolation of DNA. For example, blood cells may be treated with detergent and proteinase enzyme to disrupt the cells and DNA extracted from the aqueous layer of phenol–chloroform and precipitated with ethanol. The purification of plasmids (circular DNA) from bacteria may involve alkaline lysis and binding of DNA to an ion exchange resin prior to elution and precipitation.

It is possible to observe DNA at the molecular level using atomic force microscopy (AFM),[31] as shown in Figure 14.3 for a preparation of plasmid DNA. The DNA double helix is approximately 2 nm in diameter which is within the range seen in this image.

Solutions of double stranded DNA have high viscosity with a buoyant density of around 1.7 g cm^{-3}. DNAs rich in GC base pairs have higher buoyant

Table 14.1 Common purines, pyrimidines, nucleosides and nucleotides.[30]

Base	*Molecular formula*
Adenine	$C_5H_5N_5$
2′-Deoxyadenosine	$C_{10}H_{13}N_5O_3$
2′-Deoxyadenosine 5′-triphosphate (dATP)	$C_{10}H_{16}N_5O_{12}P_3$
Guanine	$C_5H_5N_5O$
2′-Deoxyguanosine	$C_{10}H_{13}N_5O_4$
2′-Deoxyguanosine 5′-triphosphate (dGTP)	$C_{10}H_{16}N_5O_{13}P_3$
Cytosine	$C_4H_5N_3O$
2′-Deoxycytidine	$C_9H_{13}N_3O_4$
2′-Deoxycytidine 5′-triphosphate (dCTP)	$C_9H_{16}N_3O_{13}P_3$
Thymine	$C_5H_6N_2O_2$
2′-Deoxythymidine	$C_{10}H_{14}N_2O_5$
2′-Deoxythymidine 5′-triphosphate (dTTP)	$C_{10}H_{17}N_2O_{14}P_3$
Uracil	$C_4H_4N_2O_2$
Uridine	$C_9H_{12}N_2O_6$
Uridine 5′-triphosphate (UTP)	$C_9H_{15}N_2O_{15}P_3$

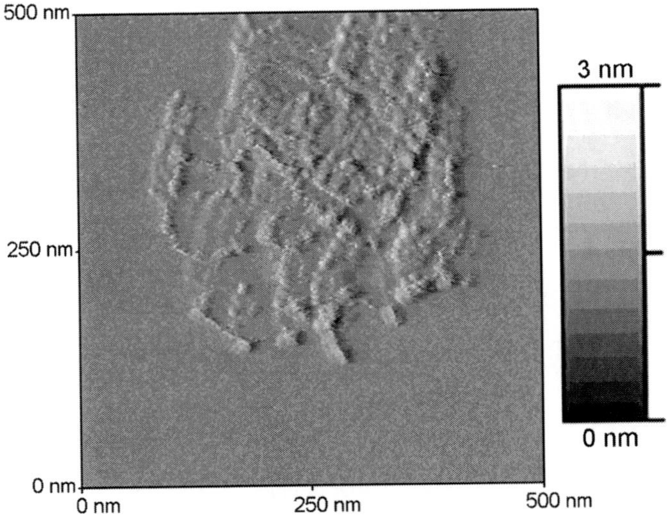

Figure 14.3 Atomic force microscope image of plasmid DNA on cleaved mica. The non-contact mode image was obtained with a Veeco Explorer AFM (courtesy of Dr Andy Wright, Advanced Materials Group, Glyndwr University, Wrexham).

densities than those of DNAs rich in AT base pairs. DNA has a UV absorption λmax of approximately 260 nm. The optical density (OD) at 260 nm can be used for its quantification and the ratio OD_{260}/OD_{280} provides an estimate of DNA purity. Heating leads to breakdown of the hydrogen bonds between bases and hence thermal denaturation or "melting" in which double stranded DNA separates into single strands at temperatures usually between 80 and 100 °C. Melting of DNA leads to an increase in the OD_{260} of the solution. The melting temperature (T_m) depends on the base composition, with DNA molecules rich in GC base pairs having higher T_m. Cooling allows complementary regions to anneal and reform double stranded DNA.

Agarose gel electrophoresis is frequently used for separation, visualization and purification of DNA of various sizes, as exemplified in Figure 14.4.

14.3.2 Chemical Synthesis of Oligonucleotides

Oligonucleotides are defined as specific sequences of nucleic acids often between 10 and 30 nucleotides long, but may also be more than 100 nucleotides. Oligonucleotides are used, for example, as primers in PCR (see 14.4.2) and may be designed to include restriction enzyme sites or mutations. Oligonucleotides may be used for hybridization studies[7] and some may have value as therapeutic agents.

Oligonucleotides can be synthesized in the laboratory by joining DNA phosphoramidite monomers (bases) one at a time, in the required order, to

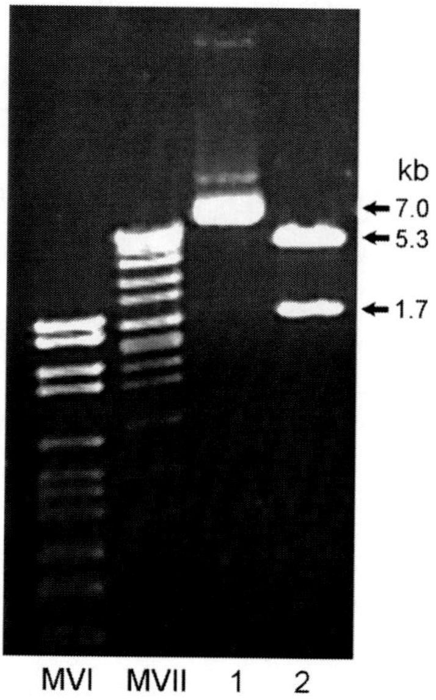

Figure 14.4 Agarose gel electrophoresis of DNA. Samples of DNA were loaded in wells at the top of the gel. An electrical potential was applied to the gel and the negatively charged DNA migrated towards the anode at the bottom of the gel. Larger molecules migrated through the gel more slowly because of greater frictional drag. In this example, DNA was visualized by intercalation with ethidium bromide which fluoresces with UV irradiation. Lanes MVI and MVII are markers containing DNA fragments of known sizes. Lane 1 contains a plasmid preparation (7 kb). The bright band is supercoiled plasmid which has migrated faster than the open circular plasmid which is the higher and fainter band. Lane 2 contains the plasmid preparation from lane 1 which has been digested with 2 restriction endonucleases to give an insert of about 1700 base pairs and a larger band of the cut plasmid (about 5300 base pairs).

elongate the nucleotide chain. Different synthetic methods are used and the synthesis is often highly automated, but the following general procedure provides a good idea of the synthetic process.

- The first base, blocked at its 5′ hydroxy group by dimethoxytrityl (DMT), is attached by an ester linkage at its 3′ hydroxyl group to a solid support (*e.g.* crosslinked beaded polystyrene) in a reaction column.
- Deblocking: the DMT protecting the 5′-hydroxyl group is removed (*e.g.* by dichloroacetic acid)
- Activation and addition (condensation): the next DNA phosphoramidite monomer to be linked to the growing oligonucleotide is activated and added, in excess, to the reaction column. A phosphite link with the active 5′-hydroxy group of the preceding base is formed.
- Blocking of unreacted 5′-hydroxyl groups: as the reaction on the solid support may not proceed to completion, any unreacted 5′-hydroxyl

groups are acetylated to prevent it to take part in the next step and form a sequence with a missing unit (or deletion).

- Stabilization of phosphite links: the phosphite link is oxidized to form a more stable phosphate bond.
- Repeat of synthetic cycle: the 5′-hydroxy group of the newly bound base is then activated and the above procedure repeated until all of the bases in required order have been added to the oligonucleotide.
- Cleavage from solid support: when the synthesis is complete, the oligonucleotide-bound solid support is treated with concentrated ammonia to detach the chain from the polymer and cleave the protecting groups from the chain.
- Purification: the synthetic oligonucleotide is then desalted, followed by analysis and purification.

See references for details of oligonucleotide synthesis and conjugations.[32–36] Many modifications to the basic procedure are possible, both in the chemistry of the synthesis and in the oligonucleotide composition. For example, other groups, such as fluorophores, biotin, amino acids and small organic groups may be conjugated to the oligonucleotide. A related interesting development is the synthesis of peptide nucleic acids (PNAs), which are stable analogues of oligonucleotides, in which the negatively charged phosphate backbone is replaced by a neutral amide linkage.

14.4 Genetic Engineering Techniques[37]

Genetic engineering or recombinant DNA technology is the application of molecular biology techniques for the formation of new combinations of heritable materials. A manipulated gene sequence may be introduced into a host organism in which it does not naturally occur and the new trait propagated and perpetuated in the host.

In addition to agarose gel electrophoresis (Figure 14.4), other common techniques in genetic engineering include site-directed mutagenesis to make specific base changes in nucleotide sequences[10] and blotting for detecting fragments of DNA (Southern blotting) or RNA (Northern blotting). In Southern and Northern blotting the nucleic acids are separated by electrophoresis and then transferred to a synthetic membrane for hybridization and detection with a complementary nucleic acid probe.[7] Nucleic acid probes are also used for *in situ* hybridization where the location of nucleotide sequences in tissue sections or cultured cells can be observed (example Figure 14.5).[38]

A further development of hybridization technology is the microarrays technique whereby a vast number of known DNA sequences are used to probe a test sample for the presence of specific nucleic acid sequences.[26] Microarrays may be used, for example, for gene expression profiling, to detect different genes in a preparation, single nucleotide polymorphisms detection and in detection of alternative splicing.

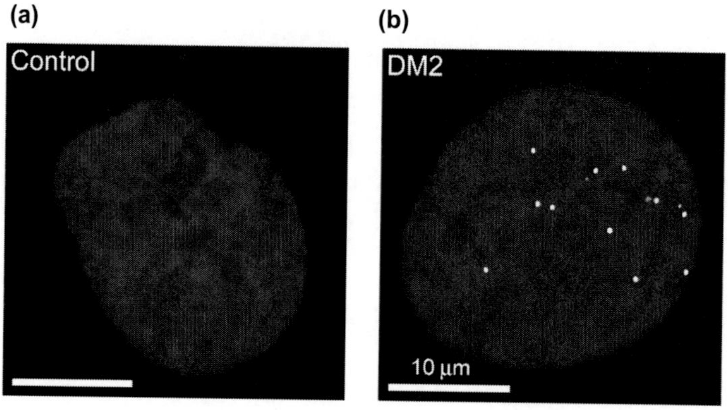

Figure 14.5 *In situ* hybridization to identify myotonic dystrophy type 2 (DM2). Transcription of the tetranucleotide (CCTG) repeats in DM2 cells gives rise to mutant RNA (with CCUG repeats) which accumulates to form nuclear foci. *In situ* hybridization with $(CAGG)_{10}$ probe with a fluorescent label fails to bind unaffected cells (a) but identifies the ribonuclear foci in DM2 cells (b). Images captured with Leica SP5 confocal microscope (see ref. 38).

Table 14.2 Some commonly used type-2 restriction endonucleases[6] with recognition sequences (5′ to 3′ strand) and cutting sites (/).

Restriction endonuclease	*Recognition sequence*
Eco RI	G/AATTC
Bam HI	G/GATCC
Not I	GC/GGCCGC
Sac I	GAGCT/C
Eco RV	GAT/ATC

14.4.1 Restriction Endonucleases

A key development in genetic manipulation procedures is the ability to cut strands of DNA at specific discrete sequences by type II restriction endonucleases (REs) (Table 14.2). These REs recognize a defined, usually palindromic, sequence between 4 and 8 nucleotides in the double stranded DNA and cleave within the sequence.[6]

Where the cutting site is offset from the axis of symmetry, a staggered cut occurs leading to fragments with protruding single stranded "sticky ends". For example, *Eco* RI, *Bam* HI and *Not* I generate 5′ extensions whereas *Sac* I gives 3′ extensions. When the cut site is in the centre of the axis of symmetry (*e.g. Eco* RV) the break is double stranded and "blunt ends" are produced. DNA that has been specifically cut with REs can ligate with other DNA (*e.g.* a vector) that has been cut in a complementary way. The DNA ligase enzyme catalyses the joining of DNA molecules.

14.4.2 Polymerase Chain Reaction

The polymerase chain reaction (PCR) is a very widely used standard laboratory technique. It is an enzymic procedure to amplify exponentially a specific sequence of DNA that may initially be present at very low levels in a complex mixture.[11,12] The method involves a cycling reaction between three different temperatures. A thermostable DNA polymerase[13] allows the whole reaction to be performed in a sealed tube without having to replenish the enzyme.

The basic reaction mixture contains two primers which are synthetic oligonucleotides, each approximately 20 to 25 nucleotides in length. The two primers, each with approximately the same melting temperature (*e.g.* $T_m \approx$ 60 °C), are designed to bind specifically to the sequence of a target DNA that is required to be amplified. One primer binds to the sense strand of the target DNA and the other binds to the antisense strand. Also present in the reaction mix are four types of deoxyribonucleotide triphosphates (dNTPs) used in the synthesis of DNA (dATP, dTTP, dGTP and dCTP) and the heat resistant DNA polymerase (*e.g.* Taq polymerase). Magnesium ions are also required in the reaction mixture. Reaction tubes are subjected to a cycling of the temperature. The precise reaction conditions such as temperature and incubation time depend on the nature of the target DNA and primers and require optimization. A typical example of conditions for the 3-stage cycling process is as follows.

1. Denaturation ($\approx$95 °C, 30 s): melting of the double stranded DNA.
2. Annealing ($\approx$55 °C, 1 min): primers hybridize to the separated sense and antisense strands on the target DNA.
3. Extension ($\approx$72 °C, 1 min): DNA synthesis occurs with extension from the 3′ end of the primers to complete the synthesis of a new set of DNA molecules.

The reaction is cycled (approximately 25 to 40 times), each cycle leading to a doubling of the number of copies of the target sequence between and including the primer sequences. Accumulation of the product and depletion of substrates leads to a cessation of amplification.

Real time PCR is the detection of PCR products as they are produced in the reaction tube, as illustrated in Figure 14.6. The determination of reaction kinetics using real time PCR enables the quantification of PCR product.

The impact of PCR in molecular biology has been profound with this technique now seen as essential for many common procedures such as cloning specific DNA fragments, screening for positive clones, investigating gene expression patterns and detecting and identifying sequences for diagnostic applications and in forensic science.

A technique that is frequently used in combination with PCR is reverse transcription (RT), where the initial target sequence is mRNA and not DNA. The first step in RT-PCR is production of complementary DNA (cDNA) from the RNA template using reverse transcriptase enzyme. Therefore, cDNA

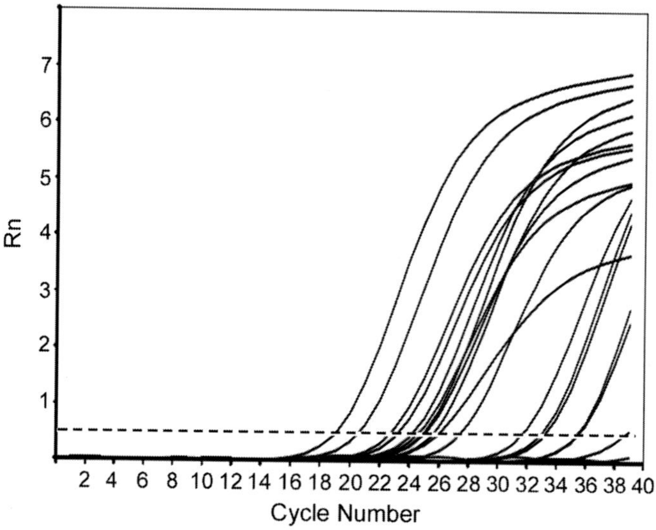

Figure 14.6 Real time quantitative RT-PCR amplification plot showing the number of cycles of PCR versus the intensity of fluorescent reporter signal (Rn). In this example, mRNA was extracted from cells that were expressing muscle glycogen phosphorylase and was converted into cDNA. Primers and probes specific for muscle glycogen phosphorylase were used in the amplification reaction (TaqMan, Applied Biosystems). The threshold level for the fluorescent signal (shown as dashed line) is in the exponential phase of the amplification and is the point at which the reaction reaches a level of detection above background. The number of cycles required to reach the threshold line is the cycle threshold (Ct) which is dependent on the number of copies present in the starting cDNA. The greater the number of copies of target cDNA in the starting material, the fewer cycles of amplification are needed for the fluorescent signal to reach the threshold. Values were obtained with an Applied Biosystems 7500 Real Time PCR System (courtesy of Dr Kathryn Wright, RJAH Orthopaedic Hospital, Oswestry, UK).

doesn't contain introns and RT-PCR is an amplification of the coding sequence of DNA.

A collection of cDNA clones may be generated from the mRNA sequences isolated from an organism or a specific tissue or cell type. The cDNA clones can be taken up and established in bacterial host cells (transformed) to make a "cDNA library", which contains a representation of all the expressed genes from the starting material.

14.4.3 Genome Sequencing and Analysis

Nucleotide sequencing was developed in the 1970s.[8,9] The first genome to be sequenced, a bacteriophage containing about 5300 bases, was completed in 1977. Major advances in automated laboratory techniques for DNA sequencing (based on the Sanger method) and computational power in the 1980s and

1990s facilitated the expansion of the science of "genomics". The completion of first draft of the entire human genome sequence, which consists of more than 2.9×10^9 base pairs, was announced in 2000, and published in 2001.[14,15] The high out-breeding in human populations leads to genetic diversity, so there is no definitive human genome. Common differences are the single-base changes that exist between individuals.

These single nucleotide polymorphisms (SNPs) have a frequency of about one in every 1000 bases. Rarely, mutations that cause specific single gene inherited disorders may occur. More commonly, SNPs may have no known effect or they may have subtle consequences such as affecting susceptibility to various diseases or direct effects such as influencing eye colour. Therefore, an important part of the human genome project is the single nucleotide polymorphism database. The majority of SNPs are in non-coding regions. SNPs may also occur within genes and occasionally (but not usually) alter amino acid sequences. Larger sequence variants also occur which may include block substitutions, insertions, deletions, inversions, duplications and copy number differences. Sequence-repeat polymorphisms are the targets for DNA fingerprinting in forensic science. The in depth analysis of relationships between genetic variation and physiology and disease remains a massive task.

Automated Sanger methods of sequencing are being replaced by next-generation sequencing technologies.[18] Increased capacity is leading to the completion of an increasing number of genome sequences. The completion of the human genome has led to the development of other disciplines such as "Functional Genomics" to determine the function of genes and gene products, "Proteomics" to study all the proteins expressed by a particular cell type and "Bioinformatics" for the computational study of sequence information.

The list of organisms whose genome has been sequenced is being added to all the time. The genomes of many bacteria, viruses and microbes that live in extreme environments have been sequenced, in addition to a number of important fungi, plants and animals including yeasts, rice, maize, fruit fly, mosquito, honey bee, nematode, zebra fish and chicken. Since the sequencing of the human genome, a number of other mammalian genomes, including mouse, rat, chimpanzee, macaque monkey, dog, cat, cow, elephant, pig and horse have been sequenced. The availability of genome sequences from different organisms allows for the study of "Comparative Genomics". The function and role of genes in human diseases may be determined by their comparison with genes in different organisms which have the corresponding functions (orthologs). Comparisons of the human, mouse and rat genomes have been reported.[39]Almost all known human-disease causing genes have rat orthologs.

14.4.4 Cloning of Individual Genes into Vectors

Once a specific gene sequence has been prepared (*e.g.* by PCR), the next step is usually to "clone" the DNA into a suitable vector, followed by confirmatory sequencing. A cloning vector is a nucleotide structure that is used to carry inserted foreign DNA for propagation or for protein production. Plasmids and phage (virus) vectors are commonly used for cloning. Phage vectors such as bacteriophage lambda often contain a linear double stranded DNA molecule. Foreign DNA can be cloned into the phage lambda chromosome and used to infect a host cell. When the DNA is injected into its host it becomes circular and may insert into the host chromosome or multiply and produce more phage.

Plasmids are frequently the cloning vector of choice. Plasmids are usually double stranded circular DNA molecules that replicate and express their proteins separately from the genomic DNA of bacteria. In nature, plasmids rapidly spread traits (such as antibiotic resistance) through populations of bacteria. Genetic engineering has allowed the alteration of some plasmids so that DNA fragments can be readily cloned into them and "transformed" into host bacterial cells. Transformation and selection of the host cell allows for the maintenance and propagation of the DNA sequence in an easily accessible form and replication of the DNA in the host cell is less prone to errors than PCR. Additional DNA sequences allow the plasmid to replicate and be expressed in mammalian cells.

14.4.5 Transfection of Mammalian Cells

A number of methods exist for the transfer of DNA to mammalian cells in culture, including infection with viruses such as adenovirus and retroviruses. Frequently, mammalian cells in culture are efficiently "transfected" with plasmids. Methods of transfection are shown in Table 14.3.

The transfection may be "transient", in which the DNA does not integrate into the host chromosome and high levels of protein expression may be achieved for a short period of time. Alternatively, the transfection may be "stable" in which the protein production is often at a lower level but permanent. An example of the transient transfection of mammalian cells is shown in Figure 14.7.

In this example, cultured fibroblast cells that are lamin A/C deficient were electroporated and transfected with a mammalian expression plasmid

Table 14.3 Methods for the plasmid transfection of mammalian cells.[a]

Transfection method
DEAE-dextran
Calcium phosphate
Cationic liposomes
Activated dendrimers
Electroporation (high voltage may form pores in membrane)
Biolostic particle delivery
Microinjection

[a]Chemical methods of transfection usually involve the interaction of positively charged molecules with the negatively charged phosphates of DNA which are taken up by the cells by endocytosis.

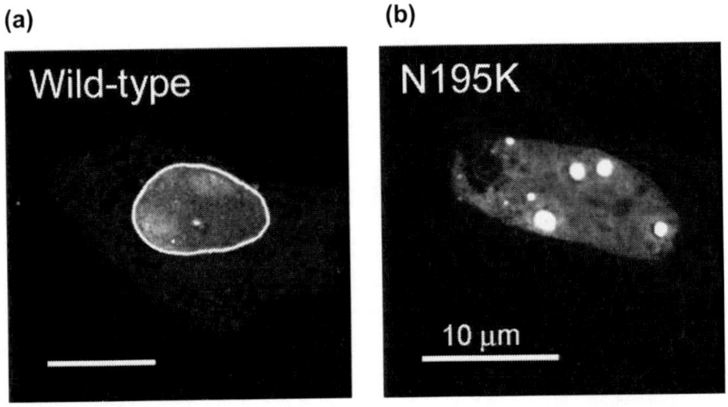

Figure 14.7 Fibroblast cells (lamin A/C deficient) transfected with (a) normal wild-type lamin A or (b) N195K mutant lamin A. Transfected wild-type lamin A is located largely at the nuclear envelope with the appearance of a normal nuclear lamina. In the example shown, the mutant lamin A undergoes abnormal lamina assembly with the formation of large nucleoplasmic foci (see ref. 40).

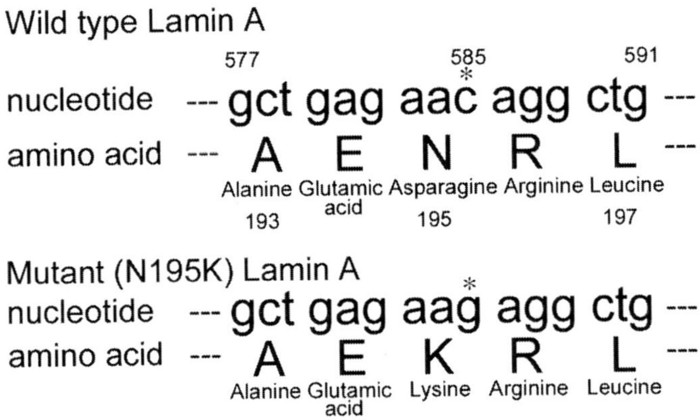

Figure 14.8 Illustration of a small region of lamin A cDNA showing part of the wild type sequence above the corresponding sequence with the c585g mutation (*). The amino acids that are translated from each DNA sequence are shown below each nucleotide triplet.

containing the human cDNA of either normal (wild-type) or mutant (N195K) lamin A.[40] The mutant was prepared by site-directed mutagenesis to make a cytosine to guanine change at base number 585 in exon 3 of the lamin A cDNA, which resulted in an asparagine to lysine change at amino acid 195 in the translated protein (Figure 14.8).

This N195K pathogenic mutation has been associated with dilated cardiomyopathy.[41] The proteins resulting from the transfections were labeled with a fluorescent tag and the cells examined microscopically. Figure 14.7 shows that the wild type lamin A has a normal appearance and is expressed at the nuclear envelope where it presumably forms a normal lamina. In contrast, mutant lamin A shows a defective assembly of the nuclear lamina with large foci of the transfected lamin A inside the nucleus.[40]

14.4.6 Transformation of Plant Cells

The production of transgenic plants often involves the maintenance of plant cells in culture. The target cells for gene transfer are often undifferentiated plant callus which is derived from explants of areas such as buds or root tips. Plant cells may have their cellulose cell wall removed to produce protoplasts which may increase the efficiency of DNA transfer.

A common method for gene transfer to plant cells involves infection with *Agrobacterium*.[42] For example, the transfer gene may be engineered into a derivative of a disarmed tumor-inducing (Ti) plasmid. Ti plasmid can enter the plant cells via infection with *Agrobacterium tumefaciens* and the transfer gene may incorporate into the plant nuclear DNA to give a stable transformation.

Alternatively, direct methods for DNA transfer may be used to give transient or stable transformations. Direct transfer may be achieved by, for example, treating the cells with polyethylene glycol, electroporation, or bombarding with DNA-coated microprojectiles. Chloroplasts, of which there may be thousands in photosynthetic cells, are DNA-containing organelles which can be targets for DNA transfer instead of genomic DNA.

Another method is the use of viruses as vectors for the delivery of nucleic acids into plants.[43] Plant virus vectors are often RNA instead of DNA, the transfection is transient and the infection may be systemic, so that tissue culture may not be required.

Other methods of transformation are also available,[44] including the use of site-specific recombination, which should make the process more predictable.[45]

If the targets for transformation are cells in culture, it is necessary to select the transformed cells. Under appropriate culture conditions, protoplasts can regenerate their cell walls and shoot and root development can occur to give rise to transgenic plants.

14.5 Applications of Recombinant DNA Technology

Recent years have witnessed a massive increase in the development and applications of recombinant DNA technology. The major players are the agricultural and health industries (Table 14.4). Other disciplines such as forensic science[46] and bioremediation[47,48] have also benefited from this technology (Table 14.4). Bioremediation is the use of microorganisms to reduce the toxicity of hazardous substances, such as the use of bacteria to decontaminate soil or water polluted with petroleum hydrocarbons. The future application of genetic manipulation for the production of biofuels could give huge benefits by reducing the reliance on fossil fuels. There may be considerable overlap between the disciplines that employ recombinant DNA technology.

Table 14.4 Applications of recombinant DNA technology in medicine and biotechnology.

Industry	*Applications*
Agriculture	Tolerance to herbicides
	Resistance to insect pests or viruses
	Delayed ripening or softening
	Improved yield
	Tolerance to biotic stresses
	Use of marginalized land
	Nutritional benefits
	Biofuels
Health care	Research and Disease models
	Diagnostics
	Therapeutic recombinant proteins
	Other biopharmaceuticals from cell cultures
	"Pharming" (Transgenic animals and plants)
	Vaccines
	Gene therapy
Forensics	Linking to crime scene
(DNA fingerprinting)	Parental testing
Bioremediation	Degradation of organic contaminants
	Immobilization of heavy metals

14.5.1 Recombinant DNA Technology in the Food Industry[49]

The commercial use of transgenic plants has increased from 1.7 million hectares in 6 countries in 1996 to 134 million hectares in 25 countries in 2009. By 2009, 46% of the global area of GM crops was being grown by developing countries. The eight countries with the largest areas of GM crops are: USA, Brazil, Argentina, India, Canada, China, Paraguay and South Africa.[49] In 2009, the top four transgenic crops (by area) were soybean, maize, cotton and canola (rape). This is likely to change with the approval given by the Chinese government in 2009 for growing GM rice.

Applications of GM technology in the food industry include the following.

- Producing tolerance to specific herbicides.
- Conferring resistance to insect pests, *e.g. Bacillus thuringiensis* (Bt) crops that contain insecticidal toxin genes.
- Increasing the shelf life of fruits and vegetables by delayed ripening or softening.
- Improved yield.
- Tolerance to other biotic stresses, *e.g.* transgenic rice plants that are resistant to rice yellow mottle virus.[50]
- Increased use of marginalized land, *e.g.* increase in salt tolerance to allow cultivation in coastal land.
- Increased nutritional benefits, *e.g.* transgenic rice with increased production of beta-carotene for vitamin A synthesis[51] or with increased iron content.

"Stacked traits", in which a plant contains more than one genetic modification are increasingly in use. Therefore, benefits of GM crops may include reduced production costs, higher yields, improved traits, reduced environmental impact (*e.g.* reduced use of pesticides and less soil erosion) and new cropping options.

A joint report by several scientific institutions in 2000[52] argued that transgenic plants are needed to meet the world food requirements. GM technology in conjunction with other developments should be used to increase the production of main staple food (rice, maize, wheat, cassava, yam, sorghum, potatoes and sweet potatoes). Increased production of other crops, including legumes, millet, cotton, rape, bananas and plantains, is also needed.[52] In 2009 it was reported that sustainable food production will need to be doubled by 2050, on approximately the same area of land, with the use of less resources and the increasing challenges of a changing climate.[49] There is also an immediate need to alleviate the poverty, hunger and malnutrition which affects more than 1 billion people. It has been argued that the most promising strategy to meet these challenges is to integrate the best of conventional crop technology and the best of crop biotechnology applications, including novel traits.[49]

However, there is opposition to the cultivation of GM crops. Concerns include the risks to human health such as potential toxicity, possible antibiotic resistance from GM crops, potential immune responses to GM food and reduced nutritional quality.[53] Other concerns include the effects of GM crops on the environment, such as the unintended transfer of transgenes to closely related wild species or subspecies through GM pollen diffusion and the horizontal gene transfer from transgenic plants to non-target organisms such as soil microbes and insects. Other environmental concerns include the effects on biodiversity, the emergence of herbicide resistant weeds and restricted access to seeds by patenting of GM plants.[54–56]

14.5.2 Recombinant DNA Technology in the Health Industry[57]

14.5.2.1 Therapeutic Proteins

An early application of genetic engineering by the health industry was the production of therapeutic recombinant human proteins. Human genes were cloned into a vector, transformed into bacteria and the bacteria cultured to produce the corresponding protein. The first therapeutic recombinant protein, insulin, was produced in 1982. In addition to bacteria, yeast, animal and plant cells are now also used as hosts, as these cells may perform beneficial post-translational modifications of the proteins.

The manipulation of DNA has allowed these procedures to advance so that recombinant proteins may be engineered to improve their properties. A promising technique is the production of regions of antibodies and chimeric antibodies by recombinant DNA technology.[58] A further method for therapeutic protein production has been "pharming", *i.e.* the creation of transgenic animals which express the recombinant protein in body fluids,[59,60] or transgenic plants which may be a source of safer and cheaper pharmaceuticals or vaccines.[61]

14.5.2.2 Disease Models for Research

Transgenic animal models, particularly mice, are used in the study of human genetic diseases. For this comparative genomics, a specific gene may be targeted and inactivated in the germ line to give rise to a mouse with a "gene knockout". An example is the targeted inactivation of the Muscleblind gene which gives a mouse with some of the features of myotonic dystrophy.[62] Also, genes may be transfected and incorporated into the germ line genome to give a mouse with a "gene knock in" as a model for a dominant gain of function disease. An example is the knock in of mutant Huntingtin for mouse models of Huntington's disease.[63]

14.5.2.3 Diagnostic Screening

Recombinant DNA technology can be used in the study and diagnosis of genetic diseases that may be divided into single gene disorders, polygenic diseases, and chromosome abnormalities.

Single Gene (Monogenic) Disorders. These are caused by a mutation in a single gene. Such mutations may be single-base "missense" mutations or small in-frame deletions or additions in which a mutant protein is produced. Alternatively, the mutations may be "nonsense", in which the reading frame of the DNA triplet code for amino acids is disrupted so that the protein product is absent or truncated. Examples are autosomal dominant Emery–Dreifuss muscular dystrophy (EDMD), which often arises due to missense mutations in lamin A/C and X-linked recessive EDMD, which frequently occurs because of nonsense mutations in the emerin gene.[64]

Figure 14.9 shows a skin biopsy from a female carrier of X-linked EDMD. Emerin protein was present in the nuclear membrane of some of the cells (from the wild type copy of the gene on the X chromosome), but was completely absent from the remainder of the cells (due to the mutant copy of the gene on the other X chromosome). As expected, staining for lamin A/C appeared normal in all the cells. Therefore, antibody staining of emerin may be used in the diagnosis of X-linked EDMD and for identification of carriers.[65]

Another type of disease is the nucleotide repeat diseases.[66] These expanded repeat sequences may occur inside a coding sequence of the gene (*e.g.* Huntington's Disease) or in non-coding regions (*e.g.* Myotonic Dystrophy and Friedreich's ataxia).

Polygenic Diseases. These diseases involve a number of genes and environmental factors and include predisposition to type 1 diabetes, certain cancers and schizophrenia. The link between a sequence of DNA and the disease to which it may contribute is often not very clear cut.

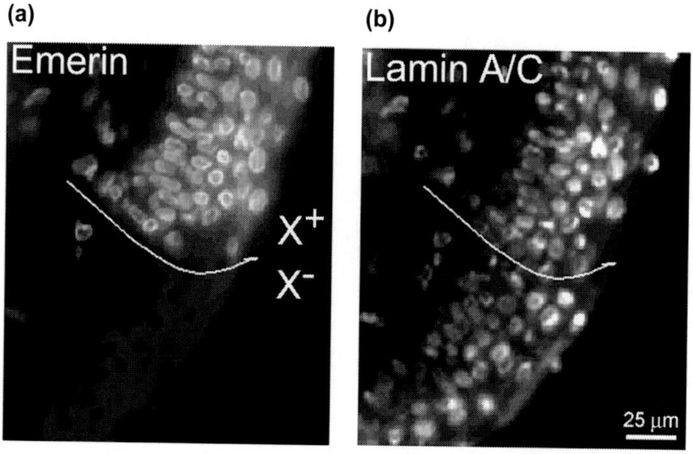

Figure 14.9 Immunolocalization of emerin (a) and lamin A/C (b) in a section of superficial skin biopsy from an X-linked EDMD carrier (reproduced from ref. 64 with permission from Elsevier).

Chromosome Abnormalities. Chromosome abnormalities involve large scale loss, gain or rearrangement of genetic material and often lead to spontaneous abortion. Chromosome abnormalities occur in less than 1% of live births and most of these are associated with clinical disorders. The most severe conditions are caused by loss or gain of whole chromosomes, such as the gain of an extra chromosome 21 (Down's syndrome) or the loss of an X chromosome in females (Turner's syndrome).

The study of patterns of inheritance and advances in recombinant DNA technology has facilitated the identification of affected genes in many genetic disorders. Procedures such as sequencing, PCR and hybridization techniques are central to diagnostic testing of gene disorders. The UK National Health Service has an integrated genetic service providing diagnosis, risk estimation, counselling and surveillance of families at high risk of serious genetic disorders.[67,68] For couples with a high risk of passing on a genetic disease to their children, *in vitro* fertilization with pre-implantation screening may be an option.

14.5.2.4 Gene Therapy[69]

Gene therapy is an attempt to alleviate or cure a genetic disorder by the modification or correction of genetic information. In contrast to the germ line alterations of DNA in mouse models, gene therapy treats somatic (non-sex) cells of the patients. Different strategies for somatic gene therapy exist which are dependent on the nature of the genetic defect and location of the target tissue.

Gene therapy is often designed to treat single gene disorders. Frequently, the aim is to replace the function of a mutant gene with a normal version of the gene to allow expression of the normal protein. A normal gene may be inserted into a non-specific location within the genome and expressed to replace the missing protein. Sometimes a gene in a plasmid is not incorporated into the genome, but maintained as a discrete entity (episome). Another approach is to replace a mutant with a normal gene through homologous recombination. This has the advantage of introducing the gene into its normal location in the genome, but the process is often very inefficient. An alternative to replacing the whole faulty gene may be targeted gene repair in which the correct sequence is introduced and incorporated during replication, thus repairing the mutation. For gene repair it is necessary to know the exact nature of the mutation.

Duchenne muscular dystrophy is a severely progressive condition affecting one in every 3500 newborn males. This condition is due to defects in the dystrophin gene on the X chromosome. This massive gene is 2.6 Mb in size and comprises 97 exons. Deletions of one or more exons of the dystrophin gene often result in absence of dystrophin protein which occurs in Duchenne muscular dystrophy. Sometimes "in-frame" deletions in dystrophin may occur, so allowing the translation of a partially functional dystrophin protein, which is characteristic of Becker muscular dystrophy (the milder form of the disease).

Figure 14.10 shows detection of dystrophin protein in sections of muscle from control (normal dystrophin), Becker muscular dystrophy and Duchenne muscular dystrophy patients.

Gene therapy methods have been used to deliver normal dystrophin-encoding DNA permanently into muscle cells.[70] Sometimes the target for therapy may be the gene transcript rather than the gene. For example, a promising approach for treatment of Duchenne muscular dystrophy is the use of antisense oligonucleotides to splice out mutation containing exons from the dystrophin pre-mRNA.[71,72] The aim of this exon skipping is to generate a translatable transcript from the mutant dystrophin gene, to produce functional

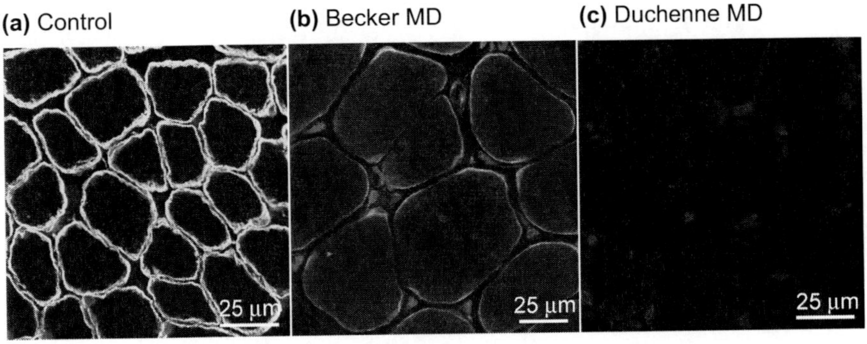

Figure 14.10 Immunohistochemical detection of dystrophin protein in sections of human skeletal muscle. (a) Control muscle (normal dystrophin), the protein is localised at the sarcolema of all the muscle fibres. (b) Becker muscular dystrophy, the level of labelling of the fibres is less than the control. (c) Duchenne muscular dystrophy, dystrophin is absent. (Note: A small proportion of Duchenne muscle fibres have dystrophin due to exon skipping in the dystrophin mRNA which gives rise to "revertant fibres" (not shown), containing dystrophin with internal deletions.)

or partially functional dystrophin protein, so that patients may develop the less severe symptoms of Becker rather than Duchenne muscular dystrophy. In order to monitor the efficiency of these therapy studies, it is important to be able to measure the amount of "repaired" dystrophin protein. For these evaluations, exon specific monoclonal antibodies are an extremely useful tool to identify amino acid sequences that can only be due to successful dystrophin repair, because these exons are deleted from the patient's genome.[73]

Instead of targeting the coding sequences directly, a slightly different approach may be to target the regulatory sequences of the gene in order to increase (or decrease) transcription of the gene. Regulatory sequences may be included with the transferred DNA in order to control expression of the therapeutic gene at a later date.[74] To treat disorders caused by a gain of gene function, the strategy would probably be one of gene inhibition. Gene inhibition may be achieved by direct interference, for example using antisense oligonucleotides or short interfering RNA[75,76] to target the gene or mRNA and ribozymes may degrade the gene transcript. The age of onset of disease is a significant factor in treatment strategy. For example, the treatment of mutations that first exert their effect before birth may be expected to have a poorer outcome than the treatment of mutations in an individual prior to the onset of pathology. Another potential therapy is the use of DNA vaccines in which antibodies against the expressed protein are produced by the immune system of the patient.[77]

Viruses to be used for delivery of gene therapy are genetically altered to make them replication defective, have any disease causing genes removed and to carry the human therapeutic gene. The therapeutic genes are delivered to the patient's target cells by the normal viral infection route. With non-viral delivery methods, the therapeutic human gene is often cloned into a plasmid. The plasmid may be engineered to promote integration into the host genome or may contain mammalian expression elements to allow replication and protein expression that is distinct from that of the genome. Some gene therapy delivery vectors are shown in Table 14.5.

Table 14.5 Gene therapy delivery mechanisms.

Vector type	*Examples*
Viral vectors (transduction)	Adenoviruses (dsDNA, *e.g.* common cold virus)
	Retroviruses (ssRNA, *e.g.* HIV)
	Herpes viruses (dsDNA, *e.g.* HSV, CMV)
	Baculoviruses (dsDNA, *e.g.* polyhedrovirus)
	Adeno-associated viruses (small ssDNA parvoviruses)
	Alphaviruses (ssRNA, *e.g.* semliki forest virus)
Plasmids (transfection)	Direct transfer
	Liposomes
	Lipofection
	Cationic polymers
Others	Targeted delivery
	Artificial human chromosome
	Bacteria

An alternative delivery method may involve direct transfer. For example, the nucleic acid may be injected into skeletal muscle and the efficiency improved by applying an electric field to the muscle.[78] The DNA may be enclosed in artificial lipid vesicles (liposomes) or nanoparticles which are able to fuse with the plasma membrane, or they may be coated with antibodies that interact with specific cell surface receptors to target them to a particular cell type.[79] DNA–lipid complexes (lipofection) and cationic polymers with DNA, can be taken into cells by endocytosis.[80,81] Bacteria may be useful for the targeted delivery and transfer of plasmid DNA to mammalian cells and tissues.[82] Another approach is the engineering and introduction of a 47th artificial human chromosome into target cells. Such structures could, in theory, carry telomeres, one or more replication origins, a centromere and large genes with the regulatory elements for appropriate expression.[83]

The location and cell type of the target tissue are major factors in the decision of which delivery method to use for gene therapy. Many gene therapy trials are for the treatment of cancers. Virus vectors may be selective for the proliferating cancer cells. Targets may be the inhibition of oncogenes,[84] the restoration of tumor suppressor genes[85] or the introduction of suicide genes.[86] For the treatment of disorders of the blood and immune system, cells can be removed for *ex vivo* gene therapy and then returned to the patient. An example of *ex vivo* gene therapy is the treatment of bone marrow stem cells from 5 boys with X-linked severe combined immunodeficiency (SCID-X1) with retrovirus-mediated immunoglobulin gamma(c) gene transfer. The bone marrow cells were returned to the patients and the immunodeficiency eradicated.[87] Unfortunately, within 3 years of treatment, 2 of the boys developed leukaemia.[88] The retrovirus vector had integrated into their genomes close to a proto-oncogene promoter. A more successful study has been the transduction of bone marrow stem cells with a retrovirus containing the ADA (adenosine deaminase) gene in order to treat ADA-SCID (ADA-severe combined immunodeficiency).[89] Eight of 10 children were essentially cured of this fatal disorder.[89]

An attractive target for gene therapy is the eye, due to the high degree of compartmentalisation and immune tolerance of this easily accessible organ.[90]

At the moment, gene therapy is an imperfect treatment for many genetic diseases. Problems include inefficiencies of gene delivery, therapeutic DNA remaining functional for short time periods, problems with integrating into the genome and the immune response which gene transfer may trigger. Viral vectors often provide efficient delivery mechanisms, but may be too unpredictable. Synthetic vectors are considered safer than viruses, but often the efficiency of transfer is too low to be clinically useful. However, significant advances and improvements continue to be made and the number and range of diseases that can be successfully treated by gene therapy are likely to increase in the near future.[69]

14.6 Synthetic Biology

Construction of the first self-replicating bacterial cell with a totally synthetic genome was reported by a team at the J. Craig Venter Institute in 2010.[91]

The synthetic DNA sequence, based on the bacterial *Mycoplasma mycoides* genome with various alterations including "watermark" sequences, was designed using a computer and constructed using chemical synthesis without any pieces of natural DNA. Specific cassettes of DNA were chemically synthesised and assembled in yeast cells by a three stage process to make the 1.08 million base pair artificial chromosome. This synthetic genome was then isolated from yeast and transplanted into the recipient cell which was a modified *Mycoplasma capricolum* bacterium. This process produced the landmark achievement of bacterial cells under the exclusive control of a synthetic genome.[91] Following this proof of concept, these techniques may now be applied to build more complex organisms with useful properties that can be utilized on an industrial scale, in areas including health, energy, agriculture and the environment.

14.7 Concluding Remarks

Since the 1970s, remarkable advances have been made in the techniques for studying and manipulating nucleic acids. These techniques have been exploited by industries including health and agriculture and their uses have raised scientific, political and ethical questions. These genetic manipulation techniques have often involved adding or altering single or low numbers of genes, out of the thousands or tens of thousands of genes present in most organisms. In the new era of synthetic biology in which it is now possible to engineer whole genomes, scientists can have complete control over design of the genome and the subsequent traits and properties of the resultant organism. The routine and highly accurate synthesis and assembly of large sequences of DNA is likely to lead to redundancy of other molecular biology techniques such as gene cloning and plasmid manipulation. Important products of synthetic biology are likely to include biofuels, pharmaceuticals, vaccines, food products and environmentally friendly pesticides. Other applications of this technology may include bioremediation, clean water treatment and synthetic photosynthesis to reduce carbon dioxide levels.[92] Therefore, synthetic biology has the potential to transform world industries, giving enormous benefits to society. However, consideration must also be given to the potential risks, with a requirement for ethical regulation to ensure that this technology is used for the benefit and not the detriment of society.

In addition to the advances in DNA synthesis, innovation in DNA sequencing has continued at pace over the past decade, which is particularly apparent with the increases in throughput that have occurred. From years 2000 to 2010, the speed of DNA sequencing increased by around 50 000 times[17] and it became possible to sequence a human genome on a single machine in one day. A large incentive which is helping to drive this innovation is the $10 million Archon X prize which will be awarded to the first team to build a device capable of accurately sequencing 100 human genomes within a 10 day period for minimal cost. The automated Sanger method of sequencing, which was used for the initial human genome sequencing projects, is now being replaced by next-generation sequencing technologies.[18] Few genome sequences of human individuals have been reported, but with improvements in the technology to give rapid, cheap and high quality sequencing of DNA, we approach the time when genome sequencing of individuals will become a routine procedure. For a complete picture, the diploid genome may be sequenced, so giving the sequence of the chromosomes from each parent. Interpretation of the genome sequence information remains at a very basic level. For better understanding, it will be necessary to sequence the genomes of a large number of individuals and to analyse this information alongside many observable characteristics of the individuals, in order to delve deeper into the relationships between human genetic variation and physiology and disease. It is the detailed interpretation of a genome sequence which remains a major challenge.

Examples of convenient sources of DNA for human genome sequencing may be white blood cells or cells in saliva. However, genetically distinct populations of cells may be present in the same individual. This "mosaicism" may occur in the germ line in early development or later in life as "somatic mosaicism" and can be caused, for example, by DNA mutations, chromosomal abnormalities or spontaneous reversion.[93] Somatic mosaicism has been implicated in a number of monogenic disorders including those affecting muscles, nerves, skin and skeletal systems and metabolic, immune and clotting disorders and in tumour suppressor genes. Cancer progression is a prominent example of somatic mosaicism, with the accumulation of somatic mutations in functional pathways leading to unregulated cell division. Therefore, as the cost of DNA sequencing continues to fall, our medical records may contain not just a single genome, but genome sequences from other cell types and tissue locations at different stages of development and disease.[17] The successful interpretation of these multiple genomes will lead to a truly personalized health care system.

Acknowledgements

Thanks to S.N. Corns, M. Holt, M.J. Marshall, G.E. Morris, C.A. Sewry, C.A. Sharp, J.H.H. Williams, A.C. Wright and K.E. Wright for their contributions and comments about this chapter.

References

1. O. T. Avery, C. M. MacLeod and M. McCarty, *J. Exp. Med.*, 1944, **79**, 137.
2. J. D. Watson and F. H. C. Crick, *Nature*, 1953, **171**, 737.
3. S. Brenner, F. Jacob and M. Meselson, *Nature*, 1961, **190**, 576.
4. F. H. C. Crick, L. Barnett, S. Brenner and R. J. Watts-Tobin, *Nature*, 1961, **192**, 1227.
5. M. Nirenberg, *Trends Biochem. Sci.*, 2004, **29**, 46.
6. R. J. Roberts, T. Vincze, J. Posfai and D. Macelis, *Nucleic Acids Res.*, 2003, **31**, 418.

7. E. M. Southern, *J. Mol. Biol.*, 1975, **98**, 503.
8. F. Sanger, S. Nicklen and A. R. Coulson, *Proc. Natl. Acad. Sci. USA*, 1977, **74**, 5463.
9. A. M. Maxam and W. Gilbert, *Proc. Natl. Acad. Sci. USA*, 1977, **74**, 560.
10. C. A. Hutchison, S. Phillips, M. H. Edgell, S. Gillam, P. Jahnke and M. Smith, *J. Biol. Chem.*, 1978, **253**, 6551.
11. R. K. Saiki, S. Scharf, F. Faloona, K. B. Mullis, G. T. Horn, H. A. Erlich and N. Arnheim, *Science*, 1985, **230**, 1350.
12. K. B. Mullis, *Sci. Am.*, 1990, **262**, 56.
13. F. C. Lawyer, S. Stoffel, R. K. Saiki, K. Myambo, R. Drummond and D. H. Gelfand, *J. Biol. Chem.*, 1989, **264**, 6427.
14. International Human Genome Sequencing Consortium, *Nature*, 2001, **409**, 860.
15. J. C. Venter, M. D. Adams, E. W. Myers, P. W. Li, R. J. Mural, G. G. Sutton and H. O. Smith, *et al.*, *Science*, 2001, **291**, 1304.
16. International Human Genome Sequencing Consortium, Nature, 2004, **431**, 931.
17. J. C. Venter, *Nature*, 2010, **464**, 676.
18. M. L. Metzker, *Nature Rev. Genet.*, 2010, **11**, 31.
19. The Double Helix-50 years, Nature, 2003, **421**, 396.
20. B. Lewin, *Genes VIII*, Prentice Hall, New Jersey, 2004.
21. P. C. Turner, A. G. McLennan, A. D. Bates and M. R. H. White, *Instant Notes in Molecular Biology*, 2nd edn, Bios Scientific Publishers, Oxford, 2000.
22. P. L. Ivanov, M. J. Wadhams, R. K. Roby, M. M. Holland, V. W. Weedn and T. J. Parsons, *Nature Genet.*, 1996, **12**, 417.
23. R. E. Green, J. Krause, A. W. Briggs, T. Maricic, U. Stenzel, M. Kircher, N. Patterson, H. Li, W. Zhai and M. H. Fritz, *et al.*, *Science*, 2010, **328**(5979), 710.
24. A. Fontana, *Biosystems*, 2010, **101**(3), 187.
25. J. Locker (Ed.), *Transcription factors*, The Human Molecular Genetics Series, BIOS Scientific Publishers, Oxford, 2001.
26. J. M. Johnson, J. Castle, P. Garrett-Engele, Z. Kan, P. M. Loerch, C. D. Armour, R. Santos, E. E. Schadt, R. Stoughton and D. D. Shoemaker, *Science*, 2003, **302**, 2141.
27. S. E. Massey, G. Moura, P. Beltrao, R. Almeida, J. R. Garey, M. F. Tuite and M. A. Santos, *Genome Res.*, 2003, **13**, 544.
28. A. Bock, K. Forchhammer, J. Heider, W. Leinfelder, G. Sawers, B. Veprek and F. Zinoni, *Mol. Microbiol.*, 1991, **5**, 515.
29. J. Sambrook and D. W. Russel, *Molecular cloning: A laboratory manual*, Cold Spring Harbor Laboratory Press, 2001.
30. R. M. C. Dawson, D. C. Elliot, W. H. Elliot and K. M. Jones, *Data for Biochemical Research*, Oxford University Press, 1987.
31. L. S. Shlyakhtenko, A. A. Gall, J. J. Weimer, D. D. Hawn and Y. L. Lyubchenko, *Biophys. J.*, 1999, **77**, 568.
32. M. H. Caruthers, *Science*, 1985, **230**, 281.

33. M. H. Caruthers, A. D. Barone, S. L. Beaucage, D. R. Dodds, E. F. Fisher, L. J. McBride, M. Matteucci, Z. Stabinsky and J. Y. Tang, *Methods Enzymol.*, 1987, **154**, 287.
34. M. J. Gait, *Curr. Opin. Biotechnol.*, 1991, **2**, 61.
35. M. H. Caruthers, G. Beaton, J. V. Wu and W. Wiesler, *Methods Enzymol.*, 1992, **211**, 3.
36. D. A. Stetsenko and M. J. Gait, *Bioconjug. Chem.*, 2001, **12**, 576.
37. S. B. Primrose, R. M. Twyman and R. W. Old, *Principles of Gene Manipulation*, Blackwell Science, Oxford, 2001.
38. I. Holt, V. Jacquemin, M. Fardaei, C. A. Sewry, G. S. Butler-Browne, D. Furling, J. D. Brook and G. E. Morris, *Am. J. Path.*, 2009, **174**, 216.
39. R. A. Gibbs, G. M. Weinstock, M. L. Metzker, D. M. Muzny, E. J. Sodergren, S. Scherer and G. Scott *et al.*, *Nature.*, 2004, **428**, 493.
40. I. Holt, C. Östlund, C. L. Stewart, T. M. Nguyen, H. J. Worman and G. E. Morris, *J. Cell Sci.*, 2003, **116**, 3027.
41. D. Fatkin, C. MacRae, T. Sasaki, M. R. Wolff, M. Porcu, M. Frenneaux, J. Atherton, H. J. Vidaillet, S. Spudich, U. De Girolami, J. G. Seidman, C. Seidman, F. Muntoni, G. Muehle, W. Johnson and B. McDonough, *N. Engl. J. Med.*, 1999, **341**, 1715.
42. S. B. Gelvin, *Microbiol. Mol. Biol. Rev.*, 2003, **67**, 16.
43. C. Porta and G. P. Lomonossoff, *Mol. Biotechnol.*, 1996, **5**, 209.
44. M. Rakoczy-Trojanowska, *Cell Mol. Biol. Lett.*, 2002, **7**, 849.
45. L. A. Lyznik, W. J. Gordon-Kamm and Y. Tao, Site-specific recombination for genetic engineering in plants, *Plant Cell Rep.*, 2003, **21**, 925–932.
46. S. J. Walsh, *Expert Rev. Mol. Diagn.*, 2004, **4**, 31.
47. D. H. Pieper and W. Reineke, *Curr. Opin. Biotechnol.*, 2000, **11**, 262.
48. D. R. Lovley, *Nature Rev. Microbiol.*, 2003, **1**, 35.
49. James C, 2009, ISAAA Brief No 41, Ithaca, NY.
50. Y. M. Pinto, R. A. Kok and D. C. Baulcombe, *Nature Biotechnol.*, 1999, **17**, 702.
51. X. Ye, S. Al-Babili, A. Kloti, J. Zhang, P. Lucca, P. Beyer and I. Potrykus, *Science*, 2000, **287**, 303.
52. Royal Society London, US National Academy of Sciences, Brazilian Academy of Sciences, Chinese Academy of Sciences, Indian National Science Academy, Mexican Academy of Sciences, Third World Academy of Sciences, Transgenic plants and world agriculture, National Academic Press, Washington DC, 2000.
53. T. Malarkey T, *Mutat. Res.*, 2003, **544**, 217.
54. M. Giovannetti, *Riv. Biol.*, 2003, **96**, 207.
55. J. P. Nap, P. L. Metz, M. Escaler and A. J. Conner, *Plant J.*, 2003, **33**, 1.
56. A. J. Conner, T. R. Glare and J. P. Nap, *Plant J.*, 2003, **33**, 19.
57. S. B. Primrose and R. M. Twyman, Genomics: Applications in Human Biology, *Blackwell Publishing, Oxford*, 2004.
58. S. M. Kipriyanov and F. Le Gall, *Mol. Biotechnol.*, 2004, **26**, 39.
59. H. Niemann, R. Halter, J. W. Carnwath, D. Herrmann, E. Lemme and D. Paul, *Transgenic Res.*, 1999, **8**, 237.

60. A. J. Harvey, G. Speksnijder, L. R. Baugh, J. A. Morris and R. Ivarie R, *Nature Biotechnol.*, 2002, **20**, 396.
61. H. Warzecha and H. S. Mason, *J. Plant Physiol.*, 2003, **160**, 755.
62. R. N. Kanadia, K. A. Johnstone, A. Mankodi, C. Lungu, C. A. Thornton, D. Esson, A. M. Timmers, W. W. Hauswirth and M. S. Swanson M S, *Science*, 2003, **302**, 1978.
63. C. Cepeda, M. A. Ariano, C. R. Calvert, J. Flores-Hernandez, S. H. Chandler, B. R. Leavitt, M. R. Hayden and M. S. Levine, *J. Neurosci. Res.*, 2001, **66**, 525.
64. G. E. Morris, *Trend. Mol. Med.*, 2001, **7**, 572.
65. S. Manilal, C. A. Sewry, T. M. Nguyen, F. Muntoni and G. E. Morris, *Neuromuscul. Disord.*, 1997, **7**, 63.
66. C. J. Cummings and H. Y. Zoghbi, *Hum. Mol. Genet.*, 2000, **9**, 909.
67. D. Donnai and R. Elles, *BMJ*, 2001, **322**, 1048.
68. Genetics White Paper. Our inheritance, our future. Realising the potential of genetics in the NHS, Her Majesty's Stationary Office, 2003.
69. R. W. Herzog, O. Cao and A. Srivastava, *Discov Med.*, 2010, **9**, 105.
70. N. B. Romero, S. Braun, O. Benveniste, F. Leturcq, J. Y. Hogrel and G. E. Morris *et al.*, *Hum. Gene Therapy*, 2004, **15**, 1065.
71. Q. L. Lu, C. J. Mann, F. Lou, G. Bou-Gharios, G. E. Morris, S. A. Xue, S. Fletcher, T. A. Partridge and S. D. Wilton S D, *Nature Med.*, 2003, **9**, 1009.
72. M. Kinali, V. Arechavala-Gomeza, L. Feng, S. Cirak, D. Hunt, C. Adkin, M. Guglieri, E. Ashton, S. Abbs and P. Nihoyannopoulos *et al.*, *Lancet Neurol.*, 2009, **8**, 918.
73. G.E. Morris and T.M. Nguyen, Neuromuscul. Disord., 2004, 14, 578, (abstract GP5.07).
74. C. Toniatti, H. Bujard, R. Cortese and G. Ciliberto, *Gene Ther.*, 2004, **11**, 649.
75. E. Devroe and P. A. Silver, *Expert Opin. Biol. Ther.*, 2004, **4**, 319.
76. T. Tuschl and A. Borkhardt, *Mol. Intervent.*, 2002, **2**, 158.
77. H. S. Garmory, K. A. Brown and R. W. Titball, *Genet. Vaccines Ther.*, 2003, **1**, 2.
78. H. Gollins, J. McMahon, K. E. Wells and D. J. Wells, *Gene Ther.*, 2003, **10**, 504.
79. P. H. Tan, M. Manunta, N. Ardjomand, S. A. Xue, D. F. Larkin, D. O. Haskard, K. M. Taylor and A. J. George, *J. Gene Med.*, 2003, **5**, 311.
80. M. Sheikh, J. Feig, B. Gee, S. Li and M. Savva, *Chem. Phys. Lipids*, 2003, **124**, 49.
81. N. Raghavachari and W. E. Fahl, *J. Pharm. Sci.*, 2002, **91**, 615.
82. H. Loessner and S. Weiss, *Expert Opin. Biol. Ther.*, 2004, **4**, 157.
83. Z. Larin and J. E. Mejia, *Trends Genet.*, 2002, **18**, 313.
84. O. G. Scharovsky, V. R. Rozados, S. I. Gervasoni and P. Matar, *J. Biomed. Sci.*, 2000, **7**, 292.
85. I. A. McNeish, S. J. Bell and N. R. Lemoine, *Gene Ther.*, 2004, **11**, 497.
86. S. O. Freytag, M. Khil, H. Stricker, J. Peabody, M. Menon, M. DePeralta-Venturina, D. Nafziger, J. Pegg, D. Paielli, S. Brown, K. Barton, M. Lu, E. Aguilar-Cordova and J. H. Kim, *Cancer Res.*, 2002, **62**, 4968.

87. S. Hacein-Bey-Abina, F. Le Deist, F. Carlier, C. Bouneaud, C. Hue, J. P. De Villartay, A. J. Thrasher, N. Wulffraat, R. Sorensen, S. Dupuis-Girod, A. Fischer, E. G. Davies, W. Kuis, L. Leiva and M. Cavazzana-Calvo, *New Engl. J. Med.*, 2002, **346**, 1185.
88. S. Hacein-Bey-Abina, C. Von Kalle, M. Schmidt, M. P. McCormack, N. Wulffraat, P. Leboulch and A. Lim *et al.*, *Science*, 2003, **302**, 415.
89. A. Aiuti, F. Cattaneo, S. Galimberti, U. Benninghoff, B. Cassani, L. Callegaro, S. Scaramuzza, G. Andolfi, M. Mirolo, I. Brigida, A. Tabucchi, F. Carlucci, M. Eibl, M. Aker, S. Slavin, H. Al-Mousa, A. Al Ghonaium, A. Ferster, A. Duppenthaler, L. Notarangelo, U. Wintergerst, R. H. Buckley, M. Bregni, S. Marktel, M. G. Valsecchi, P. Rossi, F. Ciceri, R. Miniero, C. Bordignon and M. G. Roncarolo, *New Engl. J. Med.*, 2009, **360**, 447.
90. M.M. Liu, J. Tuo and C.C. Chan, *Br. J. Ophthalmol.*, 2011, **95**, 604.
91. D. G. Gibson, J. I. Glass, C. Lartigue, V. N. Noskov, R. Y. Chuang, M. A. Algire, G. A. Benders, M. G. Montague, L. Ma, M. M. Moodie, C. Merryman, S. Vashee, R. Krishnakumar, N. Assad-Garcia, C. Andrews-Pfannkoch, E. A. Denisova, L. Young, Z. Q. Qi, T. H. Segall-Shapiro, C. H. Calvey, P. P. Parmar, C. A. Hutchison, H. O. Smith and J. C. Venter, *Science*, 2010, **329**, 52.
92. Synthetic Biology: Scope, applications and implications. The Royal Academy of Engineering, London, 2009. (www.raeng.org.uk/synbio).
93. H. Youssoufian and R. E. Pyeritz, *Nature Rev. Genet.*, 2002, **3**, 748.

Subject Index

Individual chapters in this book use either British or American English spelling. This index uses British English spelling.